CONVERSIONS BETWEEN U.S. CUSTOMARY UNITS AND SI UNITS (Continued)

U.S. Customary unit		Times conversion factor		Equals SI unit	
		Accurate	Practical		
Moment of inertia (area)					
inch to fourth power	in.^{4}	416,231	416,000	millimeter to fourth power	mm^4
inch to fourth power	in.^{4}	0.416231×10^{-6}	0.416×10^{-6}	meter to fourth power	m^4
Moment of inertia (mass)					
slug foot squared	slug-ft^2	1.35582	1.36	kilogram meter squared	kg·m^2
Power					
foot-pound per second	ft-lb/s	1.35582	1.36	watt (J/s or N·m/s)	W
foot-pound per minute	ft-lb/min	0.0225970	0.0226	watt	W
horsepower (550 ft-lb/s)	hp	745.701	746	watt	W
Pressure; stress					
pound per square foot	psf	47.8803	47.9	pascal (N/m^2)	Pa
pound per square inch	psi	6894.76	6890	pascal	Pa
kip per square foot	ksf	47.8803	47.9	kilopascal	kPa
kip per square inch	ksi	6.89476	6.89	megapascal	MPa
Section modulus					
inch to third power	in.^{3}	16,387.1	16,400	millimeter to third power	mm^3
inch to third power	in.^{3}	16.3871×10^{-6}	16.4×10^{-6}	meter to third power	m^3
Velocity (linear)					
foot per second	ft/s	0.3048*	0.305	meter per second	m/s
inch per second	in./s	0.0254*	0.0254	meter per second	m/s
mile per hour	mph	0.44704*	0.447	meter per second	m/s
mile per hour	mph	1.609344*	1.61	kilometer per hour	km/h
Volume					
cubic foot	ft^3	0.0283168	0.0283	cubic meter	m^3
cubic inch	in.^{3}	16.3871×10^{-6}	16.4×10^{-6}	cubic meter	m^3
cubic inch	in.^{3}	16.3871	16.4	cubic centimeter (cc)	cm^3
gallon (231 in.^{3})	gal.	3.78541	3.79	liter	L
gallon (231 in.^{3})	gal.	0.00378541	0.00379	cubic meter	m^3

*An asterisk denotes an *exact* conversion factor

Note: To convert from SI units to USCS units, *divide* by the conversion factor

Temperature Conversion Formulas

$$T(°C) = \frac{5}{9}[T(°F) - 32] = T(K) - 273.15$$

$$T(K) = \frac{5}{9}[T(°F) - 32] + 273.15 = T(°C) + 273.15$$

$$T(°F) = \frac{9}{5}T(°C) + 32 = \frac{9}{5}T(K) - 459.67$$

Engineering Mechanics

Dynamics

Third Edition

SI Edition

Andrew Pytel
The Pennsylvania State University

Jaan Kiusalaas
The Pennsylvania State University

SI Edition prepared by Ishan Sharma
Indian Institute of Technology, Kanpur

CENGAGE
Learning™

Australia · Brazil · Japan · Korea · Mexico · Singapore · Spain · United Kingdom · United States

CENGAGE
Learning™

**Engineering Mechanics:
Dynamics, Third Edition
Andrew Pytel and Jaan Kiusalaas**

SI Edition prepared by Ishan Sharma

Publisher, Global Engineering Program:
Christopher M. Shortt

Senior Developmental Editor:
Hilda Gowans

Acquisitions Editor, SI Edition:
Swati Meherishi

Editorial Assistant: Tanya Altieri

Team Assistant: Carly Rizzo

Associate Marketing Manager:
Lauren Betsos

Director, Content and Media Production:
Barbara Fuller-Jacobson

Content Project Manager: Jennifer
Zeigler

Production Service: RPK Editorial
Services

Copyeditor: Pat Daly

Proofreader: Martha McMaster

Indexer: Shelly Gerger-Knechtl

Compositor: Integra Software Services

Senior Art Director:
Michelle Kunkler

Internal Designer: Carmela Periera

Cover Designer: Andrew Adams

Cover Image: Copyright
iStockphoto/Olga Paslawska

Permissions Account Manager, Text:
Katie Huha

Permissions Account Manager, Images:
Deanna Ettinger

Text and Image Permissions Researcher:
Kristiina Paul

Senior First Print Buyer:
Doug Wilke

For product information and technology assistance, contact us at
Cengage Learning Customer & Sales Support, 1-800-354-9706.

For permission to use material from this text or product, submit all requests online at **www.cengage.com/permissions.**
Further permissions questions can be emailed to
permissionrequest@cengage.com.

Library of Congress Control Number: 2009936717
ISBN-13: 978-0-495-29563-1
ISBN-10: 0-495-29563-9

Cengage Learning
200 First Stamford Place, Suite 400
Stamford, CT 06902
USA

Cengage Learning is a leading provider of customized learning solutions with office locations around the globe, including Singapore, the United Kingdom, Australia, Mexico, Brazil, and Japan. Locate your local office at: **international.cengage.com/region.**

Cengage Learning products are represented in Canada by Nelson Education Ltd.

For your course and learning solutions, visit
www.cengage.com/engineering.

Purchase any of our products at your local college store or at our preferred online store **www.ichapters.com.**

Printed in Canada
2 3 4 5 6 7 13 12 11 10

To Jean, Leslie, Lori, John, Nicholas

and

To Judy, Nicholas, Jennifer, Timothy

Contents

* Indicates optional articles

Preface to the SI Edition

This edition of **Engineering Mechanics: Dynamics** has been adapted to incorporate the International System of Units (*Le Système International d'Unités* or SI) throughout the book.

Le Système International d' Unités

The United States Customary System (USCS) of units uses FPS (foot-pound—second) units (also called English or Imperial units). SI units are primarily the units of the MKS (meter-kilogram-second) system. However, CGS (centimeter-gram-second) units are often accepted as SI units, especially in textbooks.

Using SI Units in this Book

In this book, we have used both MKS and CGS units. USCS units or FPS units used in the US Edition of the book have been converted to SI units throughout the text and problems. However, in case of data sourced from handbooks, government standards, and product manuals, it is not only extremely difficult to convert all values to SI, it also encroaches upon the intellectual property of the source. Also, some quantities such as the ASTM grain size number and Jominy distances are generally computed in FPS units and would lose their relevance if converted to SI. Some data in figures, tables, examples, and references, therefore, remains in FPS units. For readers unfamiliar with the relationship between the FPS and the SI systems, conversion tables have been provided inside the front and back covers of the book.

To solve problems that require the use of sourced data, the sourced values can be converted from FPS units to SI units just before they are to be used in a calculation. To obtain standardized quantities and manufacturers' data in SI units, the readers may contact the appropriate government agencies or authorities in their countries/regions.

Instructor Resources

A Printed Instructor's Solution Manual in SI units is available on request. An electronic version of the Instructor's Solutions Manual, and PowerPoint slides of the figures from the SI text are available through www.cengage.com/engineering.

The readers' feedback on this SI Edition will be highly appreciated and will help us improve subsequent editions.

The Publishers

Preface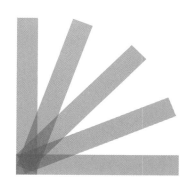

Statics and dynamics are the foundation subjects in the branch of engineering known as engineering mechanics. Engineering mechanics is, in turn, the basis of many of the traditional fields of engineering, such as aerospace engineering, civil engineering, and mechanical engineering. In addition, engineering mechanics often plays a fundamental role in such diverse fields as medicine and biology. Applying the principles of statics and dynamics to such a wide range of applications requires reasoning and practice rather than memorization. Although the principles of statics and dynamics are relatively few, they can only be truly mastered by studying and analyzing problems. Therefore, all modern textbooks, including ours, contain a large number of problems to be solved by the student. Learning the engineering approach to problem solving is one of the more valuable lessons to be learned from the study of statics and dynamics.

In this, our third edition of *Statics* and *Dynamics*, we have made every effort to improve our presentation without compromising the following principles that formed the basis of the previous editions.

- Each sample problem is carefully chosen to help students master the intricacies of engineering problem analysis.
- The selection of homework problems is balanced between "textbook" problems that illustrate the principles of engineering mechanics in a straightforward manner, and practical engineering problems that are applicable to engineering design.
- The number of problems using U.S. Customary Units and SI Units are approximately equal.
- The importance of correctly drawn free-body diagrams is emphasized throughout.
- Whenever applicable, the number of independent equations is compared to the number of unknowns before the governing equations are written.
- Numerical methods for solving problems are seamlessly integrated into the text, the emphasis being on computer applications, not on computer programming.
- Review Problems appear at the end of each chapter to encourage students to synthesize the topics covered in the chapter.

Both *Statics* and *Dynamics* contain several optional topics, which are marked with an asterisk (*). Topics so indicated can be omitted without jeopardizing the presentation of other subjects. An asterisk is also used to indicate problems that require advanced reasoning. Articles, sample problems, and problems associated with numerical methods are preceded by an icon representing a compact disk.

In this third edition of *Dynamics*, we have made what we consider to be a number of significant improvements based upon the feedback received from students and faculty who have used the previous editions. In addition, we have incorporated many of the suggestions provided by the reviewers of the second edition.

A number of articles have been reorganized, or rewritten, to make the topics easier for the student to understand. For example, the discussion of the work-energy method in Chapter 18 has been streamlined. Also, Chapter 20 (Vibrations) has been reorganized to provide a more concise presentation of the material. In addition, sections entitled Review of Equations have been added at the end of each chapter as an aid to problem solving.

The total numbers of sample problems and problems remain about the same as in the previous edition; however, the introduction of two colors improves the overall readability of the text and artwork. Compared with the previous edition, approximately one-third of the problems are new, or have been modified.

New to this edition, the Sample Problems that require numerical solutions have been solved using MATLAB©, the software program that is familiar to many engineering students.

Ancillary   *Study Guide to Accompany Pytel and Kiusalaas Engineering Mechanics, Dynamics, Third Edition*, J.L. Pytel and A. Pytel, 2009. The goals of this study guide are two-fold. First, self-tests are included to help the student focus on the salient features of the assigned reading. Second, the study guide uses "guided" problems which give the student an opportunity to work through representative problems, before attempting to solve the problems in the text.

Acknowledgments   We are grateful to the following reviewers for their valuable suggestions:

Hamid R. Hamidzadeh, Tennessee State University
Aiman S. Kuzmar, The Pennsylvania State University—Fayette,
   The Eberly Campus
Gary K. Matthew, University of Florida
Noel Perkins, University of Michigan
Corrado Poli, University of Massachusetts, Amherst

ANDREW PYTEL
JAAN KIUSALAAS

11

Introduction to Dynamics

Sir Isaac Newton (1643–1727) in his treatise Philosophiae Naturalis Principia Mathematica *established the groundwork for dynamics with his three laws of motion and the universal theory of gravitation, which are discussed in this chapter. (Time & Life Pictures/Getty Images)*

11.1 Introduction

Classical dynamics studies the motion of bodies using the principles established by Newton and Euler.[*] The organization of this text is based on the subdivisions of classical dynamics shown in Fig. 11.1.

---

[*]Sir Isaac Newton is credited with laying the foundation of classical mechanics with the publication of *Principia* in 1687. However, the laws of motion as we use them today were developed by Leonhard Euler and his contemporaries more than sixty years later. In particular, the laws for the motion of finite bodies are attributable to Euler.

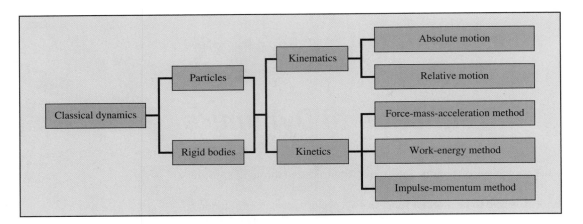

Fig. 11.1

The first part of this text deals with dynamics of particles. A *particle* is a mass point; it possesses a mass but has no size. The particle is an approximate model of a body whose dimensions are negligible in comparison with all other dimensions that appear in the formulation of the problem. For example, in studying the motion of the earth around the sun, it is permissible to consider the earth as a particle, because its diameter is much smaller than the dimensions of its orbit.

The second part of this text is devoted mainly to dynamics of *rigid bodies*. A body is said to be rigid if the distance between any two material points of the body remains constant, that is, if the body does not deform. Because any body undergoes some deformation when loads are applied to it, a truly rigid body does not exist. However, in many applications the deformation is so small (relative to the dimensions of the body) that the rigid-body idealization is a good approximation.

As seen in Fig. 11.1, the main branches of dynamics are kinematics and kinetics. *Kinematics* is the study of the geometry of motion. It is not concerned with the causes of motion. *Kinetics,* on the other hand, deals with the relationships between the forces acting on the body and the resulting motion. Kinematics is not only an important topic in its own right but is also a prerequisite to kinetics. Therefore, the study of dynamics always begins with the fundamentals of kinematics.

Kinematics can be divided into two parts as shown in Fig. 11.1: absolute motion and relative motion. The term *absolute motion* is used when the motion is described with respect to a fixed reference frame (coordinate system). Relative motion, on the other hand, describes the motion with respect to a moving coordinate system.

Figure 11.1 also lists the three main methods of kinetic analysis. The *force-mass-acceleration* (FMA) *method* is a straightforward application of the Newton-Euler laws of motion, which relate the forces acting on the body to its mass and acceleration. These relationships, called the *equations of motion,* must be integrated twice in order to obtain the velocity and the position as functions of time.

The *work-energy* and *impulse-momentum methods* are integral forms of Newton-Euler laws of motion (the equations of motion are integrated with respect to position or time). In both methods the acceleration is eliminated by the integration. These methods can be very efficient in the solution of problems concerned with velocity-position or velocity-time relationships.

The purpose of this chapter is to review the basic concepts of Newtonian mechanics: displacement, velocity, acceleration, Newton's laws, and units of measurement.

11.2 *Derivatives of Vector Functions*

A knowledge of vector calculus is a prerequisite for the study of dynamics. Here we discuss the derivatives of vectors; integration is introduced throughout the text as needed.

The vector **A** is said to be a vector function of a scalar parameter u if the magnitude and direction of **A** depend on u. (In dynamics, time is frequently chosen to be the scalar parameter.) This functional relationship is denoted by $\mathbf{A}(u)$. If the scalar variable changes from the value u to $(u + \Delta u)$, the vector **A** will change from $\mathbf{A}(u)$ to $\mathbf{A}(u + \Delta u)$. Therefore, the change in the vector **A** can be written as

$$\Delta \mathbf{A} = \mathbf{A}(u + \Delta u) - \mathbf{A}(u) \tag{11.1}$$

As seen in Fig. 11.2, $\Delta \mathbf{A}$ is due to a change in both the magnitude and the direction of the vector **A**.

The derivative of **A** with respect to the scalar u is defined as

$$\frac{d\mathbf{A}}{du} = \lim_{\Delta u \to 0} \frac{\Delta \mathbf{A}}{\Delta u} = \lim_{\Delta u \to 0} \frac{\mathbf{A}(u + \Delta u) - \mathbf{A}(u)}{\Delta u} \tag{11.2}$$

assuming that the limit exists. This definition resembles the derivative of the scalar function $y(u)$, which is defined as

$$\frac{dy}{du} = \lim_{\Delta u \to 0} \frac{\Delta y}{\Delta u} = \lim_{\Delta u \to 0} \frac{y(u + \Delta u) - y(u)}{\Delta u} \tag{11.3}$$

Caution   In dealing with a vector function, the magnitude of the derivative $|d\mathbf{A}/du|$ must not be confused with the derivative of the magnitude $d|\mathbf{A}|/du$. In general, these two derivatives will not be equal. For example, if the magnitude of a vector **A** is constant, then $d|\mathbf{A}|/du = 0$. However, $|d\mathbf{A}/du|$ will not equal zero unless the direction of **A** is also constant.

The following useful identities can be derived from the definitions of derivatives (**A** and **B** are assumed to be vector functions of the scalar u, and m is also a scalar):

$$\frac{d(m\mathbf{A})}{du} = m\frac{d\mathbf{A}}{du} + \frac{dm}{du}\mathbf{A} \tag{11.4}$$

$$\frac{d(\mathbf{A} + \mathbf{B})}{du} = \frac{d\mathbf{A}}{du} + \frac{d\mathbf{B}}{du} \tag{11.5}$$

$$\frac{d(\mathbf{A} \cdot \mathbf{B})}{du} = \mathbf{A} \cdot \frac{d\mathbf{B}}{du} + \frac{d\mathbf{A}}{du} \cdot \mathbf{B} \tag{11.6}$$

$$\frac{d(\mathbf{A} \times \mathbf{B})}{du} = \mathbf{A} \times \frac{d\mathbf{B}}{du} + \frac{d\mathbf{A}}{du} \times \mathbf{B} \tag{11.7}$$

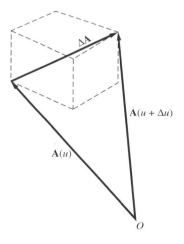

Fig. 11.2

11.3   *Position, Velocity, and Acceleration of a Particle*

a.   Position

Consider the motion of a particle along a smooth path as shown in Fig. 11.3. The position of the particle at time t is specified by the *position vector* $\mathbf{r}(t)$, which is the vector drawn from a fixed origin O to the particle. Let the location of the particle be A at time t, and B at time $t + \Delta t$, where Δt is a finite time interval. The corresponding change in the position vector of the particle,

$$\Delta \mathbf{r} = \mathbf{r}(t + \Delta t) - \mathbf{r}(t) \tag{11.8}$$

is called the *displacement vector* of the particle.

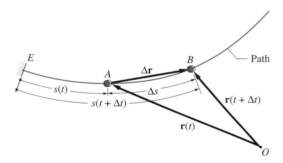

Fig. 11.3

As indicated in Fig. 11.3, the position of the particle at time t can also be specified by the *path coordinate* $s(t)$, which is the length of the path between a fixed point E and the particle. The change in path length during the time interval Δt is

$$\Delta s = s(t + \Delta t) - s(t) \tag{11.9}$$

Caution   The change in path length should not be confused with the *distance traveled* by the particle. The two are equal only if the direction of motion does not change during the time interval. If the direction of motion changes during Δt, then the distance traveled will be larger than Δs.

b.   Velocity

The velocity of the particle at time t is defined as

$$\mathbf{v}(t) = \lim_{\Delta t \to 0} \frac{\Delta \mathbf{r}}{\Delta t} = \dot{\mathbf{r}}(t) \tag{11.10}$$

where the overdot denotes differentiation with respect to time. Because the velocity is the derivative of the vector function $\mathbf{r}(t)$, it is also a vector. From Fig. 11.3

we observe that $\Delta\mathbf{r}$ becomes tangent to the path at A as $\Delta t \to 0$. Consequently, the *velocity vector is tangent to the path* of the particle.

We also deduce from Fig. 11.3 that $|\Delta\mathbf{r}| \to \Delta s$ as $\Delta t \to 0$. Therefore, the magnitude of the velocity, also known as the *speed* of the particle, is

$$v(t) = \lim_{\Delta t \to 0} \frac{|\Delta\mathbf{r}|}{\Delta t} = \lim_{\Delta t \to 0} \frac{\Delta s}{\Delta t} = \dot{s}(t) \qquad (11.11)$$

The dimension of velocity is [length/time], so the unit of velocity is m/s or ft/s.

c.   Acceleration

The velocity vectors of the particle at A (time t) and B (time $t + \Delta t$) are shown in Fig. 11.4(a). Note that both vectors are tangent to the path. The change in the velocity during the time interval Δt, shown in Fig. 11.4(b), is

$$\Delta\mathbf{v} = \mathbf{v}(t + \Delta t) - \mathbf{v}(t) \qquad (11.12)$$

The acceleration of the particle at time t is defined as

$$\mathbf{a}(t) = \lim_{\Delta t \to 0} \frac{\Delta\mathbf{v}}{\Delta t} = \dot{\mathbf{v}}(t) = \ddot{\mathbf{r}}(t) \qquad (11.13)$$

The acceleration is a vector of dimension [length/time^2]; hence its unit is m/s^2 or ft/s^2.

Caution   The acceleration vector is generally *not tangent to the path* of the particle. The direction of the acceleration coincides with $\Delta\mathbf{v}$ as $\Delta t \to 0$, which, as seen in Fig. 11.4(b), is not necessarily in the same direction as $\mathbf{v}$.

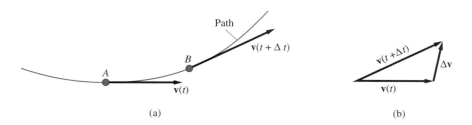

(a)   (b)

Fig. 11.4

11.4   *Newtonian Mechanics*[*]

a.   *Scope of Newtonian mechanics*

In 1687, Sir Isaac Newton (1642–1727) published his celebrated laws of motion in *Principia (Mathematical Principles of Natural Philosophy)*. Without a doubt,

---

[*]This article, which is the same as Art. 1.2 in *Statics*, is repeated here because of its relevance to our study of dynamics.

this work ranks among the most influential scientific books ever published. We should not think, however, that its publication immediately established classical mechanics. Newton's work on mechanics dealt primarily with celestial mechanics and was thus limited to particle motion. Another two hundred or so years elapsed before rigid-body dynamics, fluid mechanics, and the mechanics of deformable bodies were developed. Each of these areas required new axioms before it could assume a usable form.

Nevertheless, Newton's work is the foundation of classical, or Newtonian, mechanics. His efforts have even influenced two other branches of mechanics born at the beginning of the twentieth century: relativistic and quantum mechanics. *Relativistic mechanics* addresses phenomena that occur on a cosmic scale (velocities approaching the speed of light, strong gravitational fields, etc.). It removes two of the most objectionable postulates of Newtonian mechanics: the existence of a fixed or inertial reference frame and the assumption that time is an absolute variable, "running" at the same rate in all parts of the universe. (There is evidence that Newton himself was bothered by these two postulates.) *Quantum mechanics* is concerned with particles on the atomic or subatomic scale. It also removes two cherished concepts of classical mechanics: determinism and continuity. Quantum mechanics is essentially a probabilistic theory; instead of predicting an event, it determines the likelihood that an event will occur. Moreover, according to this theory, the events occur in discrete steps (called *quanta*) rather than in a continuous manner.

Relativistic and quantum mechanics, however, have by no means invalidated the principles of Newtonian mechanics. In the analysis of the motion of bodies encountered in our everyday experience, both theories converge on the equations of Newtonian mechanics. Thus the more esoteric theories actually reinforce the validity of Newton's laws of motion.

b. *Newton's laws for particle motion*

Using modern terminology, Newton's laws of particle motion may be stated as follows.

1. If a particle is at rest (or moving with constant velocity), it will remain at rest (or continue to move with constant velocity) unless acted on by a force.
2. A particle acted on by a force will accelerate in the direction of the force. The magnitude of the acceleration is proportional to the magnitude of the force and inversely proportional to the mass of the particle.
3. For every action, there is an equal and opposite reaction; that is, the forces of interaction between two particles are equal in magnitude and opposite in direction.

Although the first law is simply a special case of the second law, it is customary to state the first law separately because of its importance to the subject of statics.

c. *Inertial reference frames*

When applying Newton's second law, attention must be paid to the coordinate system in which the accelerations are measured. An *inertial reference* frame (also known as a Newtonian or Galilean reference frame) is defined to be any rigid coordinate system in which Newton's laws of particle motion relative to that frame are valid with an acceptable degree of accuracy. In most design applications used on the surface of the earth, an inertial frame can be approximated with sufficient accuracy by attaching the coordinate system to the earth. In the study of earth satellites, a coordinate system attached to the sun usually suffices. For interplanetary travel, it is necessary to use coordinate systems attached to the so-called fixed stars.

It can be shown that any frame that is translating with constant velocity relative to an inertial frame is itself an inertial frame. It is a common practice to omit the word *inertial* when referring to frames for which Newton's laws obviously apply.

d. Units and dimensions

The standards of measurement are called *units*. The term *dimension* refers to the type of measurement, regardless of the units used. For example, kilogram and meter/second are units, whereas mass and length/time are dimensions. The base dimensions in the *SI system* (from *Système international d'unités*) are mass $[M]$, length $[L]$, and time $[T]$, and the base units are kilogram (kg), meter (m), and second (s). All other dimensions or units are combinations of the base quantities. For example, the dimension of velocity is $[L/T]$, the unit being km/h, m/s, and so on.

A system with the base dimensions $[FLT]$ (such as the U.S. Customary system), is called a *gravitational system.* If the base dimensions are $[MLT]$ (as in the SI system), the system is known as an *absolute system.* In each system of measurement, the base units are defined by physically reproducible phenomena, or physical objects. For example, the second is defined by the duration of a specified number of radiation cycles in a certain isotope, and the kilogram is defined as the mass of a certain block of metal kept near Paris, France.

All equations representing physical phenomena must be *dimensionally homogenous;* that is, each term of the equation must have the same dimension. Otherwise, the equation will not make physical sense (it would be meaningless, for example, to add a force to a length). Checking equations for dimensional homogeneity is a good habit to learn, as it can reveal mistakes made during algebraic manipulations.

e. Mass, force, and weight

If a force **F** acts on a particle of mass m, Newton's second law states that

$$\mathbf{F} = m\mathbf{a} \tag{11.14}$$

where **a** is the acceleration vector of the particle. For a gravitational $[FLT]$ system, dimensional homogeneity of Eq. (11.14) requires the dimension of mass to be

$$[M] = \left[\frac{FT^2}{L}\right] \tag{11.15a}$$

For an absolute $[MLT]$ system of units, dimensional homogeneity of Eq. (11.4) yields for the dimension of force

$$[F] = \left[\frac{ML}{T^2}\right] \tag{11.15b}$$

The derived unit of force in the SI system is a *newton* (N), defined as the force that accelerates a 1.0-kg mass at the rate of 1.0 m/s². From Eq. (11.15b), we obtain

$$1.0\,\text{N} = 1.0\,\text{kg} \cdot \text{m/s}^2$$

Weight is the force of gravitation acting on a body. If we denote the gravitational acceleration (free-fall acceleration of the body) by g, the weight W of a body of mass m is given by Newton's second law as

$$W = mg \tag{11.16}$$

Note that mass is a constant property of a body, whereas weight is a variable that depends on the local value of g. The nominal gravitational acceleration at sea level, called *standard gravity*, is defined as $g = 9.80665$ m/s². The actual value of g varies from about 9.78 to 9.84, depending on the latitude and the proximity of large land masses. In this text, we mostly use the average value

$$g = 9.81\,\text{m/s}^2$$

in computations. However, in some cases calculation is rendered much simpler by rounding off this value to 9.8 m/s². Thus if the mass of a body is 1.0 kg, its weight on earth is $(9.81\,\text{m/s}^2)(1.0\,\text{kg}) = 9.81$ N.

f. Conversion of units

A convenient method for converting a measurement from one set of units to another set is to multiply the measurement by appropriate conversion factors. For example, to convert 180 km/h to m/s, we proceed as follows:

$$180\,\text{km/h} = 180\,\frac{\text{km}}{\text{h}} \times \frac{1.0\,\text{h}}{3600\,\text{s}} \times \frac{1000\,\text{m}}{1.0\,\text{km}} = 50\,\text{m/s}$$

we see that each conversion factor is dimensionless and of magnitude 1. Therefore, a measurement is unchanged when it is multiplied by conversion factors—only its units are altered. Note that it is permissible to cancel units during the conversion as if they were algebraic quantities.

Conversion factors applicable to mechanics are listed inside the front cover of the book.

g. *Law of gravitation*

In addition to his many other accomplishments, Newton also proposed the law of universal gravitation. Consider two particles of mass m_A and m_B that are separated by a distance R, as shown in Fig. 11.5. The law of gravitation states that the two particles are attracted to each other by forces of magnitude F that act along the line connecting the particles, where

$$F = G\frac{m_A\, m_B}{R^2}$$

(11.17)

The universal gravitational constant G is approximately 6.67×10^{-11} m^3/(kg·s^2). Although this law is valid for particles, Newton showed that it is also applicable to spherical bodies provided that their masses are distributed uniformly. (When attempting to derive this result, Newton had to develop calculus.)

If we let $m_A = M_e$ (the mass of the earth), $m_B = m$ (the mass of a body), and $R = R_e$ (the mean radius of the earth), then F in Eq. (11.17) will be the weight W of the body. Comparing $W = GM_e m/R_e^2$ with $W = mg$, we find that $g = GM_e/R_e^2$. Of course, adjustments may be necessary in the value of g for some applications in order to account for local variation of the gravitational attraction.

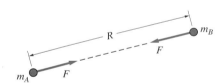

Fig. 11.5

Sample Problem 11.1

Convert 1.5 km/h to mm/s.

Solution

Using the standard conversion factors of the SI system of units, we obtain

$$1.5 \text{ km/h} = \frac{1.5 \text{ km}}{\text{h}} \times \frac{1.0 \text{ h}}{3600 \text{ s}} \times \frac{1000 \text{ m}}{1.0 \text{ km}} \times \frac{1000 \text{ mm}}{1.0 \text{ m}} = 416.66 \text{ mm/s}$$

Answer

Sample Problem 11.2

The acceleration a of a particle is related to its velocity v, its position coordinate x, and time t by the equation

$$a = Ax^3t + Bvt^2 \tag{a}$$

where A and B are constants. The dimension of the acceleration is length per unit time squared, that is, $[a] = [L/T^2]$. The dimensions of the other variables are $[v] = [L/T]$, $[x] = [L]$, and $[t] = [T]$. Derive the dimensions of A and B if Eq. (a) is to be dimensionally homogeneous. Express the units for A and B in the SI system.

Solution

For Eq. (a) to be dimensionally homogeneous, the dimension of each term on the right-hand side of the equation must be $[L/T^2]$, the same as the dimension for a. Therefore, the dimension of the first term on the right-hand side of Eq. (a) becomes

$$[Ax^3t] = [A][x^3][t] = [A][L^3][T] = \left[\frac{L}{T^2}\right] \tag{b}$$

Solving Eq. (b) for the dimension of A, we find

$$[A] = \frac{1}{[L^3][T]}\left[\frac{L}{T^2}\right] = \frac{1}{[L^2T^3]}$$

Answer

In the SI system the units of A are $\text{m}^{-2}\text{s}^{-3}$.

Performing a similar dimensional analysis on the second term on the right-hand side of Eq. (a) gives

$$[Bvt^2] = [B][v][t^2] = [B]\left[\frac{L}{T}\right][T^2] = \left[\frac{L}{T^2}\right] \tag{c}$$

Solving Eq. (c) for the dimension of B, we find

$$[B] = \left[\frac{L}{T^2}\right]\left[\frac{T}{L}\right]\left[\frac{1}{T^2}\right] = \left[\frac{1}{T^3}\right]$$

Answer

The units of B are s^{-3}.

Sample Problem 11.3

Find the gravitational force exerted by the earth on a 70-kg man whose height above the surface of the earth equals the radius of the earth. The mass and radius of the earth are $M_e = 5.9742 \times 10^{24}$ kg and $R_e = 6378$ km, respectively.

Solution

Consider a body of mass m located at a distance $2R_e$ from the center of the earth of mass M_e. The law of universal gravitation, from Eq. (11.17), states that the body is attracted to the earth by the force F given by

$$F = G\frac{mM_e}{(2R_e)^2}$$

where $G = 6.67 \times 10^{-11}$ m^3/(kg $\cdot$ s^2) is the universal gravitational constant. Substituting the values for G and the given parameters, the earth's gravitational force acting on the 70-kg man is

$$F = (6.67 \times 10^{-11})\frac{(70)(5.9742 \times 10^{24})}{[2(6378 \times 10^3)]^2} = 171.4 \text{ N} \qquad \textit{Answer}$$

Review of Equations

Differentiation formulas for vector functions

$$\frac{d(m\mathbf{A})}{du} = m\frac{d\mathbf{A}}{du} + \frac{dm}{du}\mathbf{A}$$

$$\frac{d(\mathbf{A}+\mathbf{B})}{du} = \frac{d\mathbf{A}}{du} + \frac{d\mathbf{B}}{du}$$

$$\frac{d(\mathbf{A}\cdot\mathbf{B})}{du} = \mathbf{A}\cdot\frac{d\mathbf{B}}{du} + \frac{d\mathbf{A}}{du}\cdot\mathbf{B}$$

$$\frac{d(\mathbf{A}\times\mathbf{B})}{du} = \mathbf{A}\times\frac{d\mathbf{B}}{du} + \frac{d\mathbf{A}}{du}\times\mathbf{B}$$

Position, velocity and acceleration of a particle

$$\text{Displacement}: \quad \Delta\mathbf{r} = \mathbf{r}(t+\Delta t) - \mathbf{r}(t)$$

$$\text{Change in path length}: \quad \Delta s = s(t+\Delta t) - s(t)$$

$$\text{Velocity}: \quad \mathbf{v}(t) = \lim_{\Delta t \to 0}\frac{\Delta\mathbf{r}}{\Delta t} = \dot{\mathbf{r}}(t) \quad v(t) = \dot{s}(t)$$

$$\text{Acceleration}: \quad \mathbf{a}(t) = \lim_{\Delta t \to 0}\frac{\Delta\mathbf{v}}{\Delta t} = \dot{\mathbf{v}}(t) = \ddot{\mathbf{r}}(t)$$

Newton's second law

$$\mathbf{F} = m\mathbf{a}$$

Weight-mass relationship

$$W = mg \quad g = 9.81 \text{ m/s}^2$$

Universal law of gravitation

$$F = G\frac{m_A m_B}{R^2}$$

$$G = 6.67 \times 10^{-11} \text{ m}^3/(\text{kg}\cdot\text{s}^2)$$

Problems

11.1 A person weighs 30 N on the moon, where $g = 1.6$ m/s^2. Determine (a) the mass of the person; and (b) the weight of the person on earth.

11.2 The radius and length of a steel cylinder are 60 mm and 120 mm, respectively. If the mass density of steel is 7850 kg/m^3, determine the weight of the cylinder.

11.3 Convert the following: (a) 100 kN/m^2 to lb/in.^{2}; (b) 30 m/s to km/h; (c) 800 slugs to Mg; (d) 20 lb/ft^2 to N/m^2. Use the conversion charts given on the inside cover.

11.4 Equate dimensionally Newton's second law and the universal law of gravitation and hence derive the units of the universal gravitational constant.

11.5 When a rigid body of mass m undergoes plane motion, its kinetic energy (KE) is

$$KE = \frac{1}{2}mv^2 + \frac{1}{2}mk^2\omega^2$$

where v is the velocity of its mass center, k is a constant, and ω is the angular velocity of the body in rad/s. Express the units of KE and k in terms of the base units of the SI system.

11.6 In a certain application, the acceleration a and the position coordinate x of a particle are related by

$$a = \frac{gkx}{W}$$

where g is the gravitational acceleration, k is a constant, and W is the weight of the particle. Show that this equation is dimensionally consistent if the dimension of k is $[F/L]$.

11.7 When a force F acts on a linear spring, the elongation x of the spring is given by $F = kx$, where k is called the *stiffness* of the spring. Determine the dimension of k in terms of the base dimensions of an absolute $[MLT]$ system of units.

11.8 Determine the dimensions of the following in terms of the base dimensions of a gravitational $[FLT]$ system of units: (a) mv^2; (b) mv; and (c) ma. The dimensions of the variables are $[m] = [M]$, $[v] = [L/T]$ and $a = [L/T^2]$.

11.9 A geometry textbook gives the equation of a parabola as $y = x^2$, where x and y are measured in mm. How can this equation be dimensionally correct?

11.10 The mass moment of inertia I of a homogeneous sphere about its diameter is $I = (2/5)mR^2$, where m and R are its mass and radius, respectively.

Find the dimension of I in terms of the base dimensions of (a) a gravitational $[FLT]$ system; and (b) an absolute $[MLT]$ system.

11.11   Determine the dimensions of constants A and B in the following equations, assuming each equation to be dimensionally correct: (a) $v^3 = Ax^2 + Bvt^2$; and (b) $x^2 = At^2e^{Bt^2}$. The dimensions of the variables are $[x] = [L]$, $[v] = [L/T]$ and $[a] = [L/T^2]$.

11.12   In a certain vibration problem the differential equation describing the motion of a particle of mass m is

$$m\frac{d^2x}{dt^2} + c\frac{dx}{dt} + kx = P_0 \sin \omega t$$

where x is the displacement of the particle. What are the dimensions of constants c, k, P_0, and ω in terms of the base dimensions of a gravitational $[FLT]$ system?

11.13   Using Eq. (11.17), derive the dimensions of the universal gravitational constant G in terms of the base dimensions of (a) a gravitational $[FLT]$ system; and (b) an absolute $[MLT]$ system.

11.14   A famous equation of Einstein is $E = mc^2$, where E is energy, m is mass, and c is the speed of light. Determine the dimension of energy in terms of the base dimensions of (a) a gravitational $[FLT]$ system; and (b) an absolute $[MLT]$ system.

11.15   Two 10-kg particles are placed 500 mm apart. Express the gravitational attraction acting on one of the particles as a percentage of its weight on earth.

11.16   Two identical spheres of mass 3 kg and radius 1 m are placed in contact. Find the gravitational attraction between them.

Use the following data in Problems 11.17 through 11.21:

Mass of earth $= 5.9742 \times 10^{24}$ kg
Radius of earth $= 6378$ km
Mass of moon $= 0.073483 \times 10^{24}$ kg
Radius of moon $= 1737$ km
Mass of sun $= 1.9891 \times 10^{30}$ kg
Distance between earth and sun $= 149.6 \times 10^6$ km

11.17   Find the mass of an object (in kg) that weighs 2 kN at a height of 1800 km above the earth's surface.

11.18   Prove that the weight of an object on the moon is approximately one-sixth of its weight on earth.

11.19   A man weighs 150 N on the surface of the earth. Find his weight at an elevation equal to the radius of the earth.

11.20   Determine the gravitational force exerted by the sun on a 1.0-kg object on the surface of the earth.

11.21   A spacecraft travels along the straight line connecting the earth and the sun. At what distance from earth will the gravitational forces of the earth and the sun be equal?

12

Dynamics of a Particle: Rectangular Coordinates

The fall of a skydiver is governed by the forces of gravity and aerodynamic drag. When these two forces are in balance, the skydiver is descending at a constant speed known as the terminal velocity. The determination of terminal velocity is the subject of Prob. 12.47. (Roberto Mettifogo/Photonica/Getty Images)

12.1   Introduction

In this chapter we study the dynamics (both kinematics and kinetics) of a particle in a rectangular coordinate system. The discussion is limited to a *single particle,* and the coordinate axes are assumed to be *fixed;* that is, not moving. The dynamics of two or more interacting particles and the kinematics of relative motion (translating coordinate systems) are covered in Chapter 15.

The definition of basic kinematical variables (position, velocity, and acceleration), which appeared in the previous chapter, made no reference to a coordinate system. Therefore, these definitions are applicable in any fixed reference frame. A specific coordinate system, however, is essential when we want to describe the motion. Here we employ the simplest of all reference frames: the rectangular Cartesian coordinate system. Although rectangular coordinates could be used in the solution of any problem, it is not always convenient to do so. Frequently, the curvilinear coordinate systems described in the next chapter lead to easier analyses.

Rectangular coordinates are naturally suited to the analysis of rectilinear motion (motion along a straight line) or curvilinear motion that can be described by a superposition of rectilinear motions, such as the flight of a projectile. These two applications form the bulk of this chapter.

An important problem of kinematics is introduced in the analysis of rectilinear motion: Given the acceleration of a particle, determine its velocity and position. This task, which is equivalent to integrating (solving) the second-order differential equation $\ddot{x} = f(\dot{x}, x, t)$, is encountered repeatedly throughout dynamics. Most of the differential equations encountered in this text are simple enough to be solved analytically. We do, however, include some problems that must be integrated numerically. Although these problems are optional, they are an important reminder that most practical problems do not have analytical solutions.

12.2   *Kinematics*

Figure 12.1(a) shows the path of particle A, which moves in a fixed rectangular reference frame. Letting **i**, **j**, and **k** be the base vectors (unit vectors), the position vector of the particle can be written as

$$\mathbf{r}(t) = x\mathbf{i} + y\mathbf{j} + z\mathbf{k} \qquad (12.1)$$

where x, y, and z are the time-dependent rectangular coordinates of the particle.

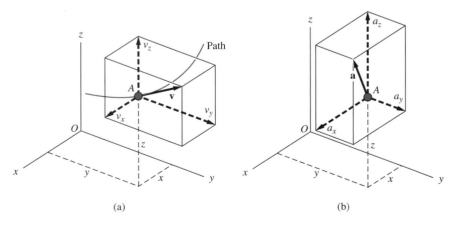

(a)   (b)

Fig. 12.1

Applying the definition of velocity, Eq. (11.10), and the chain rule of differentiation, Eq. (11.4), we obtain

$$\mathbf{v} = \frac{d\mathbf{r}}{dt} = \frac{d}{dt}(x\mathbf{i} + y\mathbf{j} + z\mathbf{k})$$

$$= x\frac{d\mathbf{i}}{dt} + \dot{x}\mathbf{i} + y\frac{d\mathbf{j}}{dt} + \dot{y}\mathbf{j} + z\frac{d\mathbf{k}}{dt} + \dot{z}\mathbf{k}$$

Because the coordinate axes are fixed,[*] the base vectors remain constant, so that $d\mathbf{i}/dt = d\mathbf{j}/dt = d\mathbf{k}/dt = \mathbf{0}$. Therefore, the velocity becomes

$$\mathbf{v} = v_x\mathbf{i} + v_y\mathbf{j} + v_z\mathbf{k} \qquad (12.2)$$

where the rectangular components, shown in Fig. 12.1(a), are

$$v_x = \dot{x} \qquad v_y = \dot{y} \qquad v_z = \dot{z} \qquad (12.3)$$

Similarly, the definition of acceleration, Eq. (11.13), yields

$$\mathbf{a} = \frac{d\mathbf{v}}{dt} = \frac{d}{dt}(v_x\mathbf{i} + v_y\mathbf{j} + v_z\mathbf{k}) = \dot{v}_x\mathbf{i} + \dot{v}_y\mathbf{j} + \dot{v}_z\mathbf{k}$$

Thus the acceleration is

$$\mathbf{a} = a_x\mathbf{i} + a_y\mathbf{j} + a_z\mathbf{k} \qquad (12.4)$$

with the rectangular components [see Fig. 12.1(b)]

$$a_x = \dot{v}_x = \ddot{x} \qquad a_y = \dot{v}_y = \ddot{y} \qquad a_z = \dot{v}_z = \ddot{z} \qquad (12.5)$$

a. Plane motion

Plane motion occurs often enough in engineering applications to warrant special attention. Figure 12.2(a) shows the path of a particle A that moves in the xy-plane. To obtain the two-dimensional rectangular components of $\mathbf{r}$, $\mathbf{v}$, and $\mathbf{a}$, we set $z = 0$ in Eqs. (12.1)–(12.5). The results are

$$\mathbf{r} = x\mathbf{i} + y\mathbf{j} \qquad \mathbf{v} = v_x\mathbf{i} + v_y\mathbf{j} \qquad \mathbf{a} = a_x\mathbf{i} + a_y\mathbf{j} \qquad (12.6)$$

where

$$v_x = \dot{x} \qquad v_y = \dot{y}$$
$$a_x = \dot{v}_x = \ddot{x} \qquad a_y = \dot{v}_y = \ddot{y} \qquad (12.7)$$

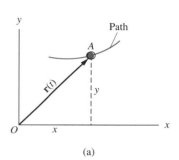

(a)

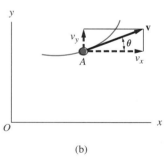

(b)

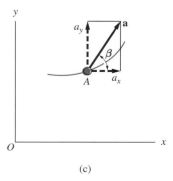

(c)

Fig. 12.2

[*]This assumption is actually overly restrictive. As we show in Chapter 16, the results remain valid if the coordinate system translates without rotating.

Figure 12.2(b) shows the rectangular components of the velocity. The angle θ, which defines the direction of **v**, can be obtained from

$$\tan \theta = \frac{v_y}{v_x} = \frac{dy/dt}{dx/dt} = \frac{dy}{dx}$$

Because the slope of the path is also equal to dy/dx, we see that **v** is tangent to the path, a result that was pointed out in the preceding chapter.

The rectangular components of **a** are shown in Fig. 12.2(c). The angle β that defines the direction of **a** can be computed from

$$\tan \beta = \frac{a_y}{a_x} = \frac{d^2y/dt^2}{d^2x/dt^2}$$

Because β is generally not equal to θ, the acceleration is not necessarily tangent to the path.

b. Rectilinear motion

If the path of a particle is a straight line, the motion is called *rectilinear.* An example of rectilinear motion, in which the particle *A* moves along the *x*-axis, is depicted in Fig. 12.3. In this case, we set $y = 0$ in Eqs. (12.6) and (12.7), obtaining **r** $= x$**i**, **v** $= v_x$**i**, and **a** $= a_x$**i**. Each of these vectors is directed along the path (i.e., the motion is one-dimensional). Because the subscripts are no longer needed, the equations for rectilinear motion along the *x*-axis are usually written as

$$\mathbf{r} = x\mathbf{i} \qquad \mathbf{v} = v\mathbf{i} \qquad \mathbf{a} = a\mathbf{i} \tag{12.8}$$

where

$$v = \dot{x} \qquad a = \dot{v} = \ddot{x} \tag{12.9}$$

In some problems, it is more convenient to express the acceleration in terms of velocity and position, rather than velocity and time. This change of variable can be accomplished by the chain rule of differentiation: $a = dv/dt = (dv/dx)(dx/dt)$. Noting that $dx/dt = v$, we obtain

$$a = v\frac{dv}{dx} \tag{12.10}$$

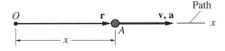

Fig. 12.3

Sample Problem   *12.1*

The position of a particle that moves along the x-axis is defined by $x = -3t^2 + 12t - 6$ m, where t is in seconds. For the time interval $t = 0$ to $t = 3$ s, (1) plot the position, velocity, and acceleration as functions of time; (2) calculate the distance traveled; and (3) determine the displacement of the particle.

Solution

Part 1

Because the motion is rectilinear, the velocity and acceleration may be calculated as follows.

$$x = -3t^2 + 12t - 6 \text{ m} \qquad \text{(a)}$$

$$v = \frac{dx}{dt} = -6t + 12 \text{ m/s} \qquad \text{(b)}$$

$$a = \frac{dv}{dt} = \frac{d^2x}{dt^2} = -6 \text{ m/s}^2 \qquad \text{(c)}$$

These functions are plotted in Figs. (a)–(c) for the prescribed time interval $t = 0$ to $t = 3$ s. Note that the plot of x is parabolic, so that successive differentiations yield a linear function for the velocity and a constant value for the acceleration. The time corresponding to the maximum (or minimum) value of x can be found by setting $dx/dt = 0$, or utilizing Eq. (b), $v = -6t + 12 = 0$, which gives $t = 2$ s. Substituting $t = 2$ s into Eq. (a), we find

$$x_{\text{max}} = -3(2)^2 + 12(2) - 6 = 6 \text{ m}$$

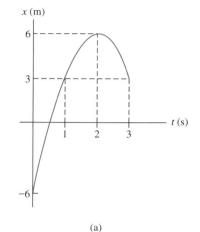

(a)

Part 2

Figure (d) shows how the particle moves during the time interval $t = 0$ to $t = 3$ s. When $t = 0$, the particle leaves A ($x = -6$ m), moving to the right. When $t = 2$ s, the particle comes to a stop at B ($x = 6$ m). Then it moves to the left, arriving at C ($x = 3$ m) when $t = 3$ s. Therefore, the distance traveled is equal to the distance that the point moves to the right ($\overline{AB}$) plus the distance it moves to the left ($\overline{BC}$), which gives

$$d = \overline{AB} + \overline{BC} = 12 + 3 = 15 \text{ m} \qquad \textbf{\textit{Answer}}$$

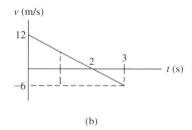

(b)

Part 3

The displacement during the time interval $t = 0$ to $t = 3$ s is the vector drawn from the initial position of the point to its final position. This vector, indicated as $\Delta \mathbf{r}$ in Fig. (d), is

$$\Delta \mathbf{r} = 9\mathbf{i} \text{ m} \qquad \textbf{\textit{Answer}}$$

Observe that the total distance traveled (15 m) is greater than the magnitude of the displacement vector (9 m) because the direction of the motion changes during the time interval.

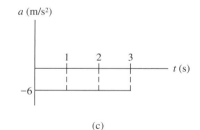

(c)

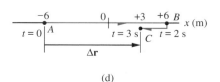

(d)

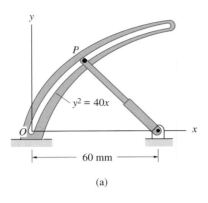

y

P

$y^2 = 40x$

O   *x*

|←   60 mm   →|

(a)

Sample Problem   *12.2*

Pin *P* at the end of the telescoping rod in Fig. (a) slides along the fixed parabolic path $y^2 = 40x$, where *x* and *y* are measured in millimeters. The *y* coordinate of *P* varies with time *t* (measured in seconds) according to $y = 4t^2 + 6t$ mm. When $y = 30$ mm, compute (1) the velocity vector of *P*; and (2) the acceleration vector of *P*.

Solution

Part 1

Substituting

$$y = 4t^2 + 6t \text{ mm} \tag{a}$$

into the equation of the path and solving for *x*, we obtain

$$x = \frac{y^2}{40} = \frac{(4t^2 + 6t)^2}{40} = 0.40t^4 + 1.20t^3 + 0.90t^2 \text{ mm} \tag{b}$$

The rectangular components of the velocity vector thus are

$$v_x = \dot{x} = 1.60t^3 + 3.60t^2 + 1.80t \text{ mm/s} \tag{c}$$

$$v_y = \dot{y} = 8t + 6 \text{ mm/s} \tag{d}$$

Setting $y = 30$ mm in Eq. (a) and solving for *t* gives $t = 2.090$ s. Substituting this value of time into Eqs. (c) and (d), we obtain

$$v_x = 34.1 \text{ mm/s} \quad \text{and} \quad v_y = 22.7 \text{ mm/s}$$

Consequently, the velocity vector at $y = 30$ mm is

$$\mathbf{v} = 34.1\mathbf{i} + 22.7\mathbf{j} \text{ mm/s} \qquad \textit{Answer}$$

The pictorial representation of this result is shown below and also in Fig. (b).

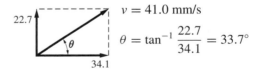

$v = 41.0$ mm/s

22.7

θ

34.1

$$\theta = \tan^{-1} \frac{22.7}{34.1} = 33.7°$$

By evaluating the slope of the path, *dy/dx*, at $y = 30$ mm, it is easy to verify that the velocity vector determined above is indeed tangent to the path.

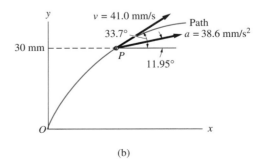

y

$v = 41.0$ mm/s

Path

33.7°

$a = 38.6$ mm/s²

30 mm

P

11.95°

O   *x*

(b)

Part 2

From Eqs. (c) and (d), we can determine the components of the acceleration vector by differentiation:

$$a_x = \dot{v}_x = 4.80t^2 + 7.20t + 1.80 \text{ mm/s}^2$$

$$a_y = \dot{v}_y = 8 \text{ mm/s}^2$$

Substituting $t = 2.090$ s, we obtain

$$a_x = 37.8 \text{ mm/s}^2 \quad \text{and} \quad a_y = 8 \text{ mm/s}^2$$

Therefore, the acceleration vector at $y = 30$ mm is

$$\mathbf{a} = 37.8\mathbf{i} + 8\mathbf{j} \text{ mm/s}^2 \qquad \textbf{\textit{Answer}}$$

The pictorial representation of **a** is

$$a = 38.6 \text{ mm/s}^2$$

$$\beta = \tan^{-1}\frac{8}{37.8} = 11.95°$$

From the drawing of the acceleration vector in Fig. (b) we see that the direction of **a** is not tangent to the path.

Sample Problem   12.3

The rigid arm OA of length R rotates about the ball-and-socket joint at O. The x- and y-coordinates describing the spatial motion of end A are

$$x = R\cos\omega t \quad y = \frac{R}{2}\sin 2\omega t$$

where ω is a constant. Find the expression for the z-coordinate of end A.

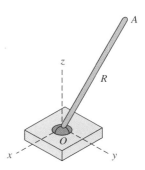

Solution

Because the arm OA is rigid, the position coordinates of end A must satisfy the equation

$$x^2 + y^2 + z^2 = R^2$$

Substituting the expressions for x and y gives

$$R^2\cos^2\omega t + \frac{R^2}{4}\sin^2 2\omega t + z^2 = R^2$$

Using the trigonometric identities $\sin 2\omega t = 2 \sin \omega t \, \cos \omega t$ and $(1 - \cos^2 \omega t) = \sin^2 \omega t$, we get

$$z^2 = R^2(1 - \cos^2 \omega t - \sin^2 \omega t \cos^2 \omega t)$$

$$= R^2(\sin^2 \omega t - \sin^2 \omega t \cos^2 \omega t)$$

$$= R^2 \sin^2 \omega t \, (1 - \cos^2 \omega t)$$

$$= R^2 \sin^4 \omega t$$

Therefore, the expression for the z-coordinate is

$$z = R \sin^2 \omega t \qquad \textit{Answer}$$

Sample Problem   **12.4**

The circular cam of radius $R = 16$ mm is pivoted at O, thus producing an eccentricity of $R/2$. Using geometry, it can be shown that the relationship between x, the position coordinate of the follower A, and the angle θ is

$$x(\theta) = \frac{R}{2}\left(\cos\theta + \sqrt{\cos^2\theta + 3}\right)$$

If the cam is rotating clockwise about O with the constant angular speed $\dot\theta = 2000$ rev/min, determine the speed the follower when $\theta = 45°$.

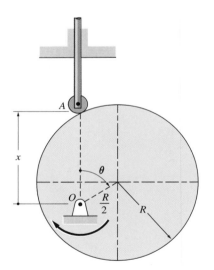

Solution

$$v = \frac{dx}{dt} = \frac{dx}{d\theta}\frac{d\theta}{dt}$$

$$= \frac{R}{2}\left[-\sin\theta + \frac{1}{2}\left(\frac{-2\cos\theta\sin\theta}{\sqrt{\cos^2\theta + 3}}\right)\right]\dot\theta$$

$$= -\frac{R}{2}\sin\theta\left(1 + \frac{\cos\theta}{\sqrt{\cos^2\theta + 3}}\right)\dot\theta$$

Substituting $R = 0.016$ m, $\theta = 45°$, and $\dot\theta = 2000(2\pi/60)$ rad/s, we get

$$v = -\frac{0.016}{2}(\sin 45°)\left(1 + \frac{\cos 45°}{\sqrt{\cos^2 45° + 3}}\right)\frac{2000(2\pi)}{60} = -1.633 \text{ m/s} \quad \textit{Answer}$$

The minus sign indicates that the follower is moving downward.

Problems

12.1   A rocket is launched vertically at time $t = 0$. The elevation of the rocket is given by

$$y = -0.13t^4 + 4.1t^3 + 0.12t^2 \text{ m}$$

where t is in seconds. Determine the maximum velocity of the rocket and the elevation at which it occurs.

12.2   When an object is tossed vertically upward on the surface of a planet, the ensuing motion in the absence of atmospheric resistance can be described by

$$x = -\frac{1}{2}gt^2 + v_0t$$

where g and v_0 are constants. (a) Derive the expressions for the velocity and acceleration of the object. Use the results to show that v_0 is the initial speed of the body and that g represents the gravitational acceleration. (b) Derive the maximum height reached by the object and the total time of flight. (c) Evaluate the results of Part (b) for $v_0 = 90$ km/h and $g = 9.8$ m/s^2 (surface of the earth).

Fig. P12.2

12.3   The position of a particle moving along the x-axis is described by

$$x = t^3 - 108t \text{ m}$$

where t is the time in seconds. For the time interval $t = 0$ to $t = 10$ s, (a) plot the position, velocity, and acceleration as functions of time; (b) find the displacement of the particle; and (c) determine the distance traveled by the particle.

12.4   The position of a particle that moves along the x-axis is given by

$$x = t^3 - 3t^2 - 45t \text{ m}$$

where t is the time in seconds. Determine the position, velocity, acceleration, and distance traveled at $t = 8$ s.

12.5   The position of a car moving on a straight highway is given by

$$x = t^2 - \frac{t^3}{90} \text{ m}$$

where t is the time in seconds. Determine (a) the distance traveled by the car before it comes to a stop; and (b) the maximum velocity reached by the car.

12.6   A body is released from rest at A and allowed to fall freely. Including the effects of air resistance, the position of the body as a function of the elapsed time is

$$x = v_0 \left(t - t_0 + t_0 e^{-t/t_0}\right)$$

where v_0 and t_0 are constants. (a) Derive the expression for the speed v of the body. Use the result to explain why v_0 is called the *terminal velocity*. (b) Derive

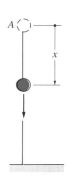

Fig. P12.6

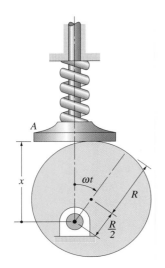

Fig. P12.9

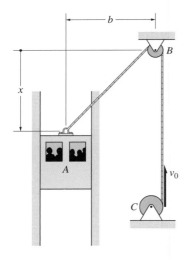

Fig. P12.10

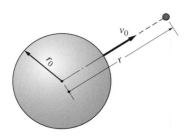

Fig. P12.11

the expressions for the acceleration a of the body as a function of t and as a function of v.

12.7   A bead moves along a straight 60-mm wire that lies along the x-axis. The position of the bead is given by

$$x = 2t^2 - 10t \text{ mm}$$

where x is measured from the center of the wire, and t is the time in seconds. Determine (a) the time when the bead leaves the wire; and (b) the distance traveled by the bead from $t = 0$ until it leaves the wire.

12.8   A particle moves along the curve $x^2 = 12y$, where x and y are measured in millimeters. The x-coordinate varies with time according to

$$x = 4t^2 - 2 \text{ mm}$$

where the time t is in seconds. Determine the magnitudes of the velocity and acceleration vectors when $t = 2$ s.

12.9   The circular cam of radius R and eccentricity $R/2$ rotates clockwise with a constant angular speed ω. The resulting vertical motion of the flat follower A can be shown to be

$$x = R\left(1 + \frac{1}{2}\cos\omega t\right)$$

(a) Obtain the velocity and acceleration of the follower as a function of t. (b) If ω were doubled, how would the maximum velocity and maximum acceleration of the follower be changed?

12.10   The elevator A is lowered by a cable that runs over pulley B. If the cable unwinds from the winch C at the constant speed v_0, the motion of the elevator is

$$x = \sqrt{(v_0 t - b)^2 - b^2}$$

Determine the velocity and acceleration of the elevator in terms of the time t.

12.11   A missile is launched from the surface of a planet with the speed v_0 at $t = 0$. According to the theory of universal gravitation, the speed v of the missile after launch is given by

$$v^2 = 2gr_0\left(\frac{r_0}{r} - 1\right) + v_0^2$$

where g is the gravitational acceleration on the surface of the planet and r_0 is the mean radius of the planet. (a) Determine the acceleration of the missile in terms of r. (b) Find the *escape velocity*, that is, the minimum value of v_0 for which the missile will not return to the planet. (c) Using the result of Part (b), calculate the escape velocity for earth, where $g = 9.8$ m/s^2 and $r_0 = 6400$ km.

12.12   The coordinates of a particle undergoing plane motion are

$$x = 15 - 2t^2 \text{ m} \qquad y = 15 - 10t + t^2 \text{ m}$$

where t is the time in seconds. Find the velocity and acceleration vectors at (a) $t = 0$ s; and (b) $t = 5$ s.

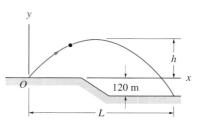

Fig. P12.13

12.13   A projectile fired at O follows a parabolic trajectory, given in parametric form by

$$x = 66t \qquad y = 86t - 4.91t^2$$

where x and y are measured in meters and t in seconds. Determine (a) the acceleration vector throughout the flight; (b) the velocity vector at O; (c) the maximum height h; and (d) the range L.

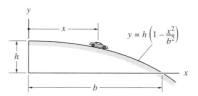

Fig. P12.14

12.14   An automobile goes down a hill that has the parabolic cross section shown. Assuming that the horizontal component of the velocity vector has a constant magnitude v_0, determine (a) the expression for the speed of the automobile in terms of x; and (b) the magnitude and direction of the acceleration.

12.15   The position of a particle in plane motion is defined by

$$x = a \cos \omega t \qquad y = b \sin \omega t$$

where $a > b$, and ω is a constant. (a) Show that the path of the particle is an ellipse. (b) Prove that the acceleration vector is always directed toward the center of the ellipse.

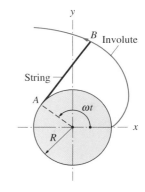

Fig. P12.16

12.16   When a taut string is unwound from a stationary cylinder, the end B of the string generates a curve known as the *involute* of a circle. If the string is unwound at the constant angular speed ω, the equation of the involute is

$$x = R \cos \omega t + R\omega t \sin \omega t \qquad y = R \sin \omega t - R\omega t \cos \omega t$$

where R is the radius of the cylinder. Find the speed of B as a function of time. Show that the velocity vector is always perpendicular to the string.

12.17   When a wheel of radius R rolls with a constant angular velocity ω, the point B on the circumference of the wheel traces out a curve known as a *cycloid*, the equation of which is

$$x = R(\omega t - \sin \omega t) \qquad y = R(1 - \cos \omega t)$$

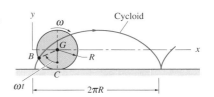

Fig. P12.17

(a) Show that the velocity vector of B is always perpendicular to $\overline{BC}$. (b) Show that the acceleration vector of B is directed along $\overline{BG}$.

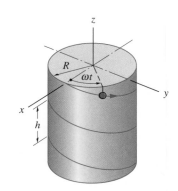

Fig. P12.18

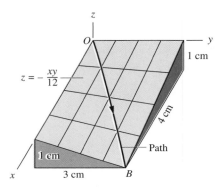

Fig. P12.19

12.18 When a particle moves along the helix shown, the components of its position vector are

$$x = R \cos \omega t \qquad y = R \sin \omega t \qquad z = -\frac{h}{2\pi} \omega t$$

where ω is constant. Show that the velocity and acceleration have constant magnitudes, and compute their values if $R = 1.2$ m, $h = 0.75$ m, and $\omega = 4\pi$ rad/s.

12.19 Path OB of a particle lies on the hyperbolic paraboloid shown. The description of motion is

$$x = \frac{4}{5} v_0 t \qquad y = \frac{3}{5} v_0 t \qquad z = -\frac{1}{25} v_0^2 t^2$$

where the coordinates are measured in inches, and v_0 is a constant. Determine (a) the velocity and acceleration when the particle is at B; and (b) the angle between the path and the xy-plane at B.

12.20 The spatial motion of a particle is described by

$$x = 3t^2 + 4t \qquad y = -4t^2 + 3t \qquad z = -6t + 9$$

where the coordinates are measured in m and the time t is in seconds. (a) Determine the velocity and acceleration vectors of the particle as functions of time. (b) Verify that the particle is undergoing plane motion (the motion is not in a coordinate plane) by showing that the unit vector perpendicular to the plane formed by $\mathbf{v}$ and $\mathbf{a}$ is constant.

12.21 The three-dimensional motion of a point is described by

$$x = R \cos \omega t \qquad y = R \sin \omega t \qquad z = \frac{R}{2} \sin 2\omega t$$

where R and ω are constants. Calculate the maximum speed and maximum acceleration of the point.

12.22 For the mechanism shown, determine (a) the velocity $\dot{x}$ of slider C in terms of θ and $\dot{\theta}$; and (b) the acceleration $\ddot{x}$ of C in terms of θ, $\dot{\theta}$, and $\ddot{\theta}$.

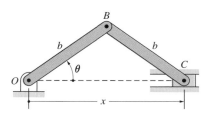

Fig. P12.22

12.23 The pin attached to the sliding collar A engages the slot in bar OB. Determine (a) the speed $\dot{y}$ of A in terms of θ and $\dot{\theta}$; and (b) the acceleration $\ddot{y}$ of A in terms of θ, $\dot{\theta}$, and $\ddot{\theta}$.

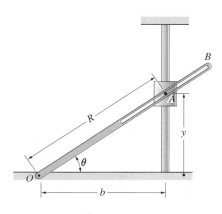

Fig. P12.23

12.24 The position coordinate of piston A can be shown to be related to the crank angle θ of the flywheel by

$$x = R\left(\cos\theta + \sqrt{9 - \sin^2\theta}\right)$$

The flywheel rotates at the constant angular speed $\dot{\theta}$. Derive the expression for the velocity $\dot{x}$ of the piston as a function of θ.

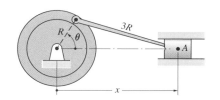

Fig. P12.24

12.25 The profile of the cam is

$$r = 55 + 10\cos\theta + 5\cos 2\theta \text{ mm}$$

If the cam rotates at the constant angular velocity of $\dot{\theta} = 1200$ rev/min, determine the maximum acceleration of follower A.

12.26 The plane C is being tracked by radar stations A and B. At the instant shown, the triangle ABC lies in the vertical plane, and the radar readings are $\theta_A = 30°$, $\theta_B = 22°$, $\dot{\theta}_A = 0.026$ rad/s, and $\dot{\theta}_B = 0.032$ rad/s. Determine (a) the altitude y; (b) the speed v; and (c) the climb angle α of the plane at this instant.

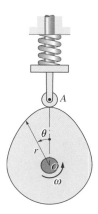

Fig. P12.25

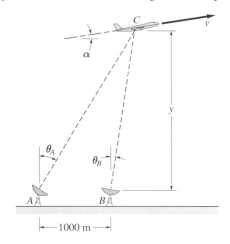

Fig. P12.26

Kinetics: Force-Mass-Acceleration Method

a. *Equations of motion*

When several forces act on a particle of mass m, Newton's second law has the form $\Sigma\mathbf{F} = m\mathbf{a}$, where $\Sigma\mathbf{F}$ is the vector sum of the forces (the resultant force), and $\mathbf{a}$ is the acceleration of the particle. The scalar representation of this vector equation in rectangular coordinates is

$$\Sigma F_x = ma_x \qquad \Sigma F_y = ma_y \qquad \Sigma F_z = ma_z \qquad (12.11)$$

Equations (12.11) are known as the *equations of motion* of the particle.

If the acceleration of the particle is known, we can use the equations of motion to find the forces. If the forces are given, the equations of motion can be solved for the accelerations. Most problems, however, are of the mixed type, where only some of the forces and some of the acceleration components are known.

We call the process of relating the forces to the acceleration of the particle by means of Eqs. (12.11) the *force-mass-acceleration (FMA) method*. Later we will learn other procedures, such as work-energy and impulse-momentum methods, that can also be used to obtain relationships between the forces and the motion.

b. *Free-body and mass-acceleration diagrams*

It is standard practice to start the FMA method by drawing two diagrams, each representing one side of Newton's second law $\Sigma \mathbf{F} = m\mathbf{a}$. The first of these is the *free-body diagram (FBD)* that shows all the forces acting on the particle. The second diagram, which we refer to as the *mass-acceleration diagram (MAD)*, displays the *inertia vector m$\mathbf{a}$* of the particle. Newton's second law can now be satisfied by requiring the two diagrams to be statically equivalent, that is, to have the same resultant.

The FBD and the MAD of a particle are shown in Fig. 12.4(a). The equal sign between the diagrams indicates static equivalence. If rectangular coordinates are employed, the inertia vector is usually represented by its rectangular components, as illustrated in Fig. 12.4(b). Once the diagrams have been drawn, it is relatively easy to write down the conditions of static equivalence, that is, the equations of motion.

The free-body diagram is as important in dynamics as it is in statics. It identifies all the forces that act on the particle in a clear and concise manner, it defines the notation used for unknown quantities, and it displays the known quantities. The mass-acceleration diagram serves a similar purpose. It also defines the notation for the unknowns, and it shows the known magnitudes and directions. But perhaps the greatest benefit of the MAD is that it focuses our attention on the kinematics required to describe the inertia vector. After all, it is kinematics that enables us to decide which components of the acceleration vector are known beforehand and which components are unknown.

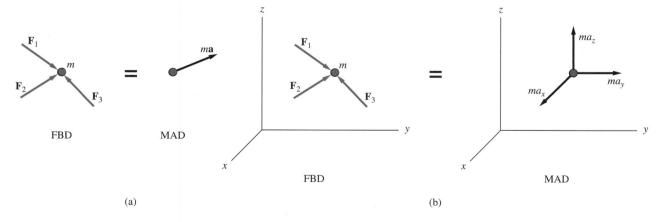

FBD   MAD

(a)

FBD

MAD

(b)

Fig. 12.4

In summary, the FMA method consists of the following steps.

Step 1: Draw the free-body diagram (FBD) of the particle that shows all forces acting on the particle.
Step 2: Use kinematics to analyze the acceleration of the particle.
Step 3: Draw the mass-acceleration diagram (MAD) for the particle that displays the inertia vector $m\mathbf{a}$, utilizing the results of Step 2.
Step 4: Referring to the FBD and MAD, relate the forces to the acceleration using static equivalence of the two diagrams.

12.4 Dynamics of Rectilinear Motion

a. Equations of motion

Figure 12.5 shows the FBD and the MAD of a particle that is in rectilinear motion along the x-axis. The corresponding equations of motion are

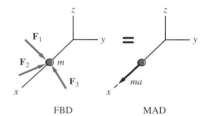

FBD   MAD

Fig. 12.5

$$\Sigma F_x = ma \qquad (12.12)$$

$$\Sigma F_y = \Sigma F_z = 0 \qquad (12.13)$$

In some problems all the forces acting on the particle are in the direction of motion (the x-direction), in which case Eqs. (12.13) are automatically satisfied. Otherwise, Eqs. (12.13) can be used in the computation of unknown forces, such as the reactions.

b. Determination of velocity and position

Let us assume that we wrote the equations of motion for an *arbitrary position* of the particle and then solved them for the acceleration a. Because the position of the particle is arbitrary, the acceleration would generally be a function of the position and velocity of the particle, and time:

$$a = f(v, x, t) \qquad (12.14)$$

An equivalent form of Eq. (12.14) is

$$\ddot{x} = f(\dot{x}, x, t)$$

which is a second-order, ordinary differential equation. The solution of this differential equation would be $x(t)$, the position as a function of time.

If all three variables (x, v, and t) appear explicitly in the expression for a in Eq. (12.14), then the chances of obtaining an analytical solution are slim. The reason is that f is usually a nonlinear function; that is, it contains nonlinear terms of the variables, such as $\sin x$ or v^2. In most cases, nonlinear differential equations can be solved only numerically. However, if f contains only one of the variables, the differential equation can be integrated in a straightforward manner, as shown below.

Case 1: $a = f(t)$ From $a = dv/dt$, we get

$$dv = a(t)\,dt \qquad (12.15)$$

Both sides of the equation can now be integrated, yielding the velocity as a function of time:

$$v(t) = \int a(t)\,dt + C_1 \qquad (12.16)$$

After the velocity has been determined, the position coordinate x can be obtained from $v = dx/dt$, or $dx = v(t)\,dt$. Integrating both sides, we get

$$x(t) = \int v(t)\,dt + C_2 \qquad (12.17)$$

The constants of integration, C_1 and C_2, can be evaluated from the initial conditions (usually the given values of x and v at $t = 0$).

Case 2: $a = f(x)$ Here we utilize Eq. (12.10): $a = v\,dv/dx$. The variables can be separated so that x and v appear on opposite sides of the equation:

$$v\,dv = a(x)\,dx \qquad (12.18)$$

The equation can now be integrated, with the result

$$\frac{1}{2}v^2 = \int a(x)\,dx + C_3$$

where C_3 is the constant of integration. Therefore,

$$v(x) = \sqrt{2\left[\int a(x)\,dx + C_3\right]} \qquad (12.19)$$

At this stage we could replace v by dx/dt in Eq. (12.19), separate the variables x and t, and integrate again to obtain $x(t)$. But the integration may not be easy due to the presence of the square root.

Case 3: $a = f(v)$ We can start with Eq. (12.18), which, after replacing $a(x)$ by $a(v)$, is

$$v\,dv = a(v)\,dx$$

Separating the variables x and v, we have

$$dx = \frac{v\,dv}{a(v)} \qquad (12.20)$$

Upon integration, we obtain x as a function of v:

$$x(v) = \int \frac{v\,dv}{a(v)} + C_4 \qquad (12.21)$$

Equation (12.21) may be inverted (solved for the velocity) if we want v as a function of x.

We could also start with Eq. (12.15): $dv = a(v)\,dt$. Rearranging terms to separate the variables leads to

$$dt = \frac{dv}{a(v)} \qquad (12.22)$$

which can be integrated, giving t in terms of v:

$$t(v) = \int \frac{dv}{a(v)} + C_5 \qquad (12.23)$$

We could now invert the result, thereby obtaining v as a function of t.

Sample Problem 12.5

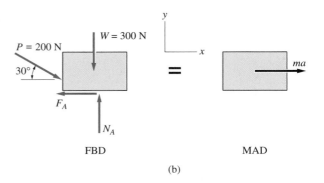

(a)

The 300-N block A in Fig. (a) is at rest on the horizontal plane when the force P is applied at $t = 0$. Find the velocity and position of the block when $t = 5$ s. The coefficients of static and kinetic friction are 0.2.

Solution

The FBD of the block is shown in Fig. (b), where N_A and F_A are the normal and friction forces exerted on the block by the plane. Figure (b) also shows the MAD. Because the motion is rectilinear, $a_y = 0$.

FBD   MAD

(b)

Referring to the FBD and MAD, we get

$$\Sigma F_y = 0 \ +\!\uparrow \qquad N_A - W - P \sin 30° = 0 \qquad \text{(a)}$$

$$\Sigma F_x = ma \ \xrightarrow{+} \qquad P \cos 30° - F_A = ma \qquad \text{(b)}$$

Equation (a) yields

$$N_A = W + P \sin 30° = 300 + 200 \sin 30° = 400 \text{ N}$$

Therefore, the friction force is

$$F_A = \mu_k N_A = 0.2 \,(400) = 80 \text{ N}$$

From Eq. (b), we obtain

$$a = \frac{1}{m}(P \cos 30° - F_A) = \frac{9.81}{300}(200 \cos 30° - 80) = 3.048 \text{ m/s}^2$$

The velocity v and position coordinate x of the block now can be found by integration as follows:

$$v = \int a \, dt = \int 3.048 \, dt = 3.048t + C_1 \qquad \text{(c)}$$

$$x = \int v \, dt = \int (3.048t + C_1)dt = 1.524t^2 + C_1 t + C_2 \qquad \text{(d)}$$

where C_1 and C_2 are constants of integration to be found from the initial conditions. The given initial velocity is zero. However, we are free to choose the origin of the x-axis. The most convenient choice is to let $x = 0$ when $t = 0$. Therefore, the initial conditions are

$$v = 0 \text{ and } x = 0 \text{ when } t = 0$$

Substituting these values into Eqs. (c) and (d) gives $C_1 = 0$ and $C_2 = 0$. Therefore, the velocity and position coordinate of the block at $t = 5$ s are

$$v = 3.048(5) = 15.24 \text{ m/s} \qquad \textit{Answer}$$

$$x = 1.524(5)^2 = 38.1 \text{ m} \qquad \textit{Answer}$$

Sample Problem 12.6

Figure (a) shows a crate of mass m resting on the bed of a dump truck. The coefficient of static friction between the crate and the bed of the truck is 0.64. In order to make the crate slide when the bed is in the position shown, the truck must accelerate to the right. Determine the smallest acceleration a for which the crate will begin to slide. Express the answer in terms of the gravitational acceleration g.

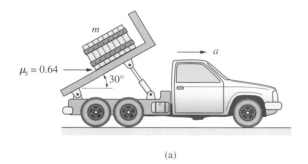

(a)

Solution

The free-body diagram (FBD) of the crate is shown in Fig. (b). In addition to the weight $W = mg$, the crate is acted on by the normal contact force N and the friction force $F = 0.64N$ (because the crate is in a state of impending sliding, F equals its maximum static value $\mu_s N$). Figure (b) also shows the mass-acceleration diagram (MAD) of the crate. Because the crate and the truck have the same acceleration before sliding occurs, the inertia vector of the crate is ma, directed horizontally.

Referring to Fig. (b), the equations of motion of the crate are

$$\Sigma F_x = ma \xrightarrow{+} \quad -N \sin 30° + 0.64N \cos 30° = ma \qquad \text{(a)}$$

$$\Sigma F_y = 0 \ +\!\uparrow \quad N \cos 30° + 0.64N \sin 30° - mg = 0 \qquad \text{(b)}$$

From Eq. (b), we obtain

$$N = \frac{mg}{\cos 30° + 0.64 \sin 30°} = 0.8432mg \qquad \text{(c)}$$

Substitution of Eq. (c) in Eq. (a) yields

$$0.8432mg(-\sin 30° + 0.64 \cos 30°) = ma$$

from which

$$a = 0.0458g \qquad \textit{Answer}$$

Note that the result is independent of m (the mass of the crate).

Sample Problem   **12.7**

The block of mass m in Fig. (a) slides on a horizontal plane with negligible friction. The position coordinate x is measured from the undeformed position of the ideal spring of stiffness k. If the block is launched at $x = 0$ with the velocity v_0 to the right, determine (1) the acceleration of the block as a function of x; (2) the velocity of the block as a function of x; and (3) the value of x when the block comes to rest for the first time.

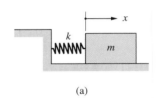

(a)

Solution

Part 1

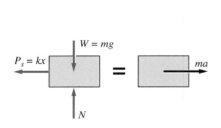

FBD   MAD

(b)

The FBD of the block for an arbitrary value of x is shown in Fig. (b), where N is the normal force exerted by the frictionless plane, and $P_s = kx$ is the force caused by the spring. Figure (b) also shows the MAD. Because the motion occurs only in the x direction, we have $a_y = 0$, and the magnitude of the inertia vector is $ma_x = ma$. Referring to the FBD and MAD, the equation of motion is

$$\Sigma F_x = ma \quad \xrightarrow{+} \quad -kx = ma$$

from which

$$a = -\frac{k}{m}x \qquad \textit{Answer} \quad \text{(a)}$$

Part 2

We can determine the velocity as a function of position by choosing x as the independent variable. Using $a = v\,dv/dx$ from Eq. (12.10), Eq. (a) becomes

$$v\frac{dv}{dx} = -\frac{k}{m}x$$

Separating the variables x and v by rearranging the terms, we find

$$v\, dv = -\frac{k}{m}x\, dx \qquad \text{(b)}$$

Integration of Eq. (b) yields

$$\frac{v^2}{2} = -\frac{kx^2}{2m} + C \qquad \text{(c)}$$

The constant of integration C can be evaluated from the initial condition: $v = v_0$ when $x = 0$, yielding $C = v_0^2/2$. Therefore, the velocity can be expressed as

$$v = \pm\sqrt{(-k/m)x^2 + v_0^2} \qquad \textbf{\textit{Answer}} \quad \text{(d)}$$

Part 3

The position of the block when it first comes to rest is found by setting $v = 0$ in Eq. (d), the result being

$$x = v_0\sqrt{m/k} \qquad \textbf{\textit{Answer}}$$

Note

With $C = v_0^2/2$, Eq. (c) can be written as $v^2 + (k/m)x^2 = v_0^2$. If we plot v versus x, called the *phase plane* plot, the result is the ellipse shown in Fig. (c). Because the plot is a closed curve, the motion of the block is *oscillatory* (the motion repeats itself), as expected.

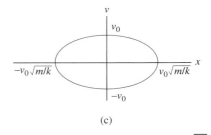

(c)

Sample Problem **12.8**

The ball shown in Fig. (a) weighs 1.5 N and is thrown upward with an initial velocity of 20 m/s. Calculate the maximum height reached by the ball if (1) air resistance is negligible; and (2) the air gives rise to a resisting force F_D, known as *aerodynamic drag*, that opposes the velocity. Assume that $F_D = cv^2$, where $c = 2 \times 10^{-3}$ N $\cdot$ s^2/m^2.

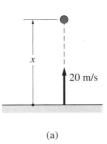

(a)

Solution

Part 1

When air resistance is neglected, the only force acting on the ball during flight is its weight W, shown in the FBD in Fig. (b). Because the motion is rectilinear, the magnitude of the inertia vector is $ma_x = ma$, as shown in the MAD in Fig. (b). Applying Newton's second law, we have

$$\Sigma F_x = ma \quad +\!\uparrow \quad -mg = ma$$

from which we find that the acceleration is

$$a = -g = -9.8 \text{ m/s}^2 \qquad \text{(a)}$$

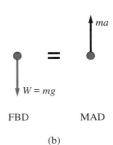

(b)

The acceleration can be integrated with respect to time to obtain the position and velocity as follows:

$$v = \int a \, dt = \int (-9.8) \, dt = -9.8t + C_1 \text{ m/s} \qquad \text{(b)}$$

$$x = \int v \, dt = \int (-9.8t + C_1) \, dt = -4.9t^2 + C_1 t + C_2 \text{ m} \qquad \text{(c)}$$

The constants of integration, C_1 and C_2, are evaluated by applying the initial conditions $x = 0$ and $v = 20$ m/s when $t = 0$, the results being $C_1 = 20$ m/s and $C_2 = 0$. Therefore, the velocity and position are given by

$$v = -9.8t + 20 \text{ m/s} \qquad \text{(d)}$$

$$x = -4.9t^2 + 20t \text{ m} \qquad \text{(e)}$$

The ball reaches its maximum height when $v = 0$. Letting $v = 0$ in Eq. (d), we obtain

$$0 = -9.8t + 20 \quad \text{or} \quad t = 2.04 \text{ s}$$

Substituting this value of time into Eq. (e), we get

$$x_{\text{max}} = -4.9(2.04)^2 + 20(2.04) = 20.4 \text{ m} \qquad \textbf{\textit{Answer}}$$

Note that in this case the acceleration, and thus the velocity and position, are independent of the weight of the ball.

Part 2

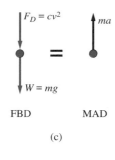

FBD   MAD

(c)

When aerodynamic drag is considered, the FBD and MAD of the ball during its upward flight are as shown in Fig. (c). Observe that the drag force F_D, which always opposes the velocity, acts downward because the positive direction for v is upward (the same as the positive direction for x). From Newton's second law, we obtain the following equation of motion:

$$\Sigma F_x = ma_x \quad +\!\uparrow \quad -mg - cv^2 = ma \qquad \text{(f)}$$

The complete solution (x and v as functions of t) of Eq. (f) is best computed numerically. However, it is possible to derive the velocity as a function of position by direct integration if the independent variable is changed from t to x. Substituting $a = v \, dv/dx$ from Eq. (12.10), Eq. (f) becomes

$$-mg - cv^2 = m\frac{dv}{dx}v$$

in which the variables x and v may be separated as follows:

$$dx = -\frac{mv \, dv}{mg + cv^2} \qquad \text{(g)}$$

Integrating both sides of this equation (using a table of integrals if necessary), we obtain

$$x = -\frac{m}{2c}\ln(mg + cv^2) + C_3 \qquad \text{(h)}$$

where C_3 is a constant of integration. Substituting the numerical values $mg = W = 1.5\,\text{N}$, $m = 1.5/9.8 = 0.153\,\text{kg}$, and $c = 2 \times 10^{-3}\,\text{N} \cdot \text{s}^2/\text{m}^2$, Eq. (h) becomes

$$x = -38.25\ \ln[1.5 + (2 \times 10^{-3})v^2] + C_3 \ \text{m} \qquad\qquad (i)$$

Applying the initial condition, $v = 20$ m/s when $x = 0$, we find that $C_3 = -31.86$ m. Therefore, the solution for x is

$$x = -38.25\ \ln[1.5 + (2 \times 10^{-3})v^2] + 31.86 \qquad\qquad (j)$$

Because the maximum height of the ball occurs when $v = 0$, we have

$$x_{\text{max}} = -38.25\ \ln 1.5 + 31.86 = 16.4 \ \text{m} \qquad\qquad \textit{Answer} \quad (k)$$

Of course, this value is smaller than the maximum height obtained in Part 1, where air resistance was neglected.

To summarize, we used the FMA approach to determine the equations of motion for both parts of this sample problem. When air resistance was neglected in Part 1, the acceleration was simply $-g$, independent of the weight of the ball. The velocity and position could be determined in terms of t by direct integration. The inclusion of aerodynamic drag in Part 2 introduced the additional term $-cv^2$ into the equation of motion with the result that the acceleration depended on c, v, and W. For this case, the solution for $x(t)$ and $v(t)$ was not determined (it would be very tedious to do so). However, $x(v)$ was readily obtained, from which we computed the maximum height of the ball.

Sample Problem   12.9

Use numerical integration to determine the time of flight and the impact velocity of the ball described in Sample Problem 12.8, Part (2).

Solution

In the solution of Sample Problem 12.8, Part (2), the acceleration of the ball during its upward flight was found to be $a = -g - (c/m)v^2$. Because air resistance always opposes the velocity, the acceleration during the downward motion is $a = -g + (c/m)v^2$. The two expressions can be combined into

$$a = -g - \frac{c}{m}\text{sgn}(v)\ v^2$$

where $\text{sgn}(v)$ denotes "the sign of v." Introducing the notation $x = x_1$, $v = x_2$, the equivalent first-order differential equations are

$$\dot{x}_1 = x_2 \quad \dot{x}_2 = -g - \frac{c}{m}\text{sgn}(x_2)\ x_2^2$$

with the initial conditions $x_1(0) = 0$, $x_2(0) = 20$ m/s. The parameters are $g = 9.8\,\text{m/s}^2$ and

$$\frac{c}{m} = \frac{2 \times 10^{-3}}{1.5/9.8} = 1.307 \times 10^{-2}\,\text{m}^{-1}$$

According to the solution of Sample Problem 12.8, the time of flight without air resistance is 2(2.04) = 4.08 s. Because air resistance will reduce this time somewhat, we chose $t = 0$ to 3.9 s as the period of integration. The MATLAB program listed here prints the numerical solution at 0.05 s intervals.

```
function example12_9
[t,x] =ode45(@f, [0:0.05:3.6], [0 20]);
printSol(t,x)
  function dxdt = f(t,x)
  dxdt = [x(2)
          -9.8-1.307e-2*sign(x(2))*x(2)^2];
  end
end
```

The arguments of the function ode45 are explained in Appendix E, which also lists the function printSol that we use to print the results. The last six lines of the printout are

```
t              x1              x2
3.6500        0.0863        -16.1165
3.7000        -0.7275       -16.4334
3.7500        -1.5569       -16.7436
3.8000        -2.4017       -17.0470
3.8500        -3.2615       -17.3437
3.9000        -4.1360       -17.6339
```

At impact with the ground, $x_1 = 0$. It is apparent from the printout that this occurs sometime between 3.65 s and 3.7 s. A more accurate value can be obtained by linear interpolation—see Eq. (E.7) in Appendix E:

$$\frac{-0.7275 - 0.0863}{3.7 - 3.65} = \frac{0 - 0.0863}{t - 3.65}$$

yielding

$$t = 3.6553 \text{ s} \qquad \qquad \textit{Answer}$$

The impact velocity is the value of x_2 at $t = 3.6553$ s. It also can be determined by linear interpolation, as follows.

$$\frac{-16.4334 - (-16.1165)}{3.7 - 3.65} = \frac{v - (-16.1165)}{3.6553 - 3.65}$$

which gives us

$$v = -16.2 \text{ m/s} \qquad \qquad \textit{Answer}$$

Problems

12.27   Calculate the force T that will lift the 50-kg crate at the speed $v = 4t$ m/s, where t is the time in seconds.

12.28   A car is traveling at 100 km/h along a straight, level road when its brakes become locked. Determine the stopping distance of the car knowing that the coefficient of kinetic friction between the tires and the road is 0.65.

12.29   Solve Prob. 12.28 if the car is traveling down a 5° incline.

12.30   A 0.1-kg block moves along the x-axis. The resultant of all forces acting on the block is $\mathbf{F} = -1.2t\mathbf{i}$ N, where t is in seconds. When $t = 0$, $x = 0$ and $\mathbf{v} = 64\mathbf{i}$ m/s. Determine the distance traveled by the particle during the time interval $t = 0$ to $t = 4$ s.

12.31   A 10-g bead slides on a wire that lies on the x-axis. The resultant of all forces acting on the bead is $\mathbf{F} = 0.04\sqrt{v}\,\mathbf{i}$ N, where the speed v is in m/s. When $t = 0$, $x = 0$, and when $t = 0.6$ s, $\mathbf{v} = 0.16\mathbf{i}$ m/s. Find x when $t = 0.8$ s.

12.32   A small ball of mass m undergoes rectilinear motion along the x-axis. The resultant of all forces acting on the ball is $\mathbf{F} = -kmv^2\mathbf{i}$, where k is a constant and v is the speed of the ball. When $t = 0$, $x = 0$ and $\mathbf{v} = v_0\mathbf{i}$. Find the speed of the ball as a function of (a) x; and (b) t.

12.33   A 4-kg block moves along the y-axis. The resultant of all forces acting on the block is $\mathbf{F} = (4t - 4)\mathbf{j}$ N, where t is in seconds. When $t = 0$, $y = 0$, and $\mathbf{v} = -8\mathbf{j}$ m/s. Find the distance traveled by the block during the time interval $t = 0$ to $t = 8$ s.

12.34   The pendulum AB is suspended from a cart that has a constant acceleration a to the right. Determine the constant angle θ of the pendulum.

12.35   A uniform cylinder is placed in a V-notched cradle. What is the largest horizontal acceleration that the cradle may have without causing the cylinder to climb out of the cradle? Neglect friction.

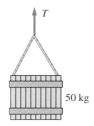

Fig. P12.27

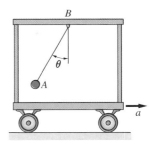

Fig. P12.34

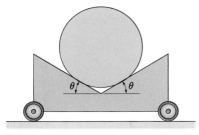

Fig. P12.35

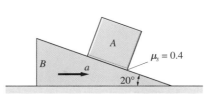

Fig. P12.36

12.36   Block A of mass m is placed on the inclined surface of wedge B. The static coefficient of friction between A and B is 0.4. Determine the smallest

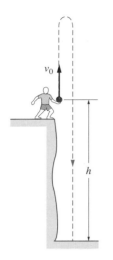

Fig. P12.37

acceleration a of the wedge that would cause the block to slide up the inclined surface.

12.37   The ball is thrown vertically upward over the edge of the cliff with the initial velocity v_0. Determine the expression for the velocity with which the ball hits the bottom of the cliff, which is the distance h below the point of release.

12.38   The cart carrying the small 2-kg package A is moving up the incline with constant acceleration a. If the package stays at rest relative to the cart in the position $\theta = 45°$, determine the value of a. Neglect friction.

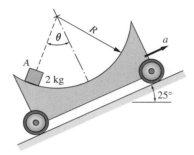

Fig. P12.38

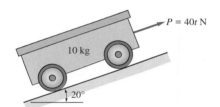

Fig. P12.39

12.39   The 10-kg cart is released from rest at time $t = 0$ on the inclined surface. The force $P = 40t$ N acts on the cart, where t is the time measured in seconds. (a) Determine the distance the cart will move down the inclined surface before reversing direction. (b) Find the velocity of the cart when it returns to the point of release.

12.40   The horizontal force $P = 40 - 10t$ N (t is the time measured in seconds) is applied to the 2-kg collar that slides along the inclined rod. At time $t = 0$, the position coordinate of the collar is $x = 0$, and its velocity is $v_0 = 3$ m/s directed down the rod. Find the time and the speed of the collar when it returns to the position $x = 0$ for the first time. Neglect friction.

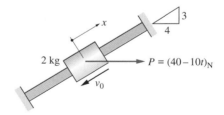

Fig. P12.40

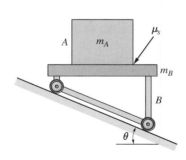

Fig. P12.41

12.41   The coefficient of static friction between the block A and the cart B is μ_s. If the assembly is released from rest on the inclined plane, determine the smallest value of μ_s that will prevent the block from sliding on the cart.

12.42   The 40-kg crate is placed on the step of an escalator. If the static coefficient of friction between the crate and the step is 0.4, determine the largest acceleration *a* of the escalator for which the crate would not slip. Assume that the direction of the acceleration is (a) up; and (b) down.

12.43   A 2000-kg rocket is launched vertically from the surface of the earth. The engine shuts off after providing a constant propulsive force of 60 kN for the first 20 seconds. Neglect the reduction in the mass of the rocket due to the burning of the fuel and the variation of the gravitational acceleration with altitude. Calculate the altitude of the rocket at the end of the powered portion of the flight.

12.44   The 300 kg rocket sled is propelled along a straight test track. The rocket engine fires for 4 seconds, producing a propulsive force of $F = 1000e^{-0.2t}$ N, and then shuts down. Assuming that the sled starts from rest time $t = 0$ and that the coefficient of kinetic friction is 0.05, determine the maximum speed reached by the sled. (g $= 10$ m/s^2)

12.45   The constant vertical force *P* is applied to the end of the rope that passes over a peg attached to the 0.2-kg block. Neglecting friction, determine *P* that would cause the block to accelerate at 1.5 m/s^2 (a) up the inclined plane; and (b) down the inclined plane.

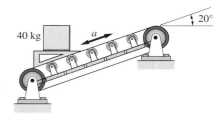

Fig. P12.42

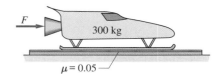

Fig. P12.44

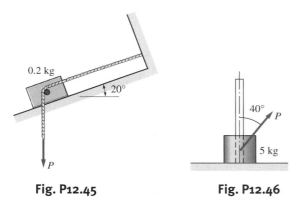

Fig. P12.45   **Fig. P12.46**

12.46   The static and kinetic coefficients of friction between the 5-kg sliding collar and the vertical guide rod are $\mu_s = 0.5$ and $\mu_k = 0.40$, respectively. If the force *P* is slowly increased until the collar starts to move, determine the initial acceleration of the collar.

12.47   The drag force acting on a 60 kg skydiver in the "spread" position shown can be approximated by $F_D = 0.0436v^2$, where F_D is in newtons and *v* is in m per second (from *Scientific and Engineering Problem-Solving with the Computer*, W. R. Bennett, Jr., Prentice Hall, New York, 1976). Assuming that the skydiver follows a vertical path, determine the terminal velocity.

12.48   When the 1.8-kg block is in the position shown, the attached spring is undeformed. If the block is released from rest in this position, determine its velocity when it hits the floor.

12.49   The spring attached to the 1.8-kg block is undeformed when the system is in the position shown. The block is pulled down until it touches the floor and

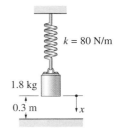

Fig. P12.47

Fig. P12.48, P12.49

then released from rest. Calculate the maximum height (measured from the floor) reached by the block.

12.50 A linear spring of stiffness k is to be designed to stop a 20-Mg railroad car traveling at 8 km/h within 400 mm after impact. Find the smallest value of k that will produce the desired result.

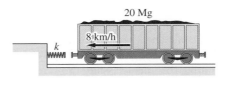

Fig. P12.50

12.51 According to the law of gravitation, the force acting on a particle of mass m located at a distance R from the center of a planet of mass M is $F = mg(R_0/R)^2$, where R_0 is the radius of the planet and g is the gravitational acceleration at its surface. If the mass m is launched vertically from the surface of the earth ($R_0 = 6400$ km and $g = 9.8$ m/s^2) with the initial velocity $v_0 = 1000$ m/s, how high above the surface of the earth will it rise if air resistance is neglected?

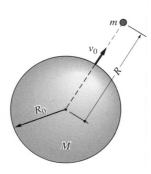

Fig. P12.51

12.52 The disk of radius R meters carries a charge with the electrostatic potential V volts. A particle of charge q coulombs lies on the axis of the disk at a distance y meters from the disk. It can be shown that the repulsive force F (in newtons) acting on the particle is

$$F = \frac{Vq}{R}\left(1 - \frac{y}{\sqrt{R^2 + y^2}}\right)$$

If a particle starts from the center of the plate with zero velocity, determine its speed at $y = R$. Neglect the effect of gravity.

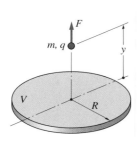

Fig. P12.52

*__12.53__ A light airplane weighing 9000 N lands with a speed of 180 km/h and coasts to a stop due to reverse propeller thrust and aerodynamic drag. Determine the stopping distance if the reverse propeller thrust is $T = 1800$ N (constant) and the drag force is $F_D = c_D v^2$, where $c_D = 0.8$ N $\cdot$ s^2/m^2 and v is the velocity in m/s.

*__12.54__ The 2 kg block slides along the inclined plane under the action of the constant force $P = 32$ N. If the block is released from rest at $x = 0$, determine the maximum velocity of the block and the value of x where it occurs. Neglect friction.

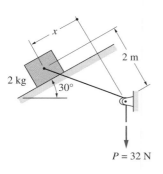

Fig. P12.54

*__12.55__ An object of mass m is released from rest and allowed to fall vertically. The aerodynamic drag force acting upon the object is $F_D = c_D v$, where v is the velocity and c_D is a constant. Derive the expression for the time required for the object to reach 90 percent of its terminal velocity after being released.

12.56   A train traveling at 20 m/s is brought to an emergency stop. During braking, the acceleration is $a = -(7/4) + (v/16)$ m/s^2, where v is the velocity in m/s. Use numerical integration to determine the stopping distance of the train and compare the result with the analytical solution $x = 241$ m.

12.57   A model ship is towed in a test basin at a speed of 20 m/s when the tow rope is released at time $t = 0$. Due to hydrodynamic resistance, the ensuing acceleration of the ship is $a = -(v^2/10)$ m/s^2, where v is the speed in m/s. Use numerical integration to determine the time when the speed of the ship is reduced to 10 m/s. Compare the answer to the analytical solution $t = 0.5$ s.

12.58   The acceleration of the skydiver described in Prob. 12.47 is $a = 9.8(1 - 32.3 \times 10^{-6}v^2)$ m/s^2, where v is the speed in m/s. Use numerical integration to find the time required for the skydiver to reach 144 km/h after jumping. Compare the answer to the analytical solution $t = 4.14$ s.

12.59   The free length of the spring attached to the 0.2 kg slider A is 5 mm. When the slider is released from rest at $x = 8$ mm, the ensuing acceleration is

$$a = -5796 \left(1 - \frac{5}{\sqrt{x^2 + 9}}\right) x \text{ mm/s}^2$$

where x is measured in mm. Use numerical integration to compute the speed of the slider when it reaches point B. Compare your answer with $v = 223$ mm/s, the value found analytically.

12.60   A 1000 N object is released from rest at 9000 m above the surface of the earth. The acceleration of the object during its fall is

$$a = -9.8 \left(1 - 6.72 \times 10^{-3}v^2 e^{-1.053 \times 10^{-4}x}\right) \text{ m/s}^2$$

where v is the speed in m/s and x is the elevation in feet. (The exponential term accounts for the variation of air density with elevation.) (a) Use numerical integration to determine the maximum speed of the object and the elevation where it occurs. (b) Plot the speed against the elevation from the time of release until the maximum speed is attained.

12.61   The static as well as the dynamic coefficient of friction between the 1.6-kg block and the horizontal surface is $\mu = 0.2$. The spring attached to the block has a stiffness of 30 N/m, and it is undeformed when $x = 0$. At time $t = 0$, the block is at $x = 0$ and moving to the right with the velocity $v = 6$ m/s. (a) Derive an expression for the acceleration of the block that is valid for both positive and negative values of v. (b) Use numerical integration to determine when the block comes to rest during the time period $t = 0$ to 1.2 s. (c) Plot the velocity against the position for the time period specified in Part (b).

12.62   The 2-kg block is at rest with the spring unstretched when the force $P(t)$ is applied at time $t = 0$. (a) Derive the expression for the acceleration of the block. (b) Using numerical integration, determine the maximum displacement and the

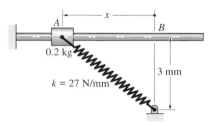

Fig. P12.59

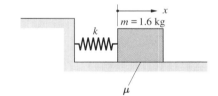

Fig. P12.61

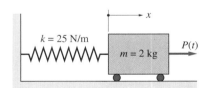

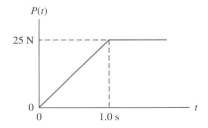

Fig. P12.62

maximum velocity of the block. (c) Plot velocity versus displacement during the time interval $0 \le t \le 3$ s.

12.63   The bubble breaks loose from the bottom of a shallow dish of water. The acceleration of the bubble, determined by its buoyancy and the viscous drag of water, is $a = 24 - 30v^{1.5}$ m/s^2, where v is the velocity in m/s. By numerical integration, determine the velocity of the bubble when it reaches the surface of the water.

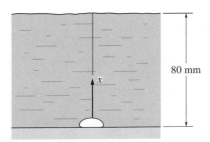

Fig. P12.63

12.5   *Curvilinear Motion*

a.   *Superposition of rectilinear motions*

Here we consider a special case of curvilinear motion that can be represented as a superposition of independent rectilinear motions. This situation occurs when the components of acceleration have the form

$$a_x = f_x(v_x, x, t) \qquad a_y = f_y(v_y, y, t) \qquad a_z = f_z(v_z, z, t) \qquad (12.24)$$

Equations (12.24) are said to be *uncoupled,* because the acceleration in any one coordinate direction is independent of the motion in the other two directions. Therefore, we can view the motions in the x-, y-, and z-directions as independent rectilinear motions that can be analyzed with the tools introduced in the previous article.

If the particle moves in a plane, say the xy-plane, its motion can be treated as a superposition of two rectilinear motions, one in the x-direction and the other in the y-direction. The flight of projectiles in a constant gravitational field falls in this category.

b.   *General curvilinear motion*

If Eqs.(12.24) are coupled, an analytical solution will be difficult or impossible. Equations of this type must be invariably solved by numerical methods—see Appendix E.3.

Sample Problem 12.10

As the 1200-kg car in Fig. (a) travels over the crest of a hill, its position is given by

$$x = v_0 t \qquad y = h \exp\left[-\left(\frac{v_0 t}{b}\right)^2\right] \qquad\qquad \text{(a)}$$

where $v_0 = 30$ m/s, $h = 10$ m, $b = 50$ m, and t is the time in seconds. Determine the contact force R between the car and the road at A.

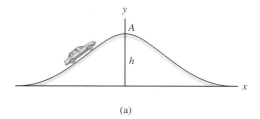

(a)

Solution

The acceleration components of the car can be obtained by differentiating Eqs. (a):

$$\dot{x} = v_0$$
$$\ddot{x} = 0$$

$$\dot{y} = -2h \left(\frac{v_0}{b}\right)^2 t \exp\left[-\left(\frac{v_0 t}{b}\right)^2\right]$$

$$\ddot{y} = -2h \left(\frac{v_0}{b}\right)^2 \exp\left[-\left(\frac{v_0 t}{b}\right)^2\right] + 4h \left(\frac{v_0}{b}\right)^4 t^2 \exp\left[-\left(\frac{v_0 t}{b}\right)^2\right]$$

At point A we have $x = 0$, and according to Eq. (a), $t = 0$. Therefore, the acceleration components at A are

$$a_x = \ddot{x} = 0$$

$$a_y = \ddot{y} = -2h \left(\frac{v_0}{b}\right)^2 = -2(10)\left(\frac{30}{50}\right)^2 = -7.2 \text{ m/s}$$

The minus sign implies that the acceleration is downward (in the negative y-direction).

The free-body diagram (FBD) and the mass-acceleration diagram (MAD) of the car at A are shown in Fig. (b), where R_x and R_y represent the components of the contact force acting on the car. The diagrams yield the following equations of motion:

$$\Sigma F_x = 0 \qquad R_x = 0$$
$$\Sigma F_y = ma_y \quad +\uparrow \quad R_y - mg = ma_y$$

Therefore,

$$R = R_y = m(g + a_y) = 1200(9.8 - 7.2) = 3120 \text{ N} \qquad \textbf{\textit{Answer}}$$

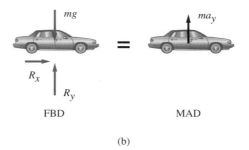

FBD　　　　　　　　　MAD

(b)

Sample Problem 12.11

As shown in Fig. (a), a projectile of weight W is launched from the origin O. The initial velocity v_0 makes an angle θ with the horizontal. The projectile lands at A, a distance R from O, as measured along the inclined plane. (1) Assuming that v_0 and θ are known, find the rectangular components of the velocity and position of the projectile as functions of time. (2) Given that $v_0 = 20$ m/s and $\theta = 30°$, determine the maximum height h and the distance R.

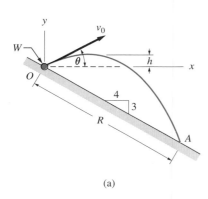

(a)

Solution

Part 1

From the free-body diagram (FBD) and mass-acceleration diagram (MAD) in Fig. (b), we obtain the equations of motion

$$\Sigma F_x = ma_x \quad \xrightarrow{+} \quad 0 = \frac{W}{g}a_x$$

$$\Sigma F_y = ma_y \quad +\uparrow \quad -W = \frac{W}{g}a_y$$

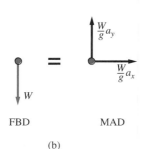

FBD　　　　　　MAD

(b)

It follows that the acceleration components are $a_x = 0$ and $a_y = -g$. Because a_x and a_y are constants, the velocity and position are readily obtained by integration, as shown in the following table.

x-direction	y-direction	
$a_x = 0$	$a_y = -g$	(a)
$v_x = \int a_x\, dt = C_1$	$v_y = \int a_y\, dt = -gt + C_3$	(b)
$x = \int v_x\, dt = C_1 t + C_2$	$y = \int v_y\, dt = -\dfrac{1}{2} g t^2 + C_3 t + C_4$	(c)

Equations (c) are parametric equations (t is the parameter) of a parabola that lies in the xy-plane. Therefore, in the absence of air resistance, a projectile follows a parabolic path.

To evaluate the constants of integration (C_1 through C_4), we must identify four conditions imposed on the motion. Examination of the problem statement and Fig. (a) reveals that there are four initial conditions. Choosing $t = 0$ as the time of launch, these conditions are

1. $x = 0$ when $t = 0$
2. $y = 0$ when $t = 0$

Substituting conditions **1** and **2** into Eqs. (c), we get $C_2 = C_4 = 0$.

3. $v_x = v_0 \cos\theta$ when $t = 0$
4. $v_y = v_0 \sin\theta$ when $t = 0$

According to Eqs. (c), conditions **3** and **4** are satisfied if $C_1 = v_0 \cos\theta$ and $C_3 = v_0 \sin\theta$. Substituting C_1 through C_4 into Eqs. (b) and (c), the rectangular components of the velocity and position are

$$v_x = v_0 \cos\theta \qquad v_y = -gt + v_0 \sin\theta \qquad \textbf{\textit{Answer}} \quad \text{(d)}$$

$$x = (v_0 \cos\theta)t \qquad y = -\frac{1}{2}gt^2 + (v_0 \sin\theta)t \qquad \textbf{\textit{Answer}} \quad \text{(e)}$$

Caution   Equations (d) and (e) are often convenient to use in solving projectile problems when air resistance is negligible. However, do not apply these equations unless the initial conditions are identical to those stated in **1** through **4** above.

Part 2

Substituting $v_0 = 20$ m/s and $\theta = 30°$ into Eqs. (d) and (e), we obtain the following description of motion:

$$v_x = 17.32 \text{ m/s} \qquad v_y = -9.8t + 10 \text{ m/s} \qquad \text{(f)}$$

$$x = 17.32t \text{ m} \qquad y = -4.905t^2 + 10t \text{ m} \qquad \text{(g)}$$

All characteristics of the motion can now be computed from Eqs. (f) and (g).

The maximum height h equals the value of y when $v_y = 0$. Letting t_1 be the time when this occurs, the second of Eqs. (f) gives us

$$0 = -9.8t_1 + 10 \quad \text{or} \quad t_1 = 1.02 \text{ s}$$

Substituting this value for t_1 in the second of Eqs. (g) yields the maximum height of the projectile

$$h = -4.905(1.02)^2 + 10(1.02) = 5.09 \text{ m} \qquad \textbf{\textit{Answer}}$$

Next we let t_2 be the time when the projectile lands at A on the inclined plane. Substituting the coordinates of A, $x = (4/5)R$ and $y = -(3/5)R$, into Eqs. (g), we obtain

$$\frac{4}{5}R = 17.32t_2 \quad \text{and} \quad -\frac{3}{5}R = -4.905t_2^2 + 10t_2$$

The solution is $t_2 = 2.5683$ s and

$$R = 55.6 \text{ m} \qquad \textbf{\textit{Answer}}$$

■

Sample Problem   12.12

A projectile of mass m is fired from point O at time $t = 0$ with the velocity v_0 as shown in Fig. (a). The aerodynamic drag F_D is proportional to the speed of the projectile: $F_D = cv$, where c is a constant. (1) Derive the equations of motion. (2) Verify that the solution of the equations of motion is

$$x = C_1 e^{-ct/m} + C_2 \quad \text{and} \quad y = C_3 e^{-ct/m} - \frac{mgt}{c} + C_4$$

where C_1 through C_4 are constants. (3) Find the maximum height h given that $W = 8$ N, $c = 0.6$ N $\cdot$ s/m, $v_0 = 30$ m/s, and $\theta = 30°$.

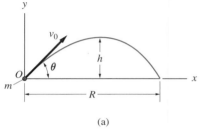

(a)

(b)

Solution

Part 1

Figure (b) shows the FBD and MAD of the projectile. The direction of the drag F_D in the FBD is opposite to the direction of the velocity vector (tangent to the path), and its rectangular components are cv_x and cv_y. The rectangular components of the inertia vector, ma_x and ma_y, are shown in the MAD. The corresponding equations of motion are

$$\Sigma F_x = ma_x \quad \xrightarrow{+} \quad -cv_x = ma_x \qquad \textbf{\textit{Answer}} \quad \text{(a)}$$

$$\Sigma F_y = ma_y \quad +\uparrow \quad -mg - cv_y = ma_y \qquad \textbf{\textit{Answer}} \quad \text{(b)}$$

Part 2

To verify that the expressions for $x(t)$ and $y(t)$ given in the problem statement satisfy the equations of motion, we must first evaluate their derivatives.

$$x = C_1 e^{-ct/m} + C_2 \qquad\qquad y = C_3 e^{-ct/m} - \frac{mgt}{c} + C_4 \qquad \text{(c)}$$

$$v_x = \dot{x} = -C_1 \frac{c}{m} e^{-ct/m} \qquad v_y = \dot{y} = -C_3 \frac{c}{m} e^{-ct/m} - \frac{mg}{c} \qquad \text{(d)}$$

$$a_x = \ddot{x} = C_1 \left(\frac{c}{m}\right)^2 e^{-ct/m} \qquad a_y = \ddot{y} = C_3 \left(\frac{c}{m}\right)^2 e^{-ct/m} \qquad \text{(e)}$$

Substitution of the above results into Eqs. (a) and (b) shows that Eqs. (c) are indeed the solution, since they satisfy the equations of motion.

Part 3

Using the given numerical values for c and W, we have $c/m = \frac{0.6(9.8)}{8} = 0.735 \, \text{s}^{-1}$, and $mg/c = 8/0.6 = 13.3$ m/s. Substituting these values into Eqs. (c) and (d), and assuming that time t is measured in seconds, we obtain

$$x = C_1 e^{-0.735t} + C_2 \text{ m} \qquad y = C_3 e^{-0.735t} - 13.3t + C_4 \text{ m} \qquad \text{(f)}$$

$$v_x = -0.735 C_1 e^{-0.735t} \text{ m/s} \qquad v_y = -0.735 C_3 e^{-0.735t} - 13.3 \text{ m/s} \qquad \text{(g)}$$

From the problem statement, we deduce that the motion must satisfy the following conditions at $t = 0$:

1. $x = 0$
2. $y = 0$
3. $v_x = 30 \cos 30° = 25.98$ m/s
4. $v_y = 30 \sin 30° = 15$ m/s

The equations obtained by substituting the four conditions into Eqs. (f) and (g) can be solved for the constants of integration. Omitting the algebraic details, the results are $C_1 = -35.35$ m ft, $C_2 = 35.35$ m, $C_3 = -38.5$ m, and $C_4 = 38.5$ m. Substituting these values into Eqs. (f) and (g), we find that

$$x = 35.35 \left(1 - e^{-0.735t}\right) \text{ m} \qquad \text{(h)}$$

$$y = 38.5 \left(1 - e^{-0.735t}\right) - 50t \text{ m} \qquad \text{(i)}$$

$$v_x = 25.98 e^{-0.735t} \text{ m/s} \qquad \text{(j)}$$

$$v_y = 28.3 e^{-0.735t} - 13.3 \text{ m/s} \qquad \text{(k)}$$

The maximum value of y occurs when $v_y = 0$. If we let this time be t_1, Eq. (k) yields

$$0 = 28.3 e^{-0.735t} - 13.3$$

from which we find

$$t_1 = -\frac{\ln(13.3/28.3)}{0.735} = -1.027 \, \text{s}$$

Substituting $t = t_1 = 1.027$ s into Eq. (i), we get for the maximum value of y

$$y_{max} = h = 38.5 \left[1 - e^{-(0.735)(1.027)}\right] - 13.3(1.027)$$

$$= 6.74 \text{ m} \qquad \qquad \textit{Answer}$$

Sample Problem 12.13

Integrate numerically the equations of motion for the projectile described in Sample Problem 12.12 using $c = 0$, 0.3, and $0.6\,\text{N} \cdot \text{s/m}$. Plot the three trajectories.

Solution

The equations of motion were derived in Sample Problem 12.12 as

$$a_x = -\frac{c}{m}v_x \qquad a_y = -g - \frac{c}{m}v_y$$

Letting

$$\begin{bmatrix} x \\ y \\ v_x \\ v_y \end{bmatrix} = \begin{bmatrix} x_1 \\ x_2 \\ x_3 \\ x_4 \end{bmatrix}$$

the equivalent first-order differential equations and the initial conditions are

$$\begin{bmatrix} \dot{x}_1 \\ \dot{x}_2 \\ \dot{x}_3 \\ \dot{x}_4 \end{bmatrix} = \begin{bmatrix} x_3 \\ x_4 \\ -(c/m)x_3 \\ -g - (c/m)x_4 \end{bmatrix}$$

$$\mathbf{x}(0) = \begin{bmatrix} 0 \\ 0 \\ 30\cos 30° = 25.98\ \text{m/s} \\ 30\sin 30° = 15\ \text{m/s} \end{bmatrix}$$

We integrated the differential equations with the MATLAB program shown below. The command `hold on` allows new curves to be added to the current plot. All other commands are explained in Appendix E. The flight time was estimated to be 2.5 s or less. By choosing 0.05 s as the time increment for the results, approximately 50 points are available for each plot.

```
function example12_13
g = 9.8; m = 8/g;
time = [0:0.05:2.5];
x0 = [0 0 25.98 15];

c = 0.6;
[t,x] = ode45(@f,time,x0);
axes('fontsize',14)
plot(x(:,1),x(:,2),'linewidth',1.5)
grid on
xlabel('x (ft)'); ylabel('y (ft)')
hold on

c = 0.3
[t,x] = ode45(@f,time,x0);
plot(x(:,1),x(:,2),'linewidth',1.5)

c = 0;
[t,x] = ode45(@f,time,x0);
plot(x(:,1),x(:,2),'linewidth',1.5)

  function dxdt = f(t,x)
  dxdt = [x(3); x(4); -c/m*x(3); -g - c/m*x(4)];
  end
end
```

The plots of y versus x are shown here. The portion of the plot below the x-axis was clipped using the "edit" facility of the MATLAB plot window.

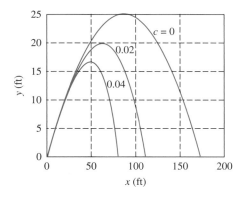

Problems

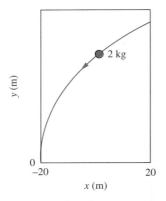

Fig. P12.65

12.64 A 0.5-kg mass moves along the path $x = \frac{1}{40}(y - 12)^2$, where x and y are measured in meters. Knowing that the y-component of the velocity is constant at 10 m/s, determine the force acting upon the mass.

12.65 A 2 kg object travels along the trajectory shown. The position coordinates of the object vary with time t (measured in seconds) as

$$x = 6 \cos\left(\frac{\pi t}{2}\right) \text{ m} \quad y = 19.6(4 - t^2) \text{ m}$$

Calculate the components of the force acting on the object at $t = 0$, 1, and 2 s.

12.66 The 60 g balancing weight A is attached to the rim of a car wheel. When the car travels at the constant speed v_0, the path of A is the curate cycloid

$$x = v_0 t - r \sin\frac{v_0 t}{R} \qquad y = R - r \cos\frac{v_0 t}{R}$$

(a) Show that the acceleration of the weight has a constant magnitude. (b) Calculate the magnitude of the force acting between the weight and the wheel if $v_0 = 90$ km/h, $R = 0.40$ m, and $r = 0.25$ m (neglect gravitational acceleration).

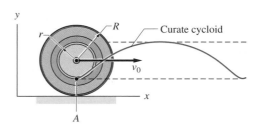

Fig. P12.66

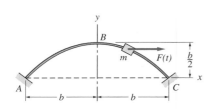

Fig. P12.67

12.67 The slider of mass $m = 0.5$ kg moves along the parabolic guide rod ABC, propelled by the horizontal force $F(t)$. The kinetic coefficient of friction between the slider and the guide rod is $\mu = 0.2$. The position of the slider is given by

$$x = b \sin\frac{2\pi t}{t_0} \qquad y = \frac{b}{4}\left(1 + \cos\frac{4\pi t}{t_0}\right)$$

where $t_0 = 0.8$ s and $b = 1.2$ m. Assuming that ABC lies in the vertical plane, determine the force F when the slider is at B.

12.68 A car of mass m travels along the cloverleaf interchange. The position of the car is given by

$$x = \frac{b}{2}\left(\sin\frac{\pi t}{4t_0} + \sin\frac{3\pi t}{4t_0}\right) \qquad y = \frac{b}{2}\left(\cos\frac{\pi t}{4t_0} - \cos\frac{3\pi t}{4t_0}\right)$$

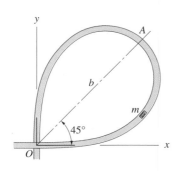

Fig. P12.68

where $b = 240$ m, and $t_0 = 12$ s is the time of travel between O and A. Determine the smallest coefficient of friction between the tires and the road that would prevent the car from skidding at A (note that $t = t_0$ when the car is at A).

12.69   The water sprinkler A is placed on sloping ground and oscillates in the xy-plane. The water leaves the sprinkler at 5 m/s and hits the ground at B, which is a distance R from the sprinkler. Derive the expression for R as a function of the angle θ. Neglect air resistance.

Fig. P12.69

12.70   A projectile, launched at A with an initial velocity of $v_0 = 10$ m/s at the angle $\theta = 65°$, impacts the vertical wall at B. Neglecting air resistance, calculate the height h.

*12.71   A projectile, launched at A with an initial velocity of 25 m/s at the angle θ, impacts the vertical wall at B. Compute the angle θ that will maximize the height h of the impact point. What is this maximum height?

12.72   The aircraft is diving at 30° from the vertical at the speed of 200 m/s. The flight path is directed toward the target at A. If the aircraft drops a package at an altitude of 1200 m, find the distance d between the point of impact and the target.

12.73   A projectile launched at A with the speed v_0 is to hit a target at B. (a) Derive the expression that determines the required angle of elevation θ. (b) Find the two solutions θ_1 and θ_2 of the expression derived in Part (a) if $v_0 = 250$ m/s and $R = 3$ km. Neglect air resistance.

12.74   The volleyball player serves the ball from point A with the speed $v_0 = 12$ m/s at the angle $\theta = 28°$. (a) Derive the equation of the trajectory (y as a function of x) of the ball. (b) Determine whether the ball clears the top of the net C and lands inside the baseline B.

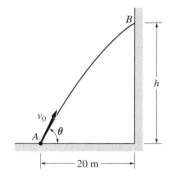

Fig. P12.70, P.12.71

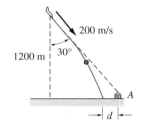

Fig. P12.72

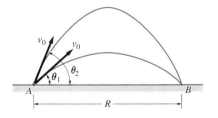

Fig. P12.73

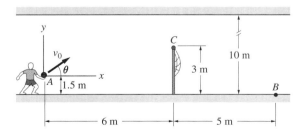

Fig. P12.74, P.12.75

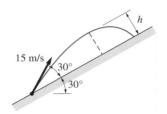

Fig. P12.76

12.75 The volleyball is served from point A with the initial speed v_0 at the angle $\theta = 70°$. Compute the largest v_0 for which the ball will not hit the ceiling.

12.76 A projectile is fired up the inclined plane with the initial velocity shown. Compute the maximum height h, measured perpendicular to the plane, that is reached by the projectile. Neglect air resistance.

12.77 A particle of mass m (kg) carrying a charge q (coulombs) enters the space between two charged plates with the horizontal velocity v_0 (m/s) as shown. Neglecting gravitational acceleration, the force acting on the particle while it is between the plates is $F = q\Delta V/2d$, where $\Delta V = V_2 - V_1$ is the electrostatic potential difference (in volts) between the plates. Derive the expression for the largest ΔV that may be applied if the particle is to miss corner A of the plate.

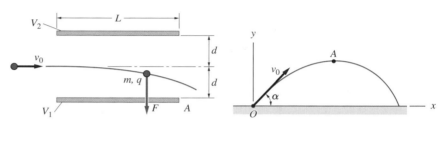

Fig. P12.77   **Fig. P12.78**

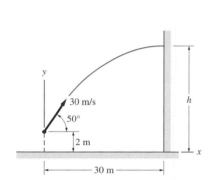

Fig. P12.79

12.78 A projectile of mass m is launched at O with the initial speed v_0 at the angle α to the horizontal. The aerodynamic drag force acting on the projectile during its flight is $\mathbf{F}_D = -c\mathbf{v}$, where c is a constant. If point A is the peak of the trajectory, derive the expressions for (a) the time required to reach A; and (b) the speed of the projectile at A.

12.79 A 0.1-kg rock is thrown at a wall from a distance of 30 m at an elevation of 2 m with the initial velocity shown. The aerodynamic drag acting on the rock is $F_D = 0.0005\,v^2$, where F_D is in newtons and the velocity v is in m/s. (a) Show that the acceleration components are

$$a_x = -0.005\,v_x\sqrt{v_x^2 + v_y^2}\ \text{m/s}^2$$

$$a_y = -0.005\,v_y\sqrt{v_x^2 + v_y^2} - 9.81\ \text{m/s}^2$$

(b) Use numerical integration to find the height h where the rock hits the wall and the speed of impact. *Note:* The analytical solution is $h = 24.0$ m, $v = 16.3$ m/s.

12.80 The 0.01-kg particle travels freely on a smooth, horizontal surface lying in the xy-plane. The force $\mathbf{F}$ acting on the particle is always directed away from the origin O, its magnitude being $F = 0.005/d^2$ N, where d is the distance in meters of the particle from O. At time $t = 0$, the position of the particle is $x = 0.3$ m, $y = 0.4$ m, and its velocity is $\mathbf{v} = -2\mathbf{j}$ m/s. (a) Derive the acceleration components of the particle, and state the initial conditions. (b) Use numerical

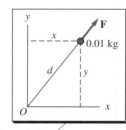

Horizontal plane

Fig. P12.80, P12.81

integration to determine the x-coordinate and the speed of the particle when it crosses the x-axis.

12.81 Solve Prob. 12.80 if the force **F** is directed toward the origin O.

12.82 The 0.25-kg ball is thrown horizontally with the velocity 36 m/s from a height of 1.8 m. The aerodynamic drag force acting on the ball is $F_D = c_D v^{1.5}$ N, where v is the speed in m/s, and $c_D = 0.03286$ N $\cdot$ (s/m)$^{1.5}$. (a) Using the coordinate system shown, derive a_x and a_y as functions of v_x and v_y, and state the initial conditions. (b) Determine the time of flight and the range R by numerical integration. (The analytic solution is $t = 0.655$ s and $R = 18.9$ m.)

12.83 A 1.0-kg ball is kicked with an initial velocity of 30 m/s into a 20-m/s headwind. The aerodynamic drag force acting on the ball is $\mathbf{F}_D = -0.5\mathbf{v}$ N. The resulting acceleration of the ball is

$$\mathbf{a} = -(10 + 0.5\,v_x)\mathbf{i} - (9.81 + 0.5v_y)\mathbf{j} \text{ m/s}^2$$

where the components of the velocity are in m/s. (a) Determine the horizontal distance of travel b and the time of flight. (b) Plot the trajectory of the ball (y vs. x).

12.84 The mass $m = 0.25$ kg, attached to a linear spring (stiffness $k = 10$ N/m, free length $L_0 = 0.5$ m), moves in the vertical plane. The spring can resist both tension and compression. The mass is released from rest at $x = 0.5$ m, $y = -0.5$ m. (a) Derive the expression for the acceleration components of the mass and state the initial conditions. (b) Integrate the acceleration components numerically for $0 \le t \le 2$ s, and plot the trajectory of the mass.

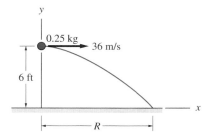

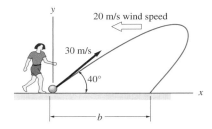

Fig. P12.83

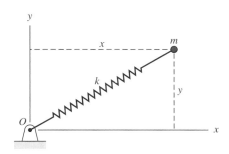

Fig. P12.84, P12.85

12.85 Solve Prob. 12.84 if the mass is released from rest at $x = y = 0.5$ m and the spring is unable to resist compression.

12.86 The plane motion of a table-tennis ball in Fig. (a) is governed by the three components of acceleration given in Fig. (b). In addition to the gravitational acceleration g, there are the effects of the aerodynamic drag a_D and aerodynamic lift a_L (the lift is caused by the Bernoulli effect, that is, the difference in air pressure caused by the spin of the ball). Realistic approximations for these accelerations are $a_D = 0.05\,v^2$ and $a_L = 0.16\omega v$, where v is in m/s and the spin ω is in rev/s.

The ball leaves the paddle at table height with the initial velocity $v_0 = 20$ m/s inclined at $\theta_0 = 60°$ with the horizontal, and a topspin $\omega = +10$ rev/s (the spin may be assumed to be constant). (a) Derive the acceleration components in the x- and y-directions. (b) Assuming that the ball lands on the table, integrate the acceleration components for the duration of the flight. Determine the time of flight and the horizontal distance traveled. (c) Plot the trajectory of the ball.

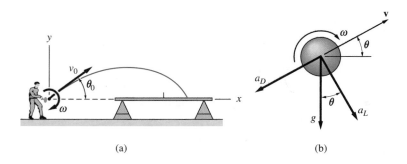

(a)   (b)

Fig. P12.86

*12.6   *Analysis of Motion by the Area Method*

This article describes the geometric relationships between the acceleration, velocity, and position diagrams of a particle undergoing rectilinear motion. These relationships then are used to develop a simple numerical method for analyzing rectilinear motion called the *area method*, which enables us to construct velocity and position diagrams from a given acceleration diagram. The area method is particularly useful in cases where the acceleration diagram is made up of straight lines.

Typical plots of acceleration, velocity, and position of a particle in rectilinear motion are shown in Fig. 12.6. Recalling that $a = dv/dt$ and $v = dx/dt$, we deduce the following relationships between the diagrams:

1. The slope of the velocity diagram at time t_i is equal to the acceleration at that time; that is, $(dv/dt)_i = a_i$, as shown in Fig. 12.6(b).
2. The slope of the position diagram at time t_i is equal to the velocity at that time, that is, $(dx/dt)_i = v_i$, as shown in Fig. 12.6(c).

Consider next the time interval that begins at time t_0 and ends at time t_n, as shown in Fig. 12.6. The initial and final values of acceleration, velocity, and position are labeled a_0, v_0, x_0 and a_n, v_n, and x_n, respectively. Rewriting $a = dv/dt$ as $dv = a\,dt$ and integrating between t_0 and t_n yields

$$v_n - v_0 = \int_{t_0}^{t_n} a(t)\,dt$$

Recognizing the right-hand side of this equation as the area of the acceleration diagram between t_0 and t_n, we have

$$v_n - v_0 = \text{area of the } a\text{-}t \text{ diagram} \Big]_{t_0}^{t_n} \qquad (12.25)$$

Similarly, rewriting $v = dx/dt$ as $dx = v\,dt$ and integrating between t_0 and t_n, we obtain

$$x_n - x_0 = \int_{t_0}^{t_n} v(t)\,dt$$

Because the right-hand side of this equation is the area of the velocity diagram between t_0 and t_n, we arrive at

$$x_n - x_0 = \text{area of the } v\text{-}t \text{ diagram} \Big]_{t_0}^{t_n} \qquad (12.26)$$

Equations (12.25) and (12.26) can be restated in the following manner.

3. The increase in velocity during a given time interval is equal to the area of the a-t diagram for that time interval [the shaded area in Fig. 12.6(a)].

4. The increase in position coordinate during a given time interval is equal to the area of the v-t diagram for that time interval [the shaded area in Fig. 12.6(b)].

The relationships **1** through **4** were stated for rectilinear motion. However, they also apply to the special case of curvilinear motion that can be described as the superposition of rectilinear motions, one along each of the coordinate axes (Art. 12.5). In the two-dimensional case, for example, the motions can be represented by two sets of diagrams: a_x-t, v_x-t, x-t, for motion in the x-direction and a_y-t, v_y-t, y-t, for motion in the y-direction.

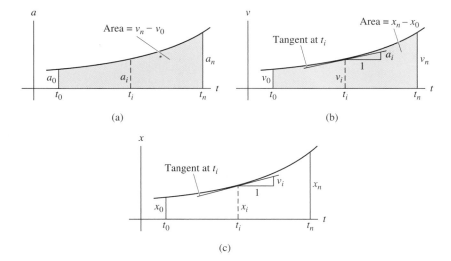

Fig. 12.6

Sample Problem   12.14

The 5-kg block in Fig. (a) is at rest at $x = 0$ and $t = 0$ when the force $P(t)$ is applied. The variation of $P(t)$ with time is shown in Fig. (b). Friction between the block and the horizontal plane can be neglected. (1) Use the area method to construct the a-t, v-t and x-t diagrams. (2) Determine the velocity and position of the block at $t = 5$ s.

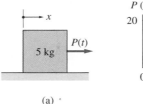

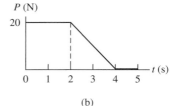

(a)   (b)

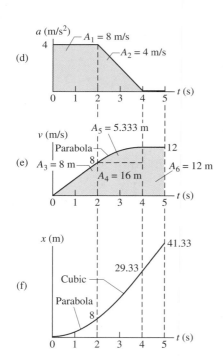

W = 5g N

FBD   MAD

(c)

(d)

(e)

(f)

Solution

Part 1

a-t Diagram   From the FBD and MAD of the block in Fig. (c), we obtain the following equation of motion:

$$\Sigma F_x = ma \quad \xrightarrow{+} \quad P = 5a$$

Therefore, the acceleration is

$$a = \frac{P}{5}$$

The resulting a-t diagram is shown in Fig. (d).

In the remainder of the solution, we will use subscripts on a, v, and x to indicate the values of these variables at various times. For example, $v_0, v_1, v_2, \ldots$ will refer to the velocities at $t = 0, 1$ s, 2 s, $\ldots$, respectively.

v-t Diagram   Before constructing the v-t diagram, we compute the areas under the a-t diagram in Fig. (d): $A_1 = 2(4) = 8$ m/s and $A_2 = (1/2)(2)(4) = 4$ m/s. The velocities $v_2, v_4,$ and v_5 are found by applying Eq. (12.25) (recall that $v_0 = 0$):

$$v_2 = v_0 + \text{area of the } a\text{-}t \text{ diagram}\Big]_{t=0}^{t=2 \text{ s}}$$

$$= v_0 + A_1 = 0 + 8 = 8 \text{ m/s}$$

$$v_4 = v_2 + \text{area of the } a\text{-}t \text{ diagram}\Big]_{t=2 \text{ s}}^{t=4 \text{ s}}$$

$$= v_2 + A_2 = 8 + 4 = 12 \text{ m/s}$$

$$v_5 = v_4 + \text{area of the } a\text{-}t \text{ diagram}\Big]_{t=4 \text{ s}}^{t=5 \text{ s}}$$

$$= v_4 + 0 = 12 \text{ m/s}$$

The values $v_0, v_2, v_4,$ and v_5 are then plotted in Fig. (e). The shape of the v-t diagram connecting these points is deduced from $a = dv/dt$, that is, the acceleration is equal to the slope of the v-t diagram.

x-t Diagram We begin by computing the areas under the v-t diagram in Fig. (e):
$A_3 = (1/2)(2)(8) = 8$ m; $A_4 = (2)(8) = 16$ m; $A_5 = (2/3)(2)(4) = 5.333$ m;
$A_6 = (1)(12) = 12$ m. The positions x_2, x_4, and x_5 are then computed from
Eq. (12.26), starting with the known value $x_0 = 0$:

$$x_2 = x_0 + \text{area of the } v\text{-}t \text{ diagram} \Big]_{t=0}^{t=2 \text{ s}}$$

$$= 0 + A_3 = 0 + 8 = 8 \text{ m}$$

$$x_4 = x_2 + \text{area of the } v\text{-}t \text{ diagram} \Big]_{t=2 \text{ s}}^{t=4 \text{ s}}$$

$$= 8 + (A_4 + A_5) = 8 + 16 + 5.333 = 29.33 \text{ m}$$

$$x_5 = x_4 + \text{area of the } v\text{-}t \text{ diagram} \Big]_{t=4 \text{ s}}^{t=5 \text{ s}}$$

$$= 29.33 + A_6 = 29.33 + 12 = 41.33 \text{ m}$$

After plotting the points x_0, x_2, x_4, and x_5 in Fig. (f), the shape of the connecting
curve can be determined from $v = dx/dt$; i.e., the slope of the x-t diagram is equal
to the velocity.

Part 2

After the diagrams in Figs. (d) through (f) have been constructed, it is a simple
matter to determine a, v, or x at a given value of time. In particular, from Figs. (e)
and (f) we see that

$$v_5 = 12 \text{ m/s} \quad \text{and} \quad x_5 = 41.3 \text{ m} \qquad \textit{Answer}$$

Sample Problem **12.15**

A golf ball is driven off a tee that is elevated 30 m above the fairway. The initial
velocity vector $\mathbf{v}_0$ of the ball is shown in Fig. (a). (1) Construct the acceleration,
velocity and position diagrams using the area method. (2) Determine the maximum height h of the ball above the tee, the range R, and the velocity vector of the
ball when it hits the fairway. Neglect air resistance.

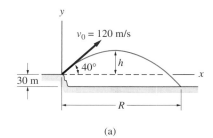

(a)

Solution

Introductory Comments

Because air resistance is neglected, the only force acting on the ball is its weight.
Therefore, $a_y = -g = -9.8$ m/s^2 and $a_x = 0$. Because the motions in the x
and y directions are uncoupled, it is convenient to consider the curvilinear motion

of the ball as a vector superposition of the motions in the x and y directions (see Art 12.5). Therefore, there will be two sets of diagrams—one set for the vertical direction and another for the horizontal direction.

For future reference, we introduce the following times: $t_0 = 0$, when the ball is hit; t_1, when the ball reaches its maximum height; and t_2, when the ball hits the fairway. In addition, subscripts 0, 1, and 2 are used to indicate the values of x, y, v_x, and v_y at these times.

From Fig. (a), we have the following five conditions imposed on the motion:

1. $x_0 = 0$
2. $y_0 = 0$
3. $y_2 = -30$ m
4. $(v_y)_0 = 120 \sin 40° = 77.13$ m/s
5. $(v_x)_0 = 120 \cos 40° = 91.93$ m/s

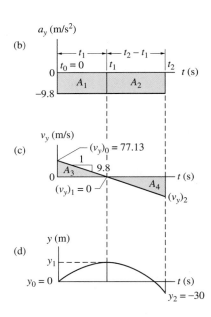

Part 1

Motion in the y-Direction   The a_y-t diagram is shown in Fig. (b), where a_y equals the constant value of -9.8 m/s^2.

From $a_y = dv_y/dt$, we conclude that the v_y-t diagram is a straight line with a slope equal to -9.8 m/s^2, as shown in Fig. (c). The initial value of this diagram is $(v_y)_0 = 77.13$ m/s. Note that $(v_y)_1 = 0$ because t_1 represents the time when the ball reaches its maximum height.

Because $v_y = dy/dt$, we deduce that the y-t diagram is a parabola as shown in Fig. (d). From the given conditions, we know that $y_0 = 0$ and $y_2 = -30$ m. Note that this curve is smooth, because there are no discontinuities in the v_y-t diagram.

Applying the area method to the diagrams in Figs. (b) through (d), we arrive at the following equations.

$(v_y)_1 = (v_y)_0 + A_1$	$0 = 77.13 - 9.8t_1$	(a)
$(v_y)_2 = (v_y)_1 + A_2$	$(v_y)_2 = 0 - 9.8(t_2 - t_1)$	(b)
$y_1 = y_0 + A_3$	$y_1 = 0 + \frac{1}{2}(77.13)t_1$	(c)
$y_2 = y_1 + A_4$	$-30 = y_1 - \frac{1}{2}(9.8)(t_2 - t_1)^2$	(d)

Solving Eqs. (a)–(d) for the four unknowns, we find $t_1 = 7.87$ s, $t_2 = 16.12$ s, $(v_y)_2 = -80.85$ m/s, and $y_1 = 303.5$ m.

Motion in the x-Direction   The a_x-t diagram, with $a_x = 0$, is shown in Fig. (e). From $a_x = dv_x/dt$, we conclude that the v_x-t diagram in Fig. (f) is a horizontal straight line with the initial value $(v_x)_0 = 91.93$ m/s. Using $v_x = dx/dt$, we find that the x-t diagram shown in Fig. (g) is an inclined straight line with the slope 91.93 m/s.

The following equations result from applying the area method to the diagrams in Figs. (e) through (f).

$(v_x)_2 = (v_x)_0 + A_5$	$(v_x)_2 = 91.93 + 0$	(e)
$x_2 = x_0 + A_6$	$x_2 = 0 + 91.93t_2$	(f)

Solving Eq. (e) gives $(v_x)_2 = 91.93$ m/s. Substituting $t_2 = 16.12$ s into Eq. (f) yields $x_2 = 1481.9$ m.

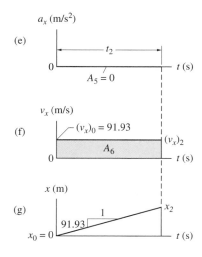

The final diagrams are shown in Fig. (h).

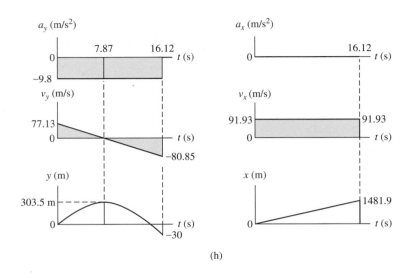

(h)

Part 2

From the position diagrams in Fig. (h) we find that

$$h = 303.5\,\text{m} \quad \text{and} \quad R = 1481.9\,\text{m} \qquad \textit{Answer}$$

According to the velocity diagram in Fig. (h), the velocity vector of the ball at impact with the fairway is

Answer

Problems

Solve the following problems using the area method. Sketch the acceleration, velocity, and position diagrams for each problem.

12.87   The figure shows the acceleration diagram for a commuter train as it travels on a straight, level track between two stations. Draw the acceleration, velocity, and position diagrams for the train. What is the distance between the two stations?

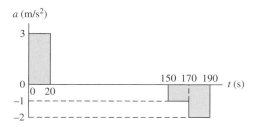

Fig. P12.87

12.88   A 2000-kg rocket is launched vertically from the surface of the earth. The engine produces a constant propulsive force of 60 kN for 20 seconds and then shuts off. Determine the altitude of the rocket at the end of the powered portion of the flight. Neclect the change in g with altitude and consider the mass of the rocket to be constant.

12.89   The volleyball player serves the ball at point A with the speed v_0 at the angle $\theta = 70°$. What is the largest v_0 for which the ball will not hit the ceiling.

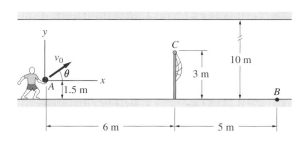

Fig. P12.89

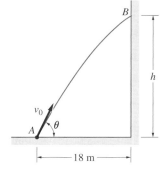

Fig. P12.90

12.90   A projectile is launched at A with the velocity $v_0 = 20$ m/s at the angle $\theta = 65°$. Find the height h of the impact point B on the vertical wall. Neglect air resistance.

12.91   A missile is launched horizontally at A with the speed $v_0 = 200$ m/s. Knowing that the range of the missile is $R = 1400$ m, calculate the launch height h and the time of flight.

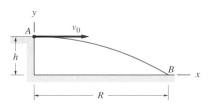

Fig. P12.91, P12.92

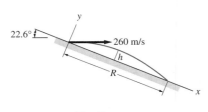

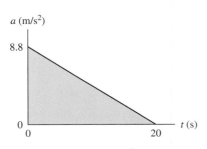

Fig. P12.93

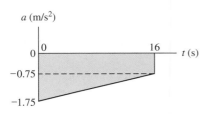

Fig. P12.95

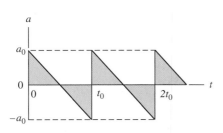

Fig. P12.97

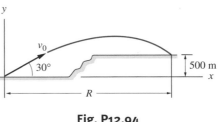

Fig. P12.99

12.92   A projectile is launched horizontally at A with the speed v_0. The time of flight is 10 s, and the path of the projectile at B is inclined at 20° with the horizontal. Determine v_0, the range R, and the launch height h. Use U.S. Customary units.

12.93   A projectile is fired horizontally at 100 m/s down the inclined plane. Draw the acceleration, velocity, and position diagrams. Use the diagrams to determine the maximum height h perpendicular to the plane, the range R along the plane, and the time of flight. Neglect air resistance.

12.94   A projectile is launched at an elevated target with initial speed $v_0 = 220$ m/s in the direction shown. Determine the time of flight and the range R.

Fig. P12.94

12.95   A car that is initially at rest accelerates along a straight, level road according to the diagram shown. Determine (a) the maximum speed; and (b) the distance traveled by the car when the maximum speed is reached.

12.96   A subway train stops at two stations that are 2 km apart. The maximum acceleration and deceleration of the train are 6.6 m/s² and 5.5 m/s², respectively, and the maximum allowable speed is 90 km/h. Find the shortest possible time of travel between the two stations.

12.97   A train is brought to an emergency stop in 16 seconds, the deceleration being as shown in the diagram. Compute the speed of the train before the brakes were applied and the stopping distance.

12.98   An airplane lands on a level runway at the speed of 40 m/s. For the first three seconds after touchdown, the reverse thrust of the propellers causes a deceleration of 3.2 m/s². For the next five seconds, the wheel brakes are applied producing an additional deceleration of 1.8 m/s². Then the reverse thrusters are shut down, and the plane is brought to a stop with only the wheel brakes. Draw the acceleration, velocity, and position diagrams. How far does the airplane travel on the runway before it comes to a stop?

12.99   A particle, at rest when $t = 0$, undergoes the periodic acceleration shown. Determine the velocity and distance traveled when (a) $t = 3t_0$; and (b) $t = 3.5t_0$.

12.100   The 8 N block is at rest on a rough surface when $t = 0$. For $t > 0$, the periodic horizontal force $P(t)$ of amplitude P_0 is applied to the block. Note that the period of $P(t)$ is 0.5 s. (a) Calculate the value of P_0 for which the average acceleration during each period is zero. (b) What is the average speed during each period in part (a)?

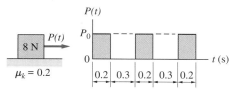

Fig. P12.100, P12.101

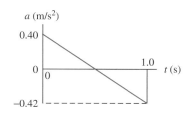

Fig. P12.102

12.101 The amplitude of the periodic force that is applied to the 8 N block is $P_0 = 6$ N. The coefficient of kinetic friction between the block and the horizontal surface is 0.2. If the velocity of the block at $t = 0$ was 2 m/s to the right, determine (a) the velocity of the block at $t = 0.7$ s; and (b) the displacement of the block from $t = 0$ to 0.7 s.

12.102 A car is traveling on a level road when it hits a small bump. The figure shows the resulting vertical acceleration of the car for the first 1.0 seconds after hitting the bump. Draw the velocity and position diagrams for the vertical motion. Determine the maximum vertical velocity and the maximum vertical displacement of the car during the 1.0-s period.

12.103 A rocket is fired vertically from the surface of the earth. The engine burns for 14 s, resulting in the acceleration shown in the diagram. Compute the maximum speed, the maximum height and the time when the maximum height occurs.

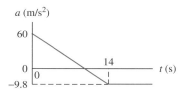

Fig. P12.103

Review of Equations

Kinematics in rectangular coordinates

$$v_x = \dot{x} \quad v_y = \dot{y} \quad v_z = \dot{z}$$
$$a_x = \dot{v}_x = \ddot{x} \quad a_y = \dot{v}_y = \ddot{y} \quad a_z = \dot{v}_z = \ddot{z}$$

Rectilinear motion

$$v = \dot{x} \quad a = \dot{v} = \ddot{x} = v\frac{dv}{dx}$$

Direct integration of equations of motion is possible in the following cases:

If $a = f(t) : dv = a(t)dt$
If $a = f(x) : vdv = a(x)dx$
If $a = f(v) : dx = \frac{vdv}{a(v)} \quad dt = \frac{dv}{a(v)}$

Force-mass-acceleration method

$$\Sigma F_x = ma_x \quad \Sigma F_y = ma_y \quad \Sigma F_z = ma_z$$

Review Problems

12.104 A particle is moving along the x-axis with the velocity $v = 2x^3 - 8x^2 + 12x$ mm/s, where x is measured in millimeters. Find the acceleration of the particle when $x = 2$ mm.

12.105 An object is undergoing rectilinear motion. During a certain six-second period, the velocity of the object changes from v_0 to 16 m/s, while its acceleration increases uniformly from 0 to 8 m/s^2. Determine v_0.

12.106 Two cars A and B are traveling in the same direction along a straight highway. At a certain instant, car A is 400 m behind car B, where the speeds of the cars are 30 m/s for A and 60 m/s for B. At the same instant, car A accelerates at the constant rate of 4 m/s^2, while car B decelerates at the constant rate of 2 m/s^2. How long will it take for car A to overtake car B?

12.107 The position coordinate of an object moving along the x-axis is given by $x = 3t^3 - 9t + 4$ mm, where the time t is measured in seconds. For the time interval $t = 0$ to $t = 2$ s, determine (a) the displacement of the object; and (b) the total distance traveled.

12.108 A boy drops a stone down a well and four seconds later hears the splash. Neglecting air resistance, compute the depth of the well. The speed of sound is 320 m/s.

12.109 A particle moving along the x-axis starts from rest at time $t = 0$ with the acceleration $a = 12t - 6t^2$ m/s^2, where t is measured in seconds. For the time interval $t = 0$ to $t = 5$ s, determine (a) displacement of the particle; and (b) the total distance traveled.

12.110 During braking, the speed of a car traveling along a straight highway varies as $v = 16 - (x/4)$ m/s where x (in meters) is the distance traveled after the brakes are applied. Determine the acceleration of the car as a function of x.

12.111 For the car described in Prob. 12.110, determine x as a function of time t (seconds). Assume that the braking begins at $t = 0$.

12.112 The projectile is launched at O with the velocity v_0 inclined at 60° to the horizontal. Determine the smallest value of v_0 for which the projectile will clear the wall AB. Neglect air resistance.

12.113 At the instant shown, the 1 kg box is sliding across the horizontal plane with a velocity of 2 m/s to the left. Find the force P that will give the block an acceleration of 10 m/s^2 to the right at this instant. The coefficient of kinetic friction between the block and the plane is 0.2.

12.114 A 1400-kg rocket is launched vertically from the surface of the earth. During the 20-second burn, the propulsive force F of the engine varies with time,

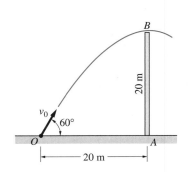

Fig. P12.112

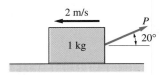

Fig. P12.113

as shown in the plot. Assuming that the gravitational acceleration and the mass of the rocket are constant, determine the elevation of the rocket at the end of the burn.

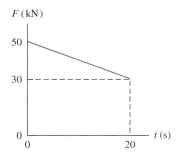

Fig. P12.114

12.115   A 0.2-kg mass moves along the x-axis. The resultant of all the forces acting on the mass $\mathbf{F} = -1.6e^{4x}\mathbf{i}$ N, where x is in meters. Knowing that $\mathbf{v} = 6\mathbf{i}$ m/s when $x = 0$, determine x when $v = 0$.

12.116   A 10-kg parcel is dropped from a height with no initial velocity. During the fall, the aerodynamic drag force acting on the parcel is $F_D = kv$, where k is a constant and v is the velocity. If the terminal velocity of the parcel is 60 m/s, determine (a) the value of k; and (b) the time when the velocity of the parcel reaches 59 m/s.

12.117   A ball is thrown down a 20° incline, as shown. Determine the initial velocity v_0 given that $\theta_0 = 25°$ and $R = 60$ m. Neglect air resistance.

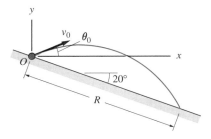

Fig. P12.117

12.118   The free length of the spring that is attached to the 0.2-kg slider A is 120 mm. If the slider is released from rest when $x = 240$ mm, calculate its initial acceleration. Neglect friction.

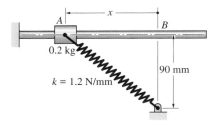

Fig. P12.118

12.119 A particle follows the path shown in the figure. The description of motion is $x = b \sin \omega t$, $y = b \exp(-\omega t/2)$, where b and ω are constants. Determine the magnitude of the acceleration at (a) point A; and (b) point B.

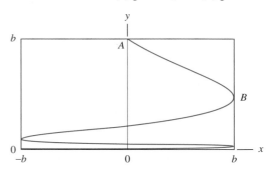

Fig. P12.119

12.120 The rope ABC passes that is attached to the 5 kg block. A constant vertical force of 10 N is applied to the end of the rope. The coefficient of kinetic friction between the block and the plane is 0.2. If the block has a velocity of 3 m/s to the left when $x = 2$ m, determine its acceleration in this position.

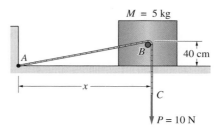

Fig. P12.120

12.121 A golf ball is hit from a tee that is elevated 8 m above a level fairway. The initial velocity of the ball is 45 m/s, inclined up at 40° to the horizontal. Determine (a) the horizontal distance traveled by the ball; and (b) the speed of the ball when it hits the fairway. Neglect air resistance.

12.122 The mass m is released with zero velocity on top of an undeformed spring of stiffness k. Derive the expressions for (a) the maximum force in the spring; and (b) the maximum velocity of the mass.

Fig. P12.122

13

Dynamics of a Particle: Curvilinear Coordinates

13.1 Introduction

The acceleration of a car traveling on a straight road is determined by its rate of change of speed. On a curved road, the acceleration also depends on the rate at which the direction of the velocity changes. This is illustrated in Prob. 13.1. (David De Lossy/Photodisc/Getty Images)

In the introduction to Chapter 12, we mentioned that curvilinear coordinates often lead to an easier description of particle motion than rectangular coordinates. In this chapter, we study two such coordinate systems: path coordinates and polar coordinates. We also consider cylindrical coordinates, which are polar coordinates with the additional axial coordinate z.

Path coordinates, also known as *normal-tangential (n-t) coordinates,* describe the motion of a particle in terms of components that are normal and tangent to its path. This is a convenient and very natural way to describe curvilinear motion if the path is known beforehand (an example is a car traveling on a curved road).

Polar (R-θ) coordinates are useful if the motion, or the forces controlling the motion, are specified in terms of the radial distance R from a fixed point, and the polar angle θ. For example, the orbital motion of a satellite is best described in terms of polar coordinates, because the gravitational force acting on the satellite depends on the distance R from the center of the earth.

This chapter has two distinct parts. The first part, which forms the bulk of the chapter, is dedicated to kinematics. Because kinematics is considerably more involved in curvilinear coordinates than in rectangular coordinates, it is also the most substantial topic in the chapter. The other part, kinetics of a particle, does not differ significantly from what was presented in the previous chapter.

13.2   Kinematics—Path (Normal-Tangential) Coordinates

a.   Plane motion

1.   Geometric Preliminaries   Figure 13.1 shows the path of a particle that moves in the xy-plane. The position of the particle is specified by the *path coordinate s*, which is the distance measured along the path from a fixed reference point. As the particle moves from A to B during an infinitesimal time interval dt, it traces an arc of radius ρ and infinitesimal length ds. The corresponding displacement of the particle is $d\mathbf{r}$, where $|d\mathbf{r}| = ds$. From Fig. 13.1 we obtain the useful relationship $ds = \rho\, d\theta$, or after division by dt,

$$\dot{s} = \rho\dot{\theta} \tag{13.1}$$

where the angle θ is measured in radians.

The radius ρ is called the *radius of curvature* of the path at A. If the equation of the path is known, its radius of curvature can be computed from

$$\rho = \frac{\left[1 + \left(\dfrac{dy}{dx}\right)^2\right]^{3/2}}{\left|\dfrac{d^2 y}{dx^2}\right|} = \frac{\left[1 + \left(\dfrac{dx}{dy}\right)^2\right]^{3/2}}{\left|\dfrac{d^2 x}{dy^2}\right|}$$

Note that either x or y can be taken as the independent variable. The inverse of ρ, that is, $1/\rho$, is known as the *curvature* of the path.

The *base vectors* $\mathbf{e}_n$ and $\mathbf{e}_t$ associated with point A on the path are shown in Fig. 13.2. Like the vectors $\mathbf{i}$ and $\mathbf{j}$ of the rectangular coordinate system, $\mathbf{e}_n$ and $\mathbf{e}_t$ are mutually perpendicular, of unit magnitude, and serve as the bases for the velocity and acceleration vectors. However, the directions of $\mathbf{e}_n$ and $\mathbf{e}_t$ are not fixed, but depend on the location A of the particle: $\mathbf{e}_t$ is tangent to the path at A and points in the direction of increasing s, whereas $\mathbf{e}_n$ is normal to the path and directed toward the *center of curvature C*. Figure 13.2 also shows how the base vectors change direction as the particle moves from point A to point B. It is customary to call $\mathbf{e}_n$ the *unit normal,* and $\mathbf{e}_t$ the *unit tangent.*

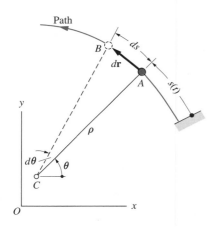

Fig. 13.1

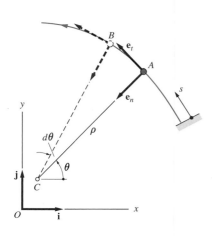

Fig. 13.2

Another useful equation can be derived from Figs. 13.1 and 13.2. Noting that $d\mathbf{r}$ is tangent to the path at A and has the magnitude ds, we can write $d\mathbf{r} = \mathbf{e}_t \, ds$. Consequently,

$$\mathbf{e}_t = \frac{d\mathbf{r}}{ds} \tag{13.2}$$

2. *Derivatives of Base Vectors* Because the directions of $\mathbf{e}_n$ and $\mathbf{e}_t$ vary with the position of the particle, their time derivatives are not zero. To obtain the derivatives, we first express the base vectors in terms of the rectangular components. Referring to Fig. 13.3, we have

$$\mathbf{e}_t = -\sin\theta\mathbf{i} + \cos\theta\mathbf{j} \qquad \mathbf{e}_n = -\cos\theta\mathbf{i} - \sin\theta\mathbf{j} \tag{13.3}$$

Differentiating with respect to time while noting that $d\mathbf{i}/dt = d\mathbf{j}/dt = \mathbf{0}$, we get

$$\dot{\mathbf{e}}_t = (-\cos\theta\mathbf{i} - \sin\theta\mathbf{j})\dot{\theta} \qquad \dot{\mathbf{e}}_n = (\sin\theta\mathbf{i} - \cos\theta\mathbf{j})\dot{\theta}$$

Comparison with Eq. (13.3) yields

$$\dot{\mathbf{e}}_t = \dot{\theta}\mathbf{e}_n \qquad \dot{\mathbf{e}}_n = -\dot{\theta}\mathbf{e}_t \tag{13.4}$$

The base vectors and their derivatives are shown in Fig. 13.4. Note that each base vector and its derivative are mutually perpendicular, which reflects the fact that only the directions of the vectors change (a change in the magnitude of a vector would be parallel to the vector).

3. *Velocity and Acceleration* We start with the definition of velocity: $\mathbf{v} = d\mathbf{r}/dt$, where $\mathbf{r}$ is the position vector of the particle. Using the chain rule of differentiation and Eq. (13.2), we can write $\mathbf{v} = (d\mathbf{r}/ds)(ds/dt) = \mathbf{e}_t(ds/dt)$, or

$$\mathbf{v} = v\mathbf{e}_t \tag{13.5}$$

where the magnitude of the velocity

$$v = \dot{s} \tag{13.6}$$

is called the *speed*. Equation (13.5) shows that the velocity is always tangent to the path (in the direction of the unit tangent $\mathbf{e}_t$).

The acceleration of the particle is obtained by differentiating the velocity with the help of Eqs. (13.4):

$$\mathbf{a} = \frac{d\mathbf{v}}{dt} = \frac{d}{dt}(v\mathbf{e}_t) = \dot{v}\mathbf{e}_t + v\dot{\mathbf{e}}_t = \dot{v}\mathbf{e}_t + v\dot{\theta}\mathbf{e}_n$$

We can eliminate $\dot{\theta}$ from the last term by substituting $\dot{\theta} = \dot{s}/\rho = v/\rho$, obtainable from Eq. (13.1). The result is

$$\mathbf{a} = a_t\mathbf{e}_t + a_n\mathbf{e}_n \tag{13.7}$$

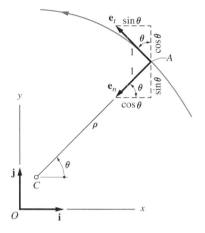

Fig. 13.3

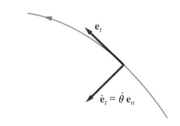

(a)

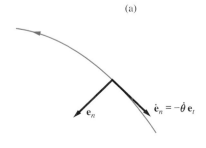

(b)

Fig. 13.4

where the normal and tangential components of the acceleration are

$$a_n = \frac{v^2}{\rho} \qquad a_t = \dot{v}$$ (13.8)

Sometimes it is advantageous to eliminate time from the expression for a_t. Using the chain rule of differentiation, we write $a_t = dv/dt = (dv/ds) \times (ds/dt)$, or

$$a_t = v\frac{dv}{ds}$$ (13.9)

Equation (13.9) is similar to $a = v\,dv/ds$, which arose in rectilinear motion. But note that a in rectilinear motion is the magnitude of the acceleration, whereas a_t in Eq. (13.9) refers to the magnitude of the tangential component of acceleration.

The velocity and acceleration vectors are shown in Fig. 13.5. It is evident that a_t is caused by a *change in the speed* of the particle. If the speed is increasing, $\mathbf{a}_t$ has the same direction as the velocity; if the speed is decreasing, $\mathbf{a}_t$ and the velocity have opposite directions. If the speed is constant, then $a_t = 0$.

The normal component a_n, sometimes called the *centripetal acceleration*, is due to a *change in the direction* of the velocity. Note that $\mathbf{a}_n$ is always directed toward the center of curvature of the path. If the path is a straight line ($1/\rho = 0$), then $a_n = 0$.

Caution   Do not confuse the following three equations:

- $\mathbf{a} = d\mathbf{v}/dt$—the definition of acceleration
- $a = dv/dt$—the magnitude of the acceleration in rectilinear motion
- $a_t = dv/dt$—the tangential component of acceleration in plane curvilinear motion

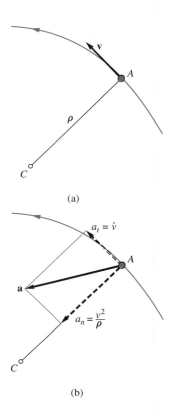

(a)

(b)

Fig. 13.5

Confusion can be avoided by meticulous use of notation: Always include subscripts and use different notation for scalars and vectors.

4.   *Motion Along a Circular Path*   The special case of a circular path plays an important role in dynamics, particularly in the kinematics of rigid bodies. If the radius of the path is R, as shown in Fig. 13.6, then Eq. (13.1) becomes $\dot{s} = R\dot{\theta}$ (note that $\rho = R$, a constant), or

$$v = R\dot{\theta}$$ (13.10)

Substituting in Eqs. (13.8), we get

$$a_n = \frac{v^2}{R} = R\dot{\theta}^2 \qquad a_t = \dot{v} = R\ddot{\theta}$$ (13.11)

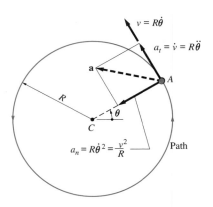

Fig. 13.6

where $\dot{\theta}$ and $\ddot{\theta}$ are known as the *angular velocity* and the *angular acceleration* of the line AC, respectively.

Figure 13.7 illustrates how the direction of the acceleration vector changes as a particle moves counterclockwise around a circle. As pointed out before, the velocity is always tangent to the path, as shown in Fig. 13.7(a). In Fig. 13.7(b), the speed is assumed to be increasing, so that $\dot{v}$ is positive. Therefore, a_t is also positive, meaning that $\mathbf{a}_t$ points in the same direction as $\mathbf{v}$. As always, $\mathbf{a}_n$ is directed toward the center of the circle.

If the speed is decreasing, as in Fig. 13.7(c), $\dot{v}$ is negative. Because a_t is now also negative, $\mathbf{a}_t$ and $\mathbf{v}$ have opposite directions. But $\mathbf{a}_n$ is still pointing to the center of the circle.

In Fig. 13.7(d), the speed is constant, that is, $\dot{v} = 0$. Consequently, $a_t = 0$, and the acceleration vector is directed toward the center of the circle.

b. *Space motion*

A description of three-dimensional particle motion using path coordinates requires a knowledge of geometry of space curves that is beyond the scope of this text. It can be shown that Eqs. (13.5) through (13.9) for velocity and acceleration are also valid for three-dimensional motion.[*] However, in space motion the "plane" of the motion is continually changing with the position of the particle.

Figure 13.8 shows a particle that is at point A at time t. The unit tangent vector is $\mathbf{e}_t = d\mathbf{r}/ds$, and the unit normal (called the *principal normal*) is $\mathbf{e}_n = \rho\, d^2\mathbf{r}/ds^2$, where $\rho = 1/|d^2\mathbf{r}/ds^2|$ is the radius of curvature of the arc ds. The third unit vector $\mathbf{e}_b$, called the *binormal*, is usually chosen so that the three unit vectors form a right-handed triad: $\mathbf{e}_b = \mathbf{e}_t \times \mathbf{e}_n$. The plane formed by $\mathbf{e}_t$ and $\mathbf{e}_n$ is called the osculating plane (after the Latin word *osculari*, "to kiss"). For two-dimensional motion, the osculating plane is fixed in space and is the plane of motion. In three-dimensional motion, the orientation of the osculating plane continually changes as the particle moves along its path. Because of the obvious geometric complexities, path coordinates are of limited use in three-dimensional motion.

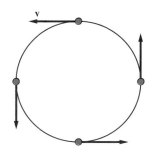

(a) Velocity

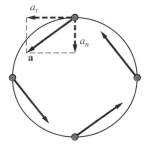

(b) Acceleration for increasing speed

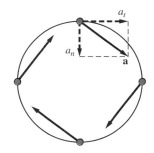

(c) Acceleration for decreasing speed

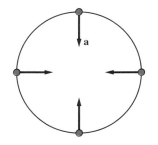

(d) Acceleration for constant speed

Fig. 13.7

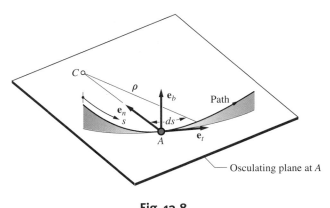

Fig. 13.8

---

[*]See *Principles of Dynamics*, Donald T. Greenwood, Prentice Hall, 1988.

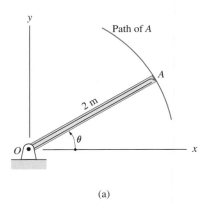

Path of A

2 m

θ

O x

(a)

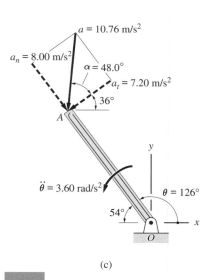

$36°$

$v = 4.00$ m/s

$\dot\theta = 2.00$ rad/s

$54°$

$\theta = 126°$

O x

(b)

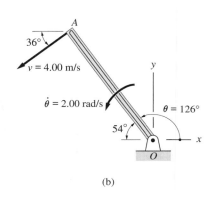

$a = 10.76$ m/s²

$a_n = 8.00$ m/s²

$\alpha = 48.0°$

$a_t = 7.20$ m/s²

$36°$

A

$\ddot\theta = 3.60$ rad/s²

$54°$

$\theta = 126°$

O x

(c)

Sample Problem   **13.1**

The angle between the 2-m bar shown in Fig. (a) and the x-axis varies according to $\theta(t) = 0.3t^3 - 1.6t + 3$ rad, where t is the time in seconds. When $t = 2$ s, (1) determine the magnitudes of the velocity and acceleration of end A; and (2) show the velocity and acceleration vectors of A on a sketch of the bar.

Solution

Part 1

It is convenient to use normal and tangential components because the path of A is a circle (centered at point O, of radius $R = 2$ m).

The angular velocity and acceleration of the bar are $\dot\theta = 0.9t^2 - 1.6$ rad/s and $\ddot\theta = 1.8t$ rad/s². At $t = 2$ s we find that

$$\dot\theta|_{t=2\,s} = 0.9(2)^2 - 1.6 = 2.00 \text{ rad/s}$$

$$\ddot\theta|_{t=2\,s} = 1.8(2) = 3.60 \text{ rad/s}^2$$

Because $\dot\theta$ and $\ddot\theta$ are positive, their directions are the same as the positive direction for θ, that is, counterclockwise.

From Eq. (13.10), the magnitude of the velocity of A is

$$v = R\dot\theta = 2(2.00) = 4.00 \text{ m/s} \qquad \textit{Answer}$$

The normal and tangential components of the acceleration of A are, using Eqs. (13.11),

$$a_n = R\dot\theta^2 = 2(2.00)^2 = 8.00 \text{ m/s}^2$$

$$a_t = R\ddot\theta = 2(3.60) = 7.20 \text{ m/s}^2$$

Therefore, the magnitude of the acceleration of A is

$$a = \sqrt{a_n^2 + a_t^2} = \sqrt{(8.00)^2 + (7.20)^2} = 10.76 \text{ m/s}^2 \qquad \textit{Answer}$$

Part 2

On substituting $t = 2$ s into the expression for $\theta(t)$, we find that the angular position of the bar at $t = 2$ s is

$$\theta|_{t=2\,s} = 0.3(2)^3 - 1.6(2) + 3 = 2.20 \text{ rad} = 126°$$

The velocity vector of end A is shown in Fig. (b). The magnitude of $\mathbf{v}$ is 4.00 m/s, as computed in Part 1, and the vector is tangent to the circular path, its direction being consistent with the direction of $\dot\theta$.

Figure (c) shows the directions of the normal and tangential components of the acceleration vector as determined in Part 1. Note that $\mathbf{a}_n$ is normal to the path and directed toward the point O, the center of the path. The direction of $\mathbf{a}_t$ is tangent to the path, consistent with the direction for $\ddot{\theta}$. The acceleration vector of magnitude 10.76 m/s^2 is also shown in Fig. (c), where the angle between $\mathbf{a}$ and $\mathbf{a}_t$ was found to be

$$\alpha = \tan^{-1}\frac{a_n}{a_t} = \tan^{-1}\frac{8.00}{7.20} = 48.0°$$

■

Sample Problem   13.2

The racing car shown in Fig. (a) is traveling at 90 km/h when it enters the semicircular curve at A. The driver increases the speed at a uniform rate, emerging from the curve at C at 144 km/h. Determine the magnitude of the acceleration when the car is at B.

Solution

Because the car follows a circular path, it is convenient to describe its motion using path coordinates. As shown in Fig. (b), we let s be the distance measured along the path from A toward C.

The magnitude of the tangential component of acceleration is *constant* between A and C, since the speed increases at a uniform rate. Therefore, integration of $a_t\,ds = v\,dv$ yields

$$\frac{v^2}{2} = a_t s + C \tag{a}$$

where C is the constant of integration. The two constants a_t and C can be evaluated using the following two conditions on the motion:

1. At A: $s = 0$, $v = 25$ m/s (90 km/h)
2. At C: $s = \pi R = 100\pi$ ft, $v = 40$ m/s (144 km/h)

Substituting condition **1** into Eq. (a), we find

$$\frac{(25)^2}{2} = 0 + C$$

from which the constant of integration is

$$C = 312.5 \text{ (m/s)}^2 \tag{b}$$

Substituting condition **2** and the value of C into Eq. (a) gives

$$\frac{(40)^2}{2} = a_t(100\pi) + 312.5$$

Solving for a_t yields

$$a_t = 1.55 \text{ m/s}^2 \tag{c}$$

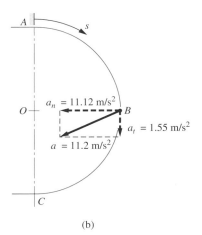

(a)

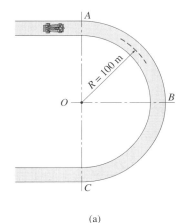

(b)

As shown in Fig. (b), the direction of a_t is downward at B, that is, in the direction of increasing speed.

On substituting the values of C and a_t into Eq. (a), the relationship between the speed v and the distance s is found to be

$$\frac{v^2}{2} = 1.55\,s + 312.5 \tag{d}$$

To compute the speed of the car at B, we substitute $s = \pi R/2 = 50\pi$ m ft into Eq. (d), the result being

$$\frac{v^2}{2} = 1.55(50\pi) + 312.5$$

$$v = 33.35 \text{ m/s}$$

From Eq. (13.11), the normal component of the acceleration at B is

$$a_n = \frac{v^2}{R} = \frac{(33.35)^2}{100} = 11.12 \text{ m/s}^2$$

directed toward the center of curvature of the path (point O), as indicated in Fig. (b).

The magnitude of the acceleration vector at B is

$$a = \sqrt{(11.12)^2 + (1.55)^2} = 11.2 \text{ m/s}^2 \qquad \textit{Answer}$$

with the direction shown in Fig. (b).

Sample Problem 13.3

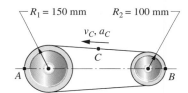

The flexible belt runs around two pulleys of different radii. At the instant shown, point C on the belt has a velocity of 5 m/s and an acceleration of 50 m/s² in the direction indicated in the figure. Compute the magnitudes of the accelerations of points A and B on the belt at this instant.

Solution

Assuming that the belt does not stretch, we conclude the following:

1. Every point on the belt has the same speed, that is, $v_A = v_B = v_C = 5$ m/s.
2. The rate of change of speed (dv/dt) of every point on the belt is the same. Therefore, $(a_A)_t = (a_B)_t = a_C = 50$ m/s².

For point A

$$(a_A)_n = \frac{v_A^2}{R_1} = \frac{(5)^2}{0.150} = 166.67 \text{ m/s}^2$$

$$a_A = \sqrt{(a_A)_n^2 + (a_A)_t^2} = \sqrt{(166.67)^2 + (50)^2} = 174.0 \text{ m/s}^2 \quad \textit{Answer}$$

For point B

$$(a_B)_n = \frac{v_B^2}{R_2} = \frac{(5)^2}{0.100} = 250.0 \text{ m/s}^2$$

$$a_B = \sqrt{(a_B)_n^2 + (a_B)_t^2} = \sqrt{(250.0)^2 + (50)^2} = 255 \text{ m/s}^2 \qquad \textbf{\textit{Answer}}$$

Sample Problem 13.4

The trolley in Fig. (a) travels at the constant speed of 90 km/h along a parabolic track described by $y = x^2/500$, where x and y are measured in meters. Compute the acceleration of the trolley when it is (1) at point O; and (2) at point A.

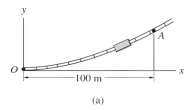

(a)

Solution

Preliminary Discussion

Because the speed of the trolley is constant, the tangential component of its acceleration is zero at all points along the track. Therefore, the acceleration has only the normal component, given by Eq. (13.8):

$$|\mathbf{a}| = a_n = \frac{v^2}{\rho} \qquad \text{(a)}$$

where ρ is the radius of curvature of the track at the point of interest. Recall that a_n is directed toward the center of curvature of the track.

The radius of curvature at any point with coordinates x and y can be computed from

$$\rho = \frac{\left[1 + \left(\dfrac{dy}{dx}\right)^2\right]^{3/2}}{\left(\dfrac{d^2y}{dx^2}\right)} \qquad \text{(b)}$$

Successive differentiations of the parabola ($y = x^2/500$) with respect to x yield

$$\frac{dy}{dx} = \frac{x}{250} \qquad \frac{d^2y}{dx^2} = \frac{1}{250} \qquad \text{(c)}$$

Substituting Eqs. (c) into Eq. (b), we find that the radius of curvature of the track is

$$\rho = 250 \left[1 + (x/250)^2\right]^{3/2} \text{ m} \tag{d}$$

$$v = \frac{90 \times 1000}{60 \times 60} = 25 \text{ m/s} \tag{e}$$

Part 1

Using Eq. (d), the radius of curvature at point O ($x_O = 0$) is

$$\rho_O = 250 \left[1 + (0/250)^2\right]^{3/2} = 250 \text{ m}$$

Therefore, the normal component of acceleration in Eq. (a) is

$$(a_n)_O = \frac{v^2}{\rho_O} = \frac{(25)^2}{250} = 2.5 \text{ m/s}^2 \qquad \textit{Answer}$$

Note that the tangent to the track at point O lies along the x-axis. Therefore, $(a_n)_O$ lies along the y-axis directed toward the center of curvature of the track, as shown in Fig. (b).

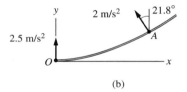

(b)

Part 2

Using Eq. (d), the radius of curvature at point A ($x_A = 100$ m) is

$$\rho_A = 250 \left[1 + (100/250)^2\right]^{3/2} = 312.3 \text{ m}$$

Therefore, the normal component of acceleration is

$$(a_n)_A = \frac{v^2}{\rho_A} = \frac{(25)^2}{312.3} = 2 \text{ m/s}^2 \qquad \textit{Answer}$$

Using the first of Eqs. (c), the slope of the track at A is

$$\left(\frac{dy}{dx}\right)_A = \frac{x_A}{250} = \frac{100}{250} = 0.4 \text{ rad} = 21.8°$$

Therefore, $(a_n)_A$ is directed as shown in Fig. (b); that is, normal to the track and directed toward its center of curvature.

Problems

13.1 A car drives through portion *AB* of the S-curve at constant speed, decelerates in *BC*, and accelerates in *CD*. Show the approximate direction of the acceleration vector at each of the five points indicated.

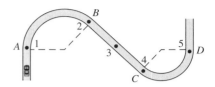

Fig. P13.1

13.2 A car is driving around a curve of radius 200 m, while increasing its speed at the rate of 0.8 m/s^2. At a certain instant, the magnitude of the total acceleration is measured to be 1.5 m/s^2. What is the speed of the car at that instant measured in km/h?

13.3 The rocket is in powered flight close to the surface of the earth. Determine the radius of curvature of the path at the instant shown if the speed of the rocket is 200 m/s. Note that the acceleration of the rocket has two components—the acceleration due to the thrust of the rocket engines and the acceleration due to gravity.

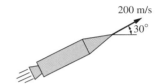

Fig. P13.3

13.4 The car is traveling at a constant speed through a dip in the road. The radius of curvature of the road at point *A*, the bottom of the dip, is 500 m. What speed of the car, measured in km/h, would result in an acceleration of magnitude $0.2g$ when the car is at *A*?

13.5 A ball is shot from the cannon at *A* with the initial velocity v_A directed at the angle θ with the horizontal. Derive the expression for the radius of curvature at *B*—the highest point on the path of the ball.

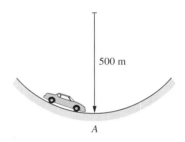

Fig. P13.4

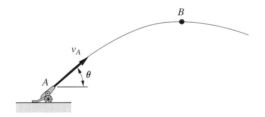

Fig. P13.5

13.6 A particle travels around a circle of radius 4 m, changing its speed at a constant rate. At a certain point *A*, the speed is 3 m/s. After traveling another quarter revolution to point *B*, the speed has increased to 6 m/s. Determine the magnitude of the acceleration of the particle at *B*.

13.7 A particle travels along a plane curve from a point *A* to a point *B*. The path length between *A* and *B* is 2 m. The speed of the particle is 4 m/s at *A* and 2 m/s at *B*. The rate of change of the speed is constant. (a) Find the tangential component of the acceleration when the particle is at *B*. (b) If the magnitude of the acceleration at *B* is 5 m/s^2, determine the radius of curvature of the path at *B*.

13.8 A particle moves along a plane curve from a point *O* to a point *B*. The path length between *O* and *B* is 2 m. The tangential component of the acceleration is

$a_t = 0.05s$ m/s², where s is the path coordinate, measured in m from point O. The speed of the particle at O is 2 m/s, and the radius of curvature of the path at B is 3 m. Determine the magnitude of the acceleration of the particle at B.

13.9   The particle passes point O at the speed of 8 m/s. Between O and B, the speed increases at the rate of $4\sqrt{v}$ m/s², where v is the speed in m/s. Determine the magnitude of the acceleration when the particle is (a) just to the left of point A; and (b) just to the right of point A.

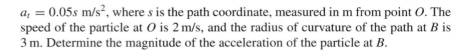

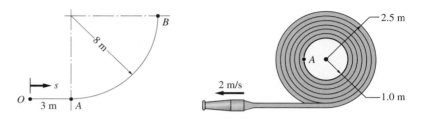

Fig. P13.9   **Fig. P13.10**

13.10   The firehose is being uncoiled from its reel at the constant speed of 2 m/s. Determine the normal component of acceleration of point A on the rim of the reel (a) at the instant shown in the figure; and (b) when almost all the hose has left the reel.

13.11   Pulley A is attached to the crankshaft of an automobile engine. If the crankshaft rotates at the constant angular speed of 2000 rev/min, determine the maximum acceleration of any point of the V-belt as it runs around the three pulleys.

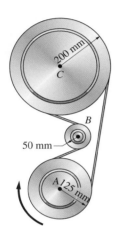

Fig. P13.11

13.12   At the instant shown, the angular speed and acceleration of rod OB are $\dot{\theta} = 8$ rad/s and $\ddot{\theta} = 24$ rad/s², respectively, both counterclockwise. Calculate (a) the velocity vectors of points A and B on the rod; and (b) the acceleration vectors of A and B.

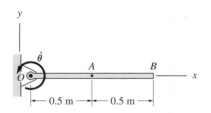

Fig. P13.12, P13.13

13.13   The angular velocity of rod OB varies as $\dot{\theta} = 8 - 12t^2$ rad/s where t is in seconds. Compute the magnitude of the acceleration of point B at (a) $t = 0$; and (b) $t = 1.0$ s.

13.14   The speed of the belt is changed at a uniform rate from 0 to 2 m/s during a time interval of 0.2 second. Calculate (a) the distance traveled by the belt during the 0.2-second interval; and (b) the maximum acceleration of any point on the belt during this interval.

13.15   The rate of change of speed of the belt is given by $0.06(10 - t)$ m/s², where t is in seconds. The speed of the belt is 0.8 m/s at $t = 0$. When the normal acceleration of a point in contact with the pulley is 40 m/s², determine (a) the speed of the belt; (b) the time required to reach that speed; and (c) the distance traveled by the belt.

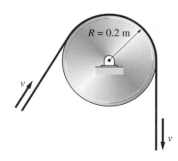

Fig. P13.14, P13.15

13.16   A motorist entering the exit ramp of a highway at 40 km/h immediately applies the brake so that the magnitude of the acceleration of the car at A is 1.5 m/s². If the tangential acceleration is maintained, how far will the car travel before coming to a stop?

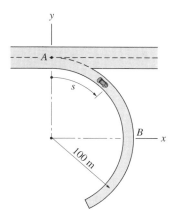

Fig. P13.16, P13.17

13.17   The tangential acceleration of a car that starts from rest at A is $(90 + s)/450$ m/s², where s is the distance in meters measured along the curve from A. Compute the acceleration vector of the car when it is at B.

13.18   A skateboarder rides down a parabolic half-pipe. The profile of the half-pipe is $y = x^2/80$ m, where x is in meters. Determine the magnitude of the rider's acceleration at point A given that $v_A = 12$ m/s and $\dot{v}_A = 4$ m/s².

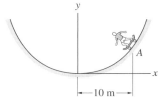

Fig. P13.18

13.19   The airplane flies along a circular path of radius 5 km that lies in the vertical plane. To simulate weightlessness, the acceleration vector of the plane is directed vertically downward and has the magnitude g. Determine the speed v of the plane and its rate of change $\dot{v}$ at point A. At the altitude of the plane, $g = 9.78$ m/s².

*13.20   A car travels around the racetrack shown. Traction between the tires and the road limits the total maximum acceleration to 5 m/s². What is the shortest possible time for the car to complete one loop around the track?

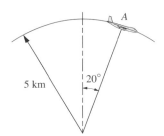

Fig. P13.19

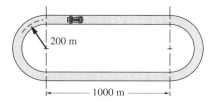

Fig. P13.20

13.21   The slot in the slider A engages the pin B that is attached to the arm OB. In the position shown, the slider is moving upward with the constant speed

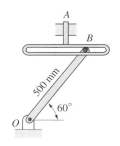

Fig. P13.21

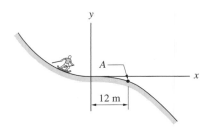

Fig. P13.22

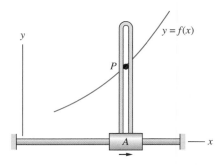

Fig. P13.23

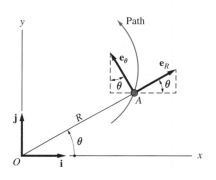

Fig. 13.9

of 150 mm/s. Determine the magnitudes of the velocity and acceleration vectors of pin *B* in this position.

*13.22 The skier glides down a slope described by $y = -x^3/3\,900$ m, where x is in m. The speed of the skier at point *A* is $v_A = 6$ m/s, and the speed is increasing at the rate $\dot{v}_A = 1$ m/s². Determine the magnitude of the skier's acceleration vector at *A*.

13.23 Pen *P* of the flatbed plotter traces the curve $y = x^3/2$, where x and y are measured in mm. When $x = 1000$ mm, the speed of slider *A* is 200,mm/s. For this position, calculate (a) the speed of *P*; and (b) the normal component of the acceleration of *P*.

13.24 A particle moves with the constant speed v_0 along the parabola $y = Ax^2 + Bx + C$. Find the maximum acceleration and the corresponding x-coordinate.

13.25 A particle moves with constant speed v_0 along the ellipse $(x/a)^2 + (y/b)^2 = 1$, where $a > b$. Determine the maximum acceleration of the particle.

13.3 | ### Kinematics—Polar and Cylindrical Coordinates

a. Plane motion (polar coordinates)

1. Geometric Preliminaries Figure 13.9 shows the polar coordinates R and θ that specify the position of particle *A* that is moving in the xy-plane. The radial coordinate R is the length of the *radial line OA*, and θ is the angle between the x-axis and the radial line. (In plane motion the polar coordinate R is equal to the magnitude of the position vector **r** of the particle.)

The base vectors $\mathbf{e}_R$ and $\mathbf{e}_\theta$ of the polar coordinate system are also shown in Fig. 13.9. The vector $\mathbf{e}_R$ is directed along the radial line, pointing away from O whereas $\mathbf{e}_\theta$ is perpendicular to $\mathbf{e}_R$, in the direction of increasing θ.

2. Derivatives of Base Vectors Note that $\mathbf{e}_R$ and $\mathbf{e}_\theta$ will rotate as the particle moves. Therefore, $\mathbf{e}_R$ and $\mathbf{e}_\theta$ are the base vectors of a rotating reference frame, similar to the path (*n-t*) coordinate system. (The fundamental difference between the two coordinate systems is that path coordinates depend on the path and the direction of motion of the particle, whereas polar coordinates are determined solely by the position of the particle.) Consequently, these base vectors possess nonzero derivatives, even though their magnitudes are constant (equal to one).

As in the preceding article, the time derivatives of the unit base vectors can be determined by first relating the vectors to the xy-coordinate system. From Fig. 13.9 we find that

$$\mathbf{e}_R = \cos\theta\mathbf{i} + \sin\theta\mathbf{j}$$
$$\mathbf{e}_\theta = -\sin\theta\mathbf{i} + \cos\theta\mathbf{j}$$

(13.12)

Differentiating with respect to time while noting that $d\mathbf{i}/dt = d\mathbf{j}/dt = \mathbf{0}$ (the xy-frame is fixed) yields

$$\frac{d\mathbf{e}_R}{dt} = (-\sin\theta\,\mathbf{i} + \cos\theta\,\mathbf{j})\dot{\theta} \qquad \frac{d\mathbf{e}_\theta}{dt} = (-\cos\theta\,\mathbf{i} - \sin\theta\,\mathbf{j})\dot{\theta}$$

Comparing these results with Eq. (13.12), we find that

$$\dot{\mathbf{e}}_R = \dot{\theta}\mathbf{e}_\theta \qquad \dot{\mathbf{e}}_\theta = -\dot{\theta}\mathbf{e}_R \qquad\qquad (13.13)$$

The variable $\dot{\theta}$ is called the *angular velocity* of the radial line. The base vectors and their derivatives are shown in Fig. 13.10. Note that $\dot{\mathbf{e}}_R$ and $\dot{\mathbf{e}}_\theta$ are perpendicular to $\mathbf{e}_R$ and $\mathbf{e}_\theta$, respectively.

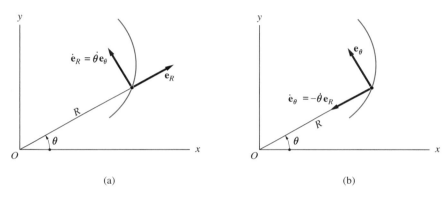

(a)   (b)

Fig. 13.10

3. *Velocity and Acceleration Vectors*   The position vector $\mathbf{r}$ of the particle can be written in polar coordinates as

$$\mathbf{r} = R\mathbf{e}_R \qquad\qquad (13.14)$$

Because the velocity vector is, by definition, $\mathbf{v} = d\mathbf{r}/dt$, we have

$$\mathbf{v} = \frac{d\mathbf{r}}{dt} = \frac{d}{dt}(R\mathbf{e}_R) = \dot{R}\mathbf{e}_R + R\dot{\mathbf{e}}_R$$

Substituting for $\dot{\mathbf{e}}_R$ from Eqs. (13.13) gives

$$\mathbf{v} = v_R\mathbf{e}_R + v_\theta\mathbf{e}_\theta \qquad\qquad (13.15)$$

where

$$v_R = \dot{R} \qquad v_\theta = R\dot{\theta} \qquad\qquad (13.16)$$

The components v_R and v_θ are called the *radial* and *transverse* components of the velocity, respectively.

The acceleration vector is computed as follows:

$$\mathbf{a} = \frac{d\mathbf{v}}{dt} = \frac{d}{dt}(\dot{R}\mathbf{e}_R + R\dot{\theta}\mathbf{e}_\theta)$$

$$= (\ddot{R}\mathbf{e}_R + \dot{R}\dot{\mathbf{e}}_R) + (\dot{R}\dot{\theta}\mathbf{e}_\theta + R\ddot{\theta}\mathbf{e}_\theta + R\dot{\theta}\dot{\mathbf{e}}_\theta)$$

The variable $\ddot{\theta}$ is called the *angular acceleration* of the radial line. Substituting for $\dot{\mathbf{e}}_R$ and $\dot{\mathbf{e}}_\theta$ from Eqs. (13.13) and rearranging terms, we obtain

$$\mathbf{a} = a_R\mathbf{e}_R + a_\theta\mathbf{e}_\theta \tag{13.17}$$

where the radial and transverse components of acceleration are given by

$$a_R = \ddot{R} - R\dot{\theta}^2 \qquad a_\theta = R\ddot{\theta} + 2\dot{R}\dot{\theta} \tag{13.18}$$

The polar components of the velocity and acceleration vectors are shown in Fig. 13.11.

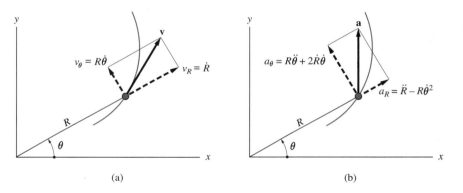

(a)   (b)

Fig. 13.11

For the special case where the path is a circle, the polar coordinate R equals the radius of the circle (a constant). Therefore, from Eqs. (13.15)–(13.18), the velocity and acceleration vectors are

$$\mathbf{v} = R\dot{\theta}\mathbf{e}_\theta$$

$$\mathbf{a} = -R\dot{\theta}^2\mathbf{e}_R + R\ddot{\theta}\mathbf{e}_\theta \tag{13.19}$$

These expressions are in agreement with Eqs. (13.10) and (13.11), where path coordinates were used. (Note that $\mathbf{e}_R = -\mathbf{e}_n$ and $\mathbf{e}_\theta = \mathbf{e}_t$ for motion on a circular path.)

b.   *Space motion (cylindrical coordinates)*

The cylindrical coordinates shown in Fig. 13.12 may be used to specify the location of a particle A that is moving in space. The three cylindrical coordinates consist of the polar coordinates R and θ, and the axial coordinate z (which is the same as the rectangular coordinate z). The unit base vectors are $\mathbf{e}_R$, $\mathbf{e}_\theta$, and $\mathbf{e}_z$, where $\mathbf{e}_z = \mathbf{k}$.

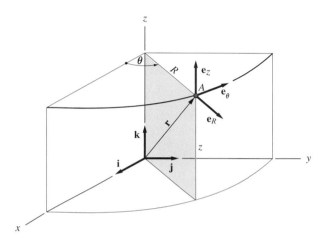

Fig. 13.12

In cylindrical coordinates, the position vector $\mathbf{r}$ of the particle in Fig. 13.12 is

$$\mathbf{r} = R\mathbf{e}_R + z\mathbf{e}_z \qquad (13.20)$$

Comparing this with $\mathbf{r} = R\mathbf{e}_R$ for polar coordinates, we conclude that the expressions for $\mathbf{v}$ and $\mathbf{a}$ in cylindrical coordinates will be the same as for polar coordinates, except for additional terms due to $z\mathbf{e}_z$. Recognizing that $\dot{\mathbf{e}}_z = \dot{\mathbf{k}} = \mathbf{0}$, Eqs. (13.15) through (13.18) can be easily modified to yield

$$\mathbf{v} = \dot{R}\mathbf{e}_R + R\dot{\theta}\mathbf{e}_\theta + \dot{z}\mathbf{e}_z$$
$$\mathbf{a} = (\ddot{R} - R\dot{\theta}^2)\mathbf{e}_R + (R\ddot{\theta} + 2\dot{R}\dot{\theta})\mathbf{e}_\theta + \ddot{z}\mathbf{e}_z \qquad (13.21)$$

Sample Problem 13.5

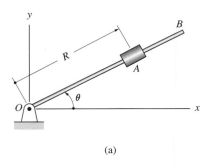

(a)

The collar A in Fig. (a) slides along the rotating rod OB. The angular position of the rod is given by $\theta = \frac{2}{3}\pi t^2$ rad, and the distance of the collar from O varies as $R = 18t^4 + 4$ m, where time t is measured in seconds. Determine the velocity and acceleration vectors of the collar at $t = 0.5$ s.

Solution

We start by determining the values of the polar coordinates of collar A and their first two derivatives at $t = 0.5$ s:

$$R = 18t^4 + 4 = 18(0.5)^4 + 4 = 5.125 \text{ m}$$

$$\dot{R} = 72t^3 = 72(0.5)^3 = 9.0 \text{ m/s}$$

$$\ddot{R} = 216t^2 = 216(0.5)^2 = 54.0 \text{ m/s}^2$$

$$\theta = \frac{2}{3}\pi t^2 = \frac{2}{3}\pi(0.5)^2 = 0.5236 \text{ rad} = 30°$$

$$\dot{\theta} = \frac{4}{3}\pi t = \frac{4}{3}\pi(0.5) = 2.094 \text{ rad/s}$$

$$\ddot{\theta} = \frac{4}{3}\pi = 4.189 \text{ rad/s}^2$$

The polar components of the velocity vector can now be calculated from Eqs. (13.16):

$$v_R = \dot{R} = 9.0 \text{ m/s} \qquad v_\theta = R\dot{\theta} = (5.125)(2.094) = 10.732 \text{ m/s}$$

Therefore, the velocity vector of the collar at $t = 0.5$ s is

$$\mathbf{v} = v_R\mathbf{e}_R + v_\theta\mathbf{e}_\theta = 9.0\mathbf{e}_R + 10.732\mathbf{e}_\theta \text{ m/s} \qquad \textit{Answer}$$

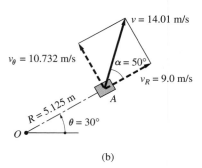

(b)

This result is shown in Fig. (b), where the magnitude of $\mathbf{v}$ and the angle α between $\mathbf{v}$ and the rod were computed from

$$v = \sqrt{v_R^2 + v_\theta^2} = \sqrt{(9.0)^2 + (10.732)^2} = 14.01 \text{ m/s}$$

$$\alpha = \tan^{-1}\frac{v_\theta}{v_R} = \tan^{-1}\frac{10.732}{9.0} = 50.0°$$

The acceleration components, obtainable from Eqs. (13.18), are

$$a_R = \ddot{R} - R\dot{\theta}^2 = 54.0 - (5.125)(2.094)^2 = 31.53 \text{ m/s}^2$$

$$a_\theta = R\ddot{\theta} + 2\dot{R}\dot{\theta} = (5.125)(4.189) + 2(9.0)(2.094) = 59.16 \text{ m/s}^2$$

The corresponding acceleration vector of the collar at $t = 0.5$ s is

$$\mathbf{a} = a_R\mathbf{e}_R + a_\theta\mathbf{e}_\theta = 31.53\mathbf{e}_R + 59.16\mathbf{e}_\theta \text{ m/s}^2 \qquad \textit{Answer}$$

which is shown in Fig. (c). The magnitude of **a** and the angle β were calculated from

$$a = \sqrt{a_R^2 + a_\theta^2} = \sqrt{(31.53)^2 + (59.16)^2} = 67.0 \text{ m/s}^2$$

$$\beta = \tan^{-1} \frac{a_\theta}{a_R} = \tan^{-1} \frac{59.16}{31.53} = 61.9°$$

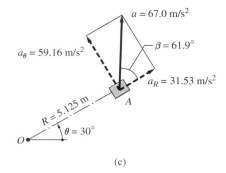

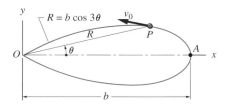

(c)

Sample Problem 13.6

As shown in the figure the particle P travels with constant speed v_0 along the path described by $R = b \cos 3\theta$. Determine the acceleration vector of the particle at point A.

Solution

Differentiating the expression for R twice, we get

$$R = b \cos 3\theta$$
$$\dot{R} = -3b\dot{\theta} \sin 3\theta$$
$$\ddot{R} = -9b\dot{\theta}^2 \cos 3\theta - 3b\ddot{\theta} \sin 3\theta$$

Substituting $\theta = 0$, we obtain at point A

$$R = b \cos 0 = b$$
$$\dot{R} = -3b\dot{\theta} \sin 0 = 0 \qquad \text{(a)}$$
$$\ddot{R} = -9b\dot{\theta}^2 \cos 0 - 3b\ddot{\theta} \sin 0 = -9b\dot{\theta}^2$$

The transverse component of the velocity is given by Eq. (13.16): $v_\theta = R\dot{\theta}$. Noting that $v_R = 0$ at A (because the tangent to the path at A is perpendicular to the radial line $\overline{OA}$), we have $v_\theta = v_0$. Therefore,

$$\dot{\theta} = \frac{v_0}{R} = \frac{v_0}{b} \qquad \text{(b)}$$

The angular acceleration $\ddot{\theta}$ of the radial line at A can be obtained from the condition that

$$v_0^2 = v_R^2 + v_\theta^2 = \dot{R}^2 + \left(R\dot{\theta}\right)^2 = \text{constant}$$

Differentiation of this expression yields

$$2\dot{R}\ddot{R} + 2R\dot{R}\dot{\theta}^2 + 2R^2\dot{\theta}\ddot{\theta} = 0$$

Since $\dot{R} = 0$ at A, this equation reduces to $0 = 2R^2\dot{\theta}\ddot{\theta}$, which can be satisfied only if

$$\ddot{\theta} = 0 \qquad \text{(c)}$$

The acceleration components in Eqs. (13.18) can now be evaluated at point A from Eqs. (a) through (c):

$$a_R = \ddot{R} - R\dot{\theta}^2 = -9b\dot{\theta}^2 - b\dot{\theta}^2 = -10b\left(\frac{v_0}{b}\right)^2 = -\frac{10v_0^2}{b}$$

$$a_\theta = R\ddot{\theta} + 2\dot{R}\dot{\theta} = 0 + 0 = 0$$

Hence, the acceleration vector of the particle at A is

$$\mathbf{a} = -\frac{10v_0^2}{b}\mathbf{e}_R = -\frac{10v_0^2}{b}\mathbf{i} \qquad \textit{Answer}$$

Sample Problem 13.7

The cable connecting the winch A to point B on the railroad car in Fig. (a) is wound in at the constant rate of 2 m. When $\theta = 60°$, determine (1) the velocity of B and $\dot{\theta}$; and (2) the acceleration of B and $\ddot{\theta}$. Neglect the radius of the winch.

Solution

From Fig. (a) we see that the length R of the cable and the angle θ are the polar coordinates of point B. At $\theta = 60°$, we have

$$R = \frac{4}{\sin\theta} = \frac{4}{\sin 60°} = 4.619 \text{ m}$$

According to the problem statement, R is being reduced at the constant rate of 5 ft/s. Therefore,

$$\dot{R} = -2 \text{ m/s} \qquad \ddot{R} = 0$$

Note that point B follows a straight, horizontal path. Consequently, its velocity and acceleration vectors will also be horizontal.

Part 1

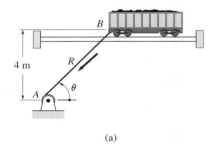

(b)

Figure (b) shows the decomposition of the velocity vector $\mathbf{v}$ of B into its radial and transverse components at $\theta = 60°$. Because $v_R = \dot{R} = -2$ m/s is negative, $\mathbf{v}_R$ is directed opposite to $\mathbf{e}_R$; that is, toward A. Knowing $\mathbf{v}_R$ and the direction of $\mathbf{v}$ (horizontal) enables us to complete the velocity diagram. From the geometry of the diagram, the speed of B at $\theta = 60°$ is

$$v = \frac{2}{\cos 60°} = 4 \text{ m/s (to the left)} \qquad \textit{Answer}$$

The velocity diagram also yields $v_\theta = 5\tan 60°$. Comparing this result with $v_R = R\dot{\theta}$ in Eqs. (13.16), we find that

$$\dot{\theta} = \frac{v_\theta}{R} = \frac{2\tan 60°}{4.619} = 0.75 \text{ rad/s (CCW)} \qquad \textit{Answer}$$

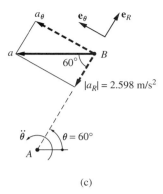

(c)

Part 2

The acceleration diagram of point B at $\theta = 60°$ is shown in Fig. (c). The radial component is, according to Eqs. (13.18),

$$a_R = \ddot{R} - R\dot{\theta}^2 = 0 - (4.619)(0.75)^2 = -2.598 \text{ m/s}^2$$

The negative sign again indicates that $\mathbf{a}_R$ is directed opposite to $\mathbf{e}_R$. Because the acceleration vector $\mathbf{a}$ is known to be horizontal, the acceleration diagram can now be completed. From the diagram, the magnitude of the acceleration at $\theta = 60°$ is

$$a = \frac{2.598}{\cos 60°} = 5.2 \text{ m/s}^2 \text{ (to the left)} \qquad \textit{Answer}$$

Referring again to the acceleration diagram, we find that $a_\theta = 2.598 \tan 60°$. Comparison with $a_\theta = R\ddot{\theta} + 2\dot{R}\dot{\theta}$ in Eqs. (13.18) yields

$$\ddot{\theta} = \frac{a_\theta - 2\dot{R}\dot{\theta}}{R} = \frac{2.598 \tan 60° - 2(-2)(0.75)}{4.619}$$

$$= 1.624 \text{ rad/s}^2 \text{ (CCW)} \qquad \textit{Answer}$$

Sample Problem   *13.8*

The passenger car of an amusement park ride is connected by the arm AB to the vertical mast OC. During a certain time interval, the mast is rotating at the constant rate $\dot{\theta} = 1.2$ rad/s while the arm is being elevated at the constant rate $\dot{\phi} = 0.3$ rad/s. Determine the cylindrical components of the velocity and acceleration of the car at the instant when $\phi = 40°$.

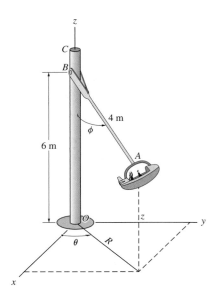

Solution

Referring to the figure, we see that the R- and z-coordinates of the car are $R = 4\sin\phi$ m and $z = 6 - 4\sin\phi$ m.

Noting that $\ddot{\phi} = 0$ ($\dot{\phi}$ is constant), we have at $\phi = 40°$

$$R = 4\sin\phi = 4\sin 40° = 2.571\,\text{m}$$

$$\dot{R} = 12\cos\phi\,\dot{\phi} = (4\cos 40°)(0.3) = 0.919\,\text{m/s}$$

$$\ddot{R} = -4\sin\phi\,\dot{\phi}^2 = -(4\sin 40°)(0.3)^2 = -0.231\,\text{m/s}^2$$

and

$$z = 6 - 4\cos\phi = 6 - 4\cos 40° = 2.936\,\text{m}$$

$$\dot{z} = 4\sin\phi\,\dot{\phi} = (4\sin 40°)(0.3) = 0.771\,\text{m/s}$$

$$\ddot{z} = 4\cos\phi\,\dot{\phi}^2 = (4\cos 40°)(0.3)^2 = 0.276\,\text{m/s}^2$$

Using Eqs. (13.21), the cylindrical components of the velocity are

$$v_R = \dot{R} = 0.919\,\text{m/s}$$

$$v_\theta = R\dot{\theta} = 2.571(1.2) = 3.085\,\text{m/s} \qquad \textit{Answer}$$

$$v_z = \dot{z} = 0.771\,\text{m/s}$$

Recalling that $\dot{\theta}$ is constant, the acceleration components in Eqs. (13.21) become

$$a_R = \ddot{R} - R\dot{\theta}^2 = -0.231 - 2.571(1.2)^2 = -3.933\,\text{m/s}^2$$

$$a_\theta = R\ddot{\theta} + 2\dot{R}\dot{\theta} = 0 + 2(0.919)(1.2) = 2.206\,\text{m/s}^2 \qquad \textit{Answer}$$

$$a_z = \ddot{z} = 0.276\,\text{m/s}^2$$

Problems

13.26   The rocket in vertical flight is being tracked by radar. Calculate the velocity and acceleration of the rocket at the instant when the radar readings are $\theta = 40°$, $R = 5\,\text{km}$, $\dot{R} = 350\,\text{m/s}$, and $\ddot{R} = 100\,\text{m/s}^2$.

13.27   The plane motion of a particle is described in polar coordinates as $R = 0.75 + 0.5t^2$ m and $\theta = \pi t^2/2$ rad, where the time t is measured in seconds. Find the magnitudes of the velocity and acceleration when $t = 2$ s.

13.28   The particle P moves along the curve $R = 4 + 2\sin\theta$ m. Knowing that $\dot{\theta} = 1.5$ rad/s (constant), determine the polar components of the velocity and acceleration of P when it is (a) at point A; and (b) at point B.

****13.29**   The particle P moves along the curve $R = 4 + 2\sin\theta$ m. At the instant that P is at point A, its speed is 4 m/s and its acceleration vector is perpendicular to the curve. Determine the magnitude of the acceleration of P at that instant.

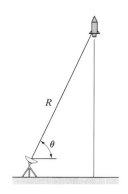

Fig. P13.26

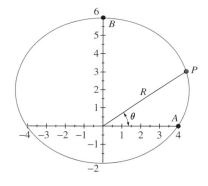

Fig. P13.28, P13.29

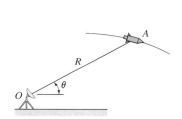

Fig. P13.30

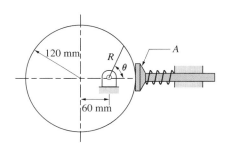

Fig. P13.31

13.30   The projectile A is being tracked by the radar at O. At a given instant, the radar readings are $\theta = 30°$, $R = 2000$ m, $\dot{R} = 200$ m/s, and $\ddot{R} = 20$ m/s^2. Determine the speed of the projectile at that instant.

13.31   A spring holds the follower A against the eccentric cam of circular profile. The equation describing the profile of the cam is $R^2 + 4R\cos\theta - 12 = 0$, where R is in mm. If the cam is rotating at the constant angular speed $\dot{\theta} = 3$ rad/s, compute the velocity and acceleration of the follower in the position $\theta = 0$.

13.32   The collar B slides on a rod that has the spiral shape described by $R = 0.3 - 0.4(\theta/\pi)$ m. The pin attached to the collar engages a slot in the arm OA, which is rotating at the constant angular speed $\dot{\theta} = 2$ rad/s. Find the polar components of the velocity and acceleration of collar B at (a) $\theta = 90°$; and (b) $\theta = 60°$.

13.33   The rod OB rotates counterclockwise about O at the constant angular speed of 30 rev/min while the collar A slides toward B with the constant speed

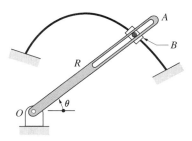

Fig. P13.32

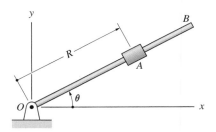

Fig. P13.33, P13.34

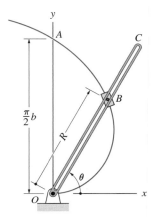

Fig. P13.36, P13.37

1 m/s, measured relative to the rod. When collar A is in the position $R = 0.2\,\text{m}$, $\theta = 0$, calculate (a) its velocity vector; and (b) its acceleration vector.

13.34   The motion of rod OB is described by $\dot{\theta} = \alpha t$, where $\alpha = 1.2\,\text{rad/s}^2$ is the constant angular acceleration of the rod. The position of the collar A on the rod is $R = v_0 t$, where $v_0 = 0.8\,\text{m/s}$ is the constant outward speed of the collar relative to the rod. Calculate the velocity and acceleration vectors of the collar as functions of time.

13.35   The plane motion of a particle described in polar coordinates is $\theta = \omega t$, $R = b\sqrt{\omega t}$, where ω and b are constants. When $\theta = \pi$, determine (a) the velocity vector of the particle; and (b) the acceleration vector of the particle.

13.36   The collar B slides along a guide rod that has the shape of the spiral $R = b\theta$. A pin on the collar slides in the slotted arm OC. If OC is rotating at the constant angular speed $\dot{\theta} = \omega$, determine the magnitude of the acceleration of the collar when it is at A.

13.37   The collar B slides along a guide rod that has the shape of the spiral $R = b\theta$. A pin on the collar slides in the slotted arm OC. If the speed of the collar is constant at v_0, determine the angular speed $\dot{\theta}$ of the arm OC in terms of v_0, b, and θ.

13.38   The slotted arm OB rotates about the pin at O. The ball A in the slot is pressed against the stationary cam C by the spring. If the angular speed of OB is $\dot{\theta} = \omega$, where ω is a constant, calculate the maximum magnitudes of (a) the velocity of A; (b) the acceleration of A; (c) $\dot{R}$ (the velocity of A relative to OB); and (d) $\ddot{R}$ (the acceleration of A relative to OB).

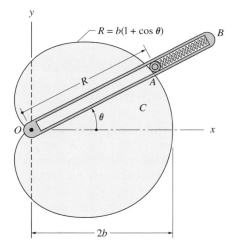

Fig. P13.38, P13.39

13.39   The slotted arm OB rotates about the pin at O. The ball A in the slot is pressed against the stationary cam C by the spring. The angular position of the arm OB depends on time t as $\theta = \pi \sin \omega t$, where ω is a constant. Determine the velocity and acceleration vectors of the ball A when $\theta = \pi/2$.

***13.40**   The curved portion of the cloverleaf highway interchange is defined by $R^2 = b^2 \sin 2\theta$, $0 \le \theta \le 90°$. If a car travels along the curve at the constant speed v_0, determine its acceleration at A.

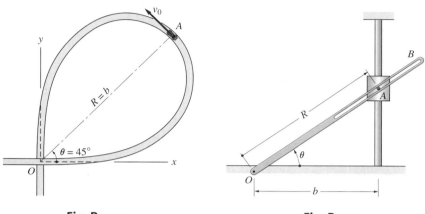

Fig. P13.40

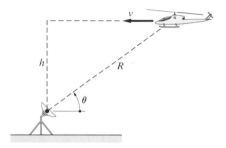

Fig. P13.41

13.41   The pin attached to the sliding collar A engages the slot in bar OB. Using polar coordinates, determine the speed of A in terms of θ and $\dot{\theta}$. (Note: The solution using rectangular coordinates was requested in Prob. 12.23.)

13.42   The helicopter is tracked by radar, which records R, θ, and $\dot{\theta}$ at regular time intervals. The readings at a certain instant are $R = 2500$ m, $\theta = 40°$, and $\dot{\theta} = 0.04$ rad/s. If the helicopter is in level flight, calculate the elevation h and speed of the helicopter at this instant.

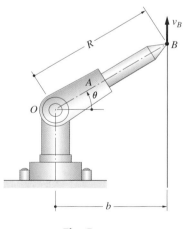

Fig. P13.42

13.43   The telescopic arm of the robot slides in the mount A, which rotates about a horizontal axis at O. End B of the arm traces the vertical line shown with the constant speed v_B. In terms of v_B, b, and θ, determine expressions for (a) $\dot{\theta}$ and $\dot{R}$; and (b) $\ddot{\theta}$ and $\ddot{R}$.

13.44   The winch D unwinds the cable BCD at the constant rate of 0.8 m/s. Determine the speed of end B of the bar AB when $R = 4$ m.

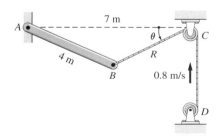

Fig. P13.44

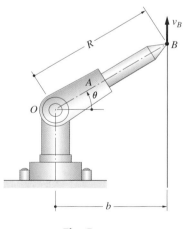

Fig. P13.43

13.45   A particle moves along a spiral described in cylindrical coordinates by $R = 0.4$ m and $z = -0.2\theta$ m, where θ is in radians. It is known that at a certain instant, $\dot{\theta} = 6$ rad/s and $\ddot{\theta} = -10$ rad/s^2. Determine the magnitudes of the velocity and acceleration vectors at this instant.

13.46   A particle moves along the path described by $R = 100$ mm and $z = 15 \sin 4\theta$ mm. If $\dot{\theta} = 0.8$ rad/s (constant), determine the magnitudes of the maximum velocity and maximum acceleration of the particle and the corresponding values of θ.

13.47   A child slides down the helical water slide AB. The description of motion in cylindrical coordinates is $R = 4$ m, $\theta = \omega^2 t^2$, and $z = h[1 - (\omega^2 t^2/\pi)]$, where $h = 3$ m and $\omega = 0.75$ rad/s. Compute the magnitudes of the velocity vector and acceleration vector when the child is at B.

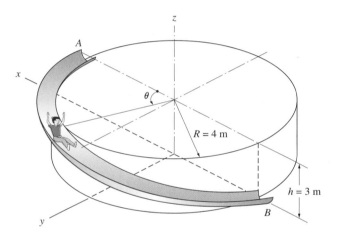

Fig. P13.47

13.48   The rod OB rotates about the z-axis with the constant angular speed $\dot{\theta} = 4$ rad/s while the slider A moves up the rod at the constant speed $\dot{s} = 2$ m/s. Determine the magnitudes of the velocity and acceleration vectors of A when $s = 1$ m.

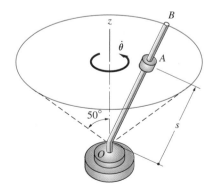

Fig. P13.48

*13.49   The rotating water sprinkler has a constant angular speed of $\dot{\theta} = 6$ rad/s about the z-axis. The speed of the water relative to the curved tube OA is 2 m/s. Compute the magnitudes of the water velocity and acceleration vectors just below the nozzle at A.

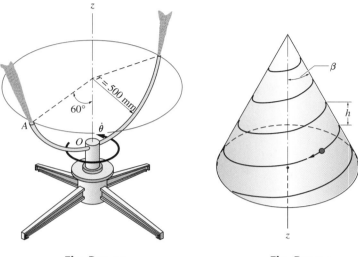

Fig. P13.49   **Fig. P13.50**

13.50   The path of the particle that is moving on the surface of a cone is defined by

$$R = \frac{h}{2\pi}\theta \tan \beta \qquad z = \frac{h}{2\pi}\theta$$

where R, θ, and z are the cylindrical coordinates. If the motion of the particle is such that $\dot{\theta} = \omega$ (constant), determine the following as functions of θ: (a) the speed of the particle; and (b) the cylindrical components of the acceleration vector.

13.4   *Kinetics: Force-Mass-Acceleration Method*

The force-mass-acceleration (FMA) method was presented in Art. 12.3 in terms of the rectangular coordinate system. When working in curvilinear coordinates, the four basic steps of the method remain unchanged: (1) draw the FBD of the particle; (2) perform kinematic analysis of the acceleration; (3) draw the MAD of the particle; and (4) derive the equations of motion from the static equivalence of the FBD and the MAD. The only significant change is in the details of the second step, namely kinematics.

a.   *Path (n-t) coordinates*

The free-body and mass-acceleration diagrams of a particle are shown in Fig. 13.13(a). Note that the MAD displays the normal and tangential (*n-t*) components of the mass-acceleration vector. The conditions for the two diagrams to be statically equivalent are

$$\Sigma F_n = ma_n \qquad \Sigma F_t = ma_t \qquad\qquad (13.22)$$

where ΣF_n and ΣF_t are *n-t* components of the resultant force acting on the particle. As mentioned before, we use path coordinates only for plane motion.

b.   *Cylindrical coordinates*

The equations of motion of a particle in terms of cylindrical components are

$$\Sigma F_R = ma_R \qquad \Sigma F_\theta = ma_\theta \qquad \Sigma F_z = ma_z \qquad (13.23)$$

where ΣF_R, ΣF_θ, and ΣF_z represent the components of the resultant force that acts on the particle. See the FBD and the MAD in Fig. 13.13(b). If the path of the particle lies in the xy plane, then the third of Eqs. (13.23) becomes $\Sigma F_z = 0$, and the motion is dependent only on the polar coordinates R and θ.

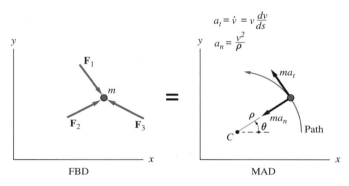

(a) Path (n-t) coordinates

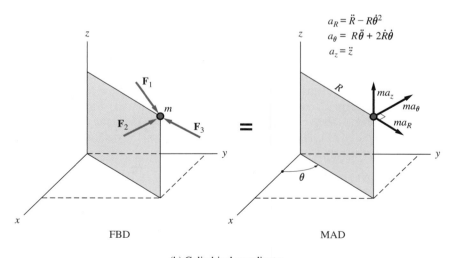

(b) Cylindrical coordinates

Fig. 13.13

Sample Problem   13.9

The strings AB and AC connect the 200-g ball A to the vertical shaft, as shown in Fig. (a). When the shaft rotates at the constant angular speed $\dot\theta$, the ball travels in a horizontal circle with the strings inclined at $\alpha = 30°$ to the shaft. Find the value of $\dot\theta$ for which the tension in string AC is 4 N.

Solution

The FBD and the MAD of the ball are shown in Fig. (b). The FBD displays the weight of the ball and the two string tensions. Because the speed of the ball is constant, the MAD contains only the normal component $a_n = R\dot\theta^2$ of the acceleration, where R is the radius of the path of A. The equation of motion in the y-direction is

$$\Sigma F_y = 0 \quad +\uparrow \quad T_{AB}\cos 30° - T_{AC}\cos 30° - 0.2 \times 9.8 = 0$$

Substituting $T_{AC} = 4$ N, we get

$$T_{AB} = \frac{1.96}{\cos 30°} + 4 = 6.26 \text{ N}$$

The second equation of motion is

$$\Sigma F_n = ma_n \quad \nwarrow+ \quad T_{AB}\sin 30° + T_{AC}\sin 30° = mR\dot\theta^2$$

With $R = 0.4\sin 30° = 0.2$ m, this equation becomes

$$(6.26 + 4)\sin 30° = 0.2 \times (0.2)\dot\theta^2$$

which yields

$$\dot\theta = 11.32 \text{ rad/s} \qquad\qquad \textit{Answer}$$

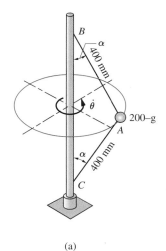

(a)

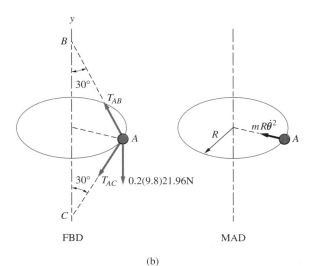

FBD   MAD

(b)

Sample Problem 13.10

The 100-g block B shown in Fig. (a) slides along the rotating bar OA. The coefficient of kinetic friction between B and OA is $\mu_k = 0.2$. In the position shown, $\dot{R} = 1$ m/s, $\dot{\theta} = 5$ rad/s, and $\ddot{\theta} = 3$ rad/s^2. For this position, determine $\ddot{R}$, the acceleration of B relative to bar OA.

Solution

Referring to the free-body diagram (FBD) in Fig. (b), we see that there are three forces acting on B: its weight of 0.98 N, the normal force N exerted by the bar OA, and the kinetic friction force $F = \mu_k N$. The direction of F is opposite to $\dot{R}$, the velocity of B relative to OA. The mass-acceleration diagram (MAD) of B is also shown in Fig. (b), where the inertia vector $m\mathbf{a}$ is described in terms of its polar components.

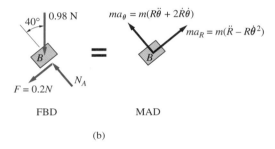

FBD   MAD

(b)

Inspection of Fig. (b) reveals that there are only two unknowns in the FBD-MAD diagrams—N_A and $\ddot{R}$; the other variables have been specified. Therefore, the two unknowns can be computed from the two equations of motion. For the direction perpendicular to the bar, we have

$$\Sigma F_\theta = m(R\ddot{\theta} + 2\dot{R}\dot{\theta})$$

$$\nwarrow\quad N_A - 0.98\cos 40° = 0.1[(0.4)(3) + (2)(1)(5)]$$

which gives

$$N_A = 1.8707 \text{ N}$$

For the radial direction, we obtain

$$\Sigma F_R = m(\ddot{R} - R\dot{\theta}^2)$$

$$\nearrow\quad -0.98\sin 40° - 0.2 N_A = 0.1[\ddot{R} - (0.4)(5)^2]$$

Substituting $N_A = 1.8707$ N and solving for $\ddot{R}$ yields

$$\ddot{R} = -0.04 \text{ m/s}^2 \qquad\qquad \textit{Answer}$$

The minus sign means that the acceleration of B relative to bar OA is directed toward point O.

Sample Problem 13.11

The 12-kg mass A in Fig. (a) slides with negligible friction in a semicircular trough of radius $R = 2$ m. The mass is launched at $\theta = 30°$ with the velocity $v_0 = 4$ m/s toward the bottom of the trough. Derive the following as functions of θ: (1) the speed of the mass; and (2) the contact force between the mass and the trough.

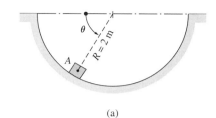

(a)

Solution

Part 1

Because the path is circular, the normal-tangential (n-t) and the polar coordinate systems could be employed with equal facility. We chose the n-t coordinates.

The free-body diagram of the mass in an arbitrary position is shown in Fig. (b). It contains the weight mg of the mass and the contact force N_A, which is normal to the surface of the trough due to the absence of friction. The mass-acceleration diagram (MAD) displays the n-t components of the inertia vector. The equation of motion in the t-direction is

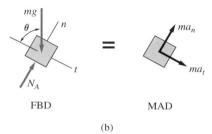

FBD   MAD

(b)

$$\Sigma F_t = ma_t \quad \searrow^+ \quad mg \cos\theta = ma_t$$

which gives

$$a_t = g \cos\theta \qquad \text{(a)}$$

Substituting $a_t = v(dv/ds)$ from Eq. (13.9), where $ds = R\,d\theta$, Eq. (a) becomes $(v/R)(dv/d\theta) = g\cos\theta$. After rearrangement, we get

$$v\,dv = gR\cos\theta\,d\theta$$

Integration yields

$$\frac{1}{2}v^2 = gR\sin\theta + C \qquad \text{(b)}$$

where C is the constant of integration. Substituting the initial condition $v = v_0 = 4$ m/s when $\theta = 30°$ in Eq. (b) gives $C = \frac{1}{2}v_0^2 - gR\sin 30° = \frac{1}{2}(4)^2 - (9.81)(2)\sin 30° = -1.810$ (m/s)^2. Therefore, the speed of the mass as a function of θ is

$$v = \pm\sqrt{2(gR\sin\theta + C)} = \pm\sqrt{2[(9.8)(2)\sin\theta - 1.810]}$$
$$= \pm\sqrt{39.2\sin\theta - 3.62}\ \text{m/s} \qquad \textit{Answer}$$

Part 2

Referring to Fig. (b), the equation of motion in the n-direction is

$$\Sigma F_n = ma_n \quad {+}\nearrow \quad N_A - mg\sin\theta = ma_n$$

Substituting $a_n = v^2/R$ and solving for N_A, we get

$$N_A = m\left(g\sin\theta + \frac{v^2}{R}\right) = 12\left(9.8\sin\theta + \frac{39.2\sin\theta - 3.62}{2}\right)$$
$$= 352.8\sin\theta - 21.7\ \text{N} \qquad \textit{Answer}$$

Sample Problem 13.12

The vertical shaft AB in Fig. (a) rotates in a bearing at A. The 0.6-kg slider P can move freely along the frictionless bar OD, which is, rigidly joined to AB at a 30° angle. At a certain instant when $r = 1.2$ m, it is known that $\dot{\theta} = 4$ rad/s, $\ddot{\theta} = 0$, and the velocity of P relative to OD is $\dot{r} = 4$ m/s. At this instant, determine the magnitude of the contact force exerted on P by OD; and $\ddot{r}$, the acceleration of P relative to OD.

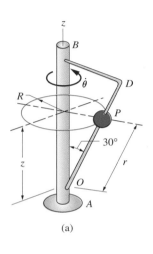

(a)

Solution

The free-body diagram (FBD) of the slider P at the instant of interest is shown in Fig. (b), where its weight is $mg = 0.6(9.81) = 5.886$ N. It is convenient to decompose the contact force exerted by OD (which is normal to OD) into two components: N_1, which is perpendicular to OD and passes through the z-axis; and N_2, which is perpendicular to both OD and N_1. The mass-acceleration diagram (MAD) of the slider P is shown in Fig. (c), where the inertia vector $m\mathbf{a}$ is expressed in terms of its cylindrical components.

From Fig. (a) we obtain $R = r \sin 30° = 1.2 \sin 30° = 0.60$ m and $z = r \cos 30°$. Differentiating with respect to time, and substituting $\dot{r} = 4$ m/s, we have

$$\dot{R} = \dot{r} \sin 30° = 4 \sin 30° = 2.00 \text{ m/s}$$

$$\ddot{R} = \ddot{r} \sin 30°$$

$$\dot{z} = \dot{r} \cos 30° = 4 \cos 30° = 3.464 \text{ m/s}$$

$$\ddot{z} = \ddot{r} \cos 30°$$

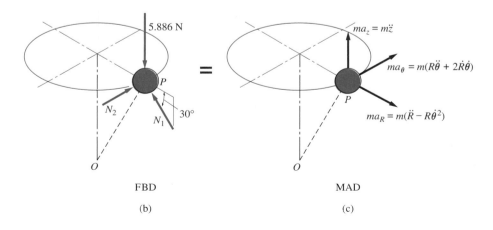

FBD

(b)

MAD

(c)

Because the values of $\dot{\theta}$, $\ddot{\theta}$, and $\dot{r}$ are given, we see that $\ddot{r}$ is the only kinematic variable that is unknown. The FBD in Fig. (b) contains two unknowns (N_1 and N_2), so that we have a problem involving three unknowns, which can be determined from the three available equations of motion for the slider.

Referring to Figs. (b) and (c), the equations of motion are

$$\Sigma F_R = ma_R = m(\ddot{R} - R\dot{\theta}^2)$$

$$\searrow^+ \quad -N_1 \cos 30° = 0.6[\ddot{r}\sin 30° - 0.60(4)^2] \qquad \text{(a)}$$

$$\Sigma F_\theta = ma_\theta = m(R\ddot{\theta} + 2\dot{R}\dot{\theta})$$

$$+\nearrow \quad N_2 = 0.6[0 + 2(2.00)(4)] = 9.600 \text{ N} \qquad \text{(b)}$$

$$\Sigma F_z = ma_z = m\ddot{z}$$

$$+\uparrow \quad N_1 \sin 30° - 5.886 = 0.6(\ddot{r}\cos 30°) \qquad \text{(c)}$$

Solving Eqs. (a) and (c) simultaneously yields

$$\ddot{r} = -3.70 \text{ m/s}^2 \qquad \text{\textit{Answer}}$$

and $N_1 = 7.931$ N. Therefore the magnitude of the contact force exerted by OD is

$$N = \sqrt{N_1^2 + N_2^2} = \sqrt{(7.931)^2 + (9.600)^2} = 12.45 \text{ N} \qquad \text{\textit{Answer}}$$

Sample Problem 13.13

The particle of mass $m = 0.3$ kg is attached to an ideal spring of stiffness $k = 28.1$ N/m. The spring is undeformed when the particle is launched on the horizontal surface with the speed $v_0 = 2$ m/s in the direction shown in Fig. (a). Friction between the particle and the surface may be neglected. (1) Derive the differential equations describing the motion of the particle in terms of the polar coordinates R and θ, and state the initial values. (2) Solve the differential equations numerically from $t = 0$ (the time of launch) to $t = 1.5$ s, and plot the trajectory of the particle.

Solution

Part 1

The free-body diagram (FBD) and the mass-acceleration diagram (MAD) of the particle at an arbitrary position are shown in Fig. (b). The only force appearing on the FBD is the spring force $F = k(R - L_0)$, where L_0 is the undeformed length of the spring. The other forces (weight of the particle and the reaction with the horizontal surface) are perpendicular to the plane of the motion; hence they do not enter the equations of motion. The MAD displays the polar components of the inertia vector, where the accelerations were obtained from Eqs. (13.18).

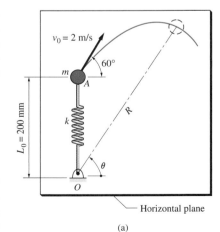

(a)

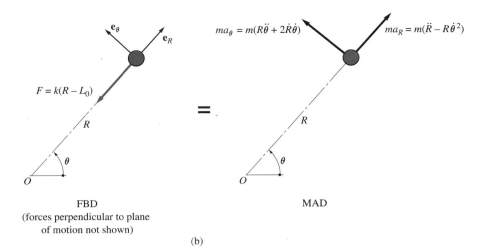

FBD
(forces perpendicular to plane
of motion not shown)

MAD

(b)

Referring to Fig. (b), the equations of motion are

$$\Sigma F_R = ma_R \quad +\nearrow \quad -k(R - L_0) = m(\ddot{R} - R\dot{\theta}^2)$$

$$\Sigma F_\theta = ma_\theta \quad \nwarrow \quad 0 = m(R\ddot{\theta} + 2\dot{R}\dot{\theta})$$

Substituting the given data—$m = 0.3\,\text{kg}$, $k = 28.1\,\text{N/m}$, and $L_0 = 0.2\,\text{m}$—and solving for $\ddot{R}$ and $\ddot{\theta}$ yield the differential equations

$$\ddot{R} = R\dot{\theta}^2 - 93.667R + 18.733 \ \text{m/s}^2 \qquad \ddot{\theta} = -\frac{2\dot{R}\dot{\theta}}{R} \ \text{rad/s}^2 \qquad \textit{Answer}$$

The initial values are

$$R_0 = L_0 = 0.2\,\text{m}$$

$$\dot{R}_0 = (v_R)_0 = 2\sin 60° = 1.73205\,\text{m/s}$$

$$\theta_0 = \frac{\pi}{2} \qquad\qquad\qquad\qquad\qquad \textit{Answer}$$

$$\dot{\theta}_0 = \frac{(v_\theta)_0}{L_0} = -\frac{2\cos 60°}{0.2} = -5.000\,\text{rad/s}$$

Part 2

Using the notation

$$\mathbf{x} = [R \ \theta \ \dot{R} \ \dot{\theta}]^T$$

the equivalent first-order equations are

$$\dot{\mathbf{x}} = \begin{bmatrix} x_3 \\ x_4 \\ x_1 x_4^2 - 93.667x_1 + 18.733 \\ -2x_3 x_4 / x_1 \end{bmatrix}$$

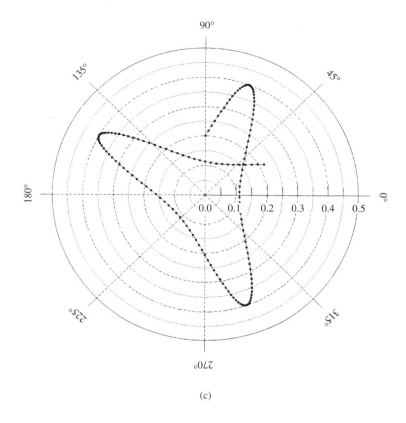

(c)

The MATLAB program that solves the equations and plots the trajectory is

```
function example13_13
time = [0:0.025:1.5];
x0 = [0.2 pi/2 1.73205 -5];
[t,x] = ode45(@f,time,x0);
axes('fontsize',14)
polar(x(:,2),x(:,1))
grid on

    function dxdt = f(t,x)
    dxdt = [x(3)
            x(4)
            x(1)*x(4)^2-93.667*x(1)+18.7333
            -2*x(3)*x(4)/x(1)];
    end
end
```

The polar plot of the trajectory is shown in Fig. (c). Note that R never exceeds 0.4 m.

Problems

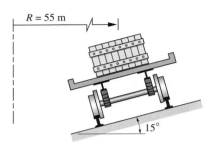

R = 55 m

15°

Fig. P13.52

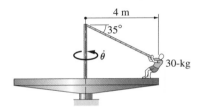

4 m

35°

θ̇

30-kg

Fig. P13.53

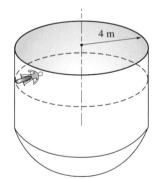

4 m

Fig. P13.54

13.51 A motorcycle travels along an unbanked curve of radius 200 m. If the static coefficient of friction between the road and the tires is 0.4, determine the largest constant speed at which the cycle will not slide.

13.52 The flatbed railway car travels at the constant speed of 60 km/h around a curve of radius 55 m and bank angle 15°. Determine the smallest static coefficient of friction between the crate and the car that would prevent the crate of mass M from sliding.

13.53 The 30-kg child holds onto a rope as the platform of the playground ride rotates about the vertical axis at the constant angular speed $\dot{\theta} = 1$ rad/s. Find the tension in the rope and the normal force between the child and the platform. Neglect friction.

13.54 The motorcyclist is riding along a horizontal circle on the inside of the cylindrical wall. If the coefficient of static friction between the tires and the wall is 0.6, what is the smallest possible speed of the motorcycle?

13.55 A car travels over the crest A of a hill where the radius of curvature is 60 m. Find the maximum speed for which the wheels will stay in contact with the road at A.

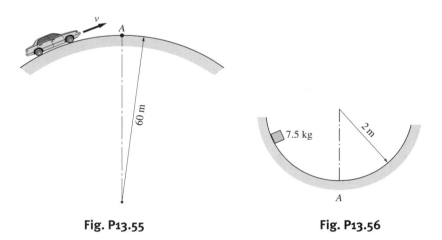

v

A

60 m

7.5 kg

2 m

A

Fig. P13.55   **Fig. P13.56**

13.56 The 7.5-kg box that is sliding down a circular chute reaches point A with a speed of 2.5 m/s. The kinetic coefficient of friction between the box and the chute is 0.3. When the box is at A, calculate (a) the normal force acting between it and the chute; and (b) its rate of change of speed.

13.57 The tension in the string of the simple pendulum is 8.5 N when $\theta = 25°$. Calculate the angular velocity and angular acceleration of the string at this instant.

13.58   The pendulum is released from rest with $\theta = 30°$. (a) Derive the equation of motion using θ as the independent variable. (b) Determine the speed of the bob as a function of θ.

13.59   The slider of mass m moves with negligible friction on a circular wire of radius R that lies in the vertical plane. The constant horizontal force F acts on the slider. If the slider starts from rest at A, determine the expression for the smallest F that would enable it to reach B.

Fig. P13.57, P13.58

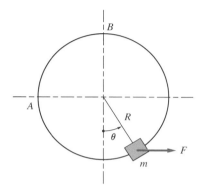

Fig. P13.59

13.60   The coin A is placed on a stationary turntable in the position $R = 0.4$ m, $\theta = 0$. The static coefficient of friction between the coin and the turntable is 0.2. If the turntable is started with the constant angular acceleration $\ddot{\theta} = 1.5 \, \text{rad/s}^2$, determine the angular velocity $\dot{\theta}$ when the coin starts slipping.

13.61   The 1-kg collar is free to slide on the smooth rod OA. The rod is rotating in the vertical plane about the pin at O at the constant angular velocity $\dot{\theta}$. Determine the minimum value of $\dot{\theta}$ for which the collar will maintain contact with the stop at A. Would friction between the collar and the rod affect your result?

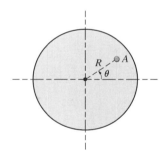

Fig. P13.60

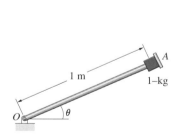

Fig. P13.61

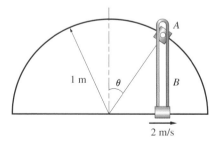

Fig. P13.62

13.62   The 500-g collar A slides on the semi-circular guide rod. A pin attached to the collar engages the vertical slot in the guide B, which is moving to the right at the constant speed of 2 m/s. Determine the force between the pin and guide B when $\theta = 45°$. Neglect friction.

13.63   A spring is connected between the 1-kg slider A and the frame. The spring has a stiffness of 5 N/m and it is undeformed when $x = 0.1$ m. Knowing

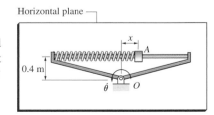

Fig. P13.63

Fig. P13.64

that the frame is rotating in the horizontal plane about O at the constant angular speed $\dot{\theta} = 2$ rad/s, determine the distance x. Neglect friction.

13.64 The spring exerts a 6 N force that presses the 0.5-kg slider A against the stop B. Assuming that the angular velocity $\dot{\theta}$ is increased slowly, find the value of $\dot{\theta}$ at which the slider loses contact with the stop B. Assume that the rotation of the assembly occurs in (a) the horizontal plane; and (b) the vertical plane.

13.65 The 0.4-kg slider A glides on the circular guide rod BC. The spring attached to the slider has a free length of 0.4 m and a stiffness of 18 N/m. The slider is launched upward from C with the velocity 2.4 m/s. Determine the magnitude of the acceleration of the slider and the contact force between the slider and the guide rod just after the launch. Neglect friction.

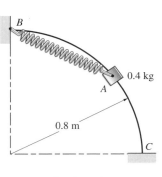

Fig. P13.65

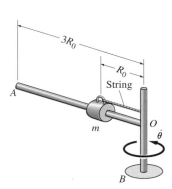

Fig. P13.66–P13.68

13.66 The rod OA carrying the sliding collar of mass $m = 2.5$ kg is rotating about the vertical axis OB with the constant angular speed $\dot{\theta} = 20$ rad/s. A string holds the collar in the position $R_0 = 1.4$ m. (a) Determine the force in the string. (b) Find $\ddot{R}$ (the acceleration of the collar relative to the rod) immediately after the string breaks. Neglect friction.

13.67 The rod OA carrying the sliding collar of mass $m = 2$ kg rotates about the vertical axis OB. The angular speed of the rod is kept constant at $\dot{\theta} = 6$ rad/s while the string pulls the collar toward O at the constant rate of 0.8 m/s. Determine the tension in the string and the contact force between the rod and the collar when the collar reaches the position $R_0 = 1.2$ m.

*****13.68** Rod OA carrying a collar of mass m rotates about the vertical axis OB. The angular speed of the assembly is kept constant at $\dot{\theta} = \omega$. If the string that holds the collar in place is cut at time $t = 0$, determine (a) the equation of motion of the collar, with R (the distance of the collar from O) as the independent variable; and (b) the speed of the collar when it reaches end A. Neglect friction.

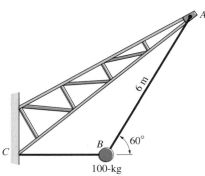

Fig. P13.69

13.69 The 100-kg wrecking ball is initially held at rest by the two cables AB and BC. Calculate the force in cable AB (a) before cable BC is released; and (b) just after cable BC is released.

13.70   The 30 mm diameter marble rolls in the groove *AOB* that has the shape of a parabola in the horizontal *xy*-plane. The cross section of the groove is a rectangle of width 20 mm Neglecting friction, find the largest speed for which the marble will stay in the groove.

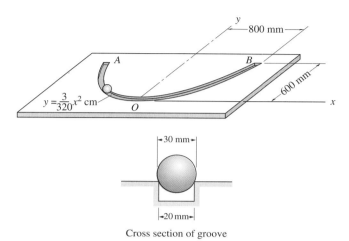

Cross section of groove

Fig. P13.70

13.71   Friction between the shoes *A* and the housing *B* enables the centrifugal clutch to transmit torque from shaft 1 to shaft 2. Centrifugal force holds the shoes against the housing, where the coefficient of static friction between the shoes and the housing is 0.8. Each shoe weighs 9.8 N, and the weights of the other parts of the shoe assemblies may be neglected. (a) Calculate the initial tension in each clutch spring *C* so that contact between the shoes and the housing occurs only when $\dot{\theta} \geq 200$ rev/min. (b) Determine the maximum torque that can be transmitted when $\dot{\theta} = 1000$ rev/min.

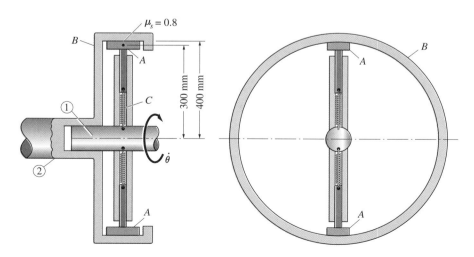

Fig. P13.71

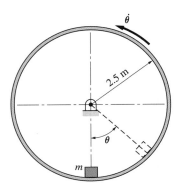

Fig. P13.72

13.72   A package of mass m is placed inside a drum that rotates in the vertical plane at the constant angular speed $\dot{\theta} = 1.36$ rad/s. If the package reaches the position $\theta = 45°$ before slipping, determine the static coefficient of friction between the package and the drum.

13.73   The path of the 3.6-kg particle P is an ellipse given by $R = R_0/(1 + e\cos\theta)$, where $R_0 = 0.5$ m and $e = 2/3$. Assuming that the angular speed of line OP is constant at 20 rad/s, calculate the polar components of the force that acts on the particle when it is at A.

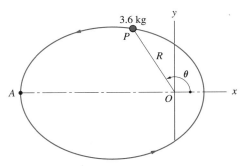

Fig. P13.73, P13.74

13.74   The 3.6-kg particle P moves along the ellipse described in Prob. 13.73. If the motion of the particle is such that $v_\theta = R\dot{\theta} = 10$ m/s (constant), determine the polar components of the force acting on the particle when it is at A.

*13.75   The 2-kg follower is attached to the end of a light telescopic rod that is pivoted at O. The follower is pressed against a frictionless spiral surface by a spring of stiffness $k = 100$ N/m and free length $L_0 = 1$ m. The equation of the spiral, which lies in the horizontal plane, is $R = b\theta/2\pi$, where $b = 0.4$ m and θ is in radians. Immediately after the rod is released from rest in position OA, determine (a) the angular acceleration $\ddot{\theta}$ of the rod; and (b) the contact force between the follower and the spiral surface.

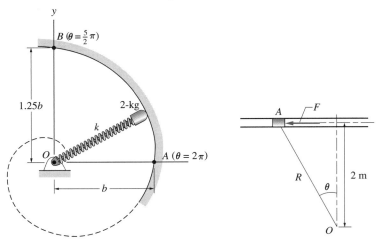

Fig. P13.75   **Fig. P13.76**

13.76   The 0.6-kg cylinder A is moving with negligible friction in the horizontal tube, propelled by the force F. If the cylinder moves so that $R\dot\theta = 4$ m/s (constant), determine F as a function of the angle θ.

13.77   The particle P moves along a curved path, its speed varying so as to keep $R^2\dot\theta$ constant. Show that the force acting on the particle always is directed along the radial line $\overline{OP}$.

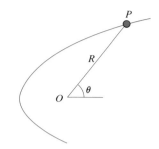

Fig. P13.77

*13.78   The 0.2-kg pellet slides in the tube OB. The tube rotates in the horizontal plane about the pin at O with the constant angular velocity $\dot\theta = 8$ rad/s. The pellet is at $R = 0.5$ m when it is fired toward B with the velocity of 2 m/s relative to the tube. (a) Determine the velocity vector of the pellet when it reaches the end of the tube at B. (b) What is the contact force between the block and the tube at B? Neglect friction.

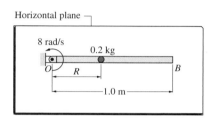

Fig. P13.78

13.79   The 0.10-kg ball A, which slides in a slot in the rotating arm OB, is kept in contact with the stationary cam C by a compression spring of stiffness k. The spring exerts a force of 2 N on the ball when the arm is stationary in position OP. If the arm rotates with the constant angular speed $\dot\theta = 20$ rad/s, calculate the minimum spring stiffness k that will maintain contact between the ball and the cam when the arm is in position OQ. Neglect friction and assume that the assembly lies in the horizontal plane.

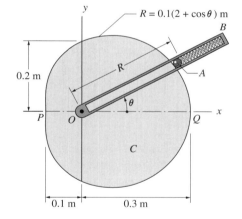

Fig. P13.79, P13.80

*13.80   The arm OB of the system described in Prob. 13.79 rotates with the constant speed $\dot\theta = 20$ rad/s. When $\theta = 60°$, the force exerted by the spring on the ball A is 12.5 N. For this position, determine the contact force between (a) the ball and the cam; and (b) the ball and the slot.

13.81   The collar A of mass m slides on the weightless rod OB, which rotates with a constant angular velocity $\dot\theta = \omega$. A pin attached to the collar engages the fixed vertical slot. Neglecting friction, determine (a) the force exerted on the pin

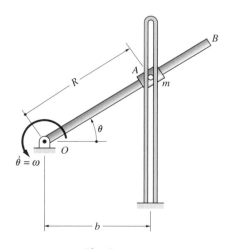

Fig. P13.81

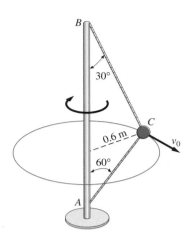

Fig. P13.82

by the slot; and (b) the force exerted on the collar by the rod. Express your answers in terms of θ, ω, m, b, and g.

13.82 The mass C is connected by two wires to the vertical shaft AB. Rotation of the shaft causes the mass to travel in the horizontal circle shown. Calculate the speed v_0 of the mass that would result in equal tensions in the wires.

13.83 The telescopic arm of the mechanical manipulator rotates about the vertical axis with the constant angular speed $\dot{\theta} = 8$ rad/s. The angle ϕ is kept constant at $45°$, but the length of the arm varies as $L = 6 + 2\sin(2\dot{\theta}t)$ m, where t is in seconds. Compute the cylindrical components of the force exerted by the arm on the 40-N manipulator head as functions of time.

13.84 The telescopic arm of the mechanical manipulator rotates about the vertical axis with the constant angular speed $\dot{\theta} = 8$ rad/s. At the same time, the arm is extended and lowered at the constant rates $\dot{L} = 1$ m/s and $\dot{\phi} = 2$ rad/s, respectively. Determine the cylindrical components of the force that the arm exerts on the 40-N manipulator head when $L = 2$ m and $\phi = 45°$.

13.85 The differential equation of motion for the simple pendulum can be shown to be $\ddot{\theta} = -(g/L)\sin\theta$. Given that $L = 9.81$ m and that the pendulum is released from rest at $\theta = 60°$, determine the time required for the pendulum to reach the position $\theta = 0$. Compare your answer with the analytical solution of 1.686 s.

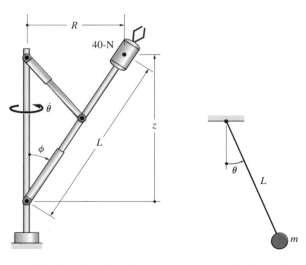

Fig. P13.83, P13.84   **Fig. P13.85**

13.86   The block of mass m is released from rest at $\theta = 0$ and allowed to slide on the circular surface. The kinetic coefficient of friction between the block and the surface is μ. (a) Show that the differential equation of motion of the block is

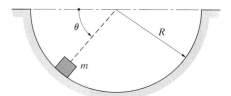

Fig. P13.86

$$\ddot{\theta} = (g/R)(\cos\theta - \mu\sin\theta) - \mu\dot{\theta}^2 \qquad (\dot{\theta} \geq 0)$$

(b) Using $R = 2$ m and $\mu = 0.3$, determine by numerical integration the value of θ where the block comes to rest for the first time.

13.87   A peg attached to the rotating disk causes the slope angle of the table OB to vary as $\theta = \theta_0 \cos\omega t$, where $\theta_0 = 15°$ and $\omega = \pi$ rad/s. At $t = 0$, the particle A is placed on the table at $R = 0.6$ m with no velocity relative to the table. (a) Derive the differential equation of motion for the particle, and state the initial conditions. (b) Integrate the equations numerically from the time of release until the particle moves off the table; plot R versus t. (c) Determine the time when A slides off the table and the velocity of A relative to the table at that instant. Neglect friction.

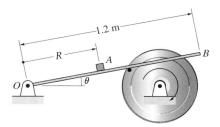

Fig. P13.87

13.88   The particle of mass m slides inside the frictionless, conical vessel. The particle is launched at $t = 0$ with the velocity $v_1 = 3$ m/s, tangent to the rim of the vessel. (a) Show that the differential equations of motion are

$$\ddot{\theta} = -\frac{2\dot{z}\dot{\theta}}{z}$$

$$\ddot{z} = \frac{z\dot{\theta}^2 \tan^2\beta - g}{1 + \tan^2\beta}$$

and state the initial conditions. (b) Using $\beta = 20°$, solve the equations of motion numerically from $t = 0$ to $t = 2$ s; plot z versus θ. (c) From the numerical solution, find the vertical distance h below which the particle will not travel.

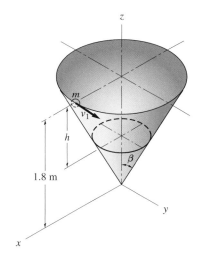

Fig. P13.88

13.89   The 0.25-kg mass, which is attached to an elastic cord of stiffness 10 N/m and free length 0.5 m, is free to move in the vertical plane. The mass is released from rest at $\theta = 0$ with the cord undeformed. (a) Derive the differential equations of motion, and state the initial conditions. (b) Solve the equations numerically from the time of release until the cord becomes vertical for the first time; plot R versus θ. (c) Find the maximum elongation of the cord.

13.90   The 0.25-kg mass in Prob. 13.89 is released from rest at $\theta = 0$ with the cord stretched by 0.25 m. (a) Derive the differential equations of motion for the mass, and state the initial conditions. (b) Solve the equations numerically from the time of release until the cord becomes slack; plot R versus θ. (c) Determine the maximum value of R, and the value of θ when the cord becomes slack.

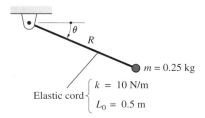

Fig. P13.89,   P13.90

Review of Equations

Kinematics in normal-tangential coordinates

$$v_n = 0 \quad v_t = v = \dot{s} \quad \text{(velocity is tangent to the path)}$$

$$a_n = v^2/\rho \quad a_t = \dot{v} = v\frac{dv}{ds}$$

$s = $ distance measured along the path of the particle
$\rho = $ radius of curvature of the path
e_n is pointed towards center of curvature
e_t is pointed in the direction of the velocity

Kinematics of motion along a circular path

$$v = R\dot{\theta}$$

$$a_n = \frac{v^2}{R} = R\dot{\theta}^2 \quad a_t = \dot{v} = R\ddot{\theta}$$

$R = $ radius of the path

Kinematics in polar and cylindrical coordinates

$$v_R = \dot{R} \quad v_\theta = R\dot{\theta} \quad v_z = \dot{z}$$

$$a_R = \ddot{R} - R\dot{\theta}^2 \quad a_\theta = R\ddot{\theta} + 2\dot{R}\dot{\theta} \quad a_z = \ddot{z}$$

e_R is pointed outward from the origin
e_θ is pointed towards increasing θ

Review Problems

13.91 The airplane flies a vertical loop of radius 400 m. The speed of the plane at the top of the loop is constant at 70 m/s. If the weight of the pilot is 600 N, determine the contact force between the pilot and his seat when the plane is at the top of the loop.

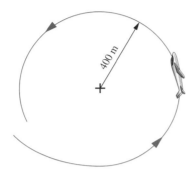

Fig. P13.91

13.92 The 200-kg car travels at the constant speed of 90 km/h on a road with the profile $y = h \cos(2\pi x/b)$ ft, where $h = 1$ m and $b = 60$ m. Determine the maximum normal force between the tires and the road.

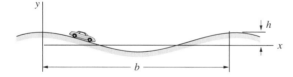

Fig. P13.92

13.93 A 600-kg satellite is in a circular orbit around a planet. The radius of the orbit is 8000 km and the period (time of a complete orbit) is 6 hours. Determine (a) the gravitational force acting on the satellite; and (b) the mass of the planet. Use $G = 6.493 \times 10^{-11}$ m^3/(kg · s^2) for the universal gravitational constant.

13.94 The bent water pipe of constant cross section is rotating about the vertical axis AB with the constant angular velocity $\dot{\theta} = 160$ rev/min. If the speed of water

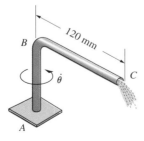

Fig. P13.94

in portion AB of the pipe is 600 mm/s (constant), determine the magnitude of the acceleration of a water particle just before it exits the pipe at C.

13.95   The car travels on a curve that has the shape of the spiral $R = (2b/\pi)\theta$, where $b = 10$ m. If $\dot{\theta} = 0.5$ rad/s (constant), determine the speed of the car and the magnitude of the acceleration when $\theta = 3\pi/2$ rad (point A).

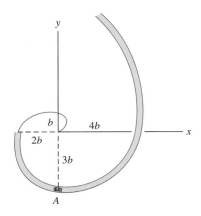

Fig. P13.95

13.96   Bar AB starts from rest at $\theta = 0$ with the constant angular acceleration $\ddot{\theta} = 6$ rad/s^2. The block of mass m begins sliding on the bar when $\theta = 45°$. Determine the coefficient of static friction between the block and the bar.

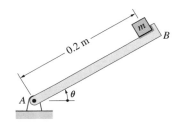

Fig. P13.96

13.97   A projectile is launched at B with the velocity v_0 inclined at angle θ_0 to the horizontal. Knowing that the radii of curvature of the trajectory at A and B are $\rho_A = 40.8$ m and $\rho_B = 63.1$ m, determine θ_0 and v_0. Neglect air resistance.

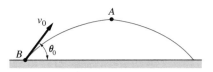

Fig. P13.97

13.98   The 5-kg package is sliding down the parabolic chute. In the position shown, the speed of the package is 2.4 m/s. Determine the normal contact force between the chute and the package in this position.

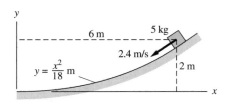

Fig. P13.98

13.99   The 0.25 m radius spool is started from rest in the position shown. The angular speed $\dot{\theta}$ of the spool varies with time, as specified on the diagram. Determine the time required to wind up the 50 m of cable.

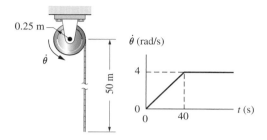

Fig. P13.99

13.100   The car travels at a constant speed on a circular, banked track of 75-m radius. If the bank angle is $\beta = 12°$ and the coefficient of static friction between the track and the tires is 0.8, determine the maximum possible speed of the car.

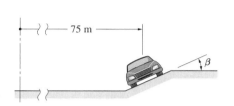

Fig. P13.100

13.101   The rope ABC is attached to the sliding collar A and passes over the peg B. The rope is kept taut by a weight attached to end C. The collar is moving to the left at the constant speed of 1.6 m/s. When the collar is in the position $\theta = 50°$, determine (a) the velocity of point C and $\dot{\theta}$; and (b) the acceleration of point C.

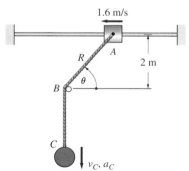

Fig. P13.101

13.102   The particle of mass m is placed on the cylindrical surface of radius R at $\theta = 0$. The particle is then displaced slightly and released from rest. Determine the speed of the particle as a function of the angle θ. At what value of θ does the particle lose contact with the surface? Neglect friction.

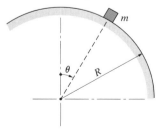

Fig. P13.102

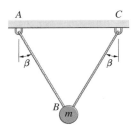

Fig. P13.103

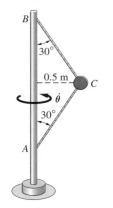

Fig. P13.105

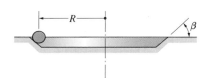

Fig. P13.108

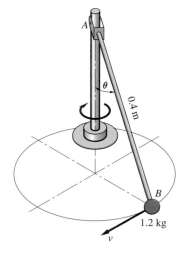

Fig. P13.109

13.103 The mass m is suspended from two wires, as shown, when wire AB is cut. If $\beta = 35°$, determine the force in wire BC (a) before AB is cut; and (b) just after AB is cut. (c) For what value of β would the results of parts (a) and (b) be the same?

13.104 A particle moves on the surface of a sphere of radius b. The description of motion in cylindrical coordinates is $\theta = \omega t$, $R = b \sin \omega t$, $z = b \cos \omega t$, where ω is a constant. Determine (a) the maximum speed; and (b) the maximum magnitude of the acceleration vector of the particle.

13.105 The mass at C is attached to the vertical pole AB by two wires. The assembly is rotating about AB at the constant angular speed $\dot\theta$. If the force in wire BC is twice the force in AC, determine the value of $\dot\theta$.

13.106 The collar A slides along the rod OB, which is rotating counterclockwise. Determine the velocity vector of A when OB is vertical, given that the speed of end B is 0.4 m/s in that position.

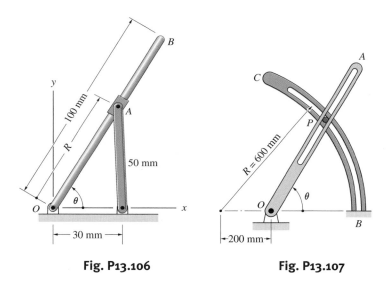

Fig. P13.106   **Fig. P13.107**

13.107 The pin P slides in slots in both the rotating arm OA and the fixed circular bar BC. If OA rotates with the constant angular speed $\dot\theta = 2$ rad/s, find the speed of P when $\theta = 60°$.

13.108 The path of a ball that rolls around a frictionless, circular track is a circle of radius R. The track is banked at the angle β. Determine the speed of the ball.

13.109 The pendulum is connected to the vertical shaft by a clevis at A. The mass of the bob B is 1.2 kg, and the mass of the arm AB is negligible. The shaft rotates with a constant angular speed, causing the bob to travel in a horizontal circle. If $\theta = 85°$, determine (a) the tensile force in AB; and (b) the speed v of the bob.

14

Work-Energy and Impulse-Momentum Principles for a Particle

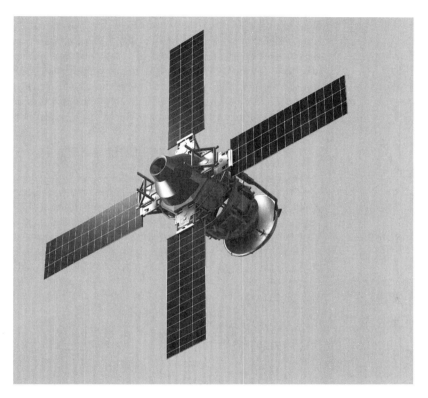

It takes a lot of energy to launch a spacecraft on an interplanetary journey. Problem 14.48 shows how to calculate the energy required to escape earth's gravitational field. (© iStockphoto.com/Konstantin Inozemtsev)

<table>
<tr><td>**14.1**</td><td>*Introduction*</td></tr>
</table>

In the force-mass-acceleration method, which we used in Chapters 12 and 13, the equations of particle motion were obtained directly from Newton's second law, $\mathbf{F} = m\mathbf{a}$. Solution of these equations required two integrations, the first to obtain the velocity, and the second to obtain the position.

Work-energy and impulse-momentum methods employ integral forms of the equations of motion. If we integrate both sides of $\mathbf{F} = m\mathbf{a}$ with respect to *position,* we obtain the equations used in the work-energy method. Integrating $\mathbf{F} = m\mathbf{a}$ with respect to *time* yields the equations of the impulse-momentum method. The integral forms of the equations of motion can be very efficient in the solution of certain types of problems. The work-energy method is useful in computing the change in speed during a displacement of the particle. The impulse-momentum method is best suited for determining the change in velocity that occurs over a time interval.

14.2   *Work of a Force*

a.   *Definition of work*

We begin by defining the work of a force, a concept that plays a fundamental role in the work-energy principle that is derived in the next article. Let point A, the point of application of a force $\mathbf{F}$, follow the path $\mathscr{L}$ shown in Fig. 14.1. The position vector of A (measured from a fixed point O) is denoted by $\mathbf{r}$ at time t and $\mathbf{r} + d\mathbf{r}$ at time $t + dt$. Note that $d\mathbf{r}$, the displacement of the point during the infinitesimal time interval dt, is tangent to the path at A. The differential work dU done by the force $\mathbf{F}$ as its point of application undergoes the displacement $d\mathbf{r}$ is defined to be

$$dU = \mathbf{F} \cdot d\mathbf{r} \tag{14.1}$$

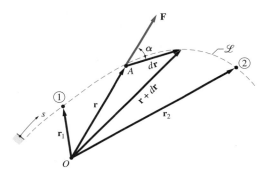

Fig. 14.1

The work done by $\mathbf{F}$ as point A moves from position 1 to 2 is obtained by integrating Eq. (14.1) along the path $\mathscr{L}$:

$$U_{1-2} = \int_{\mathscr{L}} dU = \int_{\mathbf{r}_1}^{\mathbf{r}_2} \mathbf{F} \cdot d\mathbf{r} \tag{14.2}$$

Work is a scalar quantity that may be positive, negative, or zero. Its dimension is $[FL]$ with the units being lb · ft or lb · in. in the U.S. Customary system, and N · m or *joule* (J) (1 J = 1 N · m) in the SI system.

Introducing the notation $ds = |d\mathbf{r}|$ and $F = |\mathbf{F}|$, Eq. (14.1) may be written as

$$dU = F \cos \alpha \, ds \tag{14.3}$$

where α is the angle between $\mathbf{F}$ and $d\mathbf{r}$, as shown in Fig. 14.1. Referring to Fig. 14.2(a), we see that $F \cos \alpha = F_t$ is the component of $\mathbf{F}$ that is tangent

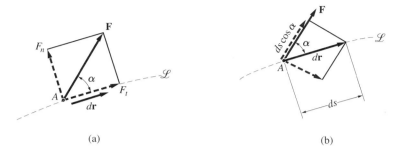

(a)   (b)

Fig. 14.2

to the path at A. Therefore, Eq. (14.2) is equivalent to

$$U_{1-2} = \int_{s_1}^{s_2} F_t \, ds \qquad (14.4)$$

where s is the path length measured from an arbitrary fixed point on the path shown in Fig. 14.1. The normal component F_n, being perpendicular to the path, does not do work ($\mathbf{F}_n \cdot d\mathbf{r} = 0$). Because the tangential component F_t does all the work, it is called the *working component* of $\mathbf{F}$. The incremental work thus can be viewed as

$$dU = F_t \, ds$$
$$= \text{(working component of } \mathbf{F}) \times \text{(magnitude of } d\mathbf{r})$$

The geometric interpretation of Eq. (14.4) is shown in Fig. 14.3—the work U_{1-2} is equal to the area under the F_t-s diagram.

Another interpretation of incremental work is obtained from Fig. 14.2(b). Note that $ds \cos \alpha$ is the component of $d\mathbf{r}$ that is parallel to $\mathbf{F}$. This component is known as the *work-absorbing component* of the differential displacement. Therefore, the incremental work is

$$dU = F(ds \cos \alpha)$$
$$= \text{(magnitude of } \mathbf{F}) \times \text{(work-absorbing component of } d\mathbf{r})$$

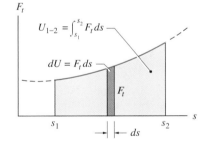

Fig. 14.3

Another useful expression for the work done by a force is obtained by writing the dot product $\mathbf{F} \cdot d\mathbf{r}$ in rectangular coordinates:

$$U_{1-2} = \int_{\mathscr{L}} (F_x \, dx + F_y \, dy + F_z \, dz)$$
$$= \int_{x_1}^{x_2} F_x \, dx + \int_{y_1}^{y_2} F_y \, dy + \int_{z_1}^{z_2} F_z \, dz \qquad (14.5)$$

where dx, dy, and dz are the components of $d\mathbf{r}$, and (x_1, y_1, z_1) and (x_2, y_2, z_2) are the coordinates of points 1 and 2, respectively.

b. Work of a constant force

If the force **F** is constant in both magnitude and direction, Eq. (14.2) becomes

$$U_{1-2} = \mathbf{F} \cdot \int_{\mathbf{r}_1}^{\mathbf{r}_2} d\mathbf{r} = \mathbf{F} \cdot \Delta\mathbf{r}$$

where $\Delta\mathbf{r}$ is the displacement vector from position 1 to position 2, as shown in Fig. 14.4.

Let α be the angle between **F** and $\Delta\mathbf{r}$, and let Δd be the displacement in the direction of **F**. Using the definition of the dot product, the work done by **F** becomes $U_{1-2} = \mathbf{F} \cdot \Delta\mathbf{r} = F|\Delta\mathbf{r}|\cos\alpha$, which may be written as[*]

$$U_{1-2} = F\Delta d \tag{14.6}$$

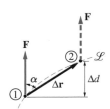

Fig. 14.4

Note that Δd is *not* the displacement of the point of application of **F**; it is the work-absorbing component of the displacement.

From Eq. (14.6) we see that the work done by a constant force depends only on the initial and final positions of its point of application; that is, the work is independent of the path $\mathscr{L}$.

If an object stays in close proximity to the surface of the earth, its weight may be considered to be a constant force, and Eq. (14.6) can be used to calculate its work. Figure 14.5 shows such an object of weight W, which moves from position 1 to 2. Observe that the increase in elevation Δh is also the work-absorbing component of the displacement. Therefore, the work done by W is

$$U_{1-2} = -W\Delta h \tag{14.7}$$

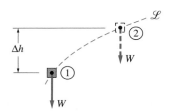

Fig. 14.5

The minus sign in this equation is due to the fact that W and the increase in elevation have opposite directions. As is the case for all constant forces, the work done by W is independent of the path.

c. Work of a central force

A *central force* has two defining characteristics: (1) it is always directed toward a fixed point, and (2) its magnitude is a function of the distance between the fixed point and the point of application of the force. Gravitational attraction and the force exerted by a spring are two common examples of a central force.

Consider the work done by the central force **F** shown in Fig. 14.6 as its point of application moves the infinitesimal distance ds along the path $\mathscr{L}$. From Eq. (14.3), the incremental work is

$$dU = -F\cos\alpha \, ds = -F \, dR$$

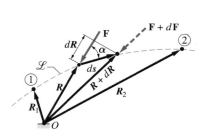

Fig. 14.6

[*]Equation (14.6) is the basis for the common "definition" of work as "force times distance." It is important to remember that this equation is valid only for a constant force; the general definition of work is given by Eq. (14.2).

where dR is the increase in the distance between the point of application of **F** and the fixed point O. The minus sign is the result of **F** and dR having opposite directions. The work of **F** during a finite displacement from position 1 to position 2 is

$$U_{1-2} = -\int_{R_1}^{R_2} F \, dR \qquad (14.8)$$

Note that the work is independent of the path $\mathscr{L}$, depending only on the initial and final values of R.

1.   *Work of a Spring Force*   An "ideal" spring has negligible weight, and its deformation (elongation or contraction) is proportional to the force that causes it. Most spiral springs closely approximate these ideal conditions. The proportionality between the force F and the resulting elongation δ is expressed as

$$F = k\delta \qquad (14.9)$$

where k is called the *stiffness* of the spring or the *spring constant*. The dimension of k is $[F/L]$; its units are N/m. Another important property of a spring is its undeformed length, also called the *free length*. Denoting the free length by L_0 and the deformed length by L, the elongation of the spring is $\delta = L - L_0$.

Figure 14.7(a) shows a spiral spring of deformed length L. Because the force F exerted by the spring is a central force, its work can be computed using Eq. (14.8). The incremental elongation $d\delta$ shown in the figure is equivalent to dR in Eq. (14.8) with both representing the incremental change in the distance between the fixed point O and the point of application of the force. Therefore, by replacing dR in Eq. (14.8) with $d\delta$, we obtain the work done by the spring as its deformation changes from δ_1 to δ_2:

$$U_{1-2} = -\int_{\delta_1}^{\delta_2} F \, d\delta = -k\int_{\delta_1}^{\delta_2} \delta \, d\delta = -\frac{1}{2}k(\delta_2^2 - \delta_1^2) \qquad (14.10)$$

Equations (14.9) and (14.10) are valid for elongation (positive δ) as well as contraction (negative δ).

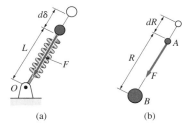

(a)   (b)

Fig. 14.7

2.   *Work of a Gravitational Force*   The work of the gravitational force (weight) acting on a body near the surface of the earth can be computed from Eq. (14.8).

Away from the surface of the earth, work must determined from Newton's law of gravitation

$$F = G\frac{m_A m_B}{R^2}$$

(11.17, repeated)

where F is the force of attraction between two bodies of masses m_A and m_B separated by the distance R, and G represents the universal gravitational constant.

The force F shown in Fig. 14.7(b) represents the gravitational force exerted by the body B on another body A. Substituting Eq. (11.17) into Eq. (14.8), the work of F becomes

$$U_{1-2} = -Gm_A m_B \int_{R_1}^{R_2} \frac{dR}{R^2} = Gm_A m_B \left(\frac{1}{R_2} - \frac{1}{R_1} \right)$$

(14.11)

where R_1 and R_2 are the initial and final distances between the bodies.

A common application of Eq. (14.11) is space mechanics, particularly the flight of satellites. In that case, $m_A = M_e$ (the mass of the earth), and R is the distance of the satellite from the center of the earth.

14.3   *Principle of Work and Kinetic Energy*

The principle of work and kinetic energy (also known as the work-energy principle) states that the work done by all forces acting on a particle (the work of the resultant force) equals the change in the kinetic energy of the particle. This principle, which forms the basis for the work-energy method of kinetic analysis, is derived by integrating Newton's second law along the path of the particle.

Consider a particle of mass m as it moves from position 1 to position 2 along the path $\mathscr{L}$ with its path coordinate changing from s_1 to s_2. Let $\Sigma\mathbf{F}$ denote the resultant force (vector sum of all the forces) acting on the particle. Applying Newton's second law in the direction tangent to the path, we get

$$\Sigma F_t = ma_t$$

(14.12)

According to Eq. (13.9), $a_t = v\,dv/ds$, which upon substitution into Eq. (14.12) yields $\Sigma F_t = mv\,dv/ds$, or

$$\Sigma F_t\,ds = mv\,dv$$

Integrating along the path $\mathscr{L}$, we get

$$\int_{s_1}^{s_2} \Sigma F_t\,ds = \int_{v_1}^{v_2} mv\,dv$$

where v_1 and v_2 are the speeds of the particle at the endpoints 1 and 2 of the path. The result of the integration is

$$U_{1-2} = \frac{1}{2}m(v_2^2 - v_1^2)$$

(14.13)

By definition, the *kinetic energy* of a particle is

$$T = \frac{1}{2}mv^2$$

(14.14)

so that Eq (14.13) becomes

$$U_{1-2} = T_2 - T_1 = \Delta T \qquad \text{(14.15)}$$

Equation (14.15) is the *work-energy principle* (or the balance of work and kinetic energy):

> Work done by the resultant force acting on a particle
> = change in the kinetic energy of the particle

When this principle is used in kinetic analysis, the method is referred to as the *work-energy method*.

Because the work-energy principle results from integrating Newton's second law, it is not an independent principle. Therefore, any problem that can be solved by the work-energy method can, in theory, be also solved by the force-mass-acceleration (FMA) method.

The main advantages of the work-energy method when compared with the FMA method are as follows:

1. In problems where the work can be calculated without integration (as in the special cases discussed in the preceding article), the change in speed of the particle may be easily obtained with a minimum of computation. (If integration must be used to compute the work, the work-energy method will, in general, have no advantage over integrating the acceleration determined by the FMA method.)
2. Only forces that do work need be considered; nonworking forces do not appear in the analysis.
3. If the final position, that is, position 2, is chosen to be an arbitrary position, the work-energy method will determine the speed as a function of position of the particle.

Caution The following points must be kept in mind when using the work-energy method:

- Work of a force is a *scalar* quantity (positive, negative, or zero) that is associated with a *change in the position* of the point of application of the force. (The phrase "work at a given position" is meaningless.)
- Kinetic energy is a *scalar* quantity (always positive) associated with the speed of a particle at a given instant of time. The units of kinetic energy are the same as the units of work: $N \cdot m$, $lb \cdot ft$, and so on.
- The work-energy principle, $U = \Delta T$, is a *scalar* equation. Although this is an obvious point, a common error is to apply the work-energy principle separately in the *x*-, *y*-, or *z*-directions, which is, of course, incorrect.
- When applying the work-energy method, an *active force diagram,* which shows only the forces that do work, may be used in place of the conventional free-body diagram. A convenient method for determining the active force diagram is to first draw the FBD of the particle in an arbitrary position and then delete or mark any forces that do not do work. In this text, we show workless forces as dashed arrows on the FBD.

Sample Problem 14.1

The collar A of mass $m = 1.8$ kg shown in Fig. (a) slides on a frictionless rod that lies in the vertical plane. A rope is attached to A and passed over a pulley at B. The constant horizontal force P is applied to the end of the rope. The collar is released from rest in position 1. (1) Determine the speed of the collar in position 2 if $P = 20$ N. (2) Find the smallest value of P for which the collar will reach position 2.

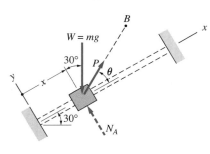

(a)

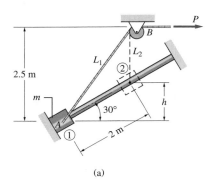

(b)

Solution

Preliminary Discussion

We are asked to determine the change in speed of the collar between two specified positions, a task that is ideally suited to the work-energy method.

The FBD of the collar is shown in Fig. (b). Only the weight W of the collar and the force P applied by the rope do work on the collar. The normal contact force N_A, being perpendicular to the rod (the path of the collar), is workless. Therefore, N_A is shown as a dashed arrow in the figure, indicating that it would not be a part of the active force diagram (recall that an active force diagram may be obtained by deleting all workless forces from the FBD).

Work of W The work done by the weight $W = mg$ of the collar can be obtained from Eq. (14.7):

$$U_{1-2} = -mgh \qquad \text{(a)}$$

where h is the change in the elevation shown in Fig. (a).

Work of P The force P in Fig. (b) is a central force (it always directed toward B) of constant magnitude. Hence, its work can be computed from Eq. (14.8):

$$U_{1-2} = -P \int_{L_1}^{L_2} dL = P(L_1 - L_2) \qquad \text{(b)}$$

where L_1 and L_2 are the lengths shown in Fig. (a).

Part 1

Denoting the speed of the collar in position 2 by v_2, we have $T_1 = 0$ (collar is at rest in position 1) and $T_2 = \frac{1}{2}mv_2^2$. Thus the work-energy principle yields

$$U_{1-2} = T_2 - T_1$$

$$-mgh + P(L_1 - L_2) = \frac{1}{2}mv_2^2 - 0 \qquad \text{(c)}$$

The work U_{1-2} was obtained by adding the contributions of W and P in Eqs. (a) and (b). From Fig. (a) we obtain $h = 2\sin 30° = 1.0$ m, $L_1 = \sqrt{(2\cos 30°)^2 + (2.5)^2} = 3.041$ m, and $L_2 = 2.5 - 1.0 = 1.5$ m. Substituting these values, together with $m = 1.8$ kg and $P = 20$ N, into Eq. (c), we obtain

$$-(1.8)(9.81)(1.0) + 20(3.041 - 1.5) = \frac{1}{2}(1.8)v_2^2$$

which yields

$$v_2 = 3.82 \text{ m/s} \qquad \textit{Answer}$$

Part 2

In this part of the problem, the collar is also at rest in position 2, so that $T_1 = T_2 = 0$. The problem is thus reduced to finding the value of P for which $U_{1-2} = 0$, that is,

$$-mgh + P(L_1 - L_2) = 0$$

The solution is

$$P = \frac{mgh}{L_1 - L_2} = \frac{(1.8)(9.81)(1.0)}{3.041 - 1.5} = 11.46 \text{ N} \qquad \textit{Answer}$$

Notes

- If the given value of P in Part 1 were less than 11.46 N, U_{1-2} would be negative, resulting in an imaginary speed (square root of a negative number) in position 2. This result would indicate that it is impossible for the collar to reach position 2.
- If friction between the rod and the collar were not negligible, we would have to add the kinetic friction force $F_k = \mu_k N_A$ on the active force diagram in Fig. (b). Because N_A varies with the position coordinate x, F_k would also be dependent on x. Therefore, integration would be required to compute the work of the friction force. Hence the work-energy method would have no advantage over the FMA method. However, in cases where the normal force is constant, the work of the friction force can be obtained without integration (see Sample Problem 14.2).

Sample Problem **14.2**

As shown in Fig (a), the block of mass $m = 1.6$ kg is placed on a horizontal plane and attached to an ideal spring. The static and kinetic coefficients of friction between the block and the plane are given in the figure. The spring has a stiffness of $k = 30$ N/m and is undeformed when $x = 0$. The block is launched at $x = 0$ with the velocity of 6 m/s to the right. (1) Determine the value of x when the block first comes to rest. (2) Show that the block does not remain at rest in the position found in Part 1. (3) Find the speed of the block when it reaches $x = 0$ for the second time.

Undeformed position

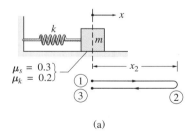

(a)

Solution

Part 1

Figure (b) shows the FBD of the block, assuming motion to the right. The weight $mg = (1.6)(9.81) = 15.696$ N and the normal contact force N_A are shown as dashed arrows, since they are workless (the forces are perpendicular to the path of the block). Nevertheless, we must compute N_A, because it determines the kinetic friction force F_k, which does perform work. From $\Sigma F_y = 0$, we get $N_A = mg = 15.696$ N. Thus $F_k = \mu_k N_A = (0.2)(15.696) = 3.139$ N. Note that F_k is constant (N_A is constant), and because it opposes the motion, its work is negative.

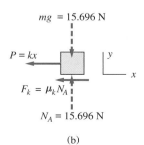

(b)

The only other force that does work is the spring force $P = kx$ (in this case, the displacement x of the block coincides with the elongation of the spring).

Applying the work-energy principle between the launch position 1 and the rest position 2, we obtain

$$U_{1-2} = T_2 - T_1$$

$$-\frac{1}{2}k\left(x_2^2 - x_1^2\right) - F_k x_2 = \frac{1}{2}mv_2^2 - \frac{1}{2}mv_1^2$$

$$-\frac{1}{2}(30)\left(x_2^2 - 0\right) - 3.139x_2 = 0 - \frac{1}{2}(1.6)(6)^2$$

The two solutions are $x_2 = 1.2850$ m and -1.4942 m. Because the launch velocity was to the right, only the positive root has physical significance; that is,

$$x_2 = 1.285 \text{ m} \qquad \textit{Answer}$$

Part 2

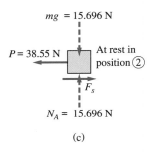

$mg = 15.696$ N

$P = 38.55$ N

At rest in position ②

F_s

$N_A = 15.696$ N

(c)

The FBD of the block at rest in position 2 is shown in Fig. (c). The spring force $P = kx_2 = (30)(1.2850) = 38.55$ N, which tends to pull the block to the left, is resisted by the static friction force F_s. Equilibrium in this position is possible only if $F_s = P = 38.55$ N. The maximum possible friction force is $F_{max} = \mu_s N_A = (0.3)(15.696) = 4.709$ N. Because $F_s > F_{max}$, equilibrium is not possible, and the block will start moving to the left.

Part 3

The position of the block when it reaches $x = 0$ for the second time is denoted by 3 in Fig (a). To obtain the speed of the block in this position, we apply the work-energy principle between positions 1 and 3 (choosing positions 2 and 3 would produce the same result). The net work of the spring force is zero, because the spring is undeformed in both positions. The work done by the kinetic friction force is $-F_k x_2$ as the block moves from 1 to 2, and $-F_k x_2$ between 2 and 3 (recall that the work of friction force is always negative) for a total of $-2F_k x_2$. Therefore, the work-energy principle becomes

$$U_{1-3} = T_3 - T_1$$

$$-2F_k x_2 = \frac{1}{2}m\left(v_3^2 - v_1^2\right)$$

$$-2(3.139)(1.2850) = \frac{1}{2}(1.6)\left(v_3^2 - 6^2\right)$$

which yields

$$v_3 = 5.09 \text{ m/s} \qquad \textit{Answer}$$

Note

If v_3 had turned out to be imaginary (square root of a negative number), we would conclude that the block was unable to reach position 3 because it had come to rest somewhere between positions 2 and 3. This rest position could be determined by another application of the work-energy principle.

Sample Problem 14.3

The 2-kg collar in Fig. (a) slides along the guide rod with negligible friction. The free length of the spring attached to the collar is $L_0 = 1.2$ m, and its stiffness is $k = 60$ N/m. If the collar is moving down the rod with the speed $v_A = 4$ m/s when it is at A, determine the speed of the collar at B.

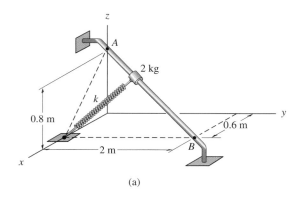

(a)

Solution

The free-body diagram (FBD) of the collar when it is an arbitrary distance s from A is shown in Fig. (b). The forces acting on the collar are the weight W, the spring force F, and the normal contact force N applied by the rod. Only W and F do work on the collar; N is workless, since it is perpendicular to the path AB of the collar.

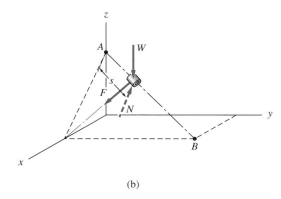

(b)

Work of W   From Eq. (14.7), the work of the weight is

$$U_{A-B} = -W(z_B - z_A) = -2(9.8)(0 - 0.8) = 15.7 \, \text{N} \cdot \text{m} \qquad \textbf{(a)}$$

Work of F   The work done by the spring force on the collar is obtained from Eq. (14.10):

$$U_{A-B} = -\frac{1}{2}k\left(\delta_B^2 - \delta_A^2\right) \qquad \textbf{(b)}$$

Letting L_A and L_B be the lengths of the spring when the collar is at A and B, respectively, the corresponding elongations of the spring are

$$\delta_A = L_A - L_0 = \sqrt{0.6^2 + 0.8^2} - 1.2 = -0.2\,\text{m}$$

$$\delta_B = L_B - L_0 = 2 - 1.2 = 0.8\,\text{m}$$

The negative sign indicates that the spring is compressed when the collar is at A. Substituting the values of δ_A and δ_B in Eq. (b), we obtain

$$U_{A-B} = -\frac{1}{2}(60)[(0.8)^2 - (-0.2)^2] = -18\,\text{N} \cdot \text{m} \qquad \text{(c)}$$

The total work done on the collar is obtained by adding Eqs. (a) and (c):

$$U_{A-B} = 15.7 - 18 = -2.3\,\text{N} \cdot \text{m}$$

Applying the work-energy principle between positions A and B, we obtain

$$U_{A-B} = T_B - T_A = \frac{1}{2}\frac{W}{g}\left(v_B^2 - v_A^2\right)$$

$$-2.3 = \frac{1}{2}\frac{(2)(9.8)}{(9.8)}\left(v_B^2 - 4^2\right)$$

which yields

$$v_B = 3.7\,\text{m/s} \qquad\qquad \textit{Answer}$$

Notes

- If the analysis had predicted an imaginary v_B (the square root of a negative number), we would conclude that the collar came to rest before reaching B.
- The ease with which v_B was obtained demonstrates the power of the work-energy method in the solution of some problems. Inspection of the FBD in Fig. (b) reveals that the relationship between the spring force F and the position coordinate s will be rather complex. It follows that if we had used the FMA method, the left-hand side of equation of motion $\Sigma F = m\ddot{s}$ would be a complicated function of s. Consequently, integration of this equation would be very difficult. However, using the work-energy method, we avoided this difficulty because the integration had been already performed in the derivation of the work-energy principle.

Problems

14.1   (a) Compute the work done by each force given in the following list as its point of application moves from 1 to 3 along the straight line connecting 1 and 3. (b) Repeat part (a) if the path consists of the straight line segments 1–2 and 2–3. (x and y are in m.)

1. $\mathbf{F} = 30\mathbf{i} - 10\mathbf{j}$ N
2. $\mathbf{F} = 3x\mathbf{i} - y\mathbf{j}$ N
3. $\mathbf{F} = 3y\mathbf{i} - x\mathbf{j}$ N

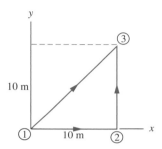

Fig. P14.1

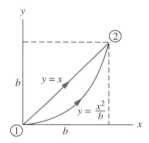

Fig. P14.2, P14.3

14.2   Compute the work of the force $\mathbf{F} = (F_0/b^3)(xy^2\mathbf{i} + x^2y\mathbf{j})$ as its point of application moves from 1 to 2 along (a) the line $y = x$; and (b) the parabola $y = x^2/b$.

14.3   Repeat Prob. 14.2 for the force $\mathbf{F} = (F_0/b^3)(x^2y\mathbf{i} + xy^2\mathbf{j})$.

14.4   The collar of weight W slides on a frictionless circular arc of radius R. The ideal spring attached to the collar has the free length $L_0 = R$ and stiffness k. When the slider moves from A to B, compute (a) the work done by the spring; and (b) the work done by the weight.

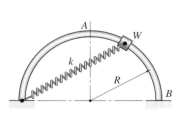

Fig. P14.4

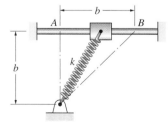

Fig. P14.5, P14.6

14.5   Derive the expression for the work done by the ideal spring on the slider when the slider moves from A to B. Assume that the free length of the spring is (a) $L_0 = b$; and (b) $L_0 = 0.8b$.

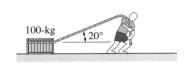

100-kg
20°

Fig. P14.7

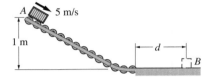

A 5 m/s

1 m

d

B

Fig. P14.8

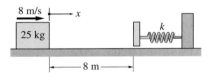

8 m/s x

25 kg

k

8 m

Fig. P14.9

14.6 The coefficient of kinetic friction between the slider and the rod is μ, and the free length of the spring is $L_0 = b$. Derive the expression for the work done by the friction force on the slider as it moves from A to B. Neglect the weight of the slider.

14.7 The man slides the 100-kg crate across the floor by pulling with a constant force of 200 N. If the crate was initially at rest, how far will the crate move before its speed is 1 m/s? The coefficient of kinetic friction between the crate and the floor is 0.18.

14.8 The 1-kg package arrives at A, the top of the inclined roller conveyor, with a speed of 5 m/s. After descending the conveyor, the package slides a distance d on the rough horizontal surface, coming to a stop at B. If the coefficient of kinetic friction between the package and the horizontal plane is 0.4, determine the distance d.

14.9 The 25-kg box is launched from the position shown along the rough horizontal plane with the velocity of 8 m/s. Determine the distance x that the box will travel before the spring stops the forward motion. The coefficient of kinetic friction between the box and the plane is $\mu_k = 0.2$, and the spring constant is $k = 150$ N/m.

14.10 The speed of the car at the base of a 10 m hill is 54 km/h. Assuming the driver keeps her foot off the brake and accelerator pedals, what will be the speed of the car at the top of the hill?

54 km/h

10 m

Fig. P14.10

14.11 The 0.8-kg slider is at rest in position 1 when the constant vertical force F is applied to the rope that is attached to the slider. What is the required magnitude of F if the slider is to reach position 2 with a speed of 6 m/s? Neglect friction.

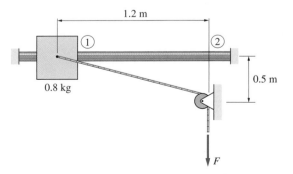

1.2 m

① ②

0.8 kg

0.5 m

F

Fig. P14.11

14.12   A crate of weight W is dragged across the floor from A to B by the constant vertical force P acting at the end of the rope. Calculate the work done on the crate by the force P. Assume that the crate does not lift off the floor.

*__14.13__   For the crate described in Prob. 14.12, determine the work done by the friction force if the kinetic coefficient of friction between the crate and floor is μ.

14.14   The 0.31-kg mass slides on a frictionless wire that lies in the vertical plane. The ideal spring attached to the mass has a free length of 80 mm and its stiffness is 120 N/m. Calculate the smallest value of the distance b if the mass is to reach the end of the wire at B after being released from rest at A.

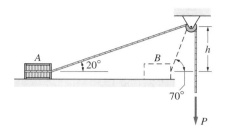

Fig. P14.12, P14.13

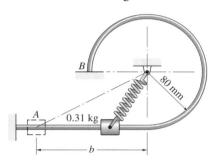

Fig. P14.14

14.15   The 1-kg collar moves from A to B along a frictionless rod. The stiffness of the spring is k and its free length is 200 mm. Compute the value of k so that the slider arrives at B with a speed of 1 m/s after being released from rest at A.

14.16   A 10-kg package, initially at rest at A, is propelled between A and B by a constant force P shown on the graph. Neglecting friction, find the smallest value of P for which the package will reach D.

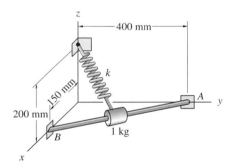

Fig. P14.15

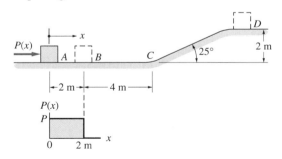

Fig. P14.16, P14.17

14.17   Solve Prob. 14.16 assuming that the coefficient of kinetic friction between the package and the contact surfaces is 0.15.

14.18   In position 1, the 0.25-kg block is held against the spring, compressing it by 150 mm. The block then is released, and the spring fires it up the cylindrical surface. Neglecting friction, find the contact force exerted on the block by the surface in position 2.

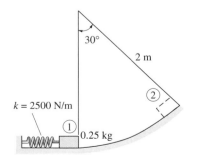

Fig. P14.18

14.19   When in the position shown, the 5-kg box is moving down the inclined plane at a speed of 6 m/s. What is the maximum force in the spring after the box hits it? The coefficient of kinetic friction between the box and the plane is $\mu_k = 0.25$, and the spring constant is $k = 4\,\text{kN/m}$.

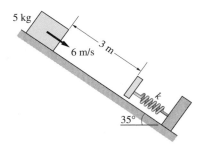

Fig. P14.19

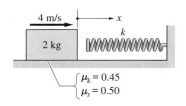

Fig. P14.20

14.20   The 2-kg block hits the spring with a speed of 4 m/s. Determine the total distance traveled by the block before it comes to a permanent stop. Use $k = 8\,\text{N/m}$ and the coefficients of friction shown.

14.21   A block of mass m is suspended from a spring of stiffness k. If the block is pulled down a distance h from its equilibrium position and released, determine its speed as it passes through the equilibrium position.

14.22   The 2-kg block slides in a frictionless slot that lies in the vertical plane. The stiffnesses of springs A and B attached to the block are $k_A = 80\,\text{N/m}$ and $k_B = 40\,\text{N/m}$. When the block is in position 1, spring A is compressed 0.2 m and spring B is undeformed. If the block is launched in position 1 with the speed v_1 down the slot, find the smallest v_1 that will enable the block to reach position 2.

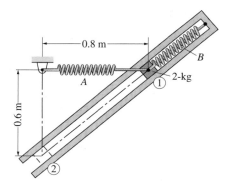

Fig. P14.22

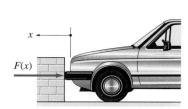

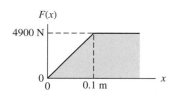

Fig. P14.23

14.23   The diagram shows the relationship between the force F and deformation x for an energy-absorbing car bumper. Determine the maximum deformation of the bumper if a 500-kg car hits a rigid wall at a speed of (a) 6 km/h; and (b) 10 km/h.

14.24   The diagram shows how the force F required to push an arrow slowly through a bale of hay varies with the distance of penetration x. Assuming that the

F-*x* plot is independent of the speed of penetration, calculate the exit speed of a 50-gm arrow if its speed at entry is 60 m/s.

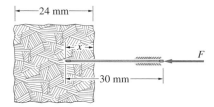

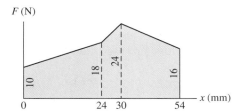

Fig. P14.24

14.25   The 2-kg weight is attached to the rim of a light wheel that is free to rotate about a vertical axis at *O*. A rope is wrapped around three-fourths of the periphery of the wheel. The wheel is at rest in the position shown, when the constant horizontal force *P* is applied to the end of the rope. Determine the smallest *P* that causes the wheel to reach the angular speed of 500 rev/min by the time the rope has been unwound. Neglect friction and the mass of the wheel.

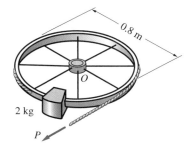

Fig. P14.25

14.4   *Conservative Forces and the Conservation of Mechanical Energy*

A force is said to be *conservative* if its work depends only on the initial and final positions of its point of application. All the specific forces discussed in the previous article are conservative, because in each case we could determine the work without having to specify the path between the end points.

It is often convenient to describe the effects of conservative forces in terms of their potential energies. Roughly speaking, potential energy is the capacity of a conservative force to do work. The principle of conservation of energy states that the total energy (the sum of all forms of energy) remains constant for a closed system. The form of the energy may change—for example, electrical energy may be converted to mechanical energy—but the total energy can neither be created nor destroyed.

In mechanics, we restrict our attention to mechanical energy, defined to be the sum of the potential and kinetic energies. If all forces acting on a particle, body, or closed system of bodies are conservative, mechanical energy is conserved, a concept known as the *principle of conservation of mechanical energy*.

This article discusses the application of the principle of conservation of mechanical energy, which is simply a restatement of the work-energy principle, $U_{1-2} = \Delta T$, for conservative force systems. Although the energy principle may be easier to apply than the work-energy method in some problems, its use is limited, because it is not valid for nonconservative forces, such as kinetic friction.

a. Conservative forces and potential energy

If the force $\mathbf{F}$ is conservative, its work

$$U_{1-2} = \int_{\mathbf{r}_1}^{\mathbf{r}_2} \mathbf{F} \cdot d\mathbf{r} \qquad \text{(14.2, repeated)}$$

is a function of the initial and final positions of its point of application. The integral in Eq. (14.2) can be a function of $\mathbf{r}_1$ and $\mathbf{r}_2$ only if the integrand is an exact differential of some scalar function $-V(\mathbf{r})$; that is, if the integrand can be written as

$$\mathbf{F} \cdot d\mathbf{r} = -dV \qquad \text{(14.16)}$$

(the minus sign is introduced by convention). The function $V(\mathbf{r})$ is called the *potential energy* of the force $\mathbf{F}$. Substituting Eq. (14.16) into Eq. (14.2), we get

$$U_{1-2} = -\int_{\mathbf{r}_1}^{\mathbf{r}_2} dV = -(V_2 - V_1) = -\Delta V \qquad \text{(14.17)}$$

where we used the notation $V_1 = V(\mathbf{r}_1)$ and $V_2 = V(\mathbf{r}_2)$.

Equation (14.17) shows that

Work of a conservative force = decrease in its potential energy

Potential energy thus can be viewed as the capacity of the force to do work. Positive work diminishes the potential for further work, and negative work increases the potential.

It is important to note that Eq. (14.7) involves the *change* in potential energy. Therefore, the baseline (or *datum*) from which V is measured may be chosen arbitrarily.

A useful property of a conservative force is that its components can be derived from its potential energy. Consider a conservative force $\mathbf{F}$ that acts at a point with rectangular coordinates (x, y, z). Using $\mathbf{F} = F_x \mathbf{i} + F_y \mathbf{j} + F_z \mathbf{k}$ and $d\mathbf{r} = dx\,\mathbf{i} + dy\,\mathbf{j} + dz\,\mathbf{k}$, we obtain

$$dV = -\mathbf{F} \cdot d\mathbf{r} = -(F_x\,dx + F_y\,dy + F_z\,dz) \qquad \text{(14.18)}$$

Because dV is an exact differential of potential energy V, it may be written as

$$dV = \frac{\partial V}{\partial x}dx + \frac{\partial V}{\partial y}dy + \frac{\partial V}{\partial z}dz \qquad (14.19)$$

Comparing Eqs. (14.18) and (14.19), the rectangular components of **F** become

$$F_x = -\frac{\partial V}{\partial x} \qquad F_y = -\frac{\partial V}{\partial y} \qquad F_z = -\frac{\partial V}{\partial z} \qquad (14.20)$$

Equation (14.20) shows that a conservative force **F** is the *negative gradient* of its potential function V. Utilizing the gradient operator ∇ (pronounced "del"), Eq. (14.20) can be written as

$$\mathbf{F} = -\nabla V \qquad (14.21)$$

Only conservative forces are derivable from a potential function in this manner. A nonconservative force, such as friction, does not possess a potential.

b. *Conservation of mechanical energy*

If all the forces acting on a particle are conservative, then their resultant is also conservative. The potential energy $V(\mathbf{r})$ of the resultant force can be obtained by summing the potential energies of all the forces acting on the particle. If the particle moves from position 1 to position 2, Eq. (14.17) can be used to relate the work of the resultant force to the change in its potential energy: $U_{1-2} = -(V_2 - V_1)$. Substituting this into the work-energy principle $U_{1-2} = \Delta T$, we obtain $-(V_2 - V_1) = T_2 - T_1$, or

$$V_1 + T_1 = V_2 + T_2 \qquad (14.22)$$

Letting the total mechanical energy[*] E be the sum of the kinetic and potential energies—that is, letting

$$E = T + V \qquad (14.23)$$

Eq. (14.22) becomes

$$E_1 = E_2 \quad \text{or} \quad \Delta E = 0 \qquad (14.24)$$

This equation is called the *principle of conservation of mechanical energy.*

Because the work done by a kinetic friction force is not independent of the path, kinetic friction is a nonconservative force. Therefore, when kinetic friction is present, the total mechanical energy is not conserved, but is reduced by the negative work done by the friction force. This energy is not lost; it is transformed into thermal energy in the form of heat. In other words, the total energy is still conserved; it is the form of the energy that has changed.

---

[*]The emphasis here is on mechanical energy. Other forms of energy, such as heat, are excluded from our discussion.

Because the total mechanical energy E is constant if all of the forces are conservative, we conclude that

$$\frac{dE}{dt} = 0 \qquad (14.25)$$

This equation is sometimes useful for obtaining the acceleration of the particle.

c. *Computation of potential energy*

The potential energies of conservative forces can be computed by comparing their work, derived by the methods of Art. 14.2, with the definition of potential energy in Eq. (14.17). The results are summarized in Table 14.1.

1. Constant force	2. Weight	3. Spring	4. Gravity
$V_f = -Fd$	$V_g = Wh$	$V_e = \frac{1}{2}k\delta^2$	$V_g = -\dfrac{Gm_A m_B}{R}$

Table 14.1 Formulas for Potential Energy

1. *Potential Energy of a Constant Force*   The work done by a constant force **F** (constant in magnitude and direction) was previously shown to be

$$U_{1-2} = F\,\Delta d \qquad \text{(14.6, repeated)}$$

where F is the magnitude of the force and Δd represents the work-absorbing displacement of its point of application, as shown in Fig. 14.4. Comparing this with $U_{1-2} = -\Delta V$, we conclude that the potential energy of a constant force **F** is

$$V_f = -Fd \qquad (14.26)$$

where d is measured from any convenient datum.

2. *Potential Energy of a Weight*   If we assume that motion is restricted to small distances from the surface of the earth, the weight of an object may be treated as a constant force. The work done by the weight W in Fig. 14.5 was given by

$$U_{1-2} = -W\,\Delta h \qquad \text{(14.7, repeated)}$$

Comparing this equation with $U_{1-2} = -\Delta V$, we conclude that the potential energy of W, called the *gravitational potential energy,* is equal to

$$V_g = Wh \quad \text{(for constant weight } W\text{)} \qquad (14.27)$$

where the positive direction of h must be vertically upward. Because only the change in potential energy is significant, the *datum* from which y is measured is arbitrary.

3. *Potential Energy of a Spring*   According to Eq. (14.10), the work done by an ideal spring[*] when its free end moves from position 1 to 2 is

$$U_{1-2} = -\frac{1}{2}k(\delta_2^2 - \delta_1^2) = \frac{1}{2}k\Delta(\delta^2)$$

where δ is the elongation of the spring, measured from its free length L_0. The potential energy of the spring, also called *elastic potential energy,*[†] is obtained by comparing this equation with $U_{1-2} = -\Delta V$. Then we see that the elastic potential energy is

$$V_e = \frac{1}{2}k\delta^2 \qquad (14.28)$$

Observe that the elastic potential energy is always positive.

4. *Gravitational Potential Energy*   Equation (14.27) can be used to calculate the gravitational potential energy of a weight only if the variation of the gravitational acceleration g is negligible. For motions that violate this restriction, the constant weight must be replaced by the force obtained from Newton's law of gravitation $F = Gm_Am_B/R^2$. The work done by F was found in Eq. (14.11) to be

$$U_{1-2} = Gm_Am_B\left(\frac{1}{R_2} - \frac{1}{R_1}\right) = Gm_Am_B\Delta(1/R)$$

where m_A and m_B are the masses being attracted toward each other. Comparing this equation with $U_{1-2} = -\Delta V$, we see that the gravitational potential energy of the system consisting of masses m_A and m_B is

$$V_g = -\frac{Gm_Am_B}{R} \qquad (14.29)$$

Observe that the gravitational potential energy is always a negative quantity that approaches zero as R approaches infinity.

---

[*]Observe that we consider the work done *by* the spring, not the work done *on* the spring. Potential energy of a spring refers to the ability of a spring to do work on the body to which it is attached, not to the effect that the body has on the spring.
[†]*Elasticity* refers to the ability of a deformed body to return to its undeformed shape when loads are removed.

Sample Problem 14.4

The figure shows a 1-kg collar that slides along the frictionless vertical rod under the actions of gravity and an ideal spring. The spring has a stiffness of 160 N/m, and its free length is 0.9 m. The collar is released from rest in position 1. Use the principle of conservation of mechanical energy to determine the speed of the collar in position 2.

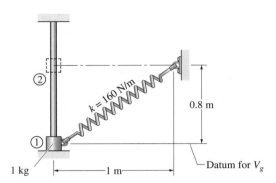

Solution

Without drawing the free-body diagram of the collar, we recognize that only the weight and the spring do work on the collar as it moves along the rod. (The normal force between the collar and the rod is a nonworking force.) Because both the weight and the spring force are conservative, we are justified in using the principle of conservation of mechanical energy.

Gravitational potential energy It is convenient to take position 1 as the datum from which the elevation h is measured. Therefore,

$$(V_g)_1 = 0 \quad (V_g)_2 = Wh_2 = (1)(9.8)(0.8) = 7.84 \text{ N-m}$$

Elastic potential energy The elongations of the spring in the two positions are

$$\delta_1 = \sqrt{0.8^2 + 1^2} - 0.9 = 0.38 \text{ m}$$
$$\delta_2 = 1 - 0.9 = 0.1 \text{ m}$$

which gives us

$$(V_e)_1 = \frac{1}{2}k\delta_1^2 = \frac{1}{2}(160)(0.38)^2 = 11.552 \text{ N-m}$$
$$(V_e)_2 = \frac{1}{2}k\delta_2^2 = \frac{1}{2}(160)(0.10)^2 = 0.8 \text{ N-m}$$

Kinetic energy Noting that the collar is at rest in position 1, the kinetic energies are

$$T_1 = 0 \qquad T_2 = \frac{1}{2}mv_2^2 = \frac{1}{2}(1)v_2^2 = 0.5\,v_2^2$$

Conservation of energy   Equating the initial and final energies, we get

$$T_1 + (V_g)_1 + (V_e)_1 = T_2 + (V_g)_2 + (V_e)_2$$

$$0 + 0 + 11.552 = 0.5\,v_2^2 + 7.84 + 0.8$$

which yields for the speed of the collar

$$v_2 = 2.4 \text{ m/s} \qquad\qquad \text{Answer}$$

Sample Problem 14.5

The figure shows a block of mass m that slides along a frictionless horizontal plane. The position coordinate x is measured from the undeformed position of the ideal spring of stiffness k. Derive the acceleration of the block as a function of x, using the principle of conservation of mechanical energy. (This problem was solved by the force-mass-acceleration method in Sample Problem 12.7.)

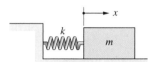

Solution

Because only the conservative spring force does work as the block moves, we are justified in using the principle of conservation of mechanical energy.

The kinetic energy T of the block is

$$T = \frac{1}{2}mv^2 = \frac{1}{2}m\dot{x}^2 \qquad\qquad \text{(a)}$$

Since the position coordinate x is measured from the undeformed position of the spring, it also corresponds to the elongation δ of the spring. Therefore, the elastic potential energy is

$$V_e = \frac{1}{2}k\delta^2 = \frac{1}{2}kx^2 \qquad\qquad \text{(b)}$$

Combining Eqs. (a) and (b), the total mechanical energy E is

$$E = T + V_e = \frac{1}{2}m\dot{x}^2 + \frac{1}{2}kx^2$$

Because the total mechanical energy is conserved, we have from Eq. (14.25)

$$\frac{dE}{dt} = m\dot{x}\ddot{x} + kx\dot{x} = 0$$

or

$$\dot{x}(m\ddot{x} + kx) = 0$$

Ignoring the static solution $\dot{x} = 0$, and recognizing that the acceleration is $a = \ddot{x}$, we get for the acceleration

$$a = -\frac{k}{m}x \qquad\qquad \text{Answer}$$

This agrees with the result obtained in Sample Problem 12.7 by the force-mass-acceleration method.

Problems

14.26   The box of weight W is held just above the top of the spring and released. Determine the maximum force in the spring during the ensuing motion.

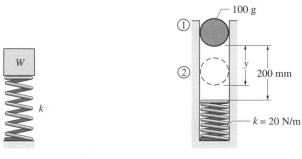

Fig. P14.26   **Fig. P14.27**

14.27   The 100-g ball is released from rest in position 1. After making contact with the spring, the ball sticks to the end of the spring and rebounds to position 2. Find the vertical distance y between positions 1 and 2.

Fig. P14.28

14.28   A linear spring of stiffness k is designed to stop the 20-Mg railroad car traveling at 8 km/h within 400 mm after impact. Find the smallest value of k that will produce the desired result. (Note: This problem was solved by the FMA method in Prob. 12.50.)

14.29   Solve Prob. 14.28 if the single spring is replaced by two springs of identical stiffness k, nested as shown. (Note that one spring is 200 mm shorter than the other.)

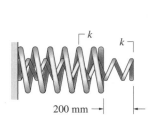

Fig. P14.29

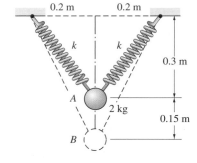

Fig. P14.30

14.30   The 2-kg weight is released from rest in position A, where the two springs of stiffness k each are undeformed. Determine the largest k for which the weight would reach position B.

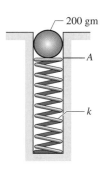

Fig. P14.31

14.31   The spring of stiffness k is undeformed in the position shown. The 200-gm ball is placed on the spring and launched vertically by compressing the spring 0.2 m and releasing it. If the ball reaches an elevation of 15 m above A, determine the value of k.

14.32 The sliding collar of weight $W = 10\,N$ is attached to two springs of stiffnesses $k_1 = 180\,N/m$ and $k_2 = 60\,N/m$. The free length of each spring is 50 cm. If the collar is released from rest in position A, determine its speed in position B. Neglect friction.

14.33 The spring attached to the 0.6-kg sliding collar has a stiffness of 200 N/m and a free length of 150 mm. If the speed of the collar in position A is 3 m/s to the right, determine the speed in position B. Neglect friction.

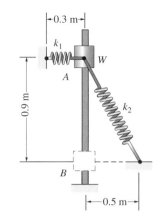

Fig. P14.32

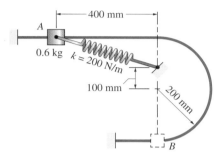

Fig. P14.33

14.34 The platform and the 12-kg block are traveling to the right at 8 m/s when the platform is brought to a sudden stop on collision with the wall. Before the collision, the spring connecting the block to the platform was undeformed. Neglecting friction, determine the speed with which the block hits the wall.

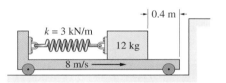

Fig. P14.34

14.35 The 1-kg collar slides with negligible friction on the circular guide rod that is attached to the platform. The collar is in position A when the platform is traveling to the right with the speed v_0. After the platform is brought to a sudden stop, the collar slides up the rod, reaching its highest position at B. Determine v_0.

14.36 The 0.5-kg pendulum oscillates with the amplitude of $\theta_{max} = 50°$. Determine the maximum force in the supporting string.

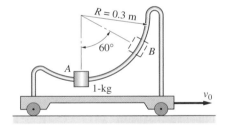

Fig. P14.35

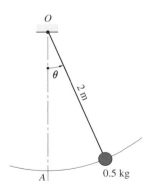

Fig. P14.36

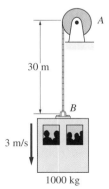

Fig. P14.37

14.37 As the steel cable unwinds from the drum A, the 1000-kg elevator descends at the constant speed of 3 m/s. If the drum suddenly stops when the elevator is in the position shown, calculate the resulting maximum force in

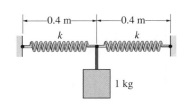

Fig. P14.39

Fig. P14.40

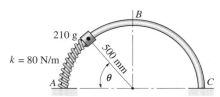

Fig. P14.41

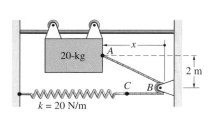

Fig. P14.44

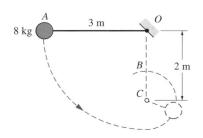

Fig. P14.45

the cable. Due to the elasticity of steel, the cable acts as an ideal spring of stiffness 800×10^3 N/m.

14.38   Solve Prob. 14.37 assuming a spring of stiffness $k = 6000$ N/m has been mounted between the elevator and the cable at B.

14.39   The 1-kg weight is attached to two springs, each of stiffness $k = 80$ N/m. The weight is being held in the position shown where each spring is stretched 4 mm. If the weight is released from this position, find its speed after a fall of 0.1 m.

14.40   The chain, 6 m long and weighing 5 N/m, is released from rest in the position shown. Neglecting friction, determine the speed of the chain at the instant when the last link leaves the table.

14.41   The semicircular rod AC lies in the vertical plane. The spring wound around the rod is undeformed when $\theta = 45°$. If the 210-g slider is pressed against the spring and released at $\theta = 30°$, determine the velocity of the slider when it passes through B. Neglect friction and assume the slider is not attached to the spring.

14.42   The particle of mass m is at rest at A when it is slightly displaced and allowed to slide down the cylindrical surface of radius R. Neglecting friction, determine (a) the speed of the particle as a function of the angle θ; and (b) the value of θ when the particle leaves the surface.

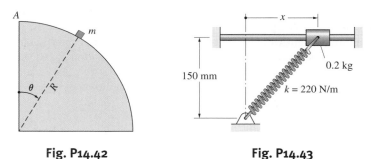

Fig. P14.42   **Fig. P14.43**

14.43   The spring attached to the 0.2-kg collar has a stiffness of 220 N/m and a free length of 150 mm. If the collar is released from rest when $x = 150$ mm, determine its acceleration as a function of x. Neglect friction.

14.44   The 20-kg sliding panel is suspended from frictionless rollers that run on the horizontal rail. The spring attached to rope ABC has a stiffness of 20 N/m and is undeformed when the panel is in the position $x = 0$. If the panel starts from rest at $x = 3$ m, determine its speed when $x = 0$.

14.45   The 8-kg pendulum is released from rest at A. Initially, the pendulum swings about O, but after the string catches the fixed peg C, rotation takes place about C. When the bob of the pendulum is at B, determine (a) its speed; and (b) the tension in portion BC of the string.

14.46   The catapult is made of two elastic bands, each 0.15 m long when unstretched. Each band behaves like an ideal spring of stiffness 100 N/m. If a 100-g rock is launched from the position shown, determine the speed of the rock when it leaves the catapult at *D*.

14.47   The 0.5-kg collar slides on the circular rod *AB* that lies in a vertical plane. The spring attached to the collar has a free length of 240 mm, and its stiffness is 80 N/m. If the collar is released from rest at *A*, determine the speed with which it arrives at *B*.

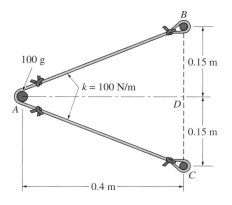

Fig. P14.46

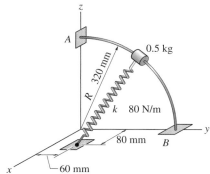

Fig. P14.47

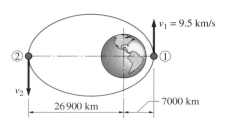

Fig. P14.49

14.48   A 1200-kg spacecraft is launched from the surface of the earth. How much energy is required for the spacecraft to escape the earth's gravitational field? Use the following data: mass of earth $= 5.974 \times 10^{24}$ kg, radius of earth $= 6378$ km, universal gravitational constant $= 6.672 \times 10^{-11}$ m$^3 \cdot$ kg$^{-1} \cdot$ s^{-2}.

14.49   A satellite is in an elliptical orbit around the earth. If the speed of the satellite at the perigee (position 1) is 9.5 km/s, calculate its speed at the apogee (position 2). Use $GM_e = 3.98 \times 10^{14}$ m^3/s^2, where M_e is the mass of the earth and G is the universal gravitational constant.

14.50   The firing of retrorockets causes the earth satellite to slow down to 3 km/s when it reaches its perigee *A*. As a result, the satellite leaves its orbit and descends toward the earth. The satellite burns up when it enters the earth's atmosphere at *B*, 100 km above the earth's surface. If the speed of the satellite was observed to be 7.7 km/s at *B*, determine the height *h* of the perigee. Use $GM_e = 3.98 \times 10^{14}$ m^3/s^2, where M_e is the mass of the earth and G is the universal gravitational constant.

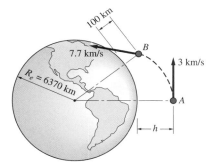

Fig. P14.50

14.5   *Power and Efficiency*

Power is defined as the *rate at which work is done*. Letting P be the power and U the work, we have

$$P = \frac{dU}{dt} \tag{14.30}$$

Power is a scalar of dimension $[FL/T]$. In the SI system, its unit is *watts* (1 W = 1 J/s = 1 N · m/s).

The *power of a force* **F** is obtained by substituting $dU = \mathbf{F} \cdot d\mathbf{r}$ from Eq. (14.1) into Eq. (14.30), yielding $P = (\mathbf{F} \cdot d\mathbf{r})/dt$. Noting that $d\mathbf{r}/dt = \mathbf{v}$ is the velocity of the particle on which **F** acts, we have

$$P = \mathbf{F} \cdot \mathbf{v} \tag{14.31}$$

Machines, being devices that do work, are also associated with power. The *input power* of a machine is the rate at which energy is supplied to the machine. The *output power* is the rate at which the machine does work. The *efficiency η* of the machine is defined as

$$\eta = \frac{output\ power}{input\ power} \times 100\% \tag{14.32}$$

The output power is always less than the input power due to loss of mechanical energy caused by friction, vibrations, and so on. Hence the efficiency of a machine is always less than 100%.

The drivetrain (transmission, drive shafts, and driving wheels) of a car is an example of a machine. The input power is supplied by the engine of the car. Some of that power is used to overcome friction in the transmission and the bearings. A lesser amount of power is converted to sound by vibrations of the drive shafts and the gears. The remaining power is the output power that is available to the driving wheels.

Sample Problem 14.6

A 1000-kg automobile accelerates under constant power from 90 km/h to 144 km/h in a distance of $^1/_4$ km on a straight, level test track. (1) Determine the horsepower delivered by the driving wheels to the car. (2) Calculate the power output of the engine if the efficiency of the drivetrain is 82%. Neglect air and rolling resistance.

Solution

Part 1

Let F be the driving force that is supplied by the driving wheels (the output end of the drivetrain) to the car, as shown in the figure. According to Newton's second law, $F = ma$, where m is the mass of the car, and a is the acceleration. Substituting $a = v(dv/dx)$, we get $F = mv(dv/dx)$. Because the velocity v and the driving force F have the same direction, the dot product $P = \mathbf{F} \cdot \mathbf{v}$ becomes

$$P = Fv = mv^2 \frac{dv}{dx}$$

Multiplying both sides by dx, we obtain

$$P\, dx = mv^2\, dv$$

Integration yields (note that P and m are constants)

$$Px = \frac{1}{3}mv^3 + C$$

where C is the constant of integration. With $m = 1000$ kg, the last equation becomes

$$Px = 333.3\,v^3 + C \qquad \text{(a)}$$

The values of P and C can be found from the given velocities at the beginning and end of the 1/4-km test. Let $x = 0$ at the start of the test, when $v = 25$ m/s (90 km/h). Substitution of these values in Eq. (a) yields

$$0 = 333.3(25)^3 + C \qquad \text{(b)}$$

At the end of the test, $x = 250\,\text{m}$ (1/4 km) and $v = 40\,\text{m/s}$ (144 km/h). Therefore, Eq. (a) becomes

$$P(250) = 333.3(40)^3 + C \tag{c}$$

The solution of Eqs. (b) and (c) is $C = -\,5.20781 \times 10^6\,\text{N-m}^2/\text{s}$ and $P = 64493.5\,\text{N}\cdot\text{m/s}$. Hence the power supplied to the car by the driving wheels is

$$P = 64493.5\,\text{N}\cdot\text{m/s} \times \frac{1\ \text{hp}}{746\ \text{N}\cdot\text{m/s}} = 86.5\ \text{hp} \qquad \textbf{\textit{Answer}}$$

Part 2

The efficiency of the drivetrain is 82%. Thus the power P' supplied to the drivetrain (the output power of the engine) is, according to Eq. (14.32),

$$P' = \frac{P}{\eta} = \frac{86.5}{0.82} = 105.5\ \text{hp} \qquad \textbf{\textit{Answer}}$$

Sample Problem   14.7

An 1800-kg delivery van is traveling at the speed of 26 m/s when the driver applies the brakes at time $t = 0$, causing all four wheels to lock. Friction between the tires and the road causes the van to skid to a stop. Determine the power of the friction force as a function of time if the kinetic coefficient of friction between the tires and the road is 0.6.

Solution

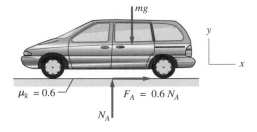

The free-body diagram of the van during braking is shown in the figure, where N_A is the resultant normal force acting on the tires and F_A is the resultant friction force. From the diagram, we obtain the following equations of motion:

$$\Sigma F_x = ma \qquad \xrightarrow{\ +\ } \qquad 0.6 N_A = ma$$

$$\Sigma F_y = 0 \qquad +\!\uparrow \qquad N_A - mg = 0$$

The solution is

$$N_A = mg = 1800(9.81) = 17\,658 \text{ N} = 17.658 \text{ kN}$$

$$a = 0.6g = 0.6(9.81) = 5.886 \text{ m/s}^2 \rightarrow$$

The velocity of the van during braking is

$$v = \int a\,dt = \int 5.886\,dt = 5.886t + C \text{ m/s}$$

The constant of integration C is found from the initial condition $v = -26$ m/s when $t = 0$, yielding $C = -26$ m/s. Therefore,

$$v = 5.886t - 26 \text{ m/s} \rightarrow \qquad\qquad (a)$$

The time when the van stops is $26/5.886 = 4.42$ s.

The power of the friction force can be calculated from Eq. (14.31). Noting that F_A and v have the same direction, we get

$$P = F_A v$$

Substituting for v from Eq. (a) and using $F_A = 0.6N_A = 0.6(17.658) = 10.595$ kN, we obtain

$$P = 10.595(5.886t - 26) = 62.4t - 275 \text{ kN} \cdot \text{m/s}$$

or

$$P = 62.4t - 275 \text{ kW} \qquad (0 \leq t \leq 4.42 \text{ s}) \qquad\qquad \textit{Answer}$$

Because P is the power supplied to the van by the friction force, it is negative, as expected.

Problems

14.51   A 50-kg man is riding a 150-kg motorcycle along a straight, level road at 72 km/h when he suddenly accelerates at full throttle. Determine his acceleration at that instant, knowing that the power output of the engine is 35 hp. Neglect any impediments to the motion, such as air resistance.

14.52   A 800-kg car accelerates from 54 km/h to 90 km/h as the engine is providing a constant 40 hp to the drive wheels. Find the distance traveled by the car during the acceleration.

14.53   Constant power of 450 kW is being delivered to the driving wheels of a 150-Mg locomotive. If the speed of the locomotive is 10 m/s at time $t = 0$, what is its speed at time $t = 60$ s?

14.54   The conveyor dumps 500-kg of coal per second into the waiting truck. Determine the power consumed by the conveyor motor if its efficiency is 70 percent.

Fig. P14.54

14.55   A 1500-kg automobile traveling along a straight, level road increases its speed from 54 km/h to 90 km/h in a distance of 100 m. The efficiency of the drive train is 85%, and the power output of the engine is constant. Determine the output horsepower of the engine during the acceleration. Neglect air and rolling resistance.

14.56   A pumping system, consisting of a pump and piping, raises water at the rate of 42 m³/min from a reservoir to a tank located 16 m above the reservoir. Determine the efficiency of the system if the output power of the motor that drives the pump is 240 hp. (Water density 1000 kg/m³)

14.57   A 1200-Mg train climbs a 3% grade (3 m rise per 100 m horizontal distance) at constant speed. If the engine of the train delivers 5000 kW of power to the driving wheels, determine the speed of the train. Neglect air and rolling resistance.

14.58   The block of mass m is at rest at time $t = 0$ when the constant force F is applied. Determine the power of F as a function of time.

14.59   The block of mass m is at rest at time $t = 0$ when the time-dependent force F is applied. If the power of F is constant, determine F as a function of time.

Fig. P14.58–P14.60

14.60   The block of mass $m = 12\,\text{kg}$ is at rest when the force $F = 60\cos\pi t$ N is applied during the time period $0 \le t \le 0.5$ s. (a) Determine the power of F as a function of t during this period. (b) Find the maximum power and the time when it occurs.

14.61   An electric hoist lifts a 500-kg mass at a constant speed of 0.68 m/s while consuming 4.2 kW of power. (a) Compute the efficiency of the hoist. (b) At what constant speed can the hoist lift an 800-kg mass with the same power consumption as in part (a)?

14.62   An electric hoist consumes 5 kW of power at a constant rate to lift the 500-kg mass. Assuming that the hoist is 80% efficient, determine the maximum speed reached by the mass.

14.63   A 60-kg man pedals a 10-kg bicycle up a 5° incline at 10 km/h. If the bicycle is 95% efficient, calculate the man's horsepower.

14.64   An 1800-kg car is traveling on a straight, level highway at the velocity v. The aerodynamic drag on the car is $F_D = 0.12v^2$ N, where the unit for v is m/s. If the maximum power available to the drive wheels is 150 kW, determine the largest possible acceleration of the car when (a) $v = 60$ km/h; and (b) $v = 120$ km/h.

14.65   The force F resisting the motion of a car depends on the speed v of the car as $F = F_0 + cv^2$, where F_0 and c are constants. If the output power of the drivetrain is 19.3 hp at the constant speed of 50 km/h and 32.3 hp at 60 km/h, determine the power requirement at 70 km/h.

14.66   The 4-kg block accelerates from rest as a result of the action of the force $F(t) = 12[1 - (t/2)]$, where F is in newtons and t is in seconds. The duration of the force is 2 seconds. Calculate (a) the power of F as a function of t; and (b) the maximum power and the time when it occurs.

*14.67   The diagram shows a typical P-v relationship for a gasoline-powered car that is accelerating at full throttle in low gear, where P is the power of the driving force F and v is the speed of the car. Determine the maximum value of F and the speed at which it occurs. (Hint: The line from the origin to a point on the curve has the slope $F = P/v$.)

Fig. P14.61, P14.62

Fig. P14.66

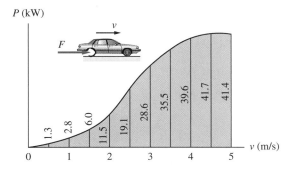

Fig. P14.67

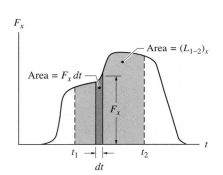

Fig. 14.8

14.6   *Principle of Impulse and Momentum*

We found the work-energy principle to be useful in determining the change in the speed of a particle during a given *displacement*. If we want to find the change in the velocity during a given *time* interval, then the principle of impulse and momentum provides a practical means of analysis. We begin with the definitions of impulse of a force and momentum of a particle,[*] and then we proceed to the derivation of the impulse-momentum principle.

a.   *Impulse of a force*

The *impulse* $\mathbf{L}_{1-2}$ of a force $\mathbf{F}$ in the time interval t_1 to t_2 is defined as

$$\mathbf{L}_{1-2} = \int_{t_1}^{t_2} \mathbf{F}\, dt \tag{14.33}$$

Impulse is a vector quantity, its rectangular components being

$$(L_{1-2})_x = \int_{t_1}^{t_2} F_x\, dt \qquad (L_{1-2})_y = \int_{t_1}^{t_2} F_y\, dt \qquad (L_{1-2})_z = \int_{t_1}^{t_2} F_z\, dt \tag{14.34}$$

where F_x, F_y, and F_z are the components of $\mathbf{F}$. As shown in Fig. 14.8, $(L_{1-2})_x$ is equal to the area under the F_x-t diagram between t_1 and t_2. Similarly, $(L_{1-2})_y$ and $(L_{1-2})_z$ are the areas under the F_y-t and F_z-t diagrams, respectively. This knowledge is very useful when computing the impulse of a force when its time dependence is given in graphical or numerical form. The dimension of impulse is $[FT]$; hence its units are N · s, lb · s, and so on.

In the work-energy methods described in the preceding articles, only forces that did work entered the analysis. Consequently, the free-body diagram could be replaced by an active-force diagram. However, free-body diagrams must always be used when calculating impulses, because a force has an impulse even if it does no work.

An important special case arises if the force $\mathbf{F}$ is *constant* in magnitude and direction. The impulse of the force then reduces to $\mathbf{L}_{1-2} = \mathbf{F} \int_{t_1}^{t_2} dt$, or

$$\mathbf{L}_{1-2} = \mathbf{F}(t_2 - t_1) = \mathbf{F}\,\Delta t \qquad (\mathbf{F}\ \text{constant}) \tag{14.35}$$

The impulse of a constant force is thus equal to the product of the force and the time interval, the impulse being in the same direction as the force.

b.   *Momentum of a particle and momentum diagrams*

The momentum $\mathbf{p}$ of a particle of mass m at an instant of time is defined as

$$\mathbf{p} = m\mathbf{v} \tag{14.36}$$

---

[*]They are more accurately called *linear* impulse and *linear* momentum, to distinguish them from *angular* impulse and *angular* momentum, which are discussed in the next article. However, it is common to omit "linear" when the meaning is clear from the context.

where **v** is the velocity vector of the particle at that instant. The momentum of a particle is a vector quantity that acts in the same direction as the velocity. The dimension of momentum is $[ML/T]$, or equivalently, $[FT]$. Therefore, the dimension of momentum is the same as the dimension of impulse.

The *momentum diagram* for a particle is a sketch of the particle showing its momentum vector $m\mathbf{v}$. Momentum diagrams are useful tools in the analysis of problems using the principle of impulse and momentum.

c. *Force-momentum relationship*

If the mass of the particle is constant, Newton's second law states that $\Sigma\mathbf{F} = m\mathbf{a}$, where $\Sigma\mathbf{F}$ is the resultant force acting on the particle. If the mass varies with time, this law takes the form

$$\Sigma\mathbf{F} = \frac{d}{dt}(m\mathbf{v})$$

Substituting $\mathbf{p} = m\mathbf{v}$ yields

$$\Sigma\mathbf{F} = \frac{d\mathbf{p}}{dt} \tag{14.37}$$

Equation (14.37) is the general form of Newton's second law: The resultant force is equal to the rate of change of the momentum. Therefore, $\Sigma\mathbf{F} = m\mathbf{a}$ should be considered as a special case that is valid for constant mass only.

d. *Impulse-momentum principle*

Multiplying both sides of Eq. (14.37) by dt and integrating from time t_1 to t_2, we obtain

$$\int_{t_1}^{t_2} \mathbf{F}\, dt = \int_{t_1}^{t_2} d\mathbf{p} = \mathbf{p}_2 - \mathbf{p}_1 \tag{14.38}$$

where $\mathbf{p}_1 = \mathbf{p}(t_1)$ and $\mathbf{p}_2 = \mathbf{p}(t_2)$ represent the momenta at t_1 and t_2, respectively. Because the left side of Eq. (14.38) is the impulse of **F** over the time interval t_1 to t_2, we have

$$\mathbf{L}_{1-2} = \mathbf{p}_2 - \mathbf{p}_1 = \Delta\mathbf{p} \tag{14.39}$$

Equation (14.39) is called the *impulse-momentum principle,* or the *balance of impulse and momentum.* When this principle is applied to the analysis of motion, the technique is called the *impulse-momentum method.*

If a rectangular coordinate system is used, Eq. (14.39) is equivalent to the following three scalar equations:

$$(L_{1-2})_x = \Delta p_x = (mv_x)_2 - (mv_x)_1$$

$$(L_{1-2})_y = \Delta p_y = (mv_y)_2 - (mv_y)_1 \qquad \text{(14.40)}$$

$$(L_{1-2})_z = \Delta p_z = (mv_z)_2 - (mv_z)_1$$

Caution   You must be careful not to confuse the work-energy principle, $U_{1-2} = T_2 - T_1$, with the impulse-momentum principle, $\mathbf{L}_{1-2} = \mathbf{p}_2 - \mathbf{p}_1$. Some of the important differences between the two principles are as follows:

- Work (U_{1-2}) is a *scalar* quantity associated with a force and a change in position.
- Impulse ($\mathbf{L}_{1-2}$) is a *vector* quantity associated with a force and a time interval.
- Kinetic energy (T) is a *scalar* quantity associated with a mass and its speed at an instant of time.
- Momentum ($\mathbf{p}$) is a *vector* quantity associated with a mass and its velocity vector at an instant of time.
- The work-energy principle, $U_{1-2} = \Delta T$, is a *scalar* relationship, whereas the impulse-momentum principle, $\mathbf{L}_{1-2} = \Delta \mathbf{p}$, is a *vector* relationship.

e.   *Conservation of momentum*

From the impulse-momentum principle, Eq. (14.39), we see that if the impulse acting on a particle is zero during a given time interval, the momentum of the particle will be conserved during that interval. In other words, if $\mathbf{L}_{1-2} = \mathbf{0}$, then

$$\mathbf{p}_1 = \mathbf{p}_2 \quad \text{or} \quad \Delta \mathbf{p} = \mathbf{0} \qquad \text{(14.41)}$$

Equation (14.41) is called the *principle of conservation of momentum.* Observe that this principle is valid only if the impulse of the resultant force acting on the particle is zero. If there is no resultant force acting on a particle, the resultant impulse will obviously be zero, and momentum will be conserved. However, it is possible for the impulse of a force—that is, its time integral—to be zero even if the force is not zero.

Because momentum is a vector quantity, it is possible for one or two of its components to be conserved, even though the total momentum itself is not conserved. For example, we see from Eq. (14.40) that the momentum in the x-direction will be conserved if $(L_{1-2})_x = 0$, regardless of the values of $(L_{1-2})_y$ and $(L_{1-2})_z$.

The conservation of momentum principle is very useful in the analysis of impact and other interactions between particles.

Sample Problem 14.8

At time $t = 0$, the velocity of the 0.5-kg particle in Fig. (a) is 10 m/s to the right. In addition to its weight (the xy-plane is vertical), the particle is acted on by the force $\mathbf{P}(t)$. The direction of $\mathbf{P}(t)$ is constant throughout the motion, but its magnitude varies with time as shown in Fig. (b). Calculate the velocity of the particle when $t = 4$ s.

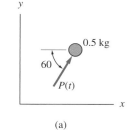

(a)

Solution

The impulse-momentum method is the most direct means of solving this problem because (1) the impulse-momentum method deals directly with the change in momentum (velocity) during a time interval, and (2) the impulses of the forces can be easily computed.

The diagrams required to solve the problem by the impulse-momentum method are shown in Figs. (c)–(e), where the subscripts 1 and 2 correspond to $t = 0$ and 4 s, respectively. The FBD of the particle at an arbitrary time t is shown in Fig. (c). Figure (d) shows the momentum diagram when $t = 0$, where $\mathbf{p}_1 = m\mathbf{v}_1 = 0.5(10)\mathbf{i} = 5\mathbf{i} \ \text{N} \cdot \text{s}$. The momentum diagram when $t = 4$ s is shown in Fig. (e), where both components of $m\mathbf{v}_2$ were drawn in the positive coordinate directions.

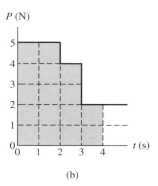

(b)

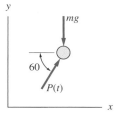

(c) FBD at time t

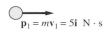

(d) Momentum diagram ($t = 0$)

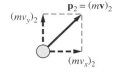

(e) Momentum diagram ($t = 4$ s)

Impulse of weight   Because the weight is a constant force, the components of its impulse are easily computed from Eq. (14.35):

$$(L_{1-2})_x = 0$$
$$(L_{1-2})_y = -mg\,\Delta t = -0.5(9.81)(4) = -19.620 \ \text{N} \cdot \text{s}$$

(a)

Impulse of P(t)   The area under the P-t diagram in Fig. (b) between 0 and 4 s is $5(2) + 4(1) + 2(1) = 16 \ \text{N} \cdot \text{s}$. Because the direction of $\mathbf{P}(t)$ is constant, the components of its impulse are

$$(L_{1-2})_x = 16(\cos 60°) = 8.0 \ \text{N} \cdot \text{s}$$
$$(L_{1-2})_y = 16(\sin 60°) = 13.856 \ \text{N} \cdot \text{s}$$

(b)

Substituting Eqs. (a) and (b) into the impulse-momentum principle, Eq. (14.40), yields for the x-direction

$$(L_{1-2})_x = (mv_x)_2 - (mv_x)_1 \quad \xrightarrow{+} \quad 8.0 = 0.5(v_x)_2 - 5$$

from which we obtain
$$(v_x)_2 = 26.00 \text{ m/s}$$

For the y-direction, we get
$$(L_{1-2})_y = (mv_y)_2 - (mv_y)_1 \quad +\uparrow \quad 13.856 - 19.620 = 0.5(v_y)_2 - 0$$

which gives us
$$(v_y)_2 = -11.53 \text{ m/s}$$

The corresponding velocity vector at time $t = 4$ s is shown below.

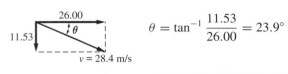

$$\theta = \tan^{-1} \frac{11.53}{26.00} = 23.9° \qquad \textbf{Answer}$$

Sample Problem **14.9**

The straight portion of a ski jump is inclined at 60° to the horizontal. After leaving the starting gate with negligible velocity, a 50-kg skier attains a speed of 25 m/s in 3.5 s. Determine the coefficient of kinetic friction between the skis and the track. Neglect air resistance.

Solution

FBD

From the FBD of the skier in the figure, we get

$$\Sigma F_y = 0 \quad +\nearrow \quad N_A - (50)(9.8)\cos 60° = 0$$

$$N_A = 245 \text{ N}$$

To obtain the friction force F, we apply the impulse-momentum principle in the x-direction. Noting that the weight of the skier and the friction force F are constant, the impulse-momentum principle yields

$$(L_{1-2})_x = m[(v_2)_x - (v_1)_x] \quad \searrow^+ \quad [(50)(9.8)\sin 60° - F]3.5 = 50(25 - 0)$$

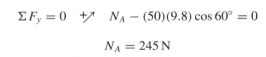

$$F = 67.2 \text{ N}$$

Therefore, the coefficient of kinetic friction is

$$\mu_k = \frac{F}{N} = \frac{67.2}{245} = 0.274 \qquad \textit{Answer}$$

Problems

14.68   The 100-gm ball hits the horizontal surface with the speed $v_0 = 20$ m/s. If 15 percent of the energy is lost during the impact, determine the impulse that acted on the ball during the impact.

Fig. P14.68

14.69   The velocity of a 2-kg particle at time $t = 0$ is $\mathbf{v} = 10\mathbf{i}$ m/s. Determine the velocity at $t = 5$ s if the particle is acted upon by the force $\mathbf{F} = 2t\mathbf{i} - 0.6t^2\mathbf{j}$ N, where t is in seconds.

14.70   A constant horizontal force **P** (not shown) acts on the 0.5-kg body as it slides along a frictionless, horizontal table. During the time interval $t = 5$ s to $t = 7.5$ s, the velocity changes as indicated in the figure. Determine the magnitude and direction of **P**.

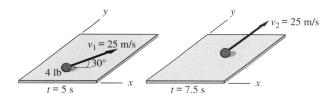

Fig. P14.70

14.71   The 0.2-kg mass moves in the vertical xy-plane. At time $t = 0$, the velocity of the mass is $8\mathbf{j}$ m/s. In addition to its weight, the mass is acted on by the force $\mathbf{F}(t) = F(t)\mathbf{i}$, where the magnitude of the force varies with time as shown in the figure. Determine the velocity vector of the mass at $t = 4$ s.

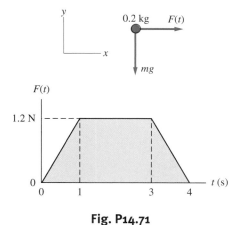

Fig. P14.71

14.72   Find the smallest time in which a 1200-kg automobile traveling at 90 km/h can be stopped on a straight road if the static coefficient of friction between the tires and the road is 0.65. Assume that the road is level.

14.73   Solve Prob. 14.72 if the car is descending a hill that is inclined at 5° with the horizontal.

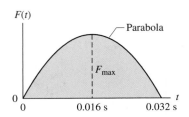

Fig. P14.74

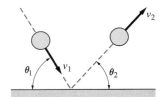

Fig. P14.75, P14.76

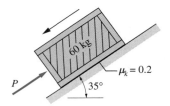

Fig. P14.77

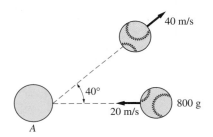

Fig. P14.78

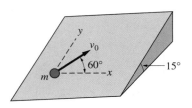

Fig. P14.80

14.74 A 0.09-kg tennis ball traveling at 15 m/s rebounds in the opposite direction at 20 m/s after being hit by a racket. During the 0.032-s period of contact, the magnitude of the force $F(t)$ exerted by the racket has the parabolic time dependence shown. Calculate F_{max}, the maximum value of $F(t)$. Because of the short time of contact, the impulse of the weight of the ball can be neglected.

14.75 A 0.05 kg ball hits a frictionless, rigid, horizontal surface with the speed $v_1 = 30$ m/s at the angle $\theta_1 = 70°$. The angle of rebound is $\theta_2 = 62°$. Compute (a) the speed of the ball immediately after the rebound; and (b) the resultant impulse acting on the ball during its time of contact with the surface.

14.76 A ball with a mass of 0.05 kg hits a frictionless, rigid, horizontal surface with the speed $v_1 = 20$ m/s at the angle $\theta_1 = 45°$. The impulse acting on the ball during its contact with the surface is 1.8 N · s. Find the rebound speed v_2 and the corresponding angle θ_2.

14.77 The 60-kg crate is sliding down the inclined plane. The coefficient of kinetic friction between the crate and the plane is 0.2, and the force P applied to the crate is constant. If the speed of the crate changes from 8 m/s to zero in 3 seconds, determine P.

14.78 The 800-g baseball is thrown horizontally toward a batter at 20 m/s. After being struck by the bat A, the velocity of the ball is 40 m/s in the direction shown. Determine the magnitude of the average force applied to the ball by the bat during the 0.02-s contact period.

14.79 The 100-kg block is at rest on a frictionless surface when the force $P(t)$ is applied. Determine (a) the maximum velocity of the block and the corresponding time; and (b) the velocity of the block at $t = 6$ s.

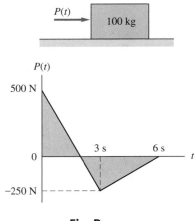

Fig. P14.79

14.80 The particle of mass m is launched on an inclined surface with the speed $v_0 = 2$ m/s in the direction shown. Neglecting friction, find the velocity vector of the particle 0.5 s later.

14.81 The 12-kg box slides down the ramp onto the horizontal surface. The impact with the corner at A reduces the speed of the box from 6 m/s to 4.8 m/s. Determine the impulse vector delivered by the corner to the box.

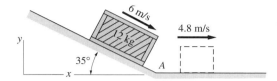

Fig. P14.81

14.82 The 35-kg cleaning "pig" is inserted into a 1-m diameter pipeline and driven forward by air pressure p which varies as $p = 64(1 - e^{-0.5t})$ N/m^2 (t is in seconds). If the "pig" starts from rest at $t = 0$, determine its speed at $t = 10$ s. Neglect friction.

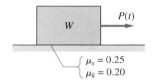

Fig. P14.82

14.83 The block of weight W is at rest on a rough surface at time $t = 0$. During the period from $t = 0$ to 2 s, the block is acted on by the sinusoidally varying force $P(t) = W \sin(\pi t/2)$ where t is in seconds. Using the coefficients of friction shown, find (a) the time when the block starts to move; and (b) the velocity of the block, measured in m/s at $t = 2$ s.

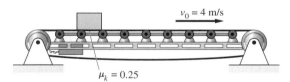

Fig. P14.83

14.84 A parcel is lowered onto a conveyor belt that is moving at 4 m/s. If the coefficient of kinetic friction between the parcel and the belt is 0.25, calculate the time that it takes for the parcel to reach the speed of the belt.

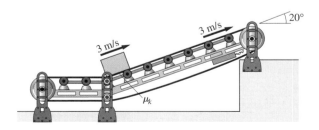

Fig. P14.84

14.85 A parcel starts up the inclined portion of the conveyor with the same velocity as the conveyor belt, namely, 3 m/s. However, the parcel slides on the belt due to insufficient friction, reaching its maximum height in 4 s. Determine the coefficient of kinetic friction between the parcel and the conveyor belt.

Fig. P14.85

Fig. P14.86

14.86 The 500-kg car starts from rest at $t = 0$ on a straight, level road. The driving force F supplied to the car by the rear wheels varies with time as

$F = 289t - 17.4t^2$ N, where t is in seconds. Calculate the speed of the car in mi/h when $t = 10$ s.

14.87 The elevator of weight W starts from rest at time $t = 0$ and reaches its operating speed in 3 s. The acceleration is accomplished by varying the tension F in cable AB (by means of a control circuit) in the manner shown on the diagram. Determine the operating speed of the elevator in m/s.

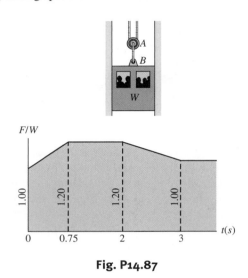

Fig. P14.87

<table>
<tr><td></td></tr>
</table>

14.7   *Principle of Angular Impulse and Momentum*

Angular impulse and angular momentum are the moments of (linear) impulse and (linear) momentum, respectively. For every equation presented in the previous article, there is an analogous equation for angular impulse and angular momentum. The angular impulse-momentum principle serves the same purpose as its linear counterpart: it relates the change in the velocity of a particle during a time interval to the forces acting on the particle.

a.   *Angular impulse of a force*

As shown in Fig. 14.9, we let **r** be the vector drawn from an arbitrary point A to the point of application of the force **F**. The *angular impulse* of **F** about point A during the time interval t_1 to t_2 is defined as

$$(\mathbf{A}_A)_{1-2} = \int_{t_1}^{t_2} \mathbf{r} \times \mathbf{F} \, dt = \int_{t_1}^{t_2} \mathbf{M}_A \, dt \qquad (14.42)$$

where $\mathbf{M}_A = \mathbf{r} \times \mathbf{F}$ is the moment of **F** about A. The angular momentum is a *vector* of dimension $[FLT]$. Therefore, its units are N · m · s, lb · ft · s, and so on.

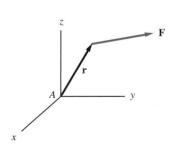

Fig. 14.9

Let A be the origin of a rectangular coordinate system, as indicated in Fig. 14.9. Then we can write $\mathbf{M}_A = M_x\mathbf{i} + M_y\mathbf{j} + M_z\mathbf{k}$, and the rectangular components of Eq. (14.42) become

$$(A_x)_{1-2} = \int_{t_1}^{t_2} M_x\, dt \qquad (A_y)_{1-2} = \int_{t_1}^{t_2} M_y\, dt \qquad (A_z)_{1-2} = \int_{t_1}^{t_2} M_z\, dt$$

$$(14.43)$$

The components of $(\mathbf{A}_A)_{1-2}$ in Eq. (14.43) are also called the angular impulses of $\mathbf{F}$ about the coordinate axes passing through point A.

If the *direction* of $\mathbf{M}_A$ is constant in the time interval t_1 to t_2, then $\mathbf{M}_A$ and $(\mathbf{A}_A)_{1-2}$ have the same direction. In that case, the magnitude of the angular impulse is $(A_A)_{1-2} = \int_{t_1}^{t_2} M_A\, dt$.

If the *direction* and the *magnitude* of $\mathbf{M}_A$ are constant, the angular impulse about A becomes $(\mathbf{A}_A)_{1-2} = \mathbf{M}_A \int_{t_1}^{t_2} dt$, which upon integration yields

$$\boxed{(\mathbf{A}_A)_{1-2} = \mathbf{M}_A(t_2 - t_1) = \mathbf{M}_A\, \Delta t \qquad (\mathbf{M}_A \text{ constant})} \qquad (14.44)$$

b.   *Angular momentum of a particle*

By definition, the *angular momentum* of a particle about a point A is

$$\boxed{\mathbf{h}_A = \mathbf{r} \times (m\mathbf{v})} \qquad (14.45)$$

where $m\mathbf{v}$ is the momentum of the particle, and $\mathbf{r}$ denotes its position vector measured from A, as shown in Fig. 14.10 (a). The angular momentum is also known as the *moment of momentum,* because Eq. (14.45) is analogous to the definition of the moment of a force: $\mathbf{M}_A = \mathbf{r} \times \mathbf{F}$. The dimension of angular momentum is $[ML^2/T]$, or equivalently, $[FLT]$, which is the same as the dimension of the angular impulse.

From the properties of the cross product, we deduce that angular momentum is a *vector* of magnitude

$$h_A = mvd \qquad (14.46)$$

where d is the *moment arm* from A to the momentum vector $m\mathbf{v}$, as shown in Fig. 14.10(a). The direction of the angular momentum vector is perpendicular to the plane shared by A and $m\mathbf{v}$. A two-dimensional view of angular momentum is shown in Fig. 14.10(b). Note that the sense of h_A (CW or CCW) is determined by the right-hand rule.

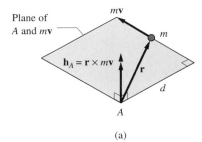

(a)

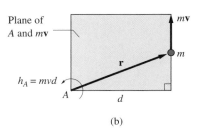

(b)

Fig. 14.10

Using the rectangular components $\mathbf{r} = x\mathbf{i} + y\mathbf{j} + z\mathbf{k}$ and $\mathbf{v} = v_x\mathbf{i} + v_y\mathbf{j} + v_z\mathbf{k}$, Eq. (14.45) becomes

$$\mathbf{h}_A = \begin{vmatrix} \mathbf{i} & \mathbf{j} & \mathbf{k} \\ x & y & z \\ mv_x & mv_y & mv_z \end{vmatrix} \tag{14.47}$$

Expanding the determinant, we get

$$\mathbf{h}_A = h_x\mathbf{i} + h_y\mathbf{j} + h_z\mathbf{k} \tag{14.48a}$$

where

$$h_x = m(yv_z - zv_y) \qquad h_y = m(zv_x - xv_z) \qquad h_z = m(xv_y - yv_x) \quad \text{(14.48b)}$$

are the angular momenta of the particle about the coordinate axes passing through A.

c.   *Moment-angular momentum relationship*

Differentiating the angular momentum of the particle in Eq. (14.45) with respect to time, we obtain

$$\dot{\mathbf{h}}_A = \frac{d}{dt}[\mathbf{r} \times (m\mathbf{v})] = \mathbf{r} \times \frac{d(m\mathbf{v})}{dt} + \frac{d\mathbf{r}}{dt} \times (m\mathbf{v})$$

According to Eq. (14.37), $d(m\mathbf{v})/dt = \Sigma\mathbf{F}$, where $\Sigma\mathbf{F}$ is the resultant force acting on the particle. Therefore,

$$\dot{\mathbf{h}}_A = \mathbf{r} \times \Sigma\mathbf{F} + \dot{\mathbf{r}} \times (m\mathbf{v}) = \mathbf{M}_A + \dot{\mathbf{r}} \times (m\mathbf{v})$$

where $\mathbf{M}_A = \mathbf{r} \times \Sigma\mathbf{F}$ is the moment of the resultant force about A. Solving for the moment, we obtain

$$\mathbf{M}_A = \dot{\mathbf{h}}_A - \dot{\mathbf{r}} \times (m\mathbf{v}) \tag{14.49}$$

Recall that $\mathbf{r}$ is the position vector of the particle measured from point A. So far, we have placed no restrictions on the choice of A; it may be either a fixed or a moving point. If A is fixed with respect to an inertial reference frame, then $\dot{\mathbf{r}}$ is the velocity $\mathbf{v}$ of the particle, and the last term of Eq. (14.49) becomes $\dot{\mathbf{r}} \times (m\mathbf{v}) = \mathbf{v} \times (m\mathbf{v}) = \mathbf{0}$ (the cross product of two parallel vectors is zero). Consequently,

$$\mathbf{M}_A = \dot{\mathbf{h}}_A \qquad (A: \text{ fixed point}) \tag{14.50}$$

When A is not fixed, then Eq. (14.50) is generally not valid, and Eq. (14.49) must be used. This case is developed further in the next chapter.

d.   *Angular impulse-momentum principle*

Integration of Eq. (14.50) over the time interval t_1 to t_2 yields

$$\int_{t_1}^{t_2} \mathbf{M}_A \, dt = \int_{t_1}^{t_2} d\mathbf{h}_A = (\mathbf{h}_A)_2 - (\mathbf{h}_A)_1$$

where $(\mathbf{h}_A)_1 = \mathbf{h}_A(t_1)$ and $(\mathbf{h}_A)_2 = \mathbf{h}_A(t_2)$. Because the left-hand-side $\int_{t_1}^{t_2} \mathbf{M}_A \, dt = (\mathbf{A}_A)_{1-2}$ is the angular impulse of the resultant force about A, we can write

$$(\mathbf{A}_A)_{1-2} = (\mathbf{h}_A)_2 - (\mathbf{h}_A)_1 = \Delta \mathbf{h}_A \qquad (A:\ \text{fixed point}) \qquad (14.51\text{a})$$

which is known as the *principle of angular impulse and angular momentum.* The rectangular components of Eq. (14.51a) are

$$(A_x)_{1-2} = (h_x)_2 - (h_x)_1$$

$$(A_y)_{1-2} = (h_y)_2 - (h_y)_1 \qquad (14.51\text{b})$$

$$(A_z)_{1-2} = (h_z)_2 - (h_z)_1$$

You can reduce the chances of mistakes in applying this principle by using a free-body diagram to compute the angular impulse of the resultant force acting on the particle.

e.   *Conservation of angular momentum*

If the angular impulse about A is zero, it follows from Eq. (14.51a) that the angular momentum of the particle about A is conserved. In other words,

$$\text{if}\quad (\mathbf{A}_A)_{1-2} = 0, \quad \text{then}\quad (\mathbf{h}_A)_1 = (\mathbf{h}_A)_2 \qquad (A:\ \text{fixed point}) \qquad (14.52)$$

which is known as the *principle of conservation of angular momentum.* If only one component of the angular impulse vanishes, then only the corresponding component of the angular momentum is conserved. For example, if $(A_x)_{1-2} = 0$ in Eq. (14.51b), then $(h_x)_1 = (h_x)_2$, even if h_y and h_z are not conserved.

Sample Problem 14.10

The circular table in Fig. (a) is being driven at a constant angular speed $\dot{\theta} = 20$ rad/s about the vertical z-axis. The block B of mass m is placed on the rotating table with zero initial velocity with the string AB taut. If the block slips for 3.11 s before reaching the speed of the table, determine the coefficient of kinetic friction between the block and the table.

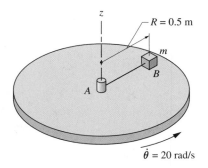

$R = 0.5$ m

m

B

A

$\dot{\theta} = 20$ rad/s

(a)

Solution

The FBD of block B is shown in Fig. (b). The tension T in the string passes through the z-axis, whereas the weight W and the normal contact force N are parallel to the axis. Hence the moment of each of these forces about the z-axis is zero. The moment of the kinetic friction force F about the z-axis is $M_z = (\mu_k N)R$. Substituting $N = mg$, obtainable from the equilibrium equation $\Sigma F_z = 0$, we get

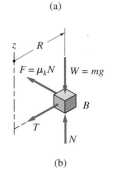

z R

$F = \mu_k N$ $W = mg$

B

T

N

(b)

$$M_z = \mu_k mg R$$

If we let $t_1 = 0$ be the time when the block is placed on the table, then $t_2 = 3.11$ s is the time when slipping stops. Because M_z is constant, the angular impulse of the friction force about the z-axis during the period of slipping is

$$+\curvearrowleft \quad (A_z)_{1-2} = M_z(t_2 - t_1) = \mu_k mg R t_2$$

The corresponding change in the angular momentum of the block about the z-axis is

$$+\curvearrowleft \quad (h_z)_2 - (h_z)_1 = mvR - 0 = mR^2\dot{\theta}$$

where we substituted $v = R\dot{\theta}$. Applying the angular impulse-momentum principle about the z-axis, we get

$$(A_z)_{1-2} = (h_z)_2 - (h_z)_1 \qquad \mu_k mg R t_2 = mR^2\dot{\theta}$$

which yields for the coefficient of kinetic friction

$$\mu_k = \frac{R\dot{\theta}}{gt_2} = \frac{(0.5)(20)}{(9.8)(3.11)} = 0.328 \qquad\qquad Answer$$

Sample Problem 14.11

The particle of mass $m = 0.3$ kg shown in Fig. (a) moves on a frictionless horizontal plane. One end of the linear spring is attached to the particle, and the other end is attached to the fixed point O. If the particle is launched from position A with the velocity $\mathbf{v}_1$ as shown, determine the spring stiffness k if the maximum distance between the path of the particle and point O is 400 mm. The spring is undeformed when the particle is at A.

Solution

The diagram in Fig. (b) shows the momentum vectors $m\mathbf{v}_1$ and $m\mathbf{v}_2$ when the particle is at points A and B, respectively. Point B is the position of the particle when the length of the spring equals its maximum value $L_2 = 400$ mm. Note that the direction of $\mathbf{v}_2$ is tangent to the path of the particle; in other words, the velocity vector is perpendicular to the spring.

Because the spring force is always directed toward O, angular momentum about O is conserved. Referring to the momentum vectors in Fig. (b), and recalling that the angular momentum equals the moment of the linear momentum ($h_O = mvd$), we obtain

$$(h_O)_1 = (h_O)_2 \quad \circlearrowleft^+ \quad (mv_1 \cos 60°)L_1 = mv_2 L_2$$

which yields

$$v_2 = \frac{v_1 \cos 60° L_1}{L_2} = \frac{2 \cos 60° (200)}{400} = 0.500 \text{ m/s}$$

Because the spring force is conservative, mechanical energy of the particle is conserved:

$$T_1 + V_1 = T_2 + V_2$$

$$\frac{1}{2}mv_1^2 + \frac{1}{2}k\delta_1^2 = \frac{1}{2}mv_2^2 + \frac{1}{2}k\delta_2^2$$

where δ_1 and δ_2 are the elongations of the spring when the particle is at A and B, respectively. Substituting $\delta_1 = 0$, $\delta_2 = L_2 - L_1 = 200$ mm $= 0.200$ m, and the values for m, v_1, and v_2, we obtain

$$\frac{1}{2}(0.3)(2)^2 + 0 = \frac{1}{2}(0.3)(0.500)^2 + \frac{1}{2}k(0.200)^2$$

from which the spring stiffness is found to be

$$k = 28.1 \text{ N/m} \qquad\qquad \textit{Answer}$$

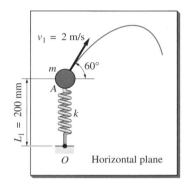

(a)

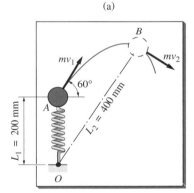

(b) Momentum diagram

Problems

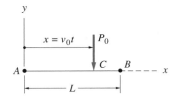

Fig. P14.88

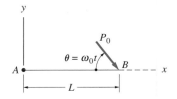

Fig. P14.89

14.88   The force P_0 has constant magnitude and direction, but its point of application C moves along the x-axis with the constant speed v_0. Determine the angular impulse of the force about point A for the time period during which C moves from A to B.

14.89   The force shown has the constant magnitude P_0 and the fixed point of application B, but its line of action rotates with the constant angular speed ω_0. Determine the angular impulse of the force about A for the time during which the force rotates from $\theta = 0$ to $\theta = 90°$.

14.90   The velocity of the 500-g particle at B is $\mathbf{v} = 2\mathbf{i} + 4\mathbf{j} + 6\mathbf{k}$ m/s. Calculate the angular momentum of the particle about point A at this instant.

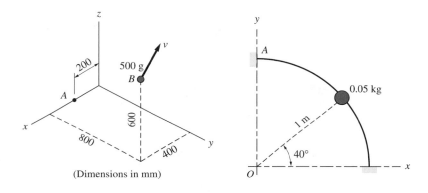

Fig. P14.90   **Fig. P14.91**

14.91   The 0.05-kg bead is sliding down the circular wire. When the bead is in the position shown, its speed is 5 m/s. For this instant, determine the angular momentum of the bead about (a) point O; and (b) point A.

14.92   The particle of mass m travels along a circular path of radius R. When the particle is in the position shown, its speed is v. What is the angular momentum of the particle about the origin O in this position?

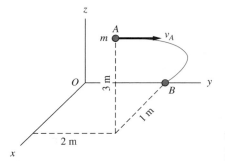

Fig. P14.93

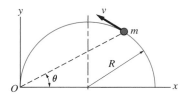

Fig. P14.92

14.93   The particle of mass m moves so that its angular momentum about point O is always conserved. When the particle is at A, its velocity is $v_A = 5$ m/s in the y-direction. At B, the speed has increased to $v_B = 15$ m/s. Determine the velocity vector of the particle at B.

14.94   The particle of mass m is restrained by a string to travel around a circular path on the horizontal table. The coefficient of kinetic friction between the particle and the table is 0.15. If the initial speed of the particle is $v_1 = 8$ m/s, determine the time that elapses before the particle stops.

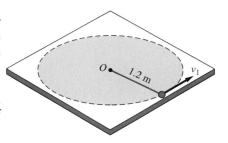

Fig. P14.94

14.95   A rope wound around the rim of the wheel is pulled by force F, which varies with time as shown. Calculate the angular impulse of F about the center of the wheel during the time interval $t = 0$ to $t = 0.5$ s.

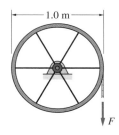
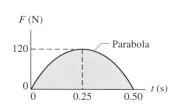

Fig. P14.95

14.96   The assembly consists of a 0.8 kg collar attached to a bent rod of negligible weight. The assembly rotates freely about the vertical axis with the angular velocity $\dot{\theta} = 10$ rad/s when the braking couple M is applied at time $t = 0$. If M varies with time as shown, how long will the assembly continue to rotate?

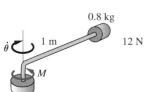
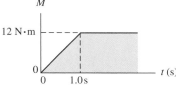

Fig. P14.96

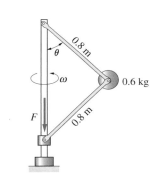

Fig. P14.97

14.97   The 0.6-kg mass is supported by two arms of negligible mass. The angle θ of the arms can be varied by changing the force F acting on the sliding collar. When $\theta = 70°$, the assembly is rotating freely about the vertical axis with the angular velocity $\omega = 15$ rad/s. Determine the angular velocity after θ is reduced to $30°$.

14.98   The path of the earth satellite is the ellipse

$$\frac{1}{R} = (119.3 \times 10^{-9})(1 + 0.161 \cos \theta)$$

where R is in meters. If the speed of the satellite at A is 6000 m/s, determine its speed at B.

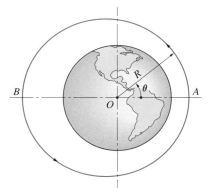

Fig. P14.98

14.99 The particle, connected by a spring to the fixed point O, slides on the frictionless, horizontal table. The particle is launched at A with the velocity v_A in the y-direction. If the velocity of the particle at B is $\mathbf{v}_B = 3.66\mathbf{i} - 5.72\mathbf{j}$ m/s, determine v_A.

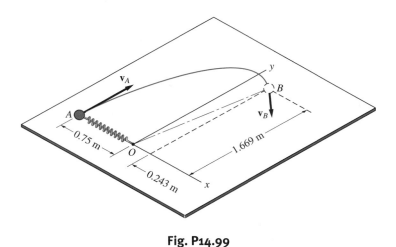

Fig. P14.99

14.100 A motor rotates the rod OB about the z-axis at the constant angular speed of 8 rad/s. The string attached to the 2-kg slider A is let out at the constant rate of 0.25 m/s. Determine the magnitude of the contact force between the slider and the rod when $R = 0.8$ m. Neglect friction and the mass of rod OB.

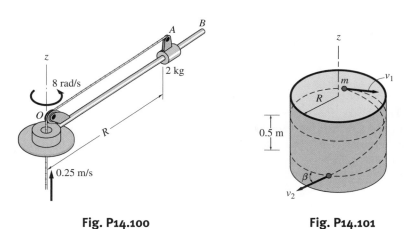

Fig. P14.100 **Fig. P14.101**

14.101 The particle of mass m is launched on the inside wall of a cylindrical vessel with the speed $v_1 = 6$ m/s directed horizontally. When the particle reaches a position 0.5 m below the launch position, determine (a) the speed v_2; and (b) the angle β between $\mathbf{v}_2$ and the horizontal. Neglect friction.

14.102   An earth satellite has an elliptical orbit with the property $R_1/R_2 = 0.25$. If the speed of the satellite at the perigee (point closest to the earth) is $v_1 = 8$ km/s, determine (a) the speed v_2 at the apogee (point farthest from the earth); and (b) the distance R_1. Use $GM_e = 3.98 \times 10^{14}$ m^3/s^2, where G is the universal gravitational constant and M_e is the mass of the earth.

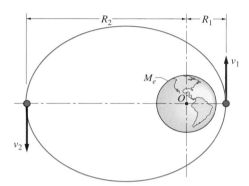

Fig. P14.102

14.103   The 0.5-kg disk slides on a frictionless, horizontal surface. The elastic cord connecting the disk to the fixed point O has a stiffness of 100 N/m and a free length of 0.75 m. The disk is given the initial velocity v_1 in the direction $\beta = 70°$ in the position shown. Determine the smallest v_1 for which the cord will always remain taut. (Hint: $\mathbf{v}$ is perpendicular to the cord when $R = R_{min}$.)

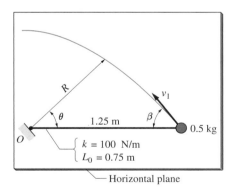

Fig. P14.103

14.104   The 2-N weight A is attached to a string that passes through the hole O in the horizontal, frictionless table. As the weight rotates about O, the end of the string is pulled downward at the constant speed of 0.05 m/s. If the angular speed $\dot{\theta}$ of the string is 8 rad/s when $R = 0.6$ m, determine the following when $R = 0.3$ m: (a) the angular speed of the string; and (b) the angular acceleration of the string.

14.105   The 0.5-kg bob of the spherical pendulum is launched in position 1 with the velocity v_1 in the direction that is horizontal and perpendicular to the string.

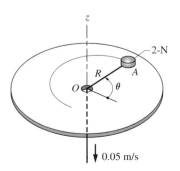

Fig. P14.104

If the lowest point reached by the bob is position 2, determine v_1. (Hint: v_2 is also horizontal and perpendicular to the string.)

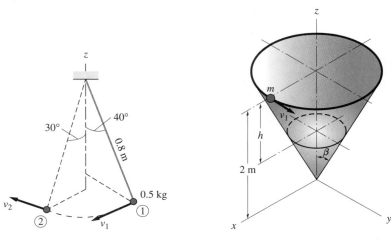

Fig. P14.105   **Fig. P14.106**

14.106 The particle of mass m slides inside a frictionless conical vessel. The initial velocity of the particle is $v_1 = 3$ m/s, tangent to the rim of the vessel. When the particle is a distance $h = 0.5$ m below the rim of the vessel, determine (a) the speed of the particle; and (b) the angle between the velocity vector and the horizontal.

*14.8 Space Motion under a Gravitational Force

Space motion under the action of a gravitational force comes under the category of central-force motion. The equations governing central-force motion can be readily derived from the principles of conservation of angular momentum and conservation of mechanical energy.

a. Central-force motion

Consider a moving particle that is acted on only by a force $\mathbf{F}$ that is always directed toward a fixed point A. For this case, called *central-force motion,* we have

$$\mathbf{M}_A = \mathbf{r} \times \mathbf{F} = \dot{\mathbf{h}}_A = \mathbf{0}$$

where $\mathbf{r}$ is the position vector of the particle, drawn from A. From Eq. (14.45), we have

$$\mathbf{h}_A = \mathbf{r} \times (m\mathbf{v}) = \text{constant} \tag{14.53}$$

Note that this equation can be valid only if $\mathbf{r}$ and $\mathbf{v}$ always lie in the same plane. Therefore, central-force motion is plane motion with constant angular momentum about point A.

b. *Motion under gravitational attraction*

The remainder of this article analyzes the motion of bodies (e.g., satellites) that move under the gravitational attraction of a planet (or the sun). We confine our attention to trajectories for which the gravitational attraction of the planet is the only force that needs to be considered. Letting m be the mass of the body and M the mass of the planet, we assume that $m \ll M$, which means that the planet may be considered to be fixed in the analysis.

Because the only force acting on the body is its weight, which is always directed toward the center of the planet, the body undergoes central-force motion, as described in Eq. (14.53). It is convenient to describe this motion, which is confined to a plane, in terms of polar coordinates R and θ, as shown in Fig. 14.11. The nonrotating xy reference frame has its origin at F, the center of the planet, also called the *focus* of the trajectory.

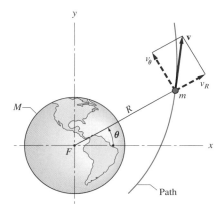

Fig. 14.11

From Eq. (14.53) we know that the angular momentum of the body about F is conserved, which means that $R(mv_\theta) = R(mR\dot{\theta})$ is constant. Letting h_0 be the angular momentum about F *per unit mass* of the body,[*] we have

$$h_0 = Rv_\theta = R^2\dot{\theta} \quad \text{(a constant)} \qquad (14.54)$$

The kinetic energy of the body is $T = \frac{1}{2}mv^2$, and its potential energy, according to Eq. (14.29), is $V_g = -GMm/R$ (where G is the universal gravitational constant). Because the gravitational attraction is a conservative force, the total energy of the body is conserved. Letting E_0 be the total energy per unit mass—that is, $E_0 = (T + V_g)/m$—we have

$$E_0 = \frac{1}{2}v^2 - \frac{GM}{R} \quad \text{(a constant)} \qquad (14.55)$$

Note that the total energy per unit mass may be positive, negative, or zero.

---

[*]The subscript 0 (the number zero) on h, and later on E, is used to denote constants that are associated with unit mass. It must not be confused with the letter O.

c.   Equation of the trajectory

We next determine the equation of the path (trajectory) of the body in the form $R = R(\theta)$. We begin by substituting $v^2 = v_R^2 + v_\theta^2$ into Eq. (14.55) to obtain

$$E_0 = \frac{1}{2}\left(v_R^2 + v_\theta^2\right) - \frac{GM}{R} \tag{a}$$

With the help of Eq. (14.54), v_R and R can be eliminated from Eq. (a). Applying the chain rule for differentiation to Eq. (14.54), we get

$$\dot{R}v_\theta + R\dot{v}_\theta = 0 \tag{b}$$

Substituting

$$\dot{R} = v_R \quad \text{and} \quad \dot{v}_\theta = \frac{dv_\theta}{d\theta}\dot{\theta} = \frac{dv_\theta}{d\theta}\frac{v_\theta}{R}$$

Eq. (b) becomes

$$v_R v_\theta = -R\left(\frac{dv_\theta}{d\theta}\frac{v_\theta}{R}\right)$$

which yields

$$v_R = -\frac{dv_\theta}{d\theta} \tag{c}$$

Substituting $R = h_0/v_\theta$ [see Eq. (14.54)] and Eq. (c) into Eq. (a), we find

$$E_0 = \frac{1}{2}\left[\left(\frac{dv_\theta}{d\theta}\right)^2 + v_\theta^2\right] - \frac{GMv_\theta}{h_0} \tag{d}$$

The solution of Eq. (d) for $d\theta$ is

$$d\theta = \pm\frac{dv_\theta}{\left(2E_0 + \dfrac{2GMv_\theta}{h_0} - v_\theta^2\right)^{1/2}} \tag{e}$$

Noting that E_0 is constant, Eq. (e) can be integrated (see a table of integrals) to yield

$$\theta = \pm\sin^{-1}\left(\frac{v_\theta - \dfrac{GM}{h_0}}{\dfrac{GM}{h_0}\left[1 + 2E_0\left(\dfrac{h_0}{GM}\right)^2\right]^{1/2}}\right) - \alpha_0 \tag{f}$$

where α_0 is a constant of integration. The dual sign means that for every point on the path associated with v_θ and θ, there is a second point associated with v_θ and $(\theta + \pi)$. Therefore, the minus sign will be omitted from here on without loss of generality.

Inverting Eq. (f), we obtain

$$\sin(\theta + \alpha_0) = \frac{v_\theta - \dfrac{GM}{h_0}}{\dfrac{GM}{h_0}e} \tag{14.56}$$

where

$$e = \sqrt{1 + 2E_0\left(\frac{h_0}{GM}\right)^2} \qquad (14.57)$$

is called the *eccentricity* of the trajectory. It can be proven that e is always a real number; that is, the term under the radical in Eq. (14.57) cannot be negative.

Solving Eq. (14.56) for v_θ, we obtain

$$v_\theta = \frac{GM}{h_0}[1 + e\sin(\theta + \alpha_0)] \qquad (14.58)$$

Because $R = h_0/v_\theta$ according to Eq. (14.54), we find that

$$R = \frac{h_0^2}{GM[1 + e\sin(\theta + \alpha_0)]} \qquad (14.59)$$

We see from Eq. (14.59) that the smallest value of R occurs when $\sin(\theta + \alpha_0) = 1$; that is, when $\theta + \alpha_0 = \pi/2$. Letting this position correspond to $\theta = 0$ (the x-axis), as shown in Fig. 14.12, the constant of integration is $\alpha_0 = \pi/2$. Because $\sin(\theta + \alpha_0) = \sin[\theta + (\pi/2)] = \cos\theta$, Eqs. (14.58) and (14.59) become

$$v_\theta = \frac{GM}{h_0}(1 + e\cos\theta) \qquad (14.60)$$

and

$$R = \frac{h_0^2}{GM(1 + e\cos\theta)} \qquad (14.61)$$

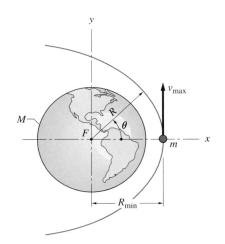

Fig. 14.12

Equation (14.61), which is the equation of the trajectory, represents a conic section (circle, ellipse, parabola, or hyperbola) in polar coordinates, where e is the eccentricity of the curve and F (the center of the planet) is the focus.

Note that Eqs. (14.60) and (14.61) are valid only if the x-axis is chosen so that R is minimized when $\theta = 0$. If this is not the case, Eqs. (14.58) and (14.59) must be used.

From Eq. (14.61), we see that the minimum value of R is

$$R_{\min} = R|_{\theta=0} = \frac{h_0^2}{GM(1 + e)} \qquad (14.62)$$

It follows that $\dot{R} = v_R = 0$ when $\theta = 0$, which means that $v = v_\theta$ at this position. Furthermore, inspection of Eq. (14.55) reveals that v is largest when R is smallest, from which we conclude that the maximum velocity is

$$v_{\max} = v_\theta|_{\theta=0} = \frac{GM(1 + e)}{h_0} \qquad (14.63)$$

Universal gravitational constant: $G = 6.673 \times 10^{-11}\,\mathrm{m}^3 \cdot \mathrm{kg}^{-1} \cdot \mathrm{s}^{-2}$		
Body	**Mean equatorial radius** **km**	**Mass** **kg**
Sun	696 000	1.9884×10^{30}
Moon	1 737	$0.073\,483 \times 10^{24}$
Mercury	2 440	$0.330\,22 \times 10^{24}$
Venus	6 052	$4.869\,0 \times 10^{24}$
Earth	6 378.14	$5.974\,2 \times 10^{24}$
Mars	3 396	$0.641\,91 \times 10^{24}$

Table 14.2 Selected Solar System Constants

directed as shown in Fig. 14.12. Substituting $h_0 = R_{\min}v_{\max}$ into Eq. (14.63) and solving for $v_{\max}$ yields

$$v_{\max} = \sqrt{\frac{GM(1+e)}{R_{\min}}} \qquad (14.64)$$

Numerical data for selected bodies of our solar system are presented in Table 14.2. These data are to be used when solving the problems at the end of this article.

d.   Classification of trajectories

It has been pointed out that Eq. (14.61) represents a conic section, which means that the trajectory must be one of the following curves, depending on the value of e.

Case I: $e = 0$ circle
Case II: $0 < e < 1$ ellipse
Case III: $e = 1$ parabola
Case IV: $e > 1$ hyperbola

If the trajectory is circular or elliptical, the body is said to be captured by the planet. The body is then known as a satellite, and its trajectory is called an orbit. For the other two cases, the gravitational pull of the planet is not strong enough to capture the body.

Case	e	Path	E_0	Velocity	Distance from focus	Drawing of path
I	$e = 0$	Circle	$E_0 < 0$	$v_{\text{circ}} = \sqrt{\dfrac{GM}{R_{\text{circ}}}}$	$R_{\text{circ}} = \dfrac{h_0^2}{GM}$	
II	$0 < e < 1$	Ellipse	$E_0 < 0$	$v_{\text{max}} = \sqrt{\dfrac{GM(1+e)}{R_{\text{min}}}}$ $v_{\text{min}} = \sqrt{\dfrac{GM(1-e)}{R_{\text{max}}}}$ $= \sqrt{\dfrac{(1-e)^2}{1+e} \cdot \dfrac{GM}{R_{\text{min}}}}$	$R_{\text{min}} = \dfrac{h_0^2}{GM(1+e)}$ $R_{\text{max}} = \dfrac{h_0^2}{GM(1-e)}$	
III	$e = 1$	Parabola	$E_0 = 0$	$v_{\text{max}} = v_{\text{esc}}$ $= \sqrt{\dfrac{2GM}{R_{\text{min}}}}$	$R_{\text{min}} = \dfrac{h_0^2}{2GM}$ $R_{\text{max}} = \infty \text{ at } \theta = \pi$	
IV	$e > 1$	Hyperbola	$E_0 > 0$	$v_{\text{max}} = \sqrt{\dfrac{GM(1+e)}{R_{\text{min}}}}$	$R_{\text{min}} = \dfrac{h_0^2}{GM(1+e)}$ $R_{\text{max}} = \infty$ $\text{at } \cos\theta_1 = -\dfrac{1}{e}$	

The velocity v_{esc} is the *escape velocity* (the minimum velocity required to escape the planet's gravitational field).

Table 14.3 Classification of Trajectories

The properties of the four types of trajectories are summarized in Table 14.3.

e.   *Properties of elliptical orbits*

An elliptical orbit that is centered at point O is shown in Fig. 14.13. Note that *perigee* and *apogee* are the names given to the locations of R_{max} and R_{min}, respectively.[*] It can be shown that the geometric interpretation of the eccentricity is

$$e = \frac{R_{\text{max}} - R_{\text{min}}}{R_{\text{max}} + R_{\text{min}}} \tag{14.65}$$

The length of the major semiaxis in Fig. 14.13 is given by

$$a = \frac{R_{\text{max}} + R_{\text{min}}}{2} = \frac{h_0^2}{GM(1 - e^2)} \tag{14.66}$$

---

[*]The terms *perigee* and *apogee* are used only for orbital motion with the earth as the focus. When the attracting body is not the earth, the corresponding terms are *periapsis* and *apoapsis,* respectively.

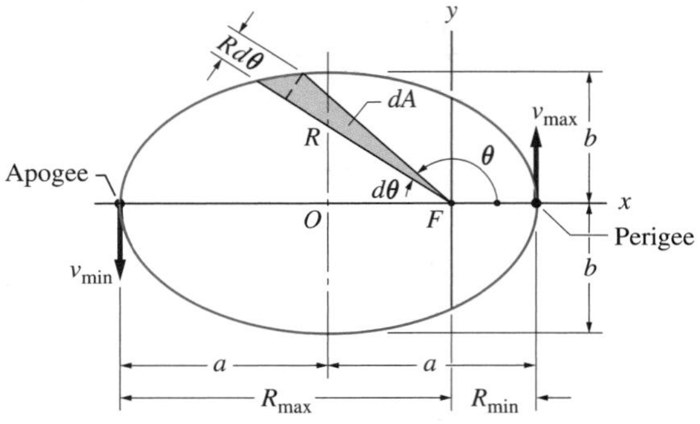

Fig. 14.13

From analytic geometry, the length of the minor semiaxis may be written as

$$b = \sqrt{R_{max}R_{min}} = \frac{h_0^2}{GM\sqrt{1 - e^2}} \qquad (14.67)$$

Using Eqs. (14.66) and (14.67), the area A of the ellipse can be written as

$$A = \pi ab = \frac{\pi}{2}(R_{max} + R_{min})\sqrt{R_{max}R_{min}} = \frac{\pi h_0^4}{(GM)^2(1 - e^2)^{3/2}} \qquad (14.68)$$

The *period* τ of an elliptical (or circular) orbit is the time required to complete one revolution around the path. The period may be related to the area of the ellipse by noting that the differential area shown in Fig. 14.13 may be expressed as $dA = (1/2)R(R d\theta)$. The rate at which the area is swept over by the line between the focus and the satellite, namely dA/dt, is called the *areal velocity*. Using Eq. (14.54), we have

$$\frac{dA}{dt} = \frac{1}{2}R^2\dot{\theta} = \frac{h_0}{2} \qquad (14.69)$$

Note that the areal velocity is constant. This is one of the celebrated laws published by Johannes Kepler (1571–1630), based on astronomical observations. Integrating with respect to time from $t = 0$ to $t = \tau$ gives $A = h_0\tau/2$. Therefore, using Eq. (14.68), the period is

$$\tau = \frac{2A}{h_0} = \frac{2\pi h_0^3}{(GM)^2(1 - e^2)^{3/2}} \qquad (14.70)$$

A special case of an elliptical path is the ballistic trajectory, where the ellipse intersects the surface of the planet. Such a trajectory is followed by all projectiles if air resistance is neglected. If the elevation of the projectile is small enough so that the variation of gravitational force with height may be neglected, then the trajectory assumes the familiar parabolic form. Hence, the parabolic trajectory is an approximation of the true elliptical path, valid for small changes of elevation.

Sample Problem *14.12*

The eccentricity of the earth's orbit around the sun is 0.017, and the period of the orbit is 365.26 days. Calculate the maximum and minimum values of (1) the earth's distance from the center of the sun; and (2) the velocity of the earth around the sun.

Solution

Preliminary Calculations

Using Table 14.1, the constant GM_s, where M_s is the mass of the sun, is found to be

$$GM_s = (6.673 \times 10^{-11})(1.9884 \times 10^{30}) = 1.3269 \times 10^{20} \text{ m}^3/\text{s}^2$$

The given period of the orbit is

$$\tau = 365.26 \text{ days} \times \frac{24 \text{ h}}{1 \text{ day}} \times \frac{3600 \text{ s}}{1 \text{ h}} = 31.56 \times 10^6 \text{ s}$$

Solving Eq. (14.70) for h_0 (the constant angular momentum per unit mass of the earth about the center of the sun) yields

$$h_0 = \left[\frac{(GM_s)^2 \tau (1 - e^2)^{3/2}}{2\pi} \right]^{1/3}$$

$$= \left[\frac{(1.3269 \times 10^{20})^2 (31.56 \times 10^6)[1 - (0.017)^2]^{3/2}}{2\pi} \right]^{1/3}$$

$$= 4.455 \times 10^{15} \text{ m}^2/\text{s}$$

Part 1

Using the formulas for Case II in Table 14.3 and the constants determined above, the maximum and minimum distances of the earth from the center of the sun are

$$\left. \begin{matrix} R_{min} \\ R_{max} \end{matrix} \right\} = \frac{h_0^2}{GM_s(1 \pm e)} = \frac{(4.455 \times 10^{15})^2}{(1.3269 \times 10^{20})(1 \pm 0.017)}$$

from which we find that

$$R_{max} = 1.522 \times 10^{11} \text{ m} \qquad R_{min} = 1.471 \times 10^{11} \text{ m} \qquad \textit{Answer}$$

Part 2

Referring to Case II of Table 14.3 again, the maximum velocity of the earth is given by

$$v_{max} = \sqrt{\frac{GM_s(1 + e)}{R_{min}}} = \sqrt{\frac{(1.3269 \times 10^{20})(1 + 0.017)}{1.471 \times 10^{11}}} \qquad \textit{Answer}$$

$$= 30.29 \times 10^3 \text{ m/s} = 30.29 \text{ km/s}$$

The minimum velocity of the earth, again from the formulas in Table 14.1, is

$$v_{min} = \sqrt{\frac{GM_s(1-e)}{R_{max}}} = \sqrt{\frac{(1.3269 \times 10^{20})(1-0.017)}{1.522 \times 10^{11}}} \qquad \textit{Answer}$$

$$= 29.28 \times 10^3 \text{ m/s} = 29.28 \text{ km/s}$$

Note that the mean velocity of the earth is $(1/2)(v_{max} + v_{min}) = (1/2) \times (30.29 + 29.28) = 29.79$ km/s, which agrees within four significant digits with the commonly accepted value.

An alternative method for computing v_{max} and v_{min} is to use the fact that h_0 is constant, that is, to solve the equations $h_0 = v_{max}R_{min} = v_{min}R_{max}$.

Problems

14.107   A space probe is launched at A with the velocity v_0, directed as shown. For what value of v_0 will the probe pass through B? What is the classification of the trajectory?

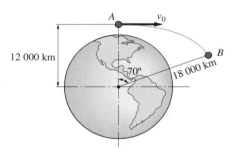

Fig. P14.107

14.108   The orbit of Ceres, a dwarf planet between Mars and Jupiter, has an eccentricity of 0.08 and a major semi-axis of 413.7×10^9 m. Determine the period of the orbit.

14.109   Assuming that the orbit of the moon around the earth is a circle (its eccentricity is actually 0.055), and knowing that the period of the orbit is 27.3 days, compute the distance between the centers of the earth and moon.

14.110   Calculate the maximum and minimum distances between the centers of the earth and moon, taking the eccentricity of the moon's orbit into account. Use the data given in Prob. 14.109.

14.111   The orbit of Phobos, a Martian moon, has an eccentricity of 0.018 and a major semi-axis of length 9380 km. Determine the orbital period of Phobos.

14.112   A 14-Mg spacecraft is in orbit around the moon. The spacecraft's maximum and minimum altitudes above the lunar surface are 340 and 140 km, respectively. Neglecting the gravitational effect of the earth, determine the minimum energy required for the spacecraft to escape lunar gravity in order to return to the earth.

14.113   A 6000-kg space capsule is in a circular orbit of radius 6688 km around the earth. When the capsule is at *A*, its speed is reduced by the firing of retro-rockets. Determine the required impulse of the retro-rockets if the capsule is to land at *B* with its flight path tangent to the earth's surface.

14.114   As the spacecraft approaches the planet Venus, its speed is $v_1 = 3000$ m/s when $x_1 = 2 \times 10^8$ m and $y_1 = 0.5 \times 10^8$ m. Determine (a) the type of trajectory that the spacecraft will travel; (b) the minimum distance between the surface of Venus and the trajectory; and (c) the maximum speed of the craft.

14.115   For the spacecraft described in Prob. 14.114, determine (a) the largest speed v_1 that would produce an elliptical orbit, assuming that an elliptical orbit were possible; and (b) whether an elliptical orbit is possible without the craft hitting the surface of Venus.

14.116   An earth satellite is inserted into its orbit at *A* with the speed $v_1 = 9200$ m/s, in the direction $\beta = 5°$. (a) Show that the trajectory is an ellipse. (b) Calculate the angle from the line *FA* to the major axis of the orbit. (c) Calculate the smallest distance between the orbit and the surface of the earth.

14.117   The speed of the spacecraft at *A* is $v_1 = 11.4$ km/s in the direction $\beta = 10°$. (a) Show that the trajectory is hyperbolic. (b) Find the terminal speed of the craft, ignoring the gravitational attraction of the sun.

14.118   A 200-kg communications satellite is traveling in a circular "parking" orbit of radius 7000 km miles (orbit 1). The satellite must traverse the elliptical orbit *AB* in order to reach the desired geosynchronous circular orbit of radius 42000 km miles (orbit 2). Determine the impulses that the satellite must receive in positions *A* and *B* during its transfer between the orbits.

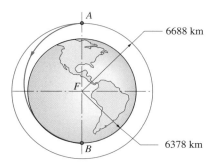

Fig. P14.113

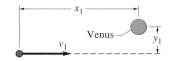

Fig. P14.114, P14.115

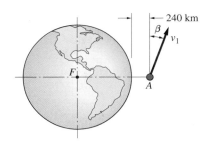

Fig. P14.116, P14.117

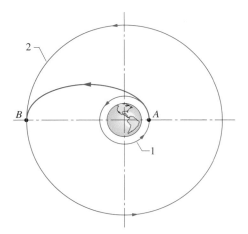

Fig. P14.118

14.119 A spacecraft is traveling in a circular orbit around the earth at an altitude of 400 km. When the craft reaches point A, its engines are fired for a short period, reducing its speed by 7.5%. The resulting path is the crash trajectory AB. Determine the angle β, measured from the nonrotating x-axis, which locates the landing site.

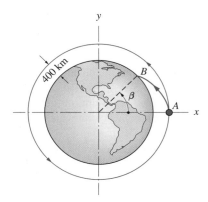

Fig. P14.119

14.120 A ballistic missile fired from the North Pole lands at the equator, after reaching a maximum altitude of 360 km above the surface of the earth. Neglecting air resistance, find the firing angle (measured from the vertical) and the initial speed of the missile.

14.121 A satellite is launched into orbit around the earth at an altitude $H_0 = 780$ km with the initial speed $v_0 = 28000$ km/h in the direction shown. (a) Derive the differential equations of motion for the satellite, and state the initial conditions. (b) Solve the equations numerically from the time of launch to the time when the satellite returns to the launch position; plot R versus t. (c) Determine the orbital period (time to execute one orbit). (d) Find the highest and lowest altitudes reached by the satellite.

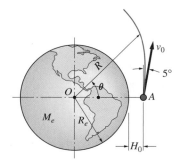

Fig. P14.121, P14.122

14.122 A spacecraft has the velocity $v_0 = 24000$ km/h in the direction shown when its altitude is $H_0 = 780$ km. (a) Derive the differential equations of motion for the spacecraft, and state the initial conditions. (b) Solve the equations numerically until the time when the satellite hits the earth; plot R versus θ. (c) Find the angle θ at the impact point.

Review of Equations

Work and potential energy of a force

	Work	Potential energy
Constant force	$U_{1-2} = F\Delta d$	$V = -Fd$
Spring force	$U_{1-2} = -\dfrac{1}{2}k(\delta_2^2 - \delta_1^2)$	$V = \dfrac{1}{2}k\delta^2$
Gravity	$U_{1-2} = Gm_A m_B \left(\dfrac{1}{R_2} - \dfrac{1}{R_1}\right)$	$V = -\dfrac{Gm_A m_B}{R}$

Principle of work and kinetic energy

$$U_{1-2} = T_2 - T_1 \quad T = \frac{1}{2}mv^2$$

Conservation of mechanical energy

$$V_1 + T_1 = V_2 + T_2$$

Power and efficiency

$$\text{Power of a force}: \quad P = \mathbf{F}\cdot\mathbf{v}$$
$$\text{Efficiency of a machine}: \quad \eta = (P_{\text{out}}/P_{\text{in}}) \times 100\%$$

Impulse and momentum

$$\text{Impulse of a force}: \quad \mathbf{L}_{1-2} = \int_{t_1}^{t_2} \mathbf{F}\,dt$$
$$\text{Momentum of a particle}: \quad \mathbf{p} = m\mathbf{v}$$
$$\text{Impulse-momentum principle}: \quad \mathbf{L}_{1-2} = \mathbf{p}_2 - \mathbf{p}_1$$

Angular impulse and angular momentum

$$\text{Angular impulse}: \quad (\mathbf{A}_A)_{1-2} = \int_{t_1}^{t_2} \mathbf{M}_A\,dt$$
$$\text{Angular momentum}: \quad \mathbf{h}_A = \mathbf{r} \times (m\mathbf{v})$$
$$\text{Angular impulse-momentum principle}:$$
$$(\mathbf{A}_A)_{1-2} = (\mathbf{h}_A)_2 - (\mathbf{h}_A)_1 \quad (A \text{ is a fixed point})$$

Review Problems

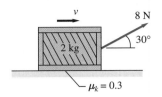

Fig. P14.123

14.123 The constant 8-N force is pulling the 2-kg crate across the horizontal surface. If the coefficient of kinetic friction between the crate and the surface is 0.3, determine the time required to increase the speed v of the crate from 3 m/s to 12 m/s.

14.124 The 0.8-kg particle slides across a frictionless, horizontal plane. The force P applied to the particle always acts in the x-direction, but its magnitude varies with time as shown in the P-t diagram. If the velocity of the particle at time $t = 0$ is 3 m/s in the direction shown, determine the speed when $t = 5$ s.

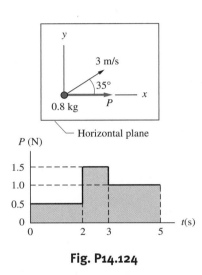

Fig. P14.124

14.125 The bicycle rider starts from rest at A and coasts down the hill. (a) Find the speed of the rider at B and at C. (b) Determine the smallest radius of curvature of the path at C for which the bicycle will not leave the ground.

Fig. P14.125

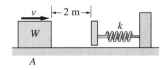

Fig. P14.126

14.126 The package of weight $W = 40$ N is sliding to the right along the rough, horizontal surface. When the package is at A, its speed is $v = 5$ m/s. The spring of stiffness $k = 200$ N/m brings the package to rest. If the coefficient of kinetic friction between the package and the surface is 0.3, determine the maximum force in the spring. Neglect the weight of the end plate that is attached to the spring.

14.127   The figure shows the elliptical orbit of an earth satellite. The minimum and maximum speeds of the satellite are $v_1 = 6$ km/s and $v_2 = 7.5$ km/s, respectively. Compute the corresponding altitudes h_1 and h_2.

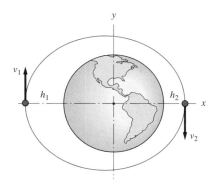

Fig. P14.127

14.128   The 0.5-kg collar is pulled along the frictionless rod by the force $P(x) = 0.3/x^2$ N, where x is in m. When $x = 0.5$ m, the velocity of the collar is 1 m/s to the right. When $x = 1.5$ m, determine (a) the velocity of the collar; and (b) the power of the force P.

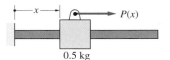

Fig. P14.128

14.129   The coefficients of static and kinetic friction between the 18-kg crate and the horizontal surface are $\mu_s = 0.25$ and $\mu_k = 0.2$, respectively. The crate is at rest when the horizontal, time-dependent force P is applied. Determine the speed of the crate at $t = 10$ s.

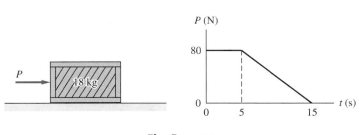

Fig. P14.129

14.130   The 2-kg collar A slides on the vertical rod with negligible friction. The spring AB has a free length of 0.8 m, and its stiffness is 20 N/m. If the mass is released from rest in the position shown, determine its speed at O.

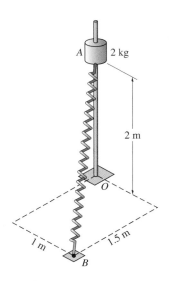

Fig. P14.130

14.131   A package of mass M is placed with zero velocity on a conveyor belt that is moving at 2 m/s. The coefficient of kinetic friction between the package and the belt is 0.4. Determine the distance that the package travels before its speed reaches the speed of the belt.

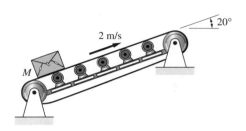

Fig. P14.131, P14.132

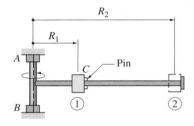

Fig P14.133

14.132   Determine the time required for the speed of the package in Prob. 14.131 to reach the speed of the conveyor belt.

14.133   The assembly shown is rotating freely about the vertical axis AB with the sliding collar C being held in position 1 by a pin. The pin subsequently falls out, allowing the collar to slide to position 2. Determine the percentage of kinetic energy that is lost in the process in terms of the distances R_1 and R_2. Neglect all masses, except the mass of collar C.

14.134   The 4-N weight slides with negligible friction on a horizontal table. The spring attached to the weight has a stiffness of 20 N/m and a free length of 0.5 m. When the weight is at A, its velocity is 10 m/s directed as shown. Determine the speed of the weight at B where the length of the spring is 0.8 m. Also, find the rate at which the spring is elongating at B.

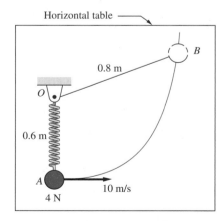

Fig. P14.134

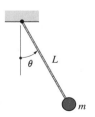

Fig P14.135

14.135   Use the moment-angular momentum relationship to show that the differential equation describing the motion of the pendulum is $\ddot{\theta} + (g/L) \sin \theta = 0$.

14.136   An unpowered space vehicle of mass m travels past the planet Venus along the trajectory AB. If the coordinates of A and B are $(-5.0, -1.0) \times 10^8$ m and $(3.960, 3.382) \times 10^8$ m, respectively, determine the speed of the vehicle at B.

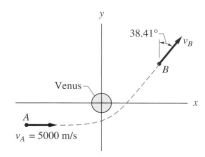

Fig. P14.136

14.137   The mass m is attached to a string and swings in a horizontal circle of radius $R = 0.25$ m when $L = 0.5$ m. The length L is then shortened by pulling the string slowly through the hole A in a table until the speed of the mass has doubled. Determine the corresponding values of (a) radius R; and (b) angle ϕ.

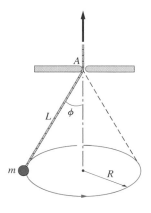

Fig. P14.137

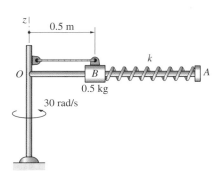

Fig. P14.138

14.138   The rod OA is rotating freely about the z-axis with the angular speed of 30 rad/s. The spring is undeformed when the cord restraining the 0.5 kg collar B breaks. If the maximum displacement of the collar relative to the rod is 0.25 m, determine the stiffness k of the spring. Neglect friction and the mass of rod OA.

14.139 The 0.6-kg mass slides with negligible friction on the cylindrical surface. The spring attached to the mass has a stiffness of 110 N/m, and its free length is 80 mm. If the mass is released from rest at A, determine its speed at B.

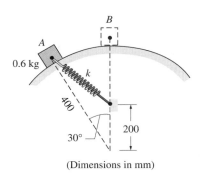

(Dimensions in mm)

Fig. P14.139

14.140 The 20-N weight is attached to a bar of negligible weight. The assembly rotates in the vertical plane about O. The spring has a free length of 1.4 m, and its stiffness is 200 N/m. If the system is released from rest when $\theta = 90°$, determine the speed of the weight when $\theta = 0$.

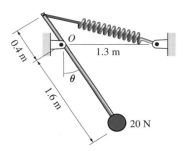

Fig. P14.140

15

Dynamics of Particle Systems

15.1   Introduction

Up to this point, our study of dynamics has focused on a single particle. Three procedures of kinetic analysis were introduced: the force-mass-acceleration, the work-energy, and the impulse-momentum methods. In this chapter, we extend these methods to systems containing two or more particles.

Before we proceed to kinetics of particle systems, it is necessary to consider another topic of kinematics, namely the concept of *relative motion*. Relative motion provides a convenient means of describing the kinematic constraints (geometric restrictions on the motion) that are usually present in a system of particles.

This chapter also introduces two new applications of dynamics: impact of particles and mass flow. Impact refers to a collision between particles and is characterized by a very short time of contact and large contact forces. The term "mass flow" is applied to problems where the mass is continuously entering or leaving the system, as in the flow of fluids through pipes and rocket propulsion.

a.   Relative motion

Our discussion of kinematics has so far been limited to *absolute* motion, where the motion of a particle is described in a fixed (inertial) reference frame. In order to emphasize the fixed nature of the reference frame, the prefix "absolute" is sometimes added to the names of kinematic variables (e.g., "absolute velocity"). Newton's second law, and the work-energy and impulse-momentum principles derived from it, apply only to absolute motion. In other words, the position, velocity, and acceleration appearing in these kinetic principles must be absolute.

In kinematics, a fixed frame is not always the most convenient reference for describing the motion of a particle. For example, the natural reference frame for observing the motion of a raindrop on the window of a moving car is the window (a moving frame), not the road (a fixed frame). A description of motion that is based on a moving frame of reference, such as the window, is termed *relative.*

Figure 15.1(a) shows the paths of two particles A and B. The position vectors of the particles, measured from the origin O of the fixed xyz coordinate system, are denoted by $\mathbf{r}_A$ and $\mathbf{r}_B$, and their velocities are $\mathbf{v}_A = \dot{\mathbf{r}}_A$ and $\mathbf{v}_B = \dot{\mathbf{r}}_B$. Because these variables are referred to a fixed reference frame, they represent the absolute position vectors and velocities of the particles. The vector $\mathbf{r}_{B/A}$, drawn from A to B, is called the position vector of B *relative* to A.

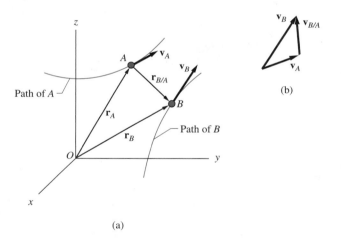

(a)

(b)

Fig. 15.1

From Fig. 15.1(a) we see that the absolute and relative position vectors are related by

$$\mathbf{r}_B = \mathbf{r}_A + \mathbf{r}_{B/A} \tag{15.1}$$

Differentiating with respect to time and introducing the notation

$$\mathbf{v}_{B/A} = \dot{\mathbf{r}}_{B/A} \tag{15.2}$$

yields

$$\mathbf{v}_B = \mathbf{v}_A + \mathbf{v}_{B/A} \qquad (15.3)$$

The vector $\mathbf{v}_{B/A}$, shown in Fig. 15.1(b), is known as the velocity of B *relative* to A.
   Differentiating both sides of Eq. (15.3), we obtain

$$\mathbf{a}_B = \mathbf{a}_A + \mathbf{a}_{B/A} \qquad (15.4)$$

where

$$\mathbf{a}_{B/A} = \dot{\mathbf{v}}_{B/A} = \ddot{\mathbf{r}}_{B/A} \qquad (15.5)$$

is called the acceleration of B *relative* to A.
   From Fig. 15.1(a), we see that the vector drawn from A to B is the negative of the vector from B to A, which leads to the following identities:

$$\mathbf{r}_{B/A} = -\mathbf{r}_{A/B} \qquad \mathbf{v}_{B/A} = -\mathbf{v}_{A/B} \qquad \mathbf{a}_{B/A} = -\mathbf{a}_{A/B} \qquad (15.6)$$

b. *Translating reference frame*

It is often convenient to describe relative motion with respect to a coordinate system that moves with the reference particle. In Fig. 15.2, the xyz axes are fixed in an inertial reference frame, whereas the $x'y'z'$ axes are attached to (and move with) the reference particle A. In this chapter, we consider only the special case in which the $x'y'z'$ axes always remain parallel to the fixed axes. In other words, the moving axes *translate* with the reference particle, but they do not rotate. Therefore, the base vectors $\mathbf{i}$, $\mathbf{j}$, and $\mathbf{k}$ of the fixed reference frame are also the base vectors of the translating frame.
   In the translating reference frame, the coordinates of particle B are x', y', and z'. Hence the relative position vector of B in this coordinate system is

$$\mathbf{r}_{B/A} = x'\mathbf{i} + y'\mathbf{j} + z'\mathbf{k} \qquad (15.7)$$

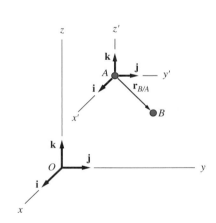

Fig. 15.2

The relative velocity and relative acceleration of B are obtained by differentiating Eq. (15.7) with respect to time (note that because the base vectors are constant, we have $d\mathbf{i}/dt = d\mathbf{j}/dt = d\mathbf{k}/dt = \mathbf{0}$):

$$\mathbf{v}_{B/A} = \frac{dx'}{dt}\mathbf{i} + \frac{dy'}{dt}\mathbf{j} + \frac{dz'}{dt}\mathbf{k} \qquad (15.8)$$

$$\mathbf{a}_{B/A} = \frac{d^2x'}{dt^2}\mathbf{i} + \frac{d^2y'}{dt^2}\mathbf{j} + \frac{d^2z'}{dt^2}\mathbf{k} \qquad (15.9)$$

Equations (15.8) and (15.9) show that $\mathbf{v}_{B/A}$ and $\mathbf{a}_{B/A}$ can be interpreted as the velocity and acceleration of particle B as seen by a *nonrotating observer* who moves with particle A.

Sample Problem 15.1

Two airplanes A and B are flying with constant velocities at the same altitude. The positions of the planes at time $t = 0$ are shown in Fig. (a) (the xy reference frame is fixed in space). Determine (1) the velocity of plane A relative to B; (2) the position vector of A relative to B as a function of time; and (3) the minimum distance between the planes and the time when this occurs.

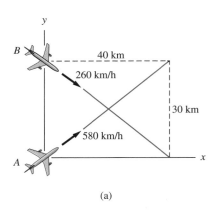

(a)

Solution

Part 1

From the geometry in Fig. (a), the velocities of the planes are

$$\mathbf{v}_A = 580 \left(\frac{40\mathbf{i} + 30\mathbf{j}}{50} \right) = 464\mathbf{i} + 348\mathbf{j} \ \text{km/h}$$

$$\mathbf{v}_B = 260 \left(\frac{40\mathbf{i} - 30\mathbf{j}}{50} \right) = 208\mathbf{i} - 156\mathbf{j} \ \text{km/h}$$

The velocity of A relative to B is

$$\mathbf{v}_{A/B} = \mathbf{v}_A - \mathbf{v}_B = (464\mathbf{i} + 348\mathbf{j}) - (208\mathbf{i} - 156\mathbf{j})$$

$$= 256\mathbf{i} + 504\mathbf{j} \ \text{km/h} \qquad \textit{Answer}$$

The magnitude and direction of the vector are

$$v_{A/B} = \sqrt{256^2 + 504^2} = 565.3 \ \text{km/h}$$

$$\theta = \tan^{-1} \frac{256}{504} = 26.93°$$

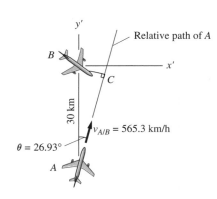

(b)

This relative velocity vector is shown in Fig. (b). Note that $\mathbf{v}_{A/B}$ is the velocity of plane A as seen by an (nonrotating) observer in plane B—that is, the velocity of A in the nonrotating $x'y'$ coordinate system attached to plane B. Since $\mathbf{v}_{A/B}$ is constant, the path of A relative to the translating $x'y'$ coordinate system is the straight line shown in Fig. (b).

Part 2

The position vector of A relative to B can be found by integrating the relative velocity:

$$\mathbf{r}_{A/B} = \int \mathbf{v}_{A/B}\, dt = \int (256\mathbf{i} + 504\mathbf{j})\, dt = (256\mathbf{i} + 504\mathbf{j})\, t + \mathbf{r}_0$$

where t is in hours and $\mathbf{r}_0$ is a constant of integration. From the initial condition, $\mathbf{r}_{A/B} = -30\mathbf{j}$ km at $t = 0$, we get $\mathbf{r}_0 = -30\mathbf{j}$ km. Therefore, the relative position vector becomes

$$\mathbf{r}_{A/B} = 256t\mathbf{i} + (504t - 30)\mathbf{j} \text{ km} \qquad \textit{Answer}$$

Part 3

Denoting the distance between the planes by s, we have

$$s^2 = |\mathbf{r}_{A/B}|^2 = (256t)^2 + (504t - 30)^2 \text{ km}^2 \qquad \text{(a)}$$

The minimum value of s occurs when $d(s^2)/dt = 0$, or

$$2(256)^2 t + 2(504t - 30)(504) = 0$$

which yields

$$t = 0.047\,32 \text{ h} = 170.3 \text{ s} \qquad \textit{Answer}$$

Substituting this value of t into Eq. (a), we get for the minimum distance between the planes

$$s_{\min} = \sqrt{[256(0.047\,32)]^2 + [504(0.047\,32) - 30]^2} = 13.59 \text{ km} \qquad \textit{Answer}$$

Note

The results in Part 3 could also be obtained from Fig. (b). The minimum distance between the planes occurs when plane A reaches position C. From triangle ABC, we obtain

$$s_{\min} = \overline{BC} = 30 \sin 26.93° = 13.59 \text{ km}$$

The time required to reach that position is

$$t = \frac{\overline{AC}}{v_{A/B}} = \frac{30 \cos 26.93°}{565.3} = 0.0473 \text{ h} = 170.3 \text{ s}$$

Problems

15.1   The two airplanes are flying at the same altitude with the velocities shown. Find the velocity of plane B as seen by a passenger in plane A.

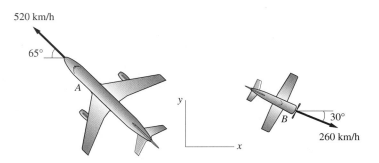

Fig. P15.1

15.2   At time $t = 0$, the two trains are 4 km apart and traveling with the velocities shown. The speed of train A is increasing at the rate of 0.5 m/s^2, whereas the speed of train B is constant. Determine the acceleration, velocity, and position of train B relative to train A as functions of t. What is the distance between the trains at $t = 120$ s?

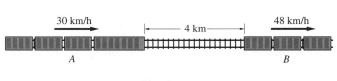

Fig. P15.2

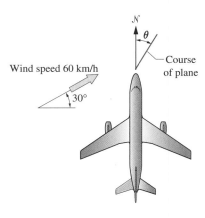

Fig. P15.3

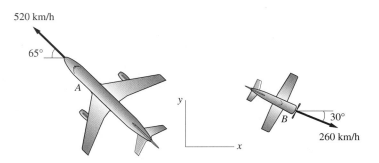

Fig. P15.4

15.3   In the position shown, the speed of car B is 16 m/s and increasing. To a passenger in car A, the acceleration of car B appears to be zero. What is the acceleration of car A in this position?

15.4   The airspeed of the plane is 560 km/h, directed north. If the wind speed is 60 km/h in the direction shown, determine the ground speed and the course (angle θ) of the plane.

15.5   The boat with a cruising speed (speed of boat relative to the water) of 24 km/h is crossing a river that has a current of 10 km/h. (a) Find the course, determined by the angle θ, that the boat must steer in order to follow a straight line from A to C. (b) Find the time required for the boat to complete the crossing.

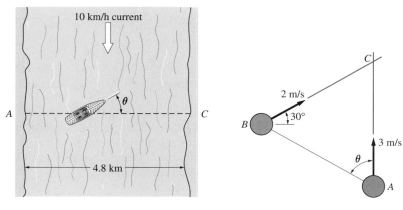

Fig. P15.5   Fig. P15.6

15.6   Two billiard balls A and B, initially at rest, are hit at the same instant and roll along the paths AC and BC. If the velocities of the balls are as shown in the figure, determine the angle θ if the balls are to collide. (Hint: $v_{A/B}$ must be directed from A toward B.)

15.7   When a stationary car is pointing into the wind, the streaks made by raindrops on the side windows are inclined at $\theta = 15°$ with the vertical. When the car is driven at 30 km/h into the wind, the angle θ increases to 75°. Find the speed of the raindrops.

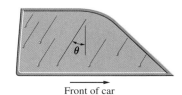

Front of car

Fig. P15.7

15.8   Two cars A and B traveling at constant speeds are in the positions shown at time $t = 0$. Determine (a) the velocity of A relative to B; (b) the position vector of A relative to B as a function of time; and (c) the minimum distance between the cars and the time when this occurs.

15.9   The crossbow is aimed at the sandbag, which is suspended from a cord. At the instant the cord is cut, the crossbow is fired. Show that the bolt will always hit the sandbag, regardless of the initial speed v_0 of the bolt. (Hint: The velocity of the bolt relative to the sandbag is constant.)

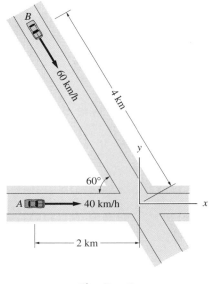

Fig. P15.8

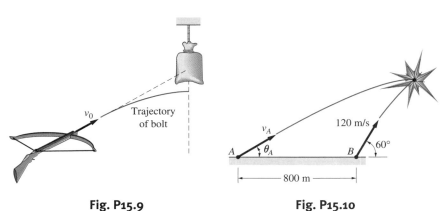

Fig. P15.9   Fig. P15.10

15.10   Two projectiles A and B are launched simultaneously in the same vertical plane with the initial positions and velocities shown in the figure. If the projectiles collide 8 s after launch, determine (a) their relative velocity at collision; and (b) the initial velocity vector of *A*. (Hint: $\mathbf{v}_{B/A}$ is constant.)

15.11   Two cars travel at the constant speeds of $v_A = 12$ m/s and $v_B = 15$ m/s around a circular track. When the cars are in the positions shown, determine the magnitudes of $\mathbf{v}_{B/A}$ and $\mathbf{a}_{B/A}$.

15.12   Cars *A* and *B* are traveling along the circular track. The speed of car *A* is constant at $v_A = 90$ km/h. In the position shown, the speed of car *B* is $v_B = 54$ km/h and is increasing at the rate of 1 m/s². Determine the magnitude of the relative acceleration between the cars in this position.

15.13   Car *A* is traveling on the circular road at the constant speed of $v_A = 70$ km/h. At the instant shown, the speed of car *B* on the overpass is $v_B = 90$ km/h, and it is decreasing at the rate of 1.8 m/s². Find the relative acceleration vector $\mathbf{a}_{B/A}$ at this instant.

15.14   Figure (a) shows a boat that is sailing in the *y*-direction at 5 km/h with the wind vane indicating the direction of the wind relative to the ship. After the ship changes course to the *x*-direction, the angle of the wind vane is as shown in Fig. (b). Determine the velocity vector of the wind.

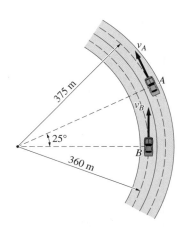

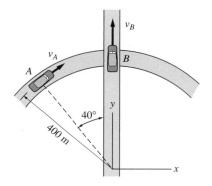

Fig. P15.11, P15.12

Fig. P15.13

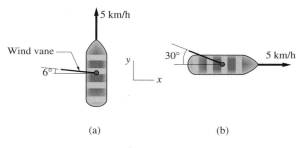

(a)   (b)

Fig. P15.14

15.3   *Kinematics of Constrained Motion*

The following terminology is used frequently when describing the kinematics of particle systems:[*]

- *Kinematic constraints*: geometric restrictions imposed on the motion of particles.
- *Equations of constraint*: mathematical expressions that describe the kinematic constraints on particles in terms of their position coordinates.
- *Kinematically independent coordinates*: position coordinates of particles that are not subject to kinematic constraints.
- *Number of degrees of freedom*: the number of kinematically independent coordinates that are required to completely describe the configuration of a system of particles.

---

[*]These terms were introduced in the discussion of virtual work in Ch. 10. The definitions are repeated here because of their importance in dynamics.

As an illustration of these concepts, consider the system shown in Fig. 15.3. The system consists of two blocks, labeled A and B, that are connected by a rope that passes over the fixed peg C. The position coordinates of the blocks are denoted by x_A and y_B. The *kinematic constraint* on the motion of the system is that the total length L of the rope does not change. The corresponding *equation of constraint* is obtained from the geometry of Fig. 15.3:

$$L = \sqrt{x_A^2 + h^2} + y_B = \text{constant} \qquad (15.10)$$

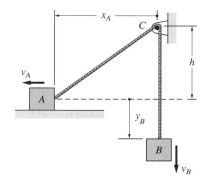

Fig. 15.3

This system has a single *degree of freedom.* We can choose either x_A or y_B as the *kinematically independent coordinate.* The position coordinate not chosen can be related to the independent coordinate using Eq. (15.10).

The constraint on the velocities v_A and v_B of the blocks can be obtained by differentiating Eq. (15.10) with respect to time. The result is

$$\dot{L} = \frac{1}{2}\frac{2x_A\dot{x}_A}{\sqrt{x_A^2 + h^2}} + \dot{y}_B = 0$$

Substituting $\dot{x}_A = v_A$ and $\dot{y}_B = v_B$, we find that the velocities are related by

$$\frac{x_A}{\sqrt{x_A^2 + h^2}}v_A + v_B = 0 \qquad (15.11)$$

If needed, the relationship between the accelerations of the blocks can be determined by differentiating Eq. (15.11) with respect to time.

An example of a system with two degrees of freedom is shown in Fig. 15.4. It is similar to the system in Fig. 15.3, but here the peg is attached to the block C that can move vertically. The three position coordinates (x_A, y_B, and y_C) of the blocks are subject to the kinematic constraint

$$L = \sqrt{x_A^2 + y_C^2} + y_B = \text{constant}$$

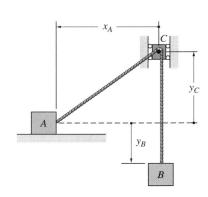

Fig. 15.4

(L is the length of the rope) so that only two of the coordinates are kinematically independent.

We note that in the above examples

(number of position coordinates) − (number of kinematic constraints)
= (number of degrees of freedom)

This is true of all but a special class of mechanical systems.[*]

---

[*]Systems that violate this rule are called *nonholonomic.* See D. T. Greenwood, *Classical Dynamics,* Prentice Hall, 1977, p. 10.

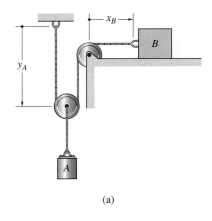

(a)

Sample Problem   15.2

Figure (a) shows a system consisting of two blocks A and B connected by an inextensible cable that runs around two pulleys. Determine the kinematic relationships between the velocities and the accelerations of the blocks.

Solution

The system shown in Fig. (a) has one degree of freedom because one coordinate (e.g., y_A or x_B), determines its configuration. It is convenient to number the pulleys and to label the fixed distance, h, as shown in Fig. (b). Letting L be the length of the cable, we have

$$L = y_A + \left(\begin{matrix} \text{length of cable wrapped} \\ \text{around pulley 1} \end{matrix} \right) + (y_A - h)$$
$$+ \left(\begin{matrix} \text{length of cable wrapped} \\ \text{around pulley 2} \end{matrix} \right) + x_B$$

Because L, h, and the lengths of cable that are wrapped around each pulley are constant, differentiation with respect to time yields

$$\frac{dL}{dt} = v_A + 0 + v_A + 0 + v_B = 0$$

which gives

$$v_B = -2v_A \qquad \qquad \textit{Answer}$$

Differentiation of this equation with respect to time yields

$$a_B = -2a_A \qquad \qquad \textit{Answer}$$

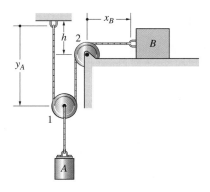

(b)

Sample Problem 15.3

The two collars A and B are joined by a rope of length L. Collar A is moving to the right with constant velocity v_A. Determine the velocity and acceleration of collar B as functions of v_A and the angle θ.

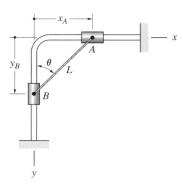

Solution

The system has a single degree of freedom because only one coordinate (x_A, y_B, or θ) determines its configuration. We choose to ignore θ for the time being and work with x_A and y_B. The equation of constraint that relates these two coordinates is

$$x_A^2 + y_B^2 = L^2 \qquad\text{(a)}$$

Differentiating this equation with respect to time (noting that $\dot{x}_A = v_A$ and $\dot{y}_B = v_B$) we obtain $2x_A v_A + 2y_B v_B = 0$, which reduces to

$$x_A v_A + y_B v_B = 0 \qquad\text{(b)}$$

Taking the time derivative of Eq. (b), we get

$$\left(x_A a_A + v_A^2\right) + \left(y_B a_B + v_B^2\right) = 0 \qquad\text{(c)}$$

where we used $\dot{v}_A = a_A$ and $\dot{v}_B = a_B$.

Solving Eq. (b) for the velocity of B yields

$$v_B = -v_A \frac{x_A}{y_B} = -v_A \frac{L \sin\theta}{L \cos\theta}$$

or

$$v_B = -v_A \tan\theta \qquad\qquad \textit{Answer} \text{ (d)}$$

Because $a_A = 0$, the acceleration of B from Eq. (c) is

$$a_B = -\frac{v_A^2 + v_B^2}{y_B}$$

Substituting for v_B from Eq. (d) and $y_B = L \cos\theta$, we obtain

$$a_B = -\frac{v_A^2 + (-v_A \tan\theta)^2}{L \cos\theta} = -\frac{v_A^2 (1 + \tan^2\theta)}{L \cos\theta}$$

Using the identity $(1 + \tan^2\theta) = 1/\cos^2\theta$, this equation reduces to

$$a_B = -\frac{v_A^2}{L \cos^3\theta} \qquad\qquad \textit{Answer}$$

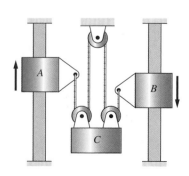

Fig. P15.15

15.15   Block *A* is moving to the right with the constant speed of 0.5 m/s. Determine the velocity of block *B*.

15.16   If block *B* is moving down with the constant speed of 0.4 m/s, find the velocity of block *A*.

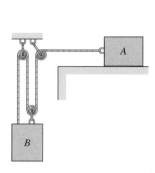

Fig. P15.16   **Fig. P15.17**

15.17   At the instant shown, the velocity of block *A* relative to block *B* is 240 mm/s directed downward. Determine the velocity of each block at this instant.

15.18   Determine the velocity of block *B* at the instant when the velocity of block *A* is 1 m/s, directed downward.

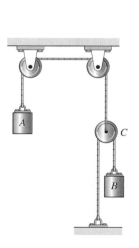

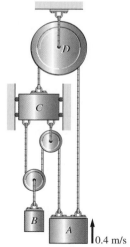

Fig. P15.18   **Fig. P15.19**

15.19   Determine the velocity of block *B* at the instant when the velocity of block *A* is 0.4 m/s, directed upward.

15.20   Collar *A* is moving upward at 1 m/s while collar *B* is moving downward at 0.4 m/s. Find the velocity of block *C*.

Fig. P15.20

15.21 Collar A is moving downward with the constant speed of 0.6 m/s. Determine the velocity of block B when $y_A = 1$ m.

15.22 Block A is moving upward at 3.6 m/s. What is the velocity of block B when $y_B = 2$ m?

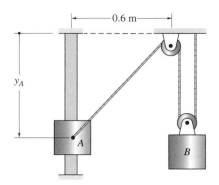

Fig. P15.21

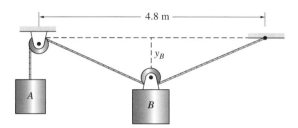

Fig. P15.22

15.23 Disk A is rotating clockwise with the angular velocity $\dot{\theta} = 2$ rad/s. Determine the velocity of block B when $\theta = 60°$.

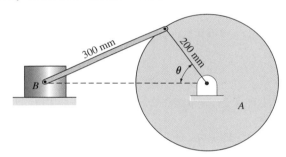

Fig. P15.23

15.24 When block C is in the position $x_C = 0.8$ m, its speed is 1 m/s to the right. Find the velocity of block A at this instant. Note that the rope runs around the pulley B and a peg attached to block C.

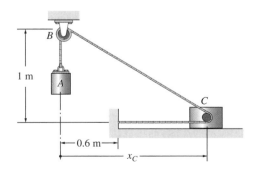

Fig. P15.24

15.25 The collars A and B are joined by the 8-in. bar. Given that $v_A = 0.3$ m/s and $a_A = -0.6$ m/s² when $x = 90$ mm, find the velocity and acceleration of B at this instant.

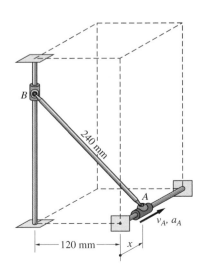

Fig. P15.25

15.26   At a certain instant, the velocity of block *A* is 0.2 m/s and the velocity of block *B* relative to block *C* is 0.6 m/s, both directed downward. Determine the velocities of *B* and *C* at this instant.

Fig. P15.26   **Fig. P15.27**

15.27   Block *C* is moving up at the constant speed of 0.3 m/s. Given that the elevations of blocks *A* and *B* are always equal, determine the velocity of *B*.

15.4   ## Kinetics: Force-Mass-Acceleration Method

Newton's second law, $\Sigma \mathbf{F} = m\mathbf{a}$, can also be used to study the motion of a particle system. The most direct approach is to use the FBD and the MAD of each particle in the system to obtain the equations of motion of the individual particles. Thus a system of *n* particles would yield the following *n* vector equations of motion: $(\Sigma \mathbf{F})_i = m\mathbf{a}_i, i = 1, 2, \ldots, n$. If the particles of the system are subject to internal kinematic constraints (due to massless connections, such as ropes joining the particles) then the forces that impose the constraints (e.g., the tensions in the ropes) appear as unknowns in these equations of motion.

In some problems it is sufficient to consider the motion of the center of mass[*] of the system, rather than the motions of individual particles. The equation of motion of the mass center is obtained by adding the equations of motion of all particles in the system. Because the constraint forces between the particles occur in equal and opposite pairs (Newton's third law), they are eliminated by the summation process.

a.   *Motion of the mass center*

Before deriving the equation of motion of the mass center, it is necessary to review the concept of mass center and to discuss forces that are external and internal to the system.

---

[*]Engineers often use *center of gravity* and *center of mass* interchangeably since these two points coincide in most applications.

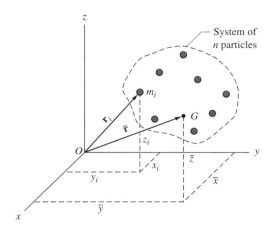

Fig. 15.5

1.   *Mass Center*   Figure 15.5 shows a system of n particles. The mass of the ith particle is denoted by m_i, and its position vector is $\mathbf{r}_i = x_i\mathbf{i} + y_i\mathbf{j} + z_i\mathbf{k}$. The mass center G of the system is defined to be the point whose position vector is

$$\bar{\mathbf{r}} = \frac{1}{m}\sum_{i=1}^{n} m_i\mathbf{r}_i \qquad (15.12a)$$

where $m = \sum_{i=1}^{n} m_i$ is the total mass of the system. The rectangular components of $\bar{\mathbf{r}}$ are[*]

$$\bar{x} = \frac{1}{m}\sum_{i=1}^{n} m_i x_i \quad \bar{y} = \frac{1}{m}\sum_{i=1}^{n} m_i y_i \quad \bar{z} = \frac{1}{m}\sum_{i=1}^{n} m_i z_i \qquad (15.12b)$$

If the xyz coordinate system is an inertial reference frame, the velocity $\bar{\mathbf{v}}$ and acceleration $\bar{\mathbf{a}}$ of the mass center are obtained by differentiating Eq. (15.12a) with respect to time:

$$\bar{\mathbf{v}} = \frac{d\bar{\mathbf{r}}}{dt} = \frac{1}{m}\sum_{i=1}^{n} m_i\mathbf{v_i} \quad \bar{\mathbf{a}} = \frac{d\bar{\mathbf{v}}}{dt} = \frac{d^2\bar{\mathbf{r}}}{dt^2} = \frac{1}{m}\sum_{i=1}^{n} m_i\mathbf{a_i} \qquad (15.13)$$

where $\mathbf{v_i}$ and $\mathbf{a_i}$ are the velocity and acceleration of the ith particle, respectively.

2.   *External and Internal Forces*   Figure 15.6(a) shows the free-body diagram of a closed system of n particles labeled 1, 2, ..., n.[†] The vector $\mathbf{F}_i$, $i = 1, 2, \ldots, n$, represents the resultant *external* force that acts on the ith particle. The external forces are caused by interaction of the particles with the external

---

[*]These equations are similar to the definition of the mass center of a rigid body in Ch. 8. The only difference is that the integrals in Ch. 8 are replaced by summations for a system of particles.
[†]A system is "closed" if no particles enter or leave the system. The reason for this restriction will become apparent later. As the particles move, the boundary of the system can be imagined to be a flexible "pouch" that always encloses the same particles.

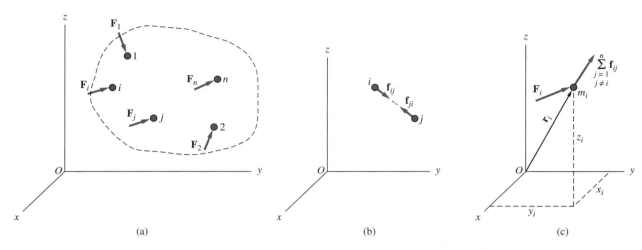

Fig. 15.6

world (i.e., their sources are external to the system). Examples of external forces that may act on a particle are its weight, its interactions with other particles that are not included in the system, and support reactions.

In addition to external forces, particles of the system may also be subjected to forces that are *internal* to the system. For example, two particles could be connected together by a spring, collide with each other, or carry electrical charges that cause them to repel or attract each other. It is not necessary to show internal forces on the FBD of the system in Fig. 15.6(a), because interactions between particles always occur as pairs of forces that are equal in magnitude, opposite in direction, and have collinear lines of action (Newton's third law). Therefore, the internal forces cancel.

Figure 15.6(b) shows the pair of internal forces that act between the *i*th and *j*th particles. The force $\mathbf{f}_{ij}$ represents the internal force acting on the *i*th particle that is caused by the *j*th particle. Similarly, $\mathbf{f}_{ji}$ is the internal force acting on the *j*th particle caused by the *i*th particle. [Each particle can have an interaction with every other particle in the system, but only one such interaction is shown in Fig.15.6 (b).] According to Newton's third law,

$$\mathbf{f}_{ij} = -\mathbf{f}_{ji} \qquad (i \neq j) \tag{15.14}$$

and the two forces are collinear.[*]

The free-body diagram of a typical (*i*th) particle of the system is shown in Fig. 15.6(c). As mentioned before, $\mathbf{F}_i$ represents the resultant external force acting on the particle (including its weight). The vector

$$\sum_{\substack{j=1 \\ j\neq i}}^{n} \mathbf{f}_{ij}$$

---

[*]Note that $i = j$ is excluded in the Eq. (15.14) because $\mathbf{f}_{ii}$, the force exerted on the *i*th particle by the *i*th particle, would be meaningless.

is the resultant internal force acting on the particle (sum of the forces applied to the ith particle by all the other particles in the system).

3. *Equation of Motion of the Mass Center* From the free-body diagram in Fig. 15.6(c), we obtain the following equation of motion for the ith particle:

$$\mathbf{F}_i + \sum_{\substack{j=1 \\ j \neq i}}^{n} \mathbf{f}_{ij} = m_i \mathbf{a}_i \qquad (i = 1, 2, \ldots, n) \tag{15.15}$$

Because there are n particles in the system, Eq. (15.15) represents n vector equations (one equation for each particle). Summing all n equations, we get

$$\sum_{i=1}^{n} \mathbf{F}_i + \sum_{i=1}^{n} \sum_{\substack{j=1 \\ j \neq i}}^{n} \mathbf{f}_{ij} = \sum_{i=1}^{n} m_i \mathbf{a}_i \tag{15.16}$$

This equation can be simplified by considering the following:

1. $\sum_{i=1}^{n} \mathbf{F}_i = \sum \mathbf{F}$ is the resultant external force acting on the system (including the weights of the particles).
2. As pointed out by Eq. (15.14), the internal forces occur in equal and opposite pairs. Therefore, their sum vanishes; that is,

$$\sum_{i=1}^{n} \sum_{\substack{j=1 \\ j \neq i}}^{n} \mathbf{f}_{ij} = \mathbf{0} \tag{15.17}$$

3. Using Eq. (15.13), the right side of Eq. (15.16) can be replaced by

$$\sum_{i=1}^{n} m_i \mathbf{a}_i = m \bar{\mathbf{a}} \tag{15.18}$$

Using the above results, Eq. (15.16) can be written as

$$\boxed{\sum \mathbf{F} = m \bar{\mathbf{a}}} \tag{15.19}$$

Comparing Eq. (15.19) and Newton's second law $\Sigma \mathbf{F} = m\mathbf{a}$ for a particle, we see that the mass center of the system moves as if it were a particle of mass equal to the total mass of the system, acted on by the resultant of the external forces acting on the system.

b. Solving equations of motion of individual particles

The equation of motion of the mass center has somewhat limited application, since it tells us nothing about the movement of individual particles or the values of internal forces. Solving the equations of motion of the individual particles suffers none of these drawbacks.

As an illustration, consider the system in Fig. 15.7(a) that consists of blocks A and B of masses m_A and m_B, respectively. The blocks are connected by a cable

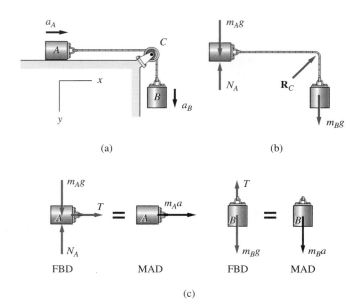

Fig. 15.7

that passes over the pulley C. The problem is to determine the force in the cable and the acceleration of each block, assuming negligible friction.

The system contains three kinematic constraints. There are two external constraints, imposed by the horizontal surface under block A and the pulley C. The corresponding constraint forces (reactions) are denoted by N_A and $\mathbf{R}_C$ in the FBD of the system in Fig. 15.7(b). The third constraint is that the length of the cable is constant. The corresponding constraint force (the cable tension), being internal to the system, does not appear in this FBD.

In this problem, no useful purpose would be served by investigating the motion of the mass center of the system. Therefore, we turn our attention to the analysis of the individual blocks, using the free-body and mass-acceleration diagrams in Fig. 15.7(c). Note that the cable force T is constant throughout the cable if the mass of the cable is negligible and friction at pulley C is ignored. The internal constraint imposed by the cable requires the accelerations of the two blocks to be equal; that is, $a_A = a_B = a$. From the diagrams, we obtain the following equations of motion:

$$\text{Block } A: \quad \sum F_x = ma \quad \xrightarrow{+} \quad T = m_A a$$

$$\text{Block } B: \quad \sum F_y = ma \quad +\!\downarrow \quad m_B g - T = m_B a$$

which yield

$$a = \frac{m_B g}{m_A + m_B} \qquad T = \frac{m_A m_B g}{m_A + m_B}$$

Sample Problem 15.4

The man shown in Fig. (a) walks from the left end to the right end of the uniform plank, which is initially at rest on a sheet of ice. Determine the distance moved by the man when he reaches the right end. The weights of the man and the plank are 60-kg and 15-kg, respectively, and friction between the plank and the ice may be neglected.

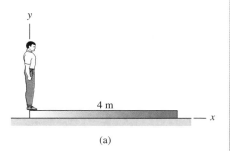

(a)

Solution

The free-body diagram of the system containing the walking man and the plank is shown in Fig. (b). The only forces that appear on this FBD are the weights of the man and the plank and the normal reaction N. The normal and the friction forces that act between the man and the plank do not appear on the FBD, because they are internal to the system.

From the FBD in Fig. (b), we see that there are no forces acting on the system in the x-direction. Therefore, according to $\sum \mathbf{F} = m\bar{\mathbf{a}}$, the mass center G of the system remains stationary, as indicated in Figs. (c) and (d).

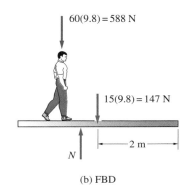

(b) FBD

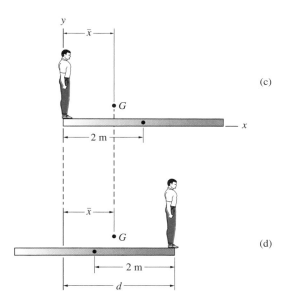

(c)

(d)

We next compute $\bar{x}$, the x-coordinate of G, when the man is at the left end of the plank. Referring to Fig. (c), we have

$$m\bar{x} = \sum_{i=1}^{n} m_i x_i$$

$$(60 + 15)\bar{x} = 60(0) + 15(2)$$

which gives $\bar{x} = 0.4$ m. Repeating the procedure when the man is at the right end of the plank, as shown in Fig. (d), we get

$$(60 + 15)\bar{x} = 60d + 15(d - 2)$$

which yields $d = \bar{x} + 0.4$. Upon substituting $\bar{x} = 0.4\,\text{m}$ (recall that $\bar{x}$ does not change as the man walks along the plank), we obtain

$$d = 0.8\,\text{m} \qquad\qquad\qquad \textit{Answer}$$

Observe that every step taken by the man results in his moving to the right and the plank moving to the left. The magnitudes of these movements are in the proper ratio to ensure that the mass center of the system does not move horizontally.

We can also solve the problem by noting that the distance between the man and the mass center G must be the same in Figs. (c) and (d), because of the symmetry of the two configurations. In other words, $\bar{x} = d - \bar{x}$, or $d = 2\bar{x}$.

Sample Problem 15.5

Figure (a) shows a 45-kg woman who is standing on a scale as she rides in an elevator that weighs 8000 N. Determine the scale reading and the corresponding acceleration of the elevator if the tension in the cable is (1) $T = 4400$ N; and (2) $T = 3600$ N. Neglect the weights of the scale and the support pulley.

Solution

Preliminary Calculations

Figure (b) shows the free-body diagram (FBD) of the system consisting of the woman and the elevator. The only external forces are the weights and cable tensions. The force acting between the woman and the scale does not appear because it is an internal force. The mass-acceleration diagram (MAD) for the system is also shown in Fig. (b). Because the woman and the elevator have the same acceleration a (assumed to be upward) the inertia vector equals the total mass of the system multiplied by a. Newton's second law yields

$$\Sigma F_y = ma_y \quad +\!\uparrow \quad 2T - 45(9.8) - 8000 = \left(45 + \frac{8000}{9.8}\right)a$$

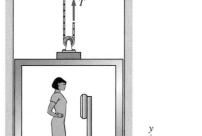

(a)

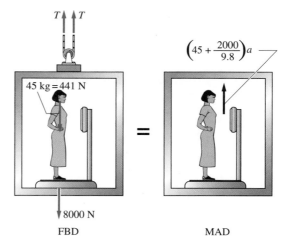

FBD   MAD

(b)

Therefore, the acceleration is

$$a = \frac{(2T - 8441)9.8}{8441} \text{ m/s}^2 \qquad \text{(a)}$$

where T is measured in pounds and $g = 9.8$ m/s^2.

To determine the force that acts between the woman and the scale, we must isolate the woman from the scale. The FBD of the woman is shown in Fig. (c), where N is the force exerted on her by the scale. This force is, of course, equal and opposite to the force exerted by the woman on the scale. The MAD of the woman is also shown in Fig. (c), where a is again directed upward, which is consistent with our previous assumption. From Newton's second law, we obtain

$$\Sigma F_y = ma_y \quad +\uparrow \quad N_A - 45(9.8) = 45a$$

from which the relation between N and a (m/s^2) is

$$N_A = 45a + 441 \text{ N} \qquad \text{(b)}$$

Substituting Eq. (a) into Eq. (b) and simplifying, we find the relationship between N and T to be

$$N_A = 0.104489T \qquad \text{(c)}$$

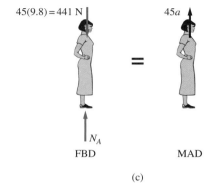

Part 1

If $T = 4400$ N, the scale reading from Eq. (c) is

$$N_A = 0.104489T = 0.104489(4400) = 459.75 \text{ N} \qquad \textit{Answer}$$

and the corresponding acceleration from Eq. (a) is

$$a = \frac{(2T - 8441)9.8}{8441} = \left[\frac{2 \times 4400 - 8441}{8441}\right]9.8 = 0.42 \text{ m/s}^2 \qquad \textit{Answer}$$

Because a is positive, it is in the assumed direction—that is, upward.

Part 2

Using $T = 3600$ N in Eqs. (a) and (c), we get

$$N_A = 0.104489\, T = 0.104489(3600) = 376.16 \text{ N} \qquad \textit{Answer}$$

and

$$a = \left[\frac{2T - 8441}{8441}\right]9.8 = \left[\frac{2 \times 3600 - 8441}{8441}\right]9.8 = -1.44 \text{ m/s}^2 \qquad \textit{Answer}$$

Because a is negative, it is directed downward.

Observe that the woman exerts a force that is greater than her weight when the acceleration is upward, and less than her weight when the acceleration is downward.

Sample Problem 15.6

The 90-N force in Fig. (a) is applied to the cable that is attached to the 60-N block A. In Fig. (b), this force is replaced by a 90-N block B. Neglecting the mass of the pulley, determine the acceleration of A and the tension in the cable for both cases.

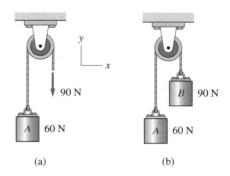

(a)   (b)

Solution

System in Fig. (a)

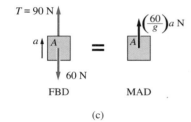

FBD   MAD

(c)

Figure (c) shows the free-body diagram (FBD) of block A. Since the mass of the pulley is to be neglected, the tension is the same throughout the cable, which gives

$$T = 90 \text{ N} \qquad \qquad \textit{Answer}$$

Figure (c) also shows the mass-acceleration diagram (MAD) for block A, where its acceleration a is assumed to be upward. Newton's second law gives

$$\Sigma F_y = ma \quad +\uparrow \quad 90 - 60 = \frac{60}{g}a$$

from which the acceleration is

$$a = \frac{g}{2} = 4.9 \text{ m/s}^2 \qquad \qquad \textit{Answer}$$

System in Fig. (b)

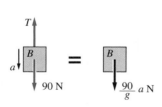

FBD   MAD

(d)

The FBDs and MADs of the blocks are shown in Fig. (d). Due to the inextensible cable, the acceleration of A is equal in magnitude to the acceleration of B, but in the opposite direction. We assumed the acceleration of A to be upward. Therefore the equation of motion of block A is

$$\Sigma F_y = ma \quad +\uparrow \quad T - 60 = \left(\frac{60}{g}\right)a$$

For block B, we have

$$\Sigma F_y = ma \quad +\uparrow \quad T - 90 = \left(\frac{90}{g}\right)a$$

Solving these two equations simultaneously, we obtain

$$T = 72 \text{ N} \quad \text{and} \quad a = \frac{g}{5} = 1.96 \text{ m/s}^2 \qquad \textbf{\textit{Answer}}$$

Note that applying a 90-N force to the end of the cable is not equivalent to attaching a 90-N weight.

Sample Problem 15.7

Figure (a) shows a system consisting of three blocks connected by an inextensible cable that runs around four pulleys. The masses of blocks A, B, and C are 60 kg, 80 kg, and 20 kg, respectively. Using the coordinates shown and neglecting the masses of the pulleys, find the acceleration of each block and the tension T in the cable.

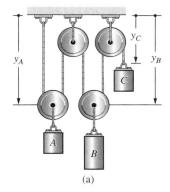

(a)

Solution

We will use the force-mass-acceleration method to derive the equation of motion for each block.

Kinematic Analysis

Letting L be the length of the cable that runs around the pulleys in Fig. (a), we have

$$L = 2y_A + 2y_B + y_C + C_1$$

where C_1 is a constant that accounts for the length of cable wrapped around the pulleys and the short cables supporting the upper two pulleys. Because the length L is constant, differentiation with respect to time yields

$$\frac{dL}{dt} = 2v_A + 2v_B + v_C = 0$$

where the v's are the velocities of the blocks. Performing another differentiation with respect to time gives the relationship between the accelerations of the blocks:

$$2a_A + 2a_B + a_C = 0 \qquad \text{(a)}$$

Kinetic Analysis

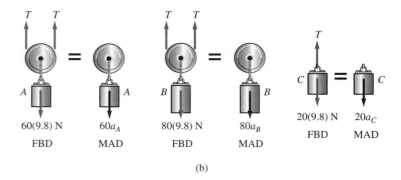

(b)

Figure (b) shows the free-body diagrams of blocks A and B (together with the massless pulleys to which they are attached), and block C. Note that the tension T is constant throughout the cable. Also shown are the corresponding mass-acceleration diagrams. Applying Newton's law $\Sigma F_y = ma$ to each block, the equations of motion are

$$+\downarrow \quad 60(9.8) - 2T = 60a_A$$

$$+\downarrow \quad 80(9.8) - 2T = 80a_B \qquad \text{(b)}$$

$$+\downarrow \quad 20(9.8) - T = 20a_C$$

Solving Eqs. (a) and (b), the cable tension and the acceleration of each block are found to be

$$T = 294 \text{ N}$$

$$a_A = 0 \qquad\qquad\qquad \textit{Answer}$$

$$a_B = 2.45 \text{ m/s}^2$$

$$a_C = -4.9 \text{ m/s}^2$$

The signs indicate that the acceleration of B is directed downward, whereas the acceleration of C is upward.

Problems

15.28   The three particles of a system move in the *xy*-plane. At a certain instant, the positions and accelerations of the particles are as shown in the figure. For this instant, determine (a) the coordinates of the mass center of the system; and (b) the acceleration of the mass center of the system.

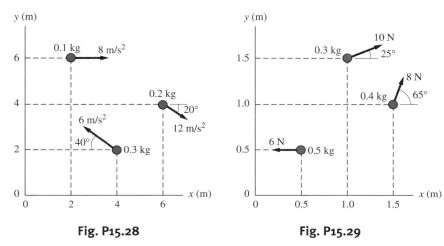

Fig. P15.28   **Fig. P15.29**

15.29   The three particles of a system move in the *xy*-plane. At the instant shown, the particles are acted on by the forces shown. For this instant, determine (a) the coordinates of the mass center of the system; and (b) the acceleration of the mass center of the system.

15.30   Two 90-kg men are seated in the 400-kg boat *A*. Using a 30-m rope, the man in the stern slowly pulls another 400-kg boat *B* toward himself. Find the distance moved by boat *A* when the two boats are about to touch. Neglect resistance from the water.

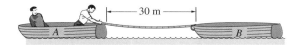

Fig. P15.30

15.31   A 0.08-kg bullet is fired from a rifle that is clamped to a trolley. The combined weight of the rifle and trolley is 20 kg. After firing, the velocity of the trolley is known to be $v_A = 4.8$ m/s to the left. Calculate (a) v_B, the velocity of the bullet after firing; and (b) the muzzle velocity (the velocity of the bullet relative to the barrel of the rifle).

15.32   A 42-kg woman *A* jumps off a 15-kg stationary cart *B*. Immediately after takeoff, the velocity of the woman relative to the cart is as shown in the figure. Determine the velocity vectors of the woman and the cart.

15.33   At the instant shown, the velocity of the 3-kg projectile *A* is 12**j** m/s when it breaks into two parts *B* and *C* weighing 2-kg and 1-kg, respectively. Part *B*

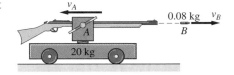

Fig. P15.31

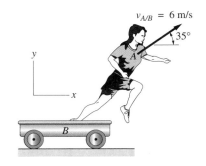

Fig. P15.32

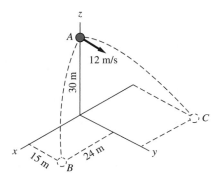

Fig. P15.33

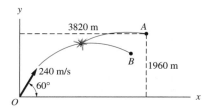

Fig. P15.34

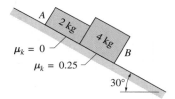

Fig. P15.36

subsequently hits the ground (the *xy*-plane) at the point (24 m, 15 m, 0). Determine the coordinates of the point where part *C* hits the ground. (Hint: Consider the motion of the mass center.)

15.34 A 60-kg projectile is launched from point *O* at $t = 0$ with the velocity shown. During flight, the projectile explodes into two parts, *A* and *B*, of masses 20 kg and 40 kg, respectively. The parts remain in the *xy*-plane. If the position of *A* at $t = 35$ s is as shown, find the position of *B* at this time. (Hint: Consider the motion of the mass center.)

15.35 Trailer *B* is hitched to the four-wheel-drive truck *A*. The coefficient of static friction between the truck tires and the road is 0.9. Determine (a) the maximum possible acceleration; and (b) the corresponding tensile force in the trailer hitch. Neglect rolling resistance of the trailer.

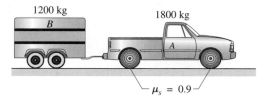

Fig. P15.35

15.36 The packages *A* and *B* slide down the inclined plane in contact with each other. Calculate the acceleration of the packages and the normal force between them.

15.37 The static coefficient of friction between the 10-kg crate and the 30-kg cart is 0.2. Find the maximum force *P* that may be applied to the crate without causing it to slip on the cart.

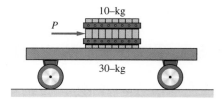

Fig. P15.37

15.38 Determine the tension in the cable connecting blocks *A* and *B* after the constant 60 N force is applied.

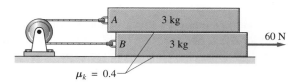

Fig. P15.38

15.39 The system consisting of blocks A and B, and the massless pulley C, is pulled upward by the constant 2.35-kN force. Determine the force in the cable joining A and B.

15.40 If the mass of block A is twice the mass of block B, find the acceleration of A in terms of the gravitational acceleration g. Neglect the masses of the pulleys.

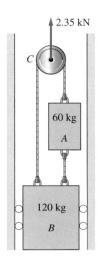

Fig. P15.39

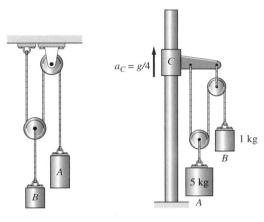

Fig. P15.40   **Fig. P15.41**

15.41 The acceleration of the sliding collar C is $g/4$, directed upward. Determine the accelerations of blocks A and B. Neglect the masses of the pulleys.

15.42 The 4-kg box A is resting on the left end of the 3-kg uniform plank B when the 50 N constant force is applied to A. Determine the distance x_A traveled by the box when it arrives at the right end of the plank. Use the kinetic coefficients of friction shown in the figure.

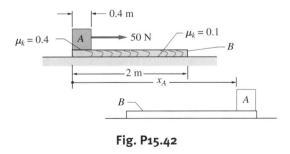

Fig. P15.42

15.43 The blocks A and B slide on the inclined plane with negligible friction. The blocks are connected by a rope that runs around the pulley C of negligible mass. Determine the acceleration of block A and the tension in the rope.

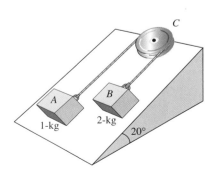

Fig. P15.43

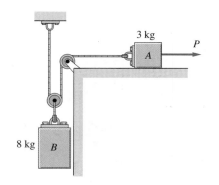

Fig. P15.44

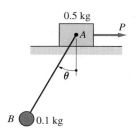

Fig. P15.45

15.44 Determine the magnitude of the force P that would cause block A to accelerate to the right at 4 m/s^2. Neglect friction and the mass of the pulley.

15.45 The block A is pulled along the horizontal surface by the constant force P. If the string connecting the bob B to the block maintains the constant angle $\theta = 35°$ to the vertical, determine the magnitude of P.

15.46 A rope connects blocks A and B as they slide down the cylindrical surface. In the position shown, the velocity of the each block is 1.2 m/s. Neglecting friction, calculate the tension in the rope for this position.

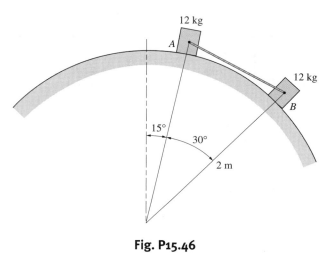

Fig. P15.46

15.47 Blocks A and B are connected by a rope as they slide down the inclined surface. The kinetic coefficient of friction between each block and the surface is shown in the figure. Determine the force in the rope.

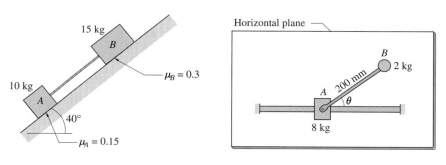

Fig. P15.47   **Fig. P15.48**

15.48 The mass B is attached to the arm that rotates in the horizontal plane about a pin in collar A. A motor in A keeps the angular speed of the arm constant at $\dot{\theta} = 2.4$ rad/s. Determine the velocity and acceleration of A as functions of the angle θ. Assume that $v_A = 0$ when $\theta = 0$. Friction and the mass of the rotating arm can be neglected.

15.49   Determine the forces in cables 1 and 2 of the pulley system shown. Neglect friction and masses of the pulleys.

*15.50   The truck A is about to tow the trailer B from a standing start. The truck has four-wheel drive, and the static coefficient of friction between its tires and the road is 0.8. Rolling resistance of the trailer is negligible. Determine the maximum possible initial acceleration of the truck if $x = 10$ m.

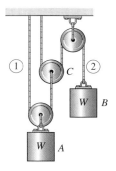

Fig. P15.49

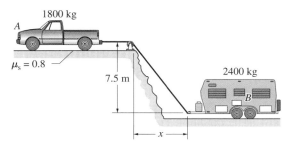

1800 kg
A
$\mu_s = 0.8$
7.5 m
2400 kg
B
x

Fig. P15.50

*15.51   Two identical blocks A and B are released from rest in the position shown. Calculate a_B and $a_{A/B}$ in terms of the gravitational acceleration g. Neglect friction.

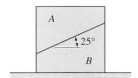

A
25°
B

Fig. P15.51

15.52   The stiffness of the spring that is attached to the two 2.5 kg blocks is 350 N/m. The system is initially at rest on the frictionless surface when the constant 1.2-lb force is applied at $t = 0$. (a) Derive the differential equation of motion for each block, and state the initial conditions. (b) Determine the speed of each block and the force P in the spring when $t = 0.1$ s. (Note: The analytical solution is $v_1 = 175.4$ mm/s, $v_2 = 44.6$ mm/s, $P = 3.01$ N.)

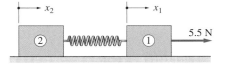

x_2 x_1
5.5 N
2   1

Fig. P15.52

15.53   The figure shows the top view of the two particles A and B that slide on a frictionless horizontal table. The particles carry identical electric charges, which give rise to the repulsive force $F = c/d^2$, where $c = 0.005$ N · m^2 and d is the distance between the particles in meters. At time $t = 0$, it is known that $d = 0.5$ m, A is at rest, and B is traveling toward A at the speed of 2 m/s. (a) Derive the differential equations of motion, and state the initial conditions. (b) Calculate the minimum value of d and the speed of each particle at that instant. (Note: The analytical solution is $d_{min} = 0.227$ m, $v_A = v_B = 800$ mm/s.)

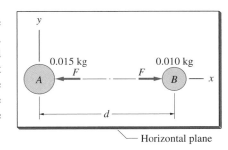

y
0.015 kg   0.010 kg
A F F B x
d
Horizontal plane

Fig. P15.53

15.54   The 0.025-kg bullet B traveling at 600 m/s hits and becomes embedded in the 15-kg block A, which was initially at rest on the frictionless surface. The force between A and B during the embedding phase is $F = 50v_{B/A}$, where F is in newtons and the relative velocity is in meters per second. (a) Determine the differential equations of motion for A and B during the embedding phase, and state the initial conditions. (b) Compute the velocity of B and the distance moved by A during 1.0 ms following the initial contact. (Note: The analytical solution is $v_B = 81.8$ m/s and $x_A = 0.567$ mm.)

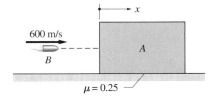

x
600 m/s
A
B
$\mu = 0.25$

Fig. P15.54

15.55   The two railroad cars are coasting with the velocities shown when they collide. The bumpers of the cars are ideal springs, the combined stiffness of two bumpers being 300 000 N/m. (a) Derive the differential equation of motion for each of the two railroad cars. (b) Assuming that the contact begins at $t = 0$, solve the equations of motion for the duration of the impact, which is approximately 0.4 s. (c) From the solution found in (b), determine the maximum value of the contact force and the time of contact.

Fig. P15.55

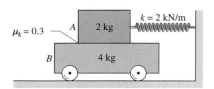

Fig. P15.56

15.56   The two blocks are released from rest at $t = 0$ with the spring stretched by 20 mm. (a) Derive the differential equations of motion for each block assuming that A slips relative to B. (b) Solve the equations of motion for the time interval $t = 0$ to $t = 0.2$ s and plot the velocity of each block versus time.

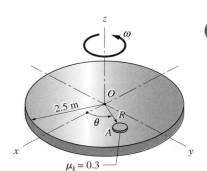

Fig. P15.57

15.57   The rough, horizontal disk of radius 2.5 m rotates at the constant angular speed $\omega = 45$ rev/min. The particle A is placed on the disk with no initial velocity at $t = 0$, $R = 1.0$ m, and $\theta = 0$. (a) Show that the differential equations of motion for the disk are

$$\ddot{R} = R\dot{\theta}^2 - \frac{\mu_k g \dot{R}}{v} \qquad \ddot{\theta} = -\frac{2\dot{R}\dot{\theta}}{R} - \frac{\mu_k g(\dot{\theta} - \omega)}{v}$$

where

$$v = \left[\dot{R}^2 + R^2\left(\dot{\theta} - \omega\right)^2\right]^{1/2}$$

and state the initial conditions. (b) Solve the equations numerically for the period of time that the particle stays on the disk. Use the numerical solution to find the speed of the particle when it is about to slide off the disk.

15.5	*Work-Energy Principles*

This article extends the work-energy methods for a single particle, presented in Arts. 14.2–14.4, to a system of particles.

a.   Work done on a system of particles

Figure 15.8 shows a typical (ith) particle belonging to a closed system of n particles. Using the notation introduced in Art. 15.4, the resultant external and internal forces acting on the ith particle are denoted by $\mathbf{F}_i$ and $\sum_{j=1}^{n} \mathbf{f}_{ij} (j \neq i)$, respectively. If the particle moves along the path $\mathscr{L}_i$ from position 1 to position 2, the work done on the particle is, according to Eq. (14.2),

$$(U_{1-2})_i = \int_{\mathscr{L}_i} \left(\mathbf{F}_i + \sum_{\substack{j=1 \\ j \neq i}}^{n} \mathbf{f}_{ij} \right) \cdot d\mathbf{r}_i \qquad (15.20)$$

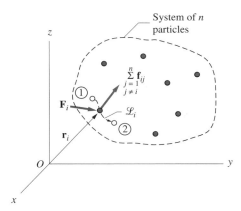

Fig. 15.8

where $\mathbf{r}_i$ is the position vector of the particle. The total work done on the system of n particles (i.e., the sum of the work done on each of the particles) can be expressed as

$$U_{1-2} = \sum_{i=1}^{n} (U_{1-2})_i = \sum_{i=1}^{n} \int_{\mathscr{L}_i} \mathbf{F}_i \cdot d\mathbf{r}_i + \sum_{i=1}^{n} \sum_{\substack{j=1 \\ j \neq i}}^{n} \int_{\mathscr{L}_i} \mathbf{f}_{ij} \cdot d\mathbf{r}_i \qquad (15.21)$$

The first term on the right side of this equation equals $(U_{1-2})_{\text{ext}}$, the work done by all of the external forces acting on the system. The second term represents $(U_{1-2})_{\text{int}}$, the total work done by the internal forces. Therefore, the total work done on the system of particles becomes

$$U_{1-2} = (U_{1-2})_{\text{ext}} + (U_{1-2})_{\text{int}} \qquad (15.22)$$

The work of the external forces, $(U_{1-2})_{\text{ext}}$, can be computed by the methods explained in Art. 14.2.

b. *Work of internal forces*

The calculation of the work done by internal forces, $(U_{1-2})_{\text{int}}$, can often be simplified as a result of the special nature of internal forces. As we have already seen in Eq. (15.14), internal forces occur in equal and opposite collinear pairs; in other words, $\mathbf{f}_{ij} = -\mathbf{f}_{ji}$. The total work done by a pair of internal forces will be zero if $d\mathbf{r}_i = d\mathbf{r}_j$, because then the work done by $\mathbf{f}_{ij}$ will cancel the work done by $\mathbf{f}_{ji}$. In this case, the internal forces are said to be *workless*.

To illustrate the difference between workless internal forces and internal forces that do work, consider the systems shown in Figs. 15.9(a) and (b), which consist of two blocks A and B that slide along a frictionless horizontal plane. Figure 15.9(a) shows the FBD of the system when the blocks are connected by an inextensible string; Fig. 15.9(b) shows the FBD of the system when the connection is an ideal spring of stiffness k. These two FBDs are identical because both systems are subject to the same external forces: W_A, W_B, N_A, and N_B. Note that in

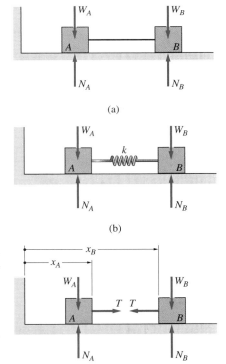

Fig. 15.9

each case $U_{\text{ext}} = 0$, because the external forces are perpendicular to the direction of motion.

To compute the work done by internal forces, it is necessary to analyze the forces acting on each block separately, as shown in the FBD of Fig. 15.9(c). The force T represents either the tension in the string for the system in Fig. 15.9(a) or the tension in the spring for the system in Fig. 15.9(b). During a differential movement of the system, the work done by T on blocks A and B is $T\,dx_A$ and $-T\,dx_B$, respectively. Therefore, the total work done by T on the system is $T(dx_A - dx_B)$. If the connection between the blocks is an inextensible string, then $dx_A = dx_B$, and the total work done by the string on the system is zero. However, if the connection is a linear spring, dx_A will generally not be equal to dx_B, because the spring can deform. We conclude that the spring force is capable of doing work on the system even though it is an internal force.

In summary, *rigid* internal connections, such as inextensible strings and pinned joints, perform equal and opposite work on the bodies they connect, which results in zero net work being performed on the system. On the other hand, *deformable* internal connections (which include springs) and friction surfaces that slide, are capable of doing work on a system.

c.   Principle of work and kinetic energy

Applying the work-energy principle, Eq. (14.15), to the arbitrary ith particle of the system, we have

$$(U_{1-2})_i = (\Delta T)_i \tag{15.23}$$

where $(U_{1-2})_i$ is the work done on the particle and $(\Delta T)_i$ is the change in its kinetic energy. If the system contains n particles, there will be n scalar equations similar to Eq. (15.23). Adding all of these equations and using Eq. (15.22), we find that

$$(U_{1-2})_{\text{ext}} + (U_{1-2})_{\text{int}} = \Delta T \tag{15.24}$$

where T, the kinetic energy of the system, is defined to be the sum of the kinetic energies of all of the particles; that is,

$$T = \sum_{i=1}^{n} T_i = \sum_{i=1}^{n} \frac{1}{2} m_i v_i^2 \tag{15.25}$$

d.   Conservation of mechanical energy

If all the forces, internal as well as external, are conservative (see Art. 14.4), the mechanical energy of the system is conserved; that is,

$$V_1 + T_1 = V_2 + T_2 \tag{15.26}$$

where V_1 and V_2 are the initial and final potential energies of the system (the sum of the potential energies of all the forces, both external and internal, that are capable of doing work on the system), and T_1 and T_2 are the initial and final kinetic energies of the system (the sum of the kinetic energies of all particles).

15.6 Principle of Impulse and Momentum

In this article we extend the impulse-momentum principles for a particle, discussed in Art. 14.6, to systems of particles.

a. Linear momentum

Consider the system of n particles in Fig. 15.10. As shown in the figure, the linear momentum of a typical (ith) particle is $\mathbf{p}_i = m_i \mathbf{v}_i$, where m_i is the mass of the particle and $\mathbf{v}_i$ is its velocity. The linear momentum (or simply the momentum) $\mathbf{p}$ of the system is defined as the vector sum of the linear momenta of all the particles in the system; that is,

$$\mathbf{p} = \sum_{i=1}^{n} \mathbf{p}_i = \sum_{i=1}^{n} m_i \mathbf{v}_i \tag{15.27}$$

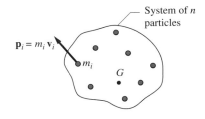

Fig. 15.10

Substituting from Eq. (15.13) $\sum_{i=1}^{n} m_i \mathbf{v}_i = m\overline{\mathbf{v}}$, where m is the total mass of the system and $\overline{\mathbf{v}}$ represents the velocity of G, the mass center of the system, we get

$$\mathbf{p} = m\overline{\mathbf{v}} \tag{15.28}$$

Thus the momentum of a system of particles of total mass m is equal to the momentum of a single particle of mass m that moves with the velocity $\overline{\mathbf{v}}$ of the mass center of the system.

b. Force-momentum relationship

Equation (15.19) stated that the motion of a closed system of particles is governed by $\Sigma \mathbf{F} = m\overline{\mathbf{a}}$, where $\Sigma \mathbf{F}$ is the resultant *external* force acting on the system, m is the total mass of the system, and $\overline{\mathbf{a}}$ is the acceleration of the mass center. Because the mass within a closed system is constant, this equation may be rewritten as

$$\Sigma \mathbf{F} = m\overline{\mathbf{a}} = m\frac{d\overline{\mathbf{v}}}{dt} = \frac{d}{dt}(m\overline{\mathbf{v}})$$

Using Eq. (15.28), the above force-momentum relationship for a closed system of particles becomes

$$\Sigma \mathbf{F} = \frac{d\mathbf{p}}{dt} \tag{15.29}$$

This equation is identical to Eq. (14.37), the force-momentum equation for a single particle.

c. Impulse-momentum principle

Multiplying both sides of Eq. (15.29) by dt and integrating between times t_1 and t_2, we find that

$$\int_{t_1}^{t_2} \Sigma \mathbf{F} \, dt = \int_{t_1}^{t_2} d\mathbf{p} = \mathbf{p}_2 - \mathbf{p}_1 \tag{15.30}$$

where $\mathbf{p}_1$ and $\mathbf{p}_2$ denote the momenta of the system at $t = t_1$ and t_2, respectively. Because the left side of Eq. (15.30) is the impulse of the *external* forces, denoted by $\mathbf{L}_{1-2}$, we can write

$$\mathbf{L}_{1-2} = \mathbf{p}_2 - \mathbf{p}_1 = \Delta\mathbf{p} \qquad (15.31)$$

which is the *impulse-momentum principle* for a system of particles. Note that this principle and Eq. (14.39), the impulse-momentum principle for a single particle, have the same form. Equation (15.31) is, of course, a vector equation that is equivalent to three scalar equations.

d. Conservation of momentum

From Eq. (15.31), we see that if the impulse of the external forces is zero, momentum of the system is conserved. In other words, if $\mathbf{L}_{1-2} = \mathbf{0}$, we obtain

$$\mathbf{p}_1 = \mathbf{p}_2 \quad \text{or} \quad \Delta\mathbf{p} = \mathbf{0} \qquad (15.32)$$

which is the *principle of conservation of momentum* for a system of particles. Equation (15.32) is identical to the principle of conservation of momentum for a particle, Eq. (14.41).

It should be noted that if the momentum of each particle of a system is conserved, then the momentum of the entire system is also conserved. However, the converse of this statement is not necessarily true: If the momentum of a system is conserved, it does not imply that the momentum of each particle is conserved.

Because Eq. (15.32) is a vector relationship, it is possible for a component of the momentum of a system to be conserved, even though the momentum vector is not conserved.

15.7 Principle of Angular Impulse and Momentum

The angular impulse-momentum principle was discussed in Art. 14.7. Here we extend the principle to a system of particles.

a. Angular momentum

Consider again a closed system of n particles. Let m_i be the mass and $\mathbf{v}_i$ the *absolute* velocity of a typical (ith) particle of the system. Recalling that the angular momentum of a particle is the moment of its linear momentum, the angular momentum of the ith particle about an arbitrary point A is $(\mathbf{h}_A)_i = \mathbf{r}_i \times (m_i\mathbf{v}_i)$, where $\mathbf{r}_i$ is the position vector of the particle *relative* to A, as shown in Fig. 15.11. The angular momentum of the system about A is obtained by adding the angular momenta of all the particles in the system about A:

$$\mathbf{h}_A = \sum_{i=1}^{n}(\mathbf{h}_A)_i = \sum_{i=1}^{n}\mathbf{r}_i \times (m_i\mathbf{v}_i) \qquad (15.33)$$

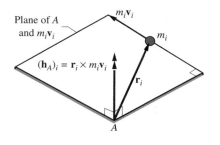

Plane of A
and $m_i\mathbf{v}_i$

$m_i\mathbf{v}_i$

m_i

$(\mathbf{h}_A)_i = \mathbf{r}_i \times m_i\mathbf{v}_i$

$\mathbf{r}_i$

A

Fig. 15.11

b. *Moment-angular momentum relationship*

Differentiating the expression for $\mathbf{h}_A$ in Eq. (15.33) with respect to time, we obtain

$$\dot{\mathbf{h}}_A = \frac{d}{dt}\left[\sum_{i=1}^{n}\mathbf{r}_i \times (m_i\mathbf{v}_i)\right] = \sum_{i=1}^{n}\mathbf{r}_i \times \frac{d(m_i\mathbf{v}_i)}{dt} + \sum_{i=1}^{n}\frac{d\mathbf{r}_i}{dt} \times (m_i\mathbf{v}_i) \qquad \text{(a)}$$

According to the force-momentum relationship in Eq. (14.37), $d(m_i\mathbf{v}_i)/dt$ is the resultant force acting on the ith particle. Therefore, the first sum on the right side of Eq. (a) represents the moment about A of all the forces that act on the particles in the system. Since the internal forces occur in equal, opposite, and collinear pairs, they contribute nothing to the resultant moment. Hence

$$\sum_{i=1}^{n}\mathbf{r}_i \times \frac{d(m_i\mathbf{v}_i)}{dt} = \Sigma\mathbf{M}_A \qquad \text{(b)}$$

where $\Sigma\mathbf{M}_A$ is the resultant moment about A of the forces that are *external* to the system.

The second sum on the right side of Eq. (a) can be simplified by recalling that $\mathbf{r}_i$ is the position vector of the ith particle relative to point A. It follows that $d\mathbf{r}_i/dt$ is the velocity of the particle relative to A; that is, $d\mathbf{r}_i/dt = \mathbf{v}_i - \mathbf{v}_A$. Therefore,

$$\sum_{i=1}^{n}\frac{d\mathbf{r}_i}{dt} \times (m_i\mathbf{v}_i) = \sum_{i=1}^{n}(\mathbf{v}_i - \mathbf{v}_A) \times (m_i\mathbf{v}_i) = \sum_{i=1}^{n}\mathbf{v}_i \times (m_i\mathbf{v}_i) - \mathbf{v}_A \times \sum_{i=1}^{n}m_i\mathbf{v}_i \qquad \text{(c)}$$

We note that $\mathbf{v}_i \times (m_i\mathbf{v}_i) = \mathbf{0}$ (the cross product of two parallel vectors vanishes). Moreover, from Eq. (15.13) we obtain $\sum_{i=1}^{n}m_i\mathbf{v}_i = m\bar{\mathbf{v}}$, where m is the total mass of the system and $\bar{\mathbf{v}}$ is the velocity of its mass center. As a result, Eq. (c) becomes

$$\sum_{i=1}^{n}\frac{d\mathbf{r}_i}{dt} \times (m_i\mathbf{v}_i) = -\mathbf{v}_A \times (m\bar{\mathbf{v}}) \qquad \text{(d)}$$

Substituting Eqs. (b) and (d) into Eq. (a) and solving for $\Sigma\mathbf{M}_A$, we get the moment-angular momentum relationship

$$\Sigma\mathbf{M}_A = \dot{\mathbf{h}}_A + \mathbf{v}_A \times (m\bar{\mathbf{v}}) \qquad \text{(15.34)}$$

If A is fixed in an inertial reference frame ($\mathbf{v}_A = \mathbf{0}$), or if A is the mass center of the system ($\mathbf{v}_A = \bar{\mathbf{v}}$), Eq. (15.34) simplifies to

$$\Sigma \mathbf{M}_A = \dot{\mathbf{h}}_A \qquad (A: \text{fixed point or mass center}) \tag{15.35}$$

c. Angular impulse-momentum principle

Multiplying each side of Eq. (15.35) by dt and integrating from t_1 and t_2, we obtain

$$\int_{t_1}^{t_2} \Sigma \mathbf{M}_A \, dt = \int_{t_1}^{t_2} d\mathbf{h}_A = (\mathbf{h}_A)_2 - (\mathbf{h}_A)_1$$

where $(\mathbf{h}_A)_1$ and $(\mathbf{h}_A)_2$ are the angular momenta about A at times t_1 and t_2, respectively. Recognizing that the left side of this equation is by definition the angular impulse of the *external* forces about A, the equation may be written as

$$(\mathbf{A}_A)_{1-2} = (\mathbf{h}_A)_2 - (\mathbf{h}_A)_1 \qquad \begin{array}{l} (A: \text{fixed point} \\ \text{or mass center}) \end{array} \tag{15.36}$$

Equation (15.36) is called the *angular impulse-momentum principle*. Note that this principle is valid only if A is a fixed point or the mass center of a closed system of particles.

d. Conservation of angular momentum

If the angular impulse of the external forces about A is zero, it follows from Eq. (15.36) that the angular momentum of the system of particles is conserved about A. In other words,

$$\text{If } (\mathbf{A}_A)_{1-2} = \mathbf{0}, \quad \text{then } (\mathbf{h}_A)_1 = (\mathbf{h}_A)_2 \qquad \begin{array}{l} (A: \text{fixed point} \\ \text{or mass center}) \end{array} \tag{15.37}$$

which is known as the *principle of conservation of angular momentum*. Observe that angular momentum about a fixed point, or about the mass center, is conserved during a given time interval if and only if the angular impulse about that point is zero throughout that time interval. Since Eq. (15.37) is a vector equation, it is possible for the angular momentum about an axis passing through A to be conserved, even though the total angular momentum about point A may not be conserved.

Sample Problem 15.8

The blocks A and B are connected by a cable that runs around two pulleys of negligible mass, as shown in Fig. (a). The kinetic coefficient of friction between the inclined plane and block A is 0.4. If the initial velocity of A is 3 m/s down the plane, determine the displacement Δs_A of block A (measured from its initial position) when the system comes to rest.

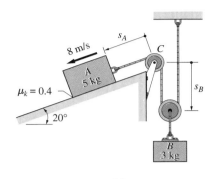

(a)

Solution

We will use the work-energy method to analyze this problem, because it is ideally suited for determining a displacement that occurs during a given change of speed. Considering the free-body diagram of the entire system in Fig. (b), we observe that the only forces that do work are the weights W_A and W_B of the blocks, and the friction force F_A beneath block A. The forces represented by the dashed arrows are workless.

1. Computation of friction force

The friction force F_A can be determined using the free-body diagram of block A in Fig. (c):

$$\Sigma F_y = 0 \quad \xrightarrow{+} \quad N_A - W_A \cos 20° = 0$$

$$F_A = \mu_k N_A = \mu_k W_A \cos 20° = 0.4(5)(9.8) \cos 20° = 18.42 \, \text{N} \qquad \text{(a)}$$

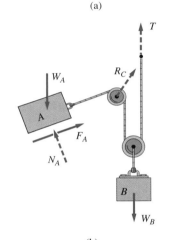

(b)

2. Kinematics

Referring to Fig. (a), we see that the kinematic constraint imposed by the constant length of the cable on the positions of the blocks is

$$s_A + 2s_B = \text{constant}$$

Therefore, the displacements (Δs_A and Δs_B) and the velocities of the two blocks are related by

$$\Delta s_A + 2\Delta s_B = 0 \qquad v_A + 2v_B = 0 \qquad \text{(b)}$$

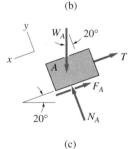

(c)

3. Work-energy principle

Let 1 and 2 denote the initial and the final (rest) positions of the system. Applying the work-energy principle, Eq. (15.24), to the system, we get

$$(U_{1-2})_{\text{ext}} + (U_{1-2})_{\text{int}} = T_2 - T_1$$

$$[(W_A \sin 20°)\Delta s_A - F_A \Delta s_A + W_B \Delta s_B] + 0 = 0 - \frac{1}{2}\left[\frac{W_A}{g}(v_A)_1^2 + \frac{W_B}{g}(v_B)_1^2\right]$$

(c)

Substituting the known values and utilizing Eqs. (b), the work-energy equation becomes

$$(5)(9.8) \sin 20 \Delta s_A - 18.42\Delta s_A + 3(9.8)\left(\frac{-\Delta s_A}{2}\right) = -\frac{1}{2}\left[\frac{49}{9.8}(3)^2 + \frac{29.4}{9.8}\left(\frac{-3}{2}\right)^2\right]$$

The solution for the displacement of block A is

$$\Delta s_A = 1.58 \text{ m} \qquad \textit{Answer}$$

Note

This problem could also be solved by using two work-energy equations, one for each block. In that case, work done by the cable tension would appear in each equation. However, when the two equations are added, the work of the cable tension would cancel, and we would finish up with Eq. (c).

Sample Problem 15.9

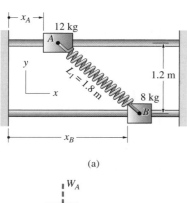

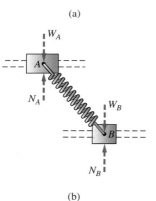

(a)

(b)

The two collars A and B shown in Fig. (a) slide along frictionless bars that lie in the same vertical plane and are 1.2 m apart. The stiffness of the spring is $k = 100$ N/m, and its free length is $L_0 = 1.2$ m. If the system is released from rest in the position shown in Fig. (a), where the spring has been stretched to the length $L_1 = 1.8$ m, calculate the maximum speed reached by each of the collars.

Solution

When the system is released from rest in the position shown in Fig. (a), which we will refer to as position 1, the tension in the spring pulls the collars toward each other. Because the free length of the spring is identical to the distance between the rails, the spring will be unstretched when A is directly above B. This position, which we shall denote as position 2, is thus the position where the speeds of the collars are maximized. After passing through position 2, the tension in the spring reduces the speeds of the collars, eventually bringing them to a temporary stop. The motion then reverses itself, with the system returning to position 1, since the system is conservative.

1. Work-energy principle

We will analyze the system consisting of the two collars and the spring. Figure (b) shows the free-body diagram of this system for an arbitrary position. The only external forces that act on this system are the weights of the collars, W_A and W_B, and the normal forces, N_A and N_B, that are provided by the rails. The force in the spring does not appear on the FBD because it is an internal force.

From the FBD in Fig. (b), we see that $(U_{1-2})_{\text{ext}} = 0$ because each external force is perpendicular to the path of the collar. Therefore, the work-energy principle,[*] Eq. (15.24), yields

$$(U_{1-2})_{\text{ext}} + (U_{1-2})_{\text{int}} = T_2 - T_1$$
$$0 + (U_{1-2})_{\text{int}} = T_2 - 0$$

where we have substituted $T_1 = 0$ for the initial kinetic energy. Using Eq. (14.10) to compute $(U_{1-2})_{\text{int}}$, the work done by the spring force, and recognizing that the

---

[*]The principle of conservation of mechanical energy, Eq. (15.26), could also have been used, because the system is conservative.

kinetic energy of the system is the sum of the kinetic energy for each collar, this equation becomes

$$-\frac{1}{2}k\left(\delta_2^2 - \delta_1^2\right) = \frac{1}{2}m_A\,(v_A)_2^2 + \frac{1}{2}m_B\,(v_B)_2^2$$

Substituting numerical values—noting that the spring deformations are $\delta_1 = L_1 - L_0 = 1.8 - 1.2 = 0.6$ m and $\delta_2 = 0$—we obtain

$$-\frac{1}{2}(100)[0 - (0.6)^2] = \frac{1}{2}(12)(v_A)_2^2 + \frac{1}{2}(8)(v_B)_2^2$$

which may be simplified to

$$6(v_A)_2^2 + 4(v_B)_2^2 = 18 \qquad\qquad \text{(a)}$$

2. Impulse-momentum principle

A second equation relating the final velocities of A and B is obtained by applying the impulse-momentum principle to the system. From the FBD in Fig. (b), we see that there are no external forces acting on the system in the x-direction. Therefore, the momentum of the system is conserved in the x-direction. (The change in momentum in the vertical direction for the system is also zero, but this is of no interest here.) Since the momentum in the x-direction in position 1 is zero, it must also be zero in position 2. Computing the momentum of the system by adding the momenta of both collars, we have

$$\xrightarrow{+}\quad (p_x)_2 = m_A(v_A)_2 + m_B(v_B)_2 = 0$$

where both velocities were assumed to be directed to the right. Substituting values for the masses, this equation becomes

$$12(v_A)_2 + 8(v_B)_2 = 0 \qquad\qquad \text{(b)}$$

Solving Eqs. (a) and (b) simultaneously, the maximum speeds of the collars are found to be

$$(v_A)_2 = \pm 1.095 \text{ m/s} \quad \text{and} \quad (v_B)_2 = \mp 1.643 \text{ m/s} \qquad \textit{Answer}$$

The duality of signs in these answers indicates that if collar A is moving to the right as it passes through position 2, collar B is moving to the left, and vice versa.

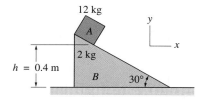

12 kg

$h = 0.4$ m

2 kg

B 30°

(a) Position ①

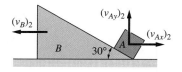

$(v_B)_2$

$(v_{Ay})_2$

B 30° A $(v_{Ax})_2$

(b) Position ②

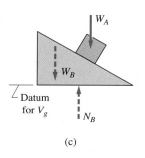

W_A

W_B

Datum
for V_g

N_B

(c)

Sample Problem   15.10

The 12-kg block A in Fig. (a) is released from rest at the top of the 2-kg wedge B (position 1). Determine the velocities of A and B when the block has reached the bottom of the inclined face of B as shown in Fig. (b) (position 2). Neglect friction.

Solution

Our solution consists of the following steps applied to the system consisting of the block A and the wedge B:

Step 1: Apply the principle of conservation of mechanical energy.
Step 2: Apply the principle of conservation of linear momentum in the x-direction.
Step 3: Relate the velocities of A and B using kinematics.
Step 4: Solve the equations that result from Steps 1 through 3.

The order in which Steps 1 through 3 are performed is immaterial. The assumed directions for the velocities of A and B in position 2 are shown in Fig. (b). The free-body diagram of the system in an arbitrary position is shown in Fig. (c).

Step 1: Conservation of Mechanical Energy

Because friction is neglected, the only force that does work on the system between positions 1 and 2 is the weight of A. This means that the system is conservative. Choosing the horizontal plane as the datum for V_g, as indicated in Fig. (c), the principle of conservation of mechanical energy,[*] Eq. (15.26), gives

$$V_1 + T_1 = V_2 + T_2$$

$$m_A g h + 0 = 0 + \frac{1}{2} m_A (v_A)_2^2 + \frac{1}{2} m_B (v_B)_2^2$$

Note that $T_1 = 0$ and $V_2 = 0$. Substituting numerical values, this equation becomes

$$12(9.81)(0.4) = \frac{1}{2}(12)(v_A)_2^2 + \frac{1}{2}(2)(v_B)_2^2$$

which after simplification can be written as

$$6(v_A)_2^2 + (v_B)_2^2 = 47.09 \tag{a}$$

Step 2: Conservation of Linear Momentum in the x-Direction

From the free-body diagram in Fig. (c) we see that there are no forces that act on the system in the x-direction. Therefore the x-component of the momentum of the system is conserved. (The normal force that acts between A and B has an x-component, but this force is internal to the system.) Because the x-component

[*]The work-energy principle, Eq. (15.24), could also have been used.

of the momentum of the system in position 1 is zero, the x-component of the momentum of the system in position 2 is also zero. Using the velocities shown in Fig. (b), we obtain

$$\xrightarrow{+} \quad (p_x)_2 = m_A(v_{Ax})_2 - m_B(v_B)_2 = 0$$

Substituting the values for m_A and m_B, we find that

$$12(v_{Ax})_2 - 2(v_B)_2 = 0 \qquad \text{(b)}$$

Step 3: Relate the Velocities of A and B Using Kinematics

The velocities of A and B must satisfy the relative velocity equation $\mathbf{v}_A = \mathbf{v}_{A/B} + \mathbf{v}_B$. The kinematic constraint is that $\mathbf{v}_{A/B}$ is directed along the inclined face of B. Assuming that $\mathbf{v}_{A/B}$ is directed down the inclined face, we have

$$\mathbf{v}_{A/B} = \;{}^{30°} \!\!\!\diagdown\!\! {}_{v_{A/B}}$$

Using the velocities shown in Fig. (b), the relative velocity equation for position 2 becomes

$$(\mathbf{v}_A)_2 = (\mathbf{v}_{A/B})_2 + (\mathbf{v}_B)_2$$
$$(v_{Ax})_2\mathbf{i} + (v_{Ay})_2\mathbf{j} = (v_{A/B})_2 \; \cos 30°\mathbf{i} - (v_{A/B})_2 \sin 30°\mathbf{j} - (v_B)_2\mathbf{i}$$

Equating the vector components gives the following two scalar equations:

$$(v_{Ax})_2 = (v_{A/B})_2 \cos 30° - (v_B)_2 \qquad \text{(c)}$$
$$(v_{Ay})_2 = -(v_{A/B})_2 \sin 30° \qquad \text{(d)}$$

Step 4: Solution of the Equations That Result from Steps 1 through 3

Inspection reveals that Eqs. (a) through (d) represent four equations that contain the four unknowns: $(v_{Ax})_2$, $(v_{Ay})_2$, $(v_B)_2$, and $(v_{A/B})_2$. Omitting the algebraic details, the solution yields

$$(v_{Ax})_2 = 0.580 \text{ m/s} \qquad (v_{Ay})_2 = -2.34 \text{ m/s}$$

Answer

$$(v_B)_2 = 3.48 \text{ m/s} \qquad (v_{A/B})_2 = 4.69 \text{ m/s}$$

Because the sign of $(v_{Ay})_2$ is negative, its direction is opposite to what is shown in Fig. (b).

The speed of A in position 2 is, therefore,

$$(v_A)_2 = \sqrt{(v_{Ax})_2^2 + (v_{Ay})_2^2} = \sqrt{(0.580)^2 + (2.34)^2}$$
$$= 2.41 \text{ m/s} \qquad \qquad \textit{Answer}$$

Sample Problem 15.11

The assembly shown in Fig. (a) consists of two small balls, each of mass m, that slide on a frictionless, rigid frame AOB of negligible mass. The support at O permits free rotation of the frame about the z-axis. The frame is initially rotating with the angular velocity ω_1 while strings hold the balls at the radial distance R_1. The strings are then cut simultaneously, permitting the balls to slide toward the stops at A and B, which are located at the radial distance R_2. Determine ω_2, the final angular velocity of the assembly, assuming that the balls do not rebound after striking the stops.

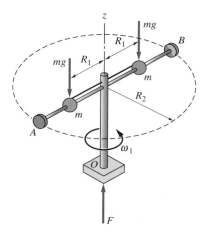

(a) Free-body diagram

Solution

The free-body diagram (FBD) of the assembly before the strings are cut is shown in Fig. (a). The only external forces acting on the assembly are the weights mg of the balls and the vertical support force F at O. (Symmetry precludes other forces or moments exerted by the support at O.) The forces acting between the balls and the rod, and the forces in the strings, are all internal forces that do not appear on the FBD of the system.

From Fig. (a) we see that the moment of the external forces about the z-axis is zero since these forces are parallel to the axis. After the strings have been cut, this FBD will change with time because the balls move away from the z-axis. However, the weights are always parallel to the z-axis, which means that the angular impulse about the z-axis continues to be zero as the balls move along the rod. When the balls hit the stops, the resulting impact forces are internal to the FBD of the assembly. Consequently, there will never be an angular impulse acting on the assembly about the z-axis, which means that angular momentum about that axis is always conserved.

The momentum diagrams of the assembly at times t_1 and t_2 are shown in Figs. (b) and (c), respectively, where t_1 is a time before the strings were cut, and t_2 is a time after the balls have come to rest relative to the rod. Only the linear momentum of each ball is shown, since the mass of the frame is negligible. The velocities of the balls are related to the angular velocities by $v_1 = R_1\omega_1$ and $v_2 = R_2\omega_2$.

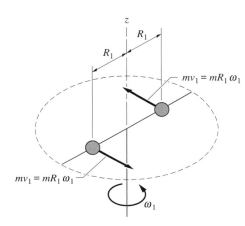

(b) Momentum diagram at $t = t_1$

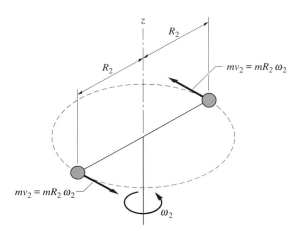

(c) Momentum diagram at $t = t_2$

Because angular momentum equals the moment of the linear momentum, conservation of angular momentum about the z-axis yields

$$(h_z)_1 = (h_z)_2$$

$$+\quad 2(m R_1 \omega_1) R_1 = 2(m R_2 \omega_2) R_2$$

from which we find

$$\omega_2 = (R_1/R_2)^2 \omega_1 \qquad\qquad \textit{Answer}$$

Note

Moving mass toward or away from the axis of rotation is an effective means of controlling angular speed (note the *square* of the ratio R_1/R_2 in the last answer). Figure skaters, for example, use the positioning of their arms to vary their rate of spin.

Problems

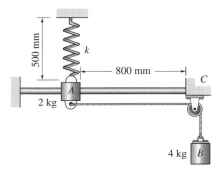

Fig. P15.58

15.58 The sliding collar A and the mass B are connected by an inextensible rope. The spring attached to A has stiffness $k = 400$ N/m, and its free length is $L_0 = 200$ mm. In the position shown, the velocity of A is v_A to the right. If A is to arrive at the stop C with zero velocity, determine v_A. Neglect friction.

15.59 In the position shown, block A is moving to the left at 6 m/s, and the spring is undeformed. Determine the spring stiffness k that would cause the system to come to rest after A has moved 1 m. The coefficient of kinetic friction between A and the horizontal surface is 0.3, and the weights of the pulleys are negligible.

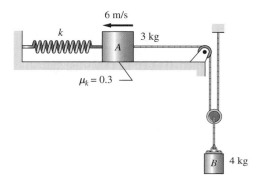

Fig. P15.59

15.60 Blocks A and B are connected by a cord that has a length of 6.5 m and passes over a small pulley C. If the system is released from rest when $x_A = 4$ m, determine the speed of A when B reaches the position shown by dashed lines. Neglect friction.

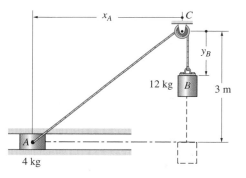

Fig. P15.60

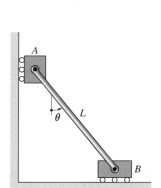

Fig. P15.61

15.61 The identical masses A and B are connected by a rigid rod of negligible mass and length L. Use the principle of conservation of mechanical energy to show that the differential equation governing the motion is $\ddot{\theta} = (g/L)\sin\theta$.

15.62 The 40-kg boy jumps from the 60-kg boat onto the dock with the velocity shown. If the boat was stationary before the boy jumped, with what velocity will the boat leave the dock?

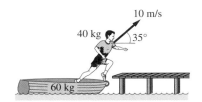

Fig. P15.62

15.63 Particle *A* of mass m_A is released from rest in the position shown and slides with negligible friction down the quarter-circular track of radius *R*. The body *B* of mass m_B containing the track can slide freely on the horizontal surface. Determine the speeds of *A* and *B* when *A* reaches the bottom of the track.

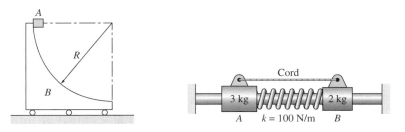

Fig. P15.63   **Fig. P15.64**

15.64 The compressive force in the spring equals 20 N when the system is at rest in the position shown. If the cord is cut, find the velocities of masses *A* and *B* when the spring force becomes zero. Neglect friction.

15.65 The package *A* lands on the stationary cart *B* with the horizontal velocity $v_0 = 2.5$ m/s. The coefficient of kinetic friction between *A* and *B* is 0.25, and the rolling resistance of *B* may be neglected. (a) Determine the velocity of *B* after the sliding of *A* relative to *B* has stopped. (b) Find the total distance that *A* slides relative to *B*.

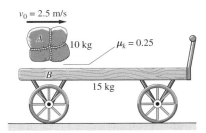

Fig. P15.65

15.66 The blocks *A* and *B* are connected by a cable that runs around two pulleys of negligible weight. Determine the time required for *A* to reach a speed of 2 m/s after the system has been released from rest. Neglect friction.

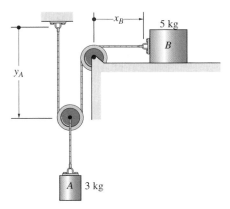

Fig. P15.66

15.67 The system is released from rest with the spring stretched 100 mm. Determine the velocity of the block relative to the cart at the instant the spring has returned to its unstretched length.

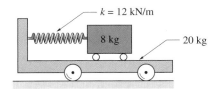

Fig. P15.67

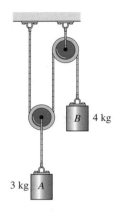

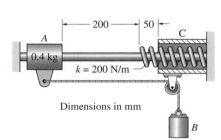

Fig. P15.68

15.68   The system consists of blocks A and B that are connected by an inextensible cable running around two pulleys. If the system is released from rest at time $t = 0$, find the velocity of A at $t = 5$ s. Neglect the weights of the pulleys.

15.69   The system is released from rest in the position shown. Neglecting friction, find the mass of B that will cause A to reach the left end of the retaining tube C with zero velocity. Note that the stiffness of the spring in the retaining tube is 200 N/m.

15.70   The block A and cart B are stationary when the constant force $P = 6$ N is applied. Find the speed of the cart when the block has moved 0.6 m relative to the cart. Neglect rolling resistance of the cart, and note that the surface between the block and cart is rough.

Fig. P15.69

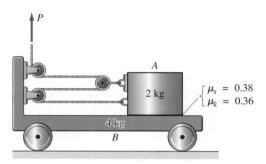

Fig. P15.70

15.71   The system is released from rest when $\theta = 0$. Determine the ratio m_A/m_B of the two masses for which the system will come to rest again when $\theta = 60°$. Neglect friction.

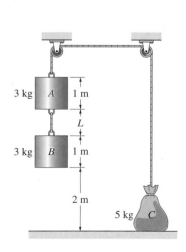

Fig. P15.73

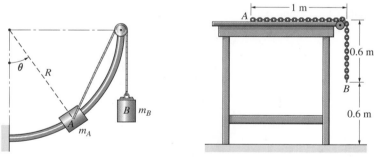

Fig. P15.71   **Fig. P15.72**

15.72   The 1.6 m chain AB weighs 4.8 N. If the chain is released from rest in the position shown, calculate the speed with which end B hits the floor. Neglect friction.

15.73   The system is released from rest in the position shown. Determine the distance L for which A would come to rest just before it hits B. (Hint: You must first compute the speed of the system when B hits the floor. Why?)

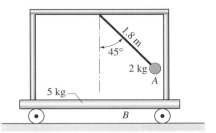

Fig. P15.75

15.74 For the two railroad cars described in Prob. 15.55, determine the maximum contact force between the bumpers during the collision. Neglect rolling resistance. (Hint: The maximum force occurs when the cars have the same velocity.)

15.75 The system is at rest when the simple pendulum is released from the position shown. When the pendulum reaches the vertical position for the first time, compute the absolute speed of (a) the pendulum's bob A; and (b) the carriage B. Neglect rolling resistance of the carriage.

15.76 The three identical masses are connected by rigid rods of negligible mass. The assembly is rotating about point O at the counterclockwise angular speed ω. Calculate the angular momentum of the assembly about its mass center G.

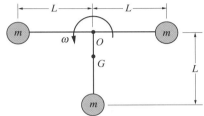

Fig. P15.76

15.77 The system consists of three 5-kg particles. At a certain instant, the positions and velocities of particles are as shown in the figure. For this instant, calculate the angular momentum of the system about (a) point O; and (b) point A.

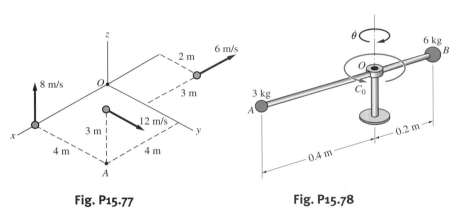

Fig. P15.77 **Fig. P15.78**

15.78 The rigid assembly, consisting of the two masses attached to a massless rod, rotates about the vertical axis at O. The assembly is initially rotating freely at the angular speed $\dot{\theta}_0 = 130$ rad/s, when the constant couple $C_0 = 6$ N · m that opposes the motion is applied. Find (a) the time required to stop the assembly; and (b) the number of revolutions made by the assembly before coming to rest.

15.79 A flyball speed governor consists of the two 0.5-kg weights and a supporting linkage of negligible weight. The position of the weights can be changed by adjusting the magnitude of the force F acting on the sliding collar. The entire assembly is initially rotating about the z-axis at 500 rev/min with the supporting arms inclined at $\alpha = 60°$. (a) Compute the initial angular momentum of the system about the z-axis. (b) Find the angular speed of the assembly when the angle α is changed to 30°. Neglect friction.

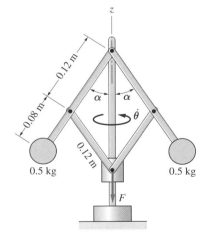

Fig. P15.79

15.80 The two blocks are released from rest simultaneously in the positions shown. The free length of the spring connecting the blocks is 0.2 m. Determine which block is the first to hit the stop at C, and find the speed of that block just before hitting. Neglect friction.

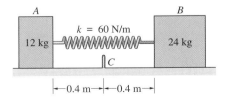

Fig. P15.80

15.81   The particles A and B, connected by a rigid rod of negligible mass, slide on the horizontal plane. When the assembly is in the position shown in Fig. (a), the velocity of each particle is v_0 in the direction indicated in the figure. Determine the velocities of the particles after the assembly has rotated through 90° to the position shown in Fig. (b). Neglect friction.

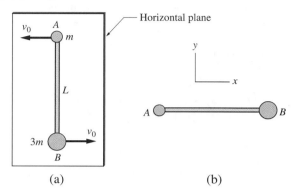

(a)   (b)

Fig. P15.81

15.82   If the system is released from rest in the position $y_B = 0$, determine the velocity of weight B when $y_B = 250$ mm. Neglect friction and the masses of the pulleys.

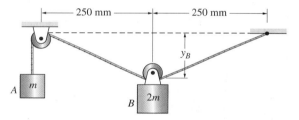

Fig. P15.82

15.83   The system consists of the electric hoist A, the crate B, and the counterweight C, all of which are suspended from the pulley D. The system is stationary

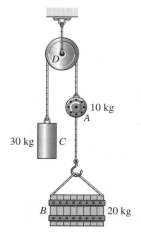

Fig. P15.83

when the hoist A is turned on, causing it to start rewinding the rope connecting A and B at the rate of 1 m/s. Determine the resulting velocities of A, B, and C.

*15.84   The assembly consists of three particles A, B, and C (each of mass m) that are connected to a light, rigid frame. The assembly is initially rotating counterclockwise about its mass center D with the angular velocity ω_0. If rod DC suddenly breaks when the assembly is in the position shown in (a), determine the velocity of masses A and B when they reach the position shown in (b).

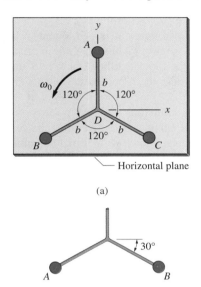

(a)

(b)

Fig. P15.84

15.85   The assembly consisting of two identical masses and a supporting frame of negligible mass rotates freely about the z-axis. Mass A is attached to the frame, but mass B is free to slide on the horizontal bar. An ideal spring of stiffness k and free length R_0 is connected between A and B. Determine all possible combinations of R and $\dot{\theta}$ for which B remains at rest relative to the frame.

15.86   Mass B of the assembly described in Prob. 15.85 is released from the position $R = R_0$ when $\dot{\theta} = \dot{\theta}_0$. (a) Find the speed of B relative to the frame when it passes point O. (b) Determine the range of $\dot{\theta}_0$ for which B will not reach O.

15.87   The mass B of the assembly described in Prob. 15.85 is released at $t = 0$ when $R = R_0$ and $\dot{\theta} = \dot{\theta}_0$. (a) Show that the differential equations of motion are

$$\ddot{R} = R\left(\dot{\theta}^2 - \frac{k}{m}\right) \qquad \ddot{\theta} = -\frac{2R\dot{R}\dot{\theta}}{R_0^2 + R^2}$$

and state the initial conditions. (b) Solve the equations numerically from $t = 0$ to 0.2 s, using the following data: $m = 0.25$ kg, $k = 300$ N/m, $R_0 = 0.4$ m, and $\dot{\theta}_0 = 30$ rad/s; plot $\dot{\theta}$ versus R. (c) Use the numerical solution to determine the range of R and $\dot{\theta}$.

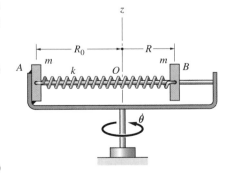

Fig. P15.85–P15.87

(a) Velocities before impact: $(v_A)_1 > (v_B)_1$

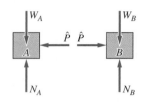

(b) FBD during impact

(c) Velocities after impact

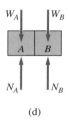

(d)

Fig. 15.12

15.8 Plastic Impact

One of the more complicated problems in dynamics is the impact, or collision, between objects. Because the magnitudes of the contact forces during an impact are usually unknown, the force-mass-acceleration method of analysis cannot be used. Furthermore, because mechanical energy is generally lost during the impact (energy may be converted into sound or heat), the work-energy approach cannot be applied directly. That leaves the impulse-momentum method as the only usable technique for the analysis of impact problems.

To explain the nature of impact, it is convenient to begin with the situation depicted in Fig. 15.12(a), where two blocks A and B are sliding to the right on a horizontal plane. Before the blocks collide, their velocities are $(v_A)_1$ and $(v_B)_1$, where $(v_A)_1 > (v_B)_1$. During the time that the blocks are in contact, equal and opposite time-dependent forces act between the blocks. These contact or *impact forces*, being caused by the collision, are equal to zero before and after the impact. One of the goals of impact analysis is to determine the velocities $(v_A)_2$ and $(v_B)_2$ of the blocks after the impact, knowing the initial velocities $(v_A)_1$ and $(v_B)_1$.

The free-body diagram of each block during the impact is shown in Fig. 15.12(b). In addition to the weights W_A and W_B and the normal forces N_A and N_B (we neglect friction), the blocks are subjected to the equal and opposite impact forces $\hat{P}$. (We will use a caret ^ above a letter to indicate an impact force.) Drawing the FBDs during the impact and identifying the impact forces are very important steps in the analysis of impact problems.

Summing forces in the y-direction for either block gives $N_A = W_A$ and $N_B = W_B$ throughout the motion. Applying the impulse-momentum principle, $(L_{1-2})_x = \Delta p_x$, to block A, and using the velocities shown in Fig. 15.12 parts (a) and (c), we obtain

$$\xrightarrow{+} \qquad -\int_{t_1}^{t_2} \hat{P}\, dt = m_A(v_A)_2 - m_A(v_A)_1 \qquad (15.38)$$

where $t = t_1$ to t_2 is the duration of the impact. Similarly, for block B we obtain

$$\xrightarrow{+} \qquad \int_{t_1}^{t_2} \hat{P}\, dt = m_B(v_B)_2 - m_B(v_B)_1 \qquad (15.39)$$

Note that the integrals in Eqs. (15.38) and (15.39) represent the impulses of the impact force $\hat{P}$.

Alternatively, we could analyze the system containing both blocks, the FBD of which is shown in Fig. 15.12(d). Because the external impulse acting on the system during the impact is zero (the impact force $\hat{P}$ is internal to the system, and the impulses of the normal forces cancel the impulses of the weights), the momentum vector of the system is conserved. The momentum balance in the x-direction for the system yields $(p_1)_x = (p_2)_x$; that is,

$$\xrightarrow{+} \qquad m_A(v_A)_1 + m_B(v_B)_1 = m_A(v_A)_2 + m_B(v_B)_2 \qquad (15.40)$$

Observe that Eq. (15.40) can also be obtained by adding Eqs. (15.38) and (15.39).

Assuming that the initial velocities are known, we see that Eq. (15.40) contains two unknown velocities: $(v_A)_2$ and $(v_B)_2$. To complete the analysis, we need another equation that takes into account the deformation characteristics of the impacting bodies[*] (the collision of steel blocks will obviously differ from the impact of rubber blocks). For the present, we will consider only the special case of *plastic impact*, where the velocities of the two blocks are the same immediately after the impact.

For plastic impact, we thus have the additional equation

$$(v_A)_2 = (v_B)_2 \qquad\qquad (15.41)$$

which, combined with Eq. (15.40), will yield the solution for $(v_A)_2$ and $(v_B)_2$. Once the final velocities have been determined, either Eq. (15.38) or (15.39) can be used to calculate the impulse of the impact force $\hat{P}$ that acts between the blocks.

In order to illustrate the analysis of plastic impact, consider the graphs of $\hat{P}$, v_A, and v_B shown in Fig. 15.13. The magnitude of the impact force $\hat{P}$ is zero except for the impact interval $\Delta t = t_2 - t_1$. The initial velocities, $(v_A)_1$ and $(v_B)_1$, are assumed to be given, and we know that $(v_A)_2 = (v_B)_2$. The graphs of $\hat{P}$ and the velocities during the period of contact are shown as dashed lines in Fig. 15.13 to emphasize that these functions are unknown. However, the area under the $\hat{P}$-t diagram can be computed, because it represents the impulse between the blocks.

Observe that the analysis presented here is concerned only with velocities immediately before and immediately after impact, and the impulse of the impact forces. It cannot deal with the variations of velocities and impact forces that occur during the impact.

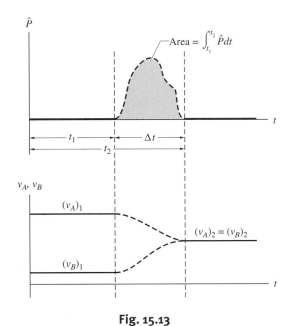

Fig. 15.13

---

[*]Note that this is one of the few situations we have encountered for which the rigid-body model must be abandoned. There is no analysis that is valid for the impact of two "rigid" objects.

Also note that the duration of the impact, Δt in Fig. 15.13, does not have to be known, because it does not enter into the analysis. However, there is a class of impact problems that requires the assumption that the duration of impact is very small. These problems are discussed in the next article.

15.9 Impulsive Motion

The analysis presented in the preceding article is independent of the duration of the impact Δt. However, there are many impact problems that can be solved only if Δt is so small that the displacement of the bodies during the impact period can be neglected. Any motion that satisfies this condition is called *impulsive motion*. Of course, the results obtained by assuming impulsive motion are only approximations. The analysis is exact only in the idealized case where $\Delta t \rightarrow 0$.

Reconsider the plastic impact of the two blocks depicted in Fig. 15.12 with the idealization $\Delta t \rightarrow 0$. Figure 15.13, which shows a finite duration of impact, is replaced by Fig. 15.14. Note that the impact interval Δt in Fig. 15.13 has become the infinitesimal time period dt in Fig. 15.14.

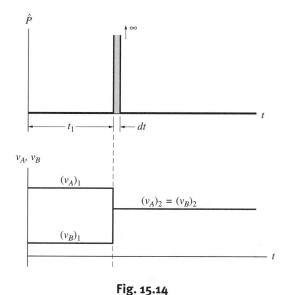

Fig. 15.14

From inspection of Fig. 15.14, we conclude the following:

1. **The magnitudes of impact forces are infinite**. The impulse of the impact force $\hat{P}$ is finite, because it is determined by the impulse-momentum equations, Eqs. (15.38) and (15.39). But because $\hat{P}$ acts over the infinitesimal time interval dt, the magnitude of $\hat{P}$ must become infinite for its impulse to remain finite. A force of infinite magnitude that acts over an infinitesimal

time interval and exerts a finite impulse (area under the force-time curve) is called an *impulsive force*.*

2. **The impulses of finite forces are negligible.** If the magnitude of a force is finite, its impulse during the impact period is negligible. For example, the weight W of a particle is an example of a finite, or nonimpulsive, force. During the time interval Δt, the impulse of W—that is, $W \Delta t$—approaches zero as $\Delta t \to 0$.

3. **The accelerations of the blocks are infinite during impact.** Because the changes in velocities are assumed to occur within an infinitesimal time period, the v-t diagrams exhibit jump discontinuities at the time of impact, as shown in Fig. 15.14. Because $a = dv/dt$, it follows that the "jumps" correspond to infinite accelerations.

4. **The blocks are in the same location before and after the impact.** Because $\Delta t \to 0$, the distances moved by the blocks during the impact are infinitesimal (the velocities are finite).

The four steps in the analysis of the impact problems are listed next. They apply to impulsive as well as nonimpulsive motion.

Step 1: Draw the FBDs of the impacting particles. Attention must be paid to identifying the impact forces—use a special symbol, such as a caret, to label each of the impact forces.

Step 2: Draw the momentum diagrams of the particles at the instant immediately before the impact. (Recall that the momentum diagram of a particle is a sketch of the particle showing its momentum vector.)

Step 3: Draw the momentum diagrams for the particles at the instant immediately after the impact.

Step 4: Using the diagrams drawn in Steps 1 through 3, derive and solve the appropriate impulse-momentum equations.

---

*An impulsive force is an example of a Dirac delta function, or a *spike*. By way of illustration, consider the rectangle of width Δx and height $1/\Delta x$, shown in (a). The area of this rectangle is 1, independent of the value of Δx. If the width Δx becomes the infinitesimal dx, the height approaches infinity, but the area is still 1. The result of this limiting procedure is the delta function shown in (b).

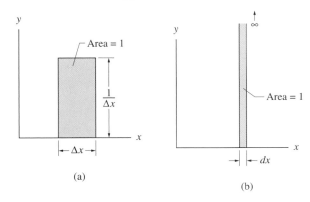

(a)

(b)

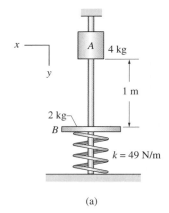

x

A | 4 kg

y

1 m

2 kg

B

$k = 49$ N/m

(a)

Sample Problem 15.12

The slider A and the plate B in Fig. (a) slide with negligible friction on the vertical guide rod. The plate is resting on a spring of stiffness $k = 49$ N/m when the slider is released in the position shown. The resulting impact between A and B is plastic and the duration of the impact is negligible. Determine (1) the velocities of A and B immediately after the impact; (2) the percentage of mechanical energy lost during the impact; (3) the impulse of the contact force during the impact; and (4) the maximum deflection of B after the impact.

Solution

There are four positions of the system that are relevant to the analysis:

Position 1 The release position.
Position 2 The position immediately before the impact.
Position 3 The position immediately after the impact.
Position 4 The position of maximum deflection of plate B.

Because the duration of the impact is infinitesimal, slider A and plate B are in essentially the same locations in positions 2 and 3.

Part 1

Motion from 1 to 2 The speed of slider A just before the impact can be determined by applying the principle of work and kinetic energy to slider A between positions 1 and 2:

$$U_{1-2} = T_2 - T_1$$

$$W_A \Delta y_A = \frac{1}{2} m_A (v_A)_2^2 - 0$$

$$4(9.8)(1) = \frac{1}{2}(4)(v_A)_2^2$$

which yields $(v_A)_2 = 4.43$ m/s.

Motion from 2 to 3 The velocities of A and B just after the impact are found from the impulse-momentum principle between positions 2 and 3. Because the duration of impact is negligible, the motion during the impact can be classified as *impulsive*. The forces that act on A and B during the impact are shown on the free-body diagrams in Fig. (b). The forces that have finite magnitudes are weights W_A and W_B, and the spring force F (because $F = k\delta$, F is finite as long as the deformation δ of the spring remains finite). The contact force $\hat{P}$ between A and B is an impulsive force; that is, its magnitude is (theoretically) infinite.

Because the analysis of impulsive motion neglects the contributions of finite forces, it is convenient to redraw the free-body diagrams displaying only the impulsive forces, as in Fig. (c). The momentum diagrams for positions 2 and 3 are shown in Figs. (d) and (e), respectively. Note that after the *plastic* impact the two masses have the same velocity; in other words, $(v_A)_3 = (v_B)_3 = v_3$.

From Fig. (c) we see that the net impulse acting on the *system* consisting of A and B is zero because the impulses of $\hat{P}$ acting on A and B cancel each other.

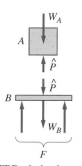

W_A

A

$\hat{P}$

$\hat{P}$

B

W_B

F

FBDs during impact

(b)

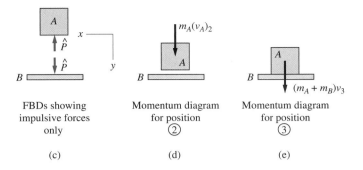

FBDs showing impulsive forces only

(c)

Momentum diagram for position ②

(d)

Momentum diagram for position ③

(e)

Hence the momentum of the system is conserved in the y-direction:

$$(p_y)_2 = (p_y)_3$$

$$+\downarrow \quad m_A(v_A)_2 = (m_A + m_B)v_3$$

$$4(4.43) = (4 + 2)v_3$$

giving for the velocity immediately after the impact

$$v_3 = 2.95 \text{ m/s} \qquad \textit{Answer}$$

Part 2

The change in the kinetic energy during the impact is

$$\Delta T = T_3 - T_2$$

$$= \frac{1}{2}(m_A + m_B)v_3^2 - \frac{1}{2}m_A(v_A)_2^2$$

$$= \frac{1}{2}(4 + 2)(2.95)^2 - \frac{1}{2}(4)(4.43)^2$$

$$= 26.11 - 39.25 = -13.14 \text{ N} \cdot \text{m}$$

The percentage of energy lost during impact is

$$\% \text{ loss} = -\frac{\Delta T}{T_2} \times 100\% = \frac{13.14}{39.25} \times 100\% = 33.5\% \qquad \textit{Answer}$$

Part 3

The impulse of $\hat{P}$ is found by applying the impulse-momentum equation to either A or B. Choosing A, we obtain from Figs. (c)–(e)

$$(L_{2-3})_y = (p_y)_3 - (p_y)_2$$

$$+\downarrow \quad -\int_{t_2}^{t_3} \hat{P} \, dt = m_A[v_3 - (v_A)_2]$$

$$= 4(2.95 - 4.43)$$

which yields

$$\int_{t_2}^{t_3} \hat{P}\, dt = 5.92 \text{ N} \cdot \text{s} \qquad\qquad \textit{Answer}$$

Application of the impulse-momentum equation to B would, of course, produce the same result. Note that we determined only the *impulse* of $\hat{P}$ during the impact; $\hat{P}$ itself is indeterminate.

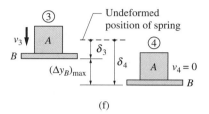

(f)

Part 4

Motion from 3 to 4   As seen in Fig. (f), the maximum displacement of plate B is $(\Delta y_B)_{\text{max}} = \delta_4 - \delta_3$, where δ_3 and δ_4 are the deformations of the spring in positions 3 and 4, respectively. Noting that δ_3 is the initial deformation of the spring caused by the weight of B, we have $\delta_3 = W_B/k = \frac{2(9.8)}{49} = 0.4$ m. Referring to Fig. (f), the work-energy principle between positions 3 and 4 yields

$$U_{3-4} = T_4 - T_3$$

$$(W_A + W_B)(\delta_4 - \delta_3) - \frac{1}{2}k\left(\delta_4^2 - \delta_3^2\right) = 0 - \frac{1}{2}(m_A + m_B)v_3^2$$

$$(4 + 2)(\delta_4 - 0.4) - \frac{1}{2}(49)\left[\delta_4^2 - (0.4)^2\right] = -\frac{1}{2}(4 + 2)(2.95)^2$$

The positive root of this equation is $\delta_4 = 1.19$ m. Therefore, the maximum deflection of plate B is

$$(\Delta y_B)_{\text{max}} = \delta_4 - \delta_3 = 1.19 - 0.4 = 0.79 \text{ m} \qquad\qquad \textit{Answer}$$

Caution   A common mistake is to apply the work-energy equation between positions 1 and 4: $U_{1-4} = T_4 - T_1$. Because the work-energy principle (or the principle of conservation of mechanical energy) does not take into account the loss of kinetic energy during an impact, it can be used only between positions that do not involve impacts.

Sample Problem 15.13

The 10-kg wedge B shown in Fig. (a) is held at rest by the stop at C when it is struck by the 50-g bullet A, which is traveling horizontally with the speed $(v_A)_1 = 900$ m/s. The duration of the impact is negligible. Assuming that all impacts are plastic and neglecting friction, calculate (1) the velocity with which the wedge starts up the incline; and (2) the impulse of each impulsive force.

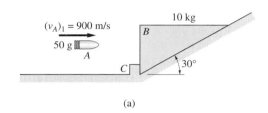

(a)

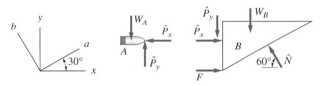

(b) Free-body diagrams during impact

Solution

Preliminary Discussion

The two positions of interest are:

Position 1   The position immediately before the impact.
Position 2   The position immediately after the impact.

Because the duration of the impact is negligible, motion is *impulsive*, with the bullet and wedge each occupying the same position before and after the impact. Observe that there are actually two impacts in this problem—the impact between A and B, and the impact between B and the incline. Assuming plastic impacts is equivalent to stating that (a) the bullet becomes embedded in the block; and (b) the wedge stays in contact with the incline.

Figure (b) shows the free-body diagrams of A and B during impact. Also shown are the xy- and ab-axes that will be used in our analysis. The forces appearing on these FBDs are W_A and W_B—the weights of A and B; F—the force at C that prevents B from sliding down the incline; $\hat{P}_x$ and $\hat{P}_y$—the components of the impact force $\hat{\mathbf{P}}$ that acts between A and B (from Newton's third law, these components occur in equal and opposite pairs); and $\hat{N}$—the normal impact force exerted on B by the incline. The impact forces $\hat{\mathbf{P}}$ and $\hat{N}$ are the only impulsive forces.

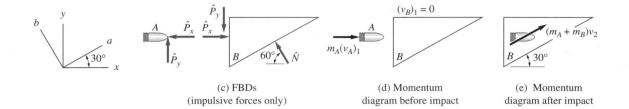

(c) FBDs
(impulsive forces only)

(d) Momentum
diagram before impact

(e) Momentum
diagram after impact

The FBDs showing only the impulsive forces are shown in Fig. (c). The momentum diagrams displaying the momenta of A and B immediately before and immediately after impact are shown in Figs. (d) and (e). In Fig. (e), observe that the direction of v_2, the common velocity of A and B after the impact, is directed up the incline.

Examination of Figs. (c) through (e) reveals that there are a total of four unknowns: the impulse of $\hat{P}_x$, the impulse of $\hat{P}_y$, the impulse of $\hat{N}$, and the final velocity v_2. Equating the impulse vectors to the change in the momentum vectors for A and B individually will yield four scalar equations that can be solved for the four unknowns. However, a more efficient solution is obtained by initially considering the system consisting of both A and B.

Part 1

From the FBDs in Fig. (c), we note that $\hat{N}$ is the only impulsive force that exerts an external impulse on the system consisting of A and B ($\hat{\mathbf{P}}$ is an internal force). Because $\hat{N}$ acts in the b-direction, momentum of the system is conserved in the a-direction. Referring to Figs. (d) and (e), we have

$$(p_a)_1 = (p_a)_2$$
$$\stackrel{+\nearrow}{} \quad m_A(v_A)_1 \cos 30° = (m_A + m_B)v_2$$
$$0.050(900) \cos 30° = (0.050 + 10)v_2$$

from which the common velocity of A and B immediately after the impact is

$$v_2 = 3.88 \text{ m/s} \qquad \qquad \textbf{\textit{Answer}}$$

Part 2

The impulse of $\hat{N}$ can be found by considering the change in the b-component of the momentum of the system. The momentum of the system in the b-direction before impact is due only to the b-component of $m_A(\mathbf{v}_A)_1$; after impact, the system has no momentum in the b-direction. Therefore, the b-component of the vector equation $\mathbf{L}_{1-2} = \Delta\mathbf{p}$ is

$$\stackrel{+\nwarrow}{} \quad \int_{t_1}^{t_2} \hat{N}\, dt = (p_b)_2 - (p_b)_1$$
$$= 0 - [-m_A(v_A)_1 \sin 30°] = 0.050(900) \sin 30°$$
$$= 22.5 \text{ N} \cdot \text{s} \qquad \qquad \textbf{\textit{Answer}}$$

The impulses of $\hat{P}_x$ and $\hat{P}_y$ can be computed by solving the two scalar impulse-momentum equations for bullet A only. Referring to Figs. (c) through (e), the x-component of the impulse-momentum equation for A is

$$(L_{1-2})_x = (p_x)_2 - (p_x)_1$$

$$\xrightarrow{+} \quad -\int_{t_1}^{t_2} \hat{P}_x \, dt = m_A v_2 \cos 30° - m_A (v_A)_1$$

$$= 0.050(3.88) \cos 30° - 0.050(900)$$

from which

$$\int_{t_1}^{t_2} \hat{P}_x \, dt = 44.8 \text{ N} \cdot \text{s} \qquad \textit{Answer}$$

Similarly, the y-component of the impulse-momentum equation for bullet A gives

$$(L_{1-2})_y = (p_y)_2 - (p_y)_1$$

$$+\uparrow \quad \int_{t_1}^{t_2} \hat{P}_y \, dt = m_A v_2 \sin 30° - 0$$

$$= 0.050(3.88) \sin 30°$$

$$= 0.0970 \text{ N} \cdot \text{s} \qquad \textit{Answer}$$

We may check our solutions by considering the impulse-momentum equation for wedge B only. For example, from Figs. (d) and (e) we note that the b-component of the momentum of B is zero, both before and after impact. Therefore, using Fig. (c) to obtain the b-component of the impulse acting on B, we obtain

$$(L_{1-2})_b = (p_b)_2 - (p_b)_1 = 0 - 0$$

$$\nwarrow \quad \int_{t_1}^{t_2} \hat{N} \, dt - \int_{t_1}^{t_2} \hat{P}_x \, dt \sin 30° - \int_{t_1}^{t_2} \hat{P}_y \, dt \cos 30° = 0$$

The fact that our answers for the impulses satisfy this equation provides a check on our solution.

Problems

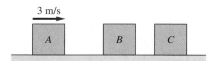

Fig. P15.88

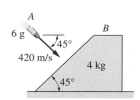

Fig. P15.89

Fig. P15.90

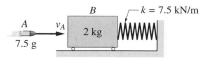

Fig. P15.91

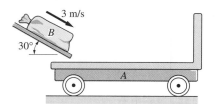

Fig. P15.94

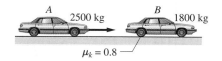

Fig. P15.95

15.88 The three identical 2-kg blocks slide on the horizontal surface with negligible friction. Initially A is moving to the right at 3 m/s while B and C are at rest. Assuming all collisions to be plastic, determine the velocities of the blocks after (a) the first collision; and (b) the second collision.

15.89 The 6-g bullet A traveling at 420 m/s hits the inclined face of the stationary block B. Assuming all impacts are plastic and neglecting friction, what will the speed of block B be immediately after the bullet hits?

15.90 The two railroad cars A and B are coasting with the speeds shown. After A collides with B, the cars become coupled together. Determine the final speed of the cars and the percentage of kinetic energy lost during the coupling procedure.

15.91 The 2-kg block B is initially at rest on the horizontal plane. After the 7.5-g bullet A is fired into the block, the maximum compression in the spring is observed to be 49 mm. Determine the initial velocity v_A of the bullet, assuming the bullet becomes embedded in the block.

15.92 The 1-kg package A is released from rest in the position shown and slides down the frictionless chute onto the 2-kg pallet B. The package and the pallet come to rest after sliding a distance d across the floor. If the coefficient of kinetic friction between the pallet and the floor is 0.2, determine the distance d.

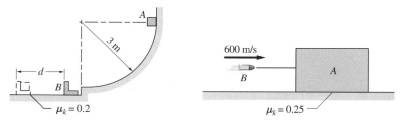

Fig. P15.92   **Fig. P15.93**

15.93 The 20-g bullet B hits the 10-kg stationary block A with a horizontal velocity of 600 m/s. The kinetic coefficient of friction between the block and the horizontal surface is 0.25. Determine (a) the total distance moved by the block after the impact; and (b) the percentage of mechanical energy lost during the impact. Assume that the bullet becomes embedded in the block.

15.94 The 10-kg bag B slides down a chute and lands on the 30-kg stationary cart A with a velocity of 3 m/s directed as shown. Neglecting rolling resistance of the cart, determine (a) the speed of the cart after the bag comes to rest on it; and (b) the percentage of mechanical energy lost during the impact.

15.95 The 1800-kg car B is stationary with its brakes locked when it is rear-ended by the 2500-kg car A. After the impact, the cars become hooked together and slide 8 m before coming to rest. The coefficient of kinetic friction between the road and the tires of *B* is 0.8. At what speed was *A* traveling immediately before the impact? Assume the brakes of *A* were not applied.

15.96   The 6-kg spiked pendulum is released in position 1. When the pendulum reaches position 2, the spike "spears" the 2-kg parcel and carries it to position 3 before stopping momentarily. Determine the angle θ in position 3.

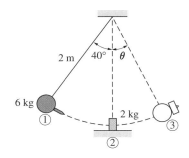

Fig. P15.96

15.97   The identical shopping carts *A* and *B* each weigh 20-kg. The 14-kg package *C* is free to slide in cart *A*. If cart *A* is pushed into the stationary cart *B* with the speed $v = 5$ m/s, determine (a) the speed of the carts immediately after the initial impact; and (b) the final speed of the carts. Neglect friction and assume all impacts are plastic.

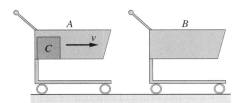

Fig. P15.97

15.98   The cars *A* and *B* collide at velocities v_A and v_B, directed as shown. After the collision, the cars become hooked together and skid 8 m in the direction shown before coming to a stop. Knowing that the coefficient of kinetic friction between the road and the tires is 0.65 for each car, find v_A and v_B.

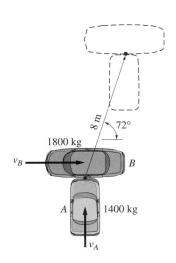

Fig. P15.98

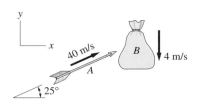

Fig. P15.99

15.99   The 4-kg sandbag B is falling vertically when it is hit by the 0.1 kg arrow A traveling at 40 m/s in the direction shown. At the time of impact, the speed of the sandbag is 4 m/s. Determine the velocity of the sandbag just after the impact, assuming that the arrow becomes embedded in the bag.

15.100   The 30-kg sack A is dropped on a spring scale from the height $h = 0$. The platform B of the scale weighs 6-kg, and the combined stiffness of the springs is 2000 N/m. Determine the maximum reading of the scale.

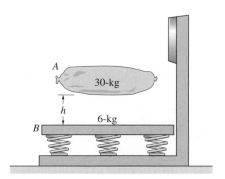

Fig. P15.100

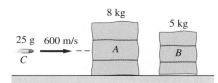

Fig. P15.101

15.101   The 25-g bullet C is fired at the hay bales A and B at a speed of 600 m/s. The bullet passes through A and becomes embedded in B. Immediately after the impacts, the velocities of the two hay bales are observed to be equal. Determine the speed of the bullet between the hay bales. Neglect friction.

15.102   The collars A and B slide with negligible friction along the wire that lies in the vertical plane. The spring attached to A has a stiffness of $k = 40$ N/m, and its undeformed length is 0.3 m. Both collars are at rest when A is released in the position shown. After the impact, A and B stick together and move the distance d before stopping momentarily. Determine d.

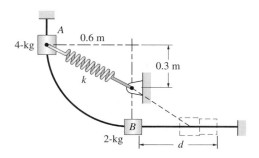

Fig. P15.102

15.103   The 60-kg cradle A is carrying the 12-kg pipe B at the speed of 8 m/s when it hits the rigid wall. Assuming all the impacts are plastic, determine the speed of the pipe immediately after the impacts. Neglect friction.

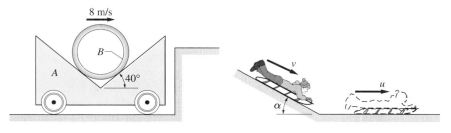

Fig. P15.103   **Fig. P15.104**

15.104   The speed of the sled just before it reaches the sharp corner at the bottom of the hill is v. When the sled hits the corner, it receives a vertical impulse from the ground that changes its speed to u, directed horizontally. In terms of v, the mass m of the sled and the rider, and the slope angle α, determine (a) u; and (b) the energy lost due to the impact.

15.105   The 12-kg pendulum is stationary in the position shown when it is hit by the 20-g bullet traveling horizontally ($\alpha = 0$). After the impact, the pendulum and the embedded bullet swing through the angle $\theta = 36°$. Calculate the initial speed of the bullet.

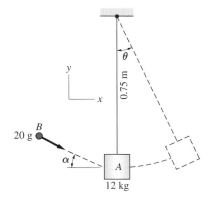

Fig. P15.105, P15.106

15.106   The 20-g bullet B is traveling at 980 m/s at an angle $\alpha = 20°$ when it becomes embedded in the 12-kg bob A of the pendulum. Before the impact, the pendulum was stationary in the vertical position ($\theta = 0$). Find the velocity vector of A immediately after the impact, assuming that A is suspended from (a) a rigid rod; and (b) an elastic (deformable) rope.

*15.107   The pendulum C is suspended from the block B that is free to move on the horizontal surface. The assembly is at rest when the bullet A hits the block with the speed of $v_1 = 300$ m/s and becomes embedded in the block. Determine the maximum angular displacement θ of the pendulum after the impact, and the velocity of the block at that instant.

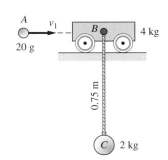

Fig. P15.107

15.10	*Elastic Impact*

Most bodies possess the ability to return either totally or partially to their original shape when released from a deformed position, a property known as *elasticity*. Therefore, there are two stages in the impact of elastic particles. Initially there is the deformation stage wherein the particle is compressed by the impact force. This stage is followed by a recovery stage during which the particle returns totally or partially to its undeformed shape. When two elastic particles collide, the recovery stage causes the particles to rebound, or move apart, after the impact (the recovery stage is absent during plastic impact). If the particles are perfectly elastic, they will return to their original shape with no energy lost during the impact. A more common occurrence is for the impacting particles to be left partially deformed by the relatively large impact forces, in which case a fraction of the initial kinetic energy is lost due to the permanent deformation. Kinetic energy can also be lost in the generation of heat and sound during the impact. Before characterizing the impact of two elastic particles, it is necessary to distinguish between direct and oblique impact.

Figure 15.15 shows the impact between two circular disks A and B that are sliding on a frictionless horizontal plane. The line that is perpendicular to the contact surface (the x-axis) is called the *line of impact*. The velocities of the particles before the impact are denoted by $(v_A)_1$ and $(v_B)_1$. When both initial velocities are directed along the line of impact, as shown in Fig. 15.15(a), the impact is

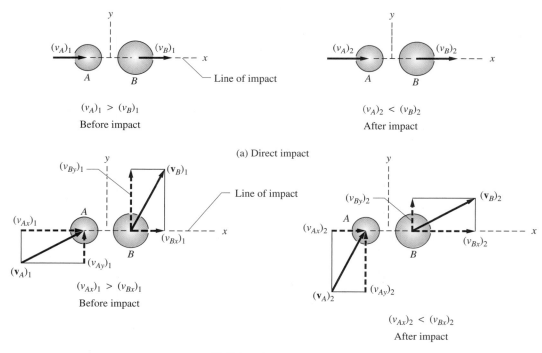

(a) Direct impact

(b) Oblique impact

Fig. 15.15

called *direct impact.* Otherwise, the impact is referred to as *oblique,* as depicted in Fig. 15.15(b). Thus direct impact is equivalent to a head-on collision, and oblique impact refers to a glancing blow. Note that the linear impulse acting on the systems is zero for both cases shown in Fig. 15.15, since no external forces act on either system.

For the direct impact in Fig. 15.15(a), momentum balance in the *x*-direction gives

$$\xrightarrow{+} \quad m_A(v_A)_1 + m_B(v_B)_1 = m_A(v_A)_2 + m_B(v_B)_2 \qquad (15.42)$$

where $(v_A)_2$ and $(v_B)_2$ are the velocities after impact.

The *coefficient of restitution e* is an experimental constant that characterizes the "elasticity" of the impacting bodies. For direct impact it is defined as

$$e = \frac{v_{\text{sep}}}{v_{\text{app}}} \qquad (15.43)$$

where

$v_{\text{sep}} = (v_B)_2 - (v_A)_2$ is the *velocity of separation* (the rate at which the distance between the particles increases after the impact)

$v_{\text{app}} = (v_A)_1 - (v_B)_1$ is the *velocity of approach* (the rate at which the distance between the particles decreases before the impact)

For an impact to occur in Fig. 15.15(a), we must have $v_{\text{app}} > 0$; if the particles are to separate after the impact (omitting the possibility that *A* could pass through *B*), we have $v_{\text{sep}} > 0$. Thus *e* will generally be a nonnegative number.

From Eq. (15.43) we see that $e = 0$ ($v_{\text{sep}} = 0$) corresponds to plastic impact. If $e = 1$, the impact is called *perfectly elastic,* a situation for which no energy is lost during the impact (see Prob. 15.108). For most impacts, the values of *e* lie between 0 and 1. (A negative coefficient of restitution indicates that one body has passed through the other, such as a bullet passing through a plate.)

Oblique impact is analyzed by assuming that *e* has the same value as for direct impact; however, in the defining equation for *e*, the velocities are replaced by their components that are directed along the line of impact. Therefore, if the line of impact coincides with the *x*-axis as shown in Fig. 15.15(b), we have

$$v_{\text{sep}} = (v_{Bx})_2 - (v_{Ax})_2$$
$$v_{\text{app}} = (v_{Ax})_1 - (v_{Bx})_1 \qquad (15.44)$$

If friction between *A* and *B* is negligible, the impact force acting on either particle will be directed along the *x*-axis, with no *y*-component. For this case, the *y*-components of the velocities of both *A* and *B* will not change during the impact.

The value of *e* is usually considered to be constant, depending only on the material of the colliding particles. But this is only an approximation to reality. Experimental evidence indicates that the coefficient of restitution actually depends on many factors (the magnitude of the relative velocity of approach, the condition of the impacting surfaces, etc.).

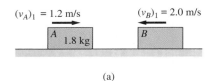

$(v_A)_1 = 1.2$ m/s $\qquad (v_B)_1 = 2.0$ m/s

(a)

Sample Problem 15.14

The 1.8-kg block A shown in Fig. (a) is sliding toward the right with the velocity 1.2 m/s when it hits block B, which is moving to the left with the velocity 2.0 m/s. The impact causes block A to stop. If the coefficient of restitution for the impact is 0.5, determine (1) the velocity of B after the impact; and (2) the mass of B. Neglect friction.

Solution

Part 1

The velocity of B after impact, $(v_B)_2$, can be calculated using the defining equation for the coefficient of restitution. Referring to Fig. (a), we see that the velocity of approach is $v_{app} = 1.2 + 2.0 = 3.2$ m/s. Noting that block A is at rest after the impact, the velocity of separation is $v_{sep} = (v_B)_2$. Therefore, $e = v_{sep} / v_{app}$ becomes

$$0.5 = \frac{(v_B)_2}{3.2}$$

from which we find

$$(v_B)_2 = 1.6 \text{ m/s} \qquad \qquad \textit{Answer}$$

Part 2

The free-body diagrams of blocks A and B during impact, shown in Fig. (b), contain the following forces: the weights W_A and W_B, the contact forces N_A and N_B, and the impact force $\hat{P}$. Since $\hat{P}$ is internal to the system of both blocks, we conclude that the net impulse acting on the system in the x-direction is zero.

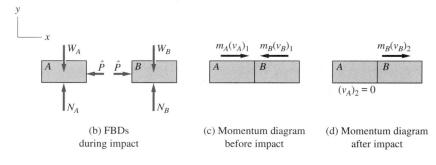

| (b) FBDs during impact | (c) Momentum diagram before impact | (d) Momentum diagram after impact |

Referring to Figs. (c) and (d), conservation of linear momentum in the x-direction for the system yields

$$(p_x)_1 = (p_x)_2$$

$$\xrightarrow{+} \quad m_A(v_A)_1 - m_B(v_B)_1 = m_B(v_B)_2$$

Substitution of the numerical values gives

$$1.8(1.2) - m_B(2.0) = m_B(1.6)$$

from which we find

$$m_B = 0.6 \text{ kg} \qquad \qquad \textit{Answer}$$

Sample Problem 15.15

Two identical disks A and B, weighing 2-kg each, are sliding across a horizontal tabletop when they collide with the velocities $(v_A)_1 = 3$ m/s and $(v_B)_1 = 2$ m/s, directed as shown in Fig. (a). If the coefficient of restitution for the impact is $e = 0.8$, calculate the velocity vectors of the disks immediately after the impact. Neglect friction.

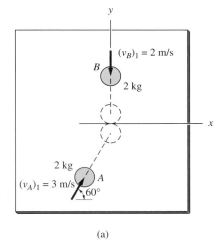

(a)

Solution

As shown in the free-body diagrams in Fig. (b), the only forces acting in the xy-plane during the impact are the impact forces $\hat{P}$, which are oppositely directed on A and B. (The weights of the disks and the normal forces exerted by the tabletop are perpendicular to the horizontal xy-plane.) In the absence of friction, $\hat{P}$ will be directed along the y-axis, which is the line of impact. Assuming that the duration of the impact is negligible, the motion is impulsive and the disks will be located at the origin of the coordinate system just before and just after the impact.

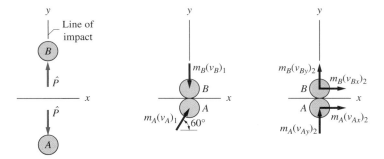

| (b) FBDs during impact (forces acting in xy-plane only) | (c) Momentum diagram before impact | (d) Momentum diagram after impact |

The momentum diagrams that display the momentum vectors for A and B immediately before and after the impact are shown in Figs. (c) and (d). Note that in Fig. (d) all components of the final velocities have been assumed to act in positive coordinate directions.

From Figs. (b)–(d), we see that there are five unknowns: $\int \hat{P} \, dt$ (the impulse of the impact force), $(v_{Ax})_2$, $(v_{Ay})_2$, $(v_{Bx})_2$, and $(v_{By})_2$. Furthermore, we see that there are five independent equations: two components of the vector impulse-momentum equation, $\mathbf{L}_{1-2} = \Delta \mathbf{p}$, for each of the two disks (a total of four equations), plus the coefficient of restitution equation, Eq. (15.43). These five equations could, of course, be used to solve for the five unknowns. However, because we are not required to find $\int \hat{P} \, dt$, a more efficient solution is obtained by also analyzing the system that contains both disks. Thus our solution consists of the following four parts.

1. Apply $(L_{1-2})_x = \Delta p_x$ to Disk A

From Fig. (b) we see that there is no impulse on A in the x-direction. Therefore, the x-component of the momentum of A is conserved. Referring to the momentum diagrams for disk A in Figs. (c) and (d), we obtain

$$(p_x)_1 = (p_x)_2$$
$$\xrightarrow{+} \quad m_A(v_A)_1 \cos 60° = m_A(v_{Ax})_2$$

Canceling m_A and substituting $(v_A)_1 = 8$ ft/s, this equation yields

$$(v_{Ax})_2 = (v_A)_1 \cos 60° = 3 \cos 60° = 1.5 \text{ m/s} \qquad \text{(a)}$$

2. Apply $(L_{1-2})_x = \Delta p_x$ to Disk B

Using the same argument as given above for disk A, we find that

$$(v_{Bx})_2 = (v_{Bx})_1 = 0 \qquad \text{(b)}$$

(Observe that the x-components of the velocities of both A and B are unchanged by the impact because $\hat{P}$ is directed along the y-axis.)

3. Apply $(L_{1-2})_y = \Delta p_y$ to the System Containing Both Disks

From Fig. (b) we see that there are no external forces, and thus no external impulses, that act on the system during impact ($\hat{P}$ is an internal force). Therefore, the momentum of the system is conserved. Referring to Figs. (c) and (d), conservation of the y-component of momentum for the system gives

$$(p_y)_1 = (p_y)_2$$
$$+\uparrow \quad m_A(v_A)_1 \sin 60° - m_B(v_B)_1 = m_A(v_{Ay})_2 + m_B(v_{By})_2$$
$$2(3) \sin 60° - 2(2) = 2(v_{Ay})_2 + 2(v_{By})_2$$

which can be reduced to

$$(v_{Ay})_2 + (v_{By})_2 = 0.598 \qquad \text{(c)}$$

4. Use the Coefficient of Restitution

Noting that the line of impact is the y-axis, the velocity of approach [see Fig. (a)] is $v_{\text{app}} = 3 \sin 60° + 2 = 4.598$ m/s. Referring to Fig. (d), we deduce that the velocity of separation is $v_{\text{sep}} = (v_{By})_2 - (v_{Ay})_2$. Therefore, $e = v_{\text{sep}}/v_{\text{app}}$ becomes

$$0.8 = \frac{(v_{By})_2 - (v_{Ay})_2}{4.598}$$

which, after simplification, may be written as

$$(v_{By})_2 - (v_{Ay})_2 = 3.678 \qquad \text{(d)}$$

Solving Eqs. (c) and (d) simultaneously, we obtain

$$(v_{Ay})_2 = -1.54 \text{ m/s} \quad \text{and} \quad (v_{By})_2 = 2.14 \text{ m/s} \qquad \text{(e)}$$

From the values given in Eqs. (a), (b), and (e), the velocities of disks A and B after impact are as follows.

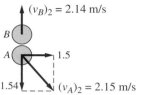

$(v_B)_2 = 2.14$ m/s

B

A 1.5

1.54 $(v_A)_2 = 2.15$ m/s

Answer

If needed, the impulse of the impact force, $\int \hat{P}\, dt$, could be computed by applying the equation $(L_{1-2})_y = \Delta p_y$ to either disk A or disk B.

Sample Problem   **15.16**

As shown in Fig. (a), the ball A of mass $m = 0.2$ kg is dropped on a rigid sur-face inclined at $\theta = 30°$ to the horizontal. The coefficient of restitution for the impact is 0.8, and friction between the ball and the surface is negligible. Deter-mine (1) the rebound speed v_2 and the rebound angle α; and (2) the percentage of kinetic energy lost during the impact.

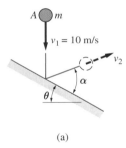

A m

$v_1 = 10$ m/s

v_2

α

θ

(a)

Solution

Part 1

As shown in the free-body diagram in Fig. (b), the forces acting on the ball during impact are its weight W and the impact force $\hat{N}$, which is normal to the friction-less plane. The y-axis, which is perpendicular to the plane, is the line of impact.

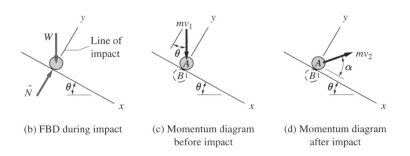

(b) FBD during impact   (c) Momentum diagram before impact   (d) Momentum diagram after impact

We will neglect the duration of the impact, which means that the motion is impulsive. It follows that the impulse of W, which is a finite force, can be neglected, and that the ball is in the same location immediately before and immediately after the impact.

Because the description of impact requires two particles, we imagine that a second particle B is embedded in the rigid plane, as shown in the momentum diagrams in Figs. (c) and (d). Because the velocity of B is always zero, the momentum diagrams before and after impact contain only the momentum of ball A.

From Fig. (b) we see that there will be no impulse acting on the ball in the x-direction during the impact (remember that the impulse of W is being neglected). Therefore, the x-component of the momentum of A is conserved. Referring to the momentum diagrams in Figs. (c) and (d), we obtain

$$(p_x)_1 = (p_x)_2$$

$$\searrow^+ \quad mv_1 \sin\theta = mv_2 \cos\alpha$$

which yields

$$v_2 \cos\alpha = v_1 \sin\theta = 10 \sin 30° = 5.0 \text{ m/s} \tag{a}$$

From Figs. (c) and (d) we see that the velocities of approach and separation are $v_{app} = v_1 \cos\theta$ and $v_{sep} = v_2 \cos\alpha$. Therefore, the definition of the coefficient of restitution, $v_{sep} = ev_{app}$, yields

$$v_2 \sin\alpha = ev_1 \cos\theta = (0.8)(10) \cos 30° = 6.928 \text{ m/s} \tag{b}$$

Solving Eqs. (a) and (b), we obtain

$$v_2 = 8.544 \text{ m/s} \quad \text{and} \quad \alpha = 54.18° \qquad \textit{Answer}$$

Part 2

The change in the kinetic energy of the ball during the impact is

$$\Delta T = T_2 - T_1 = \frac{1}{2}m\left(v_2^2 - v_1^2\right) = \frac{1}{2}(0.2)[(8.544)^2 - (10)^2]$$

$$= 7.30 - 10.0 = -2.70 \text{ N} \cdot \text{m}$$

The percentage of kinetic energy lost is

$$\% \text{ loss} = -\frac{\Delta T}{T_1} = \frac{2.70}{10.0} \times 100\% = 27.0\% \qquad \textit{Answer}$$

Problems

15.108 Prove that no energy is lost during direct impact of two particles if $e = 1$.

15.109 The two blocks slide on a horizontal surface with the speeds $(v_A)_1$ and $(v_B)_1$, where $(v_A)_1 > (v_B)_1$. Show that the speeds of the blocks after impact are

$$(v_A)_2 = \frac{(v_A)_1(m_A/m_B - e) + (v_B)_1(1 + e)}{1 + m_A/m_B}$$

$$(v_B)_2 = \frac{(v_A)_1(1 + e) + (v_B)_1(m_B/m_A - e)}{1 + m_B/m_A}$$

where e is the coefficient of restitution. Neglect friction.

Fig. P15.109

15.110 The three identical blocks ($m_A = m_B = m_C$) are at rest on a horizontal surface when block A is given the initial velocity v_0. Determine the speed of each block after all the impacts have occurred. The coefficient of restitution for each impact is 0.5. Use the formulas given in Prob. 15.109 and neglect friction.

15.111 Solve Prob. 15.110 assuming that the masses of the blocks are $m_A = m_C = m$ and $m_B = 0.6m$.

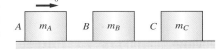

Fig. P15.110, P15.111

15.112 The three identical pendulums of mass m each are suspended so that their bobs are almost touching. After pendulum A is displaced and released, it hits B with the speed v_0. Determine the speed of bob C immediately after the impact. Assume all impacts to be perfectly elastic ($e = 1$).

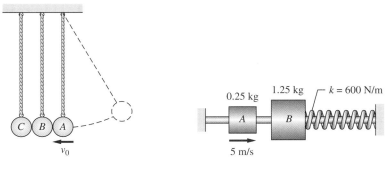

Fig. P15.112   **Fig. P15.113**

15.113 After the sliding collar A hits the stationary collar B with the speed of 5 m/s, it rebounds with the speed of 2 m/s, directed to the left. Determine the coefficient of restitution for the impact.

15.114 Two identical coins are placed on a rough, horizontal surface as shown in (a). After coin A is propelled into the stationary coin B with the initial

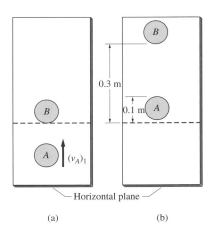

Fig. P15.114

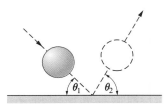

Fig. P15.115

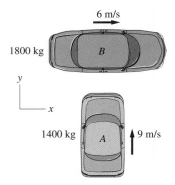

Fig. P15.116

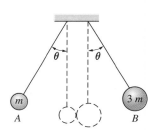

Fig. P15.119

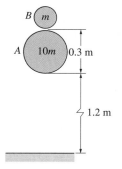

Fig. P15.120

velocity $(v_A)_1$, the coins come to rest in the positions shown in (b). Determine the coefficient of restitution for the impact between the coins.

15.115   The elastic ball is bounced off a rigid surface. Show that the relationship between the angle of incidence and the rebound angle is $\tan\theta_2 = e \tan\theta_1$, where e is the coefficient of restitution. Neglect friction.

15.116   Two cars traveling with the velocities shown collide at an intersection. The coefficient of restitution is 0.2 for the impact, and the contacting surfaces are frictionless. Calculate the velocity of each car after the impact.

15.117   The two disks A and B lie on a horizontal surface. Disk A is propelled into B, which is initially stationary, with the velocity shown. If $e = 0.85$, calculate the velocity of each disk after the impact. Neglect friction.

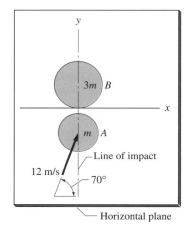

Fig. P15.117

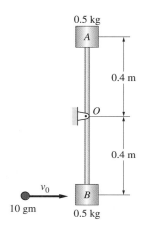

Fig. P15.118

15.118   The 0.5-kg blocks A and B are joined by of a rod of negligible weight. The assembly, which is initially at rest, is free to rotate about the pin at O. If the 10 gm pellet is fired at B with the velocity $v_0 = 100$ m/s, determine the angular velocity of the rod immediately after the impact. The coefficient of restitution for the impact is 0.75.

15.119   The two pendulums are released simultaneously from rest in the positions shown. After the bobs collide, pendulum A swings back to its release position. Calculate the coefficient of restitution for the impact.

15.120   The balls A and B are dropped simultaneously from the positions shown, with a small gap between the balls. Determine the maximum height reached by B after impacting with A. The coefficient of restitution is 0.85 for all impacts. Note that the mass of A is ten times the mass of B.

*__**15.121**__   A ball is dropped from a height h_0 onto a rigid floor. If the coefficient of restitution is 0.985, find the number of bounces made by the ball before its height of rebound is reduced to $h_0/2$.

15.122 The two disks moving along parallel paths collide with the velocities shown. The radius of A is 50 mm, and its mass is 0.4 kg. Disk B has a radius of 100 mm and a mass of 0.8 kg. Determine the speed of each disk immediately after the impact if the coefficient of restitution is 0.7. Neglect friction.

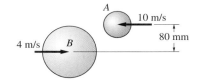

Fig. P15.122

*15.11   Mass Flow

a.   Control volume

Up to this point, we have restricted our attention to systems that always contained the same particles. In other words, we assumed that the system was closed in the sense that mass did not enter or leave the system. In this article, we apply the impulse-momentum principle to *mass flow,* where the particles continuously move through a spatial region, called the *control volume.*

An example is the flow of water through a section of pipe, as shown in Fig. 15.16(a). Here a convenient choice for the control volume V is the interior of the pipe. The water enters the control volume with the velocity $\mathbf{v}_{in}$ and exits with the velocity $\mathbf{v}_{out}$. Because the momentum of the water in the pipe is changed, a force equal to the rate of change of this momentum must exist between the surface of the control volume and the flowing water.

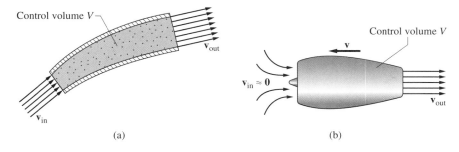

Fig. 15.16

A second example of mass flow is the jet engine of an aircraft shown in Fig. 15.16(b). The engine takes in air ($\mathbf{v}_{in} \approx \mathbf{0}$), which is mixed with fuel and ignited. The combustion gases are expelled at the velocity $\mathbf{v}_{out}$. The control volume V is the interior of the engine, which is itself moving with the velocity $\mathbf{v}$. In this case, two changes occur within the control volume: The velocity of the air is increased from zero to $\mathbf{v}_{out}$, and the velocity of the combusted fuel is changed from $\mathbf{v}$ to $\mathbf{v}_{out}$. The corresponding rate of change of momentum gives rise to a force between the engine and the flowing mass called the *thrust.*

Strictly speaking, the analysis of mass flow belongs in the realm of *fluid mechanics,* which is beyond the scope of this text. We will restrict our discussion to problems that do not require the specialized knowledge and techniques of fluid mechanics. In particular, we assume throughout that the inflow velocity ($\mathbf{v}_{in}$) and outflow velocity ($\mathbf{v}_{out}$) are constant across the inlet and outlet areas of the control volume, respectively.

b. *Impulse-momentum principle*

We will now formulate the impulse-momentum principle, based on the concept of a control volume.

Figure 15.17(a) shows a control volume V consisting of the region inside a vessel, such as a pipe. In general, any region of space may be selected as the control volume, with the best choice determined by the problem being considered. When the shape of the control volume is maintained by a vessel, as in Fig. 15.17(a), the vessel may also be included as part of the system, if convenient. However, you must clearly identify the control volume at the beginning of the analysis, because it determines the equations that will enter into the solution.

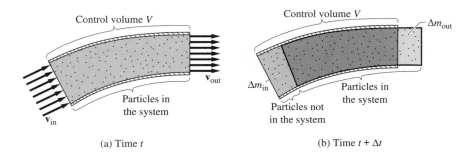

(a) Time t (b) Time $t + \Delta t$

Fig. 15.17

We now consider the momentum of the system consisting of all the particles that are inside the control volume V at time t, as indicated in Fig. 15.17(a). Note that, unlike the control volume, the system always contains the same particles: that is, the mass of the system is constant. Let the momentum of the system at time t be denoted by $\mathbf{p}$, and let the momentum of the mass within the control volume at time t be $\mathbf{p}_V$. Because the system is inside V at time t, we have $\mathbf{p} = \mathbf{p}_V$.

Consider next the momentum of the system at time $t + \Delta t$, where Δt is a small time interval. As shown in Fig. 15.17(b), some of the particles of the system have left the control volume V, and other particles, not part of the system, have entered V. The masses of these particles are labeled as Δm_{out} and Δm_{in}, respectively. Since the outlet and inlet velocities are assumed to be constant across the inlet and outlet areas, the momenta of the particles leaving and entering the control volume are, respectively,

$$\Delta \mathbf{p}_{\text{out}} = \Delta m_{\text{out}} \mathbf{v}_{\text{out}} \tag{a}$$

$$\Delta \mathbf{p}_{\text{in}} = \Delta m_{\text{in}} \mathbf{v}_{\text{in}} \tag{b}$$

The momentum of the system has changed from $\mathbf{p}$ to $\mathbf{p} + \Delta \mathbf{p}$, and the momentum of the particles that are now in V is $\mathbf{p}_V + \Delta \mathbf{p}_V$. Because the momentum of the particles leaving V may not be the same as the momentum of the particles entering V, $\Delta \mathbf{p}$ is not necessarily equal to $\Delta \mathbf{p}_V$. From Fig. 15.17(b), we see that the momentum of the system equals the momentum of the particles within V plus the momentum of the particles leaving V (these are part of the system) minus the momentum of the particles entering V (these are not part of the system); in other words,

$$\mathbf{p} + \Delta \mathbf{p} = \mathbf{p}_V + \Delta \mathbf{p}_V + \Delta \mathbf{p}_{\text{out}} - \Delta \mathbf{p}_{\text{in}} \tag{c}$$

Substituting from Eqs. (a) and (b), and recalling that $\mathbf{p} = \mathbf{p}_V$, we obtain

$$\Delta\mathbf{p} = \Delta\mathbf{p}_V + \Delta m_{\text{out}}\mathbf{v}_{\text{out}} - \Delta m_{\text{in}}\mathbf{v}_{\text{in}} \tag{d}$$

The rate at which the momentum of the system changes at time t is given by

$$\dot{\mathbf{p}} = \lim_{\Delta t \to 0} \frac{\Delta\mathbf{p}}{\Delta t}$$

which yields, in conjunction with Eq. (d),

$$\dot{\mathbf{p}} = \dot{\mathbf{p}}_V + \dot{m}_{\text{out}}\mathbf{v}_{\text{out}} - \dot{m}_{\text{in}}\mathbf{v}_{\text{in}} \tag{15.45}$$

where

$$\dot{m}_{\text{out}} = \lim_{\Delta t \to 0} \frac{\Delta m_{\text{out}}}{\Delta t} \quad \text{and} \quad \dot{m}_{\text{in}} = \lim_{\Delta t \to 0} \frac{\Delta m_{\text{in}}}{\Delta t}$$

are the mass flow rates out of and into the control volume, respectively. (The units for mass flow rate are mass per unit time, e.g., slugs/s or kg/s.)

Equation (15.45) is a form of *Reynolds' transport theorem:*

The rate at which the momentum of a system changes equals the rate at which the momentum inside the control volume changes plus the net rate at which the momentum is flowing out of the control volume.

If $\Sigma\mathbf{F}$ is the resultant force acting on the system at time t (when the entire system is contained within the control volume), then the impulse-momentum principle states that $\Sigma\mathbf{F} = \dot{\mathbf{p}}$, which, using Eq. (15.45), becomes

$$\Sigma\mathbf{F} = \dot{\mathbf{p}}_V + \dot{m}_{\text{out}}\mathbf{v}_{\text{out}} - \dot{m}_{\text{in}}\mathbf{v}_{\text{in}} \tag{15.46}$$

The following two examples of mass flow represent applications of Eq. (15.46) that deserve special attention because of their practical importance.

c. Deflection of a steady fluid stream

Consider a fluid stream that is deflected by a stationary vane, as shown in Fig. 15.18. The control volume V is taken to be the spatial region shown, with $\mathbf{v}_{\text{in}}$ and $\mathbf{v}_{\text{out}}$ being the velocities with which the stream enters and leaves the control volume, respectively. The force $\Sigma\mathbf{F}$ is the resultant force acting on the fluid within the control volume.

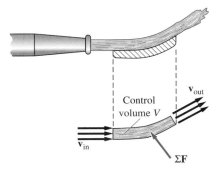

Fig. 15.18

For steady flow, $\dot{m}_\text{in} = \dot{m}_\text{out} = \dot{m}$ (a constant), which means that there is no accumulation of mass in the control volume V. Furthermore, the momentum of the fluid in the control volume is also constant; that is, $\dot{\mathbf{p}}_V = \mathbf{0}$. Making these substitutions into Eq. (15.46) gives

$$\Sigma \mathbf{F} = \dot{m}(\mathbf{v}_\text{out} - \mathbf{v}_\text{in}) \tag{15.47}$$

Observe that the resultant force $\Sigma \mathbf{F}$ acting on the fluid is constant in the case of steady flow.

d.   *Rocket propulsion*

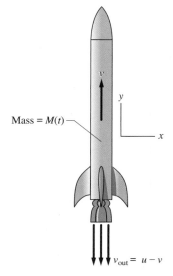

Mass = $M(t)$

$v_\text{out} = u - v$

(a)

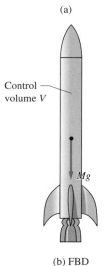

Control volume V

Mg

(b) FBD

Fig. 15.19

Figure 15.19(a) shows a rocket in vertical flight. The mass of the rocket, including its contents, is denoted by $M(t)$, and its velocity is $\mathbf{v}(t) = v(t)\mathbf{j}$, where t is time. The rocket is expelling gases at the rate $\dot{m}$ with the constant nozzle velocity $\mathbf{u} = -u\mathbf{j}$ *relative to the rocket.*

We choose the control volume V to be the rocket and its interior. Because the rocket consumes only the fuel that it carries (there is no air intake), the rates of mass flow are

$$\dot{m}_\text{out} = \dot{m} \quad \text{and} \quad \dot{m}_\text{in} = 0 \tag{h}$$

The momentum of the mass within the control volume is $\mathbf{p}_V = Mv\mathbf{j}$, which on differentiation with respect to time yields

$$\dot{\mathbf{p}}_V = \dot{M}v\mathbf{j} + M\dot{v}\mathbf{j} = -\dot{m}v\mathbf{j} + M\dot{v}\mathbf{j} \tag{i}$$

where we substituted $\dot{M} = -\dot{m}$ (this is the rate at which fuel is consumed). The velocity of the expelled gas can be written as

$$\mathbf{v}_\text{out} = (v - u)\mathbf{j} \tag{j}$$

The free-body diagram (FBD) of the rocket and its contents is shown in Fig. 15.19(b). The only force acting on the rocket is its total weight (we neglect air resistance and the pressure of the gases at the exit from the nozzle). Substituting $\Sigma \mathbf{F} = -Mg\mathbf{j}$ and Eqs. (h)–(j) into Eq. (15.46), we obtain

$$-Mg\mathbf{j} = -\dot{m}v\mathbf{j} + M\dot{v}\mathbf{j} + \dot{m}(v - u)\mathbf{j}$$

which simplifies to

$$\dot{m}u - Mg = M\dot{v} \tag{15.48}$$

Substituting $\dot{v} = a$, where a is the acceleration of the rocket, and letting

$$T = \dot{m}u \tag{15.49}$$

Eq. (15.48) becomes

$$T - Mg = Ma \tag{15.50}$$

Equation (15.50) shows that T represents a force, called the *thrust*, that tends to propel the rocket forward.

Sample Problem 15.17

Figure (a) shows water entering a 60° horizontal bend in a pipe with the velocity $v_{in} = 6$ m/s. As the water passes through the bend, its pressure drops from $p_{in} = 15$ kN/m² to $p_{out} = 10$ kN/m², and the pipe diameter increases from $d_{in} = 0.3$ m to $d_{out} = 0.4$ m. Determine the force exerted on the bend by the water. (Water weighs 1000 kg/m³.)

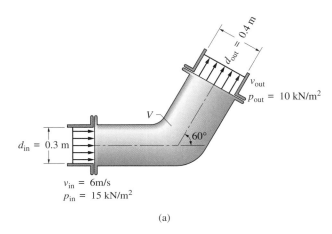

(a)

Solution

Because this problem involves steady flow, it can be solved using Eq. (15.47):

$$\Sigma \mathbf{F} = \dot{m}(\mathbf{v}_{out} - \mathbf{v}_{in}) \qquad \text{(a)}$$

As indicated in Fig. (a), we choose for the control volume V the interior of the bend. The free-body diagram of the water in the control volume is shown in Fig. (b), where P_{in} and P_{out} are the forces due to the inlet and outlet pressures, and $\mathbf{R}$ is the force exerted on the water by the wall of the bend (the weight of the water would also be included if the plane of the bend were not horizontal). The

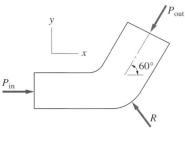

(b) FBD

261

force $\Sigma\mathbf{F}$ in Eq. (a) is the resultant of the forces appearing on the FBD. Assuming the pressure distribution to be constant over each cross section, we get

$$P_{in} = p_{in}A_{in} = 15 \times 10^3 \left[\frac{\pi(0.3)^2}{4}\right] = 1060.3 \text{ N}$$

$$P_{out} = p_{out}A_{out} = 10 \times 10^3 \left[\frac{\pi(0.4)^2}{4}\right] = 1256.6 \text{ N}$$

where A denotes the cross-sectional area of the pipe.

Because the flow is steady—that is, there is no accumulation of water in V—we have $A_{in}v_{in} = A_{out}v_{out}$. Therefore, the speed of water leaving the control volume is

$$v_{out} = v_{in}\frac{A_{in}}{A_{out}} = v_{in}\left(\frac{d_{in}}{d_{out}}\right)^2 = 6\left(\frac{0.3}{0.4}\right)^2 = 3.375 \text{ m/s}$$

The rate of mass flow through the pipe is $\dot{m} = \rho Av$, where ρ is the mass density of water, and v represents the speed of water. Using the values at the inlet of the bend (the outlet values could also be used), we obtain

$$\dot{m} = \rho A_{in}v_{in} = 1000\left[\frac{\pi}{4}(0.3)^2\right]6 = 424 \text{ kg/s}$$

Referring to Fig. (b), the left side of Eq. (a) is

$$\Sigma\mathbf{F} = P_{in}\mathbf{i} - P_{out}(\cos 60°\mathbf{i} + \sin 60°\mathbf{j}) + \mathbf{R}$$
$$= 1060.3\mathbf{i} - 1256.6(\cos 60°\mathbf{i} + \sin 60°\mathbf{j}) + \mathbf{R} = 432\mathbf{i} - 1088.3\mathbf{j} + \mathbf{R} \text{ N}$$

$$(b)$$

The right side of Eq. (a) is

$$\dot{m}(\mathbf{v}_{out} - \mathbf{v}_{in}) = 424[3.375(\cos 60°\mathbf{i} + \sin 60°\mathbf{j}) - 6\mathbf{i}] = -1828.5\mathbf{i} + 1239.3\mathbf{j} \text{ N}$$

$$(c)$$

Equating Eqs. (b) and (c), we get

$$432\mathbf{i} - 1088.3\mathbf{j} + \mathbf{R} = -1828.5\mathbf{i} + 1239.3\mathbf{j}$$

which yields for the force applied by the bend on the water

$$\mathbf{R} = -2260.5\mathbf{i} + 2327.6\mathbf{j} \text{ N}$$

Therefore, the force exerted by the water on the bend is

$$-\mathbf{R} = 2260.5\mathbf{i} - 2327.6\mathbf{j} \text{ N} \qquad\qquad Answer$$

Sample Problem *15.18*

A rocket in vertical flight has a total mass M_0 at lift-off and consumes fuel (including oxidizer) at the constant rate $\dot{m}$. The velocity u of the exhaust gases relative to the rocket is also constant. Derive the expression for the velocity of the rocket as a function of time t, where t is measured from the time of the lift-off.

Solution

The equation of motion of the rocket is given by Eq. (15.48):

$$\dot{m}u - Mg = M\dot{v} \tag{a}$$

The mass of the rocket at time t after lift-off is $M = M_0 - \dot{m}t$. Substitution in Eq. (a) yields

$$\dot{m}u - (M_0 - \dot{m}t)g = (M_0 - \dot{m}t)\dot{v}$$

Solving for $\dot{v}$, we get

$$\dot{v} = \frac{\dot{m}u}{M_0 - \dot{m}t} - g$$

Integration with respect to time gives us

$$v = \int \dot{v}\,dt = -u\ln(M_0 - \dot{m}t) - gt + C \tag{b}$$

where C is the constant of integration. The initial condition is $v = 0$ when $t = 0$, which, on substitution into Eq. (b), yields

$$0 = -u\ln M_0 + C$$

Therefore,

$$C = u\ln M_0$$

and Eq. (b) becomes

$$v = u\ln\left(\frac{M_0}{M_0 - \dot{m}t}\right) - gt \qquad\qquad \textit{Answer}$$

Problems

15.123 A 2.5-Mg rocket is in vertical flight above earth's atmosphere, where the gravitational acceleration is 9.5 m/s^2. The engine consumes fuel at the rate of 98 kg/s and expels the exhaust at the velocity of 500 m/s relative to the rocket. Determine (a) the thrust of the engine; and (b) the acceleration of the rocket.

15.124 A Saturn V rocket weighs 3.2×10^6 N at liftoff. Its first stage develops 7.5×10^6 N of thrust while consuming 2×10^6 N of fuel during its 2.5-min burn. Determine (a) the velocity of the exhaust gas relative to the rocket; and (b) the velocity of the rocket at the end of the burn. Assume vertical flight, and neglect air resistance and the change of gravity with altitude. (Hint: See Sample Problem 15.18.)

15.125 The V-2 rocket of World War II fame weighed 14 tons at liftoff, including 9.5 tons of fuel (alcohol and oxygen). The powered flight lasted 65 s, and the rocket reached a maximum velocity of 7200 km/h in vertical flight. Calculate the thrust of the rocket engine. (Hint: See Sample Problem 15.18.)

15.126 The space probe weighing 2400 N is traveling at a constant speed of 36 000 km/h when its thruster is fired for 100 s. During the firing, the fuel consumption is 1 kg/s, and the gases are expelled at 5400 km/h relative to the vehicle. If the line of thrust is inclined at 25° to the initial direction of travel, determine the final velocity of the probe. Neglect the effect of gravity.

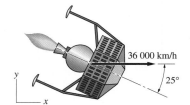

Fig. P15.126

15.127 The firehose is discharging a jet of water against the flat plate. The diameter of the jet is $d = 40$ mm, and the speed of the water is 50 m/s. Calculate the force exerted by the jet on the plate if the plate is (a) stationary; and (b) moving to the right at 4 m/s. (The density of water is 1000 kg/m^3.)

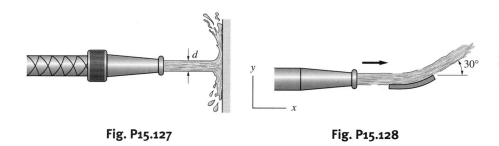

Fig. P15.127   **Fig. P15.128**

15.128 The jet of water is deflected by a stationary vane through 30° as shown. The flow rate is 0.030 m^3/min, and the speed of the jet is 20 m/s. Determine the force exerted by the jet on the vane. (The density of water is 1000 kg/m^3.)

15.129 The 0.15-kg ball is held stationary by a jet of water as shown. At the nozzle, the speed of the jet is 6 m/s, and its diameter is $d = 12$ mm. Find (a) the velocity with which the jet hits the ball; and (b) the height h of the ball. (The density of water is 1000 kg/m^3.)

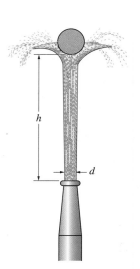

Fig. P15.129

15.130   Water runs over the spillway at a rate of 60 m³/ min. If the speed of the water at the top of the spillway is 3 m/s, calculate the horizontal force applied to the spillway by the running water. Assume that the mechanical energy of the water is conserved. (The density of water is 1000 kg/m³.)

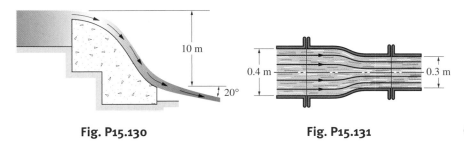

Fig. P15.130   **Fig. P15.131**

15.131   Water enters the reducer section of the pipe at the speed of 2 m/s and a gage pressure of 9 kN/m². If the gage pressure on exit is 2 kN/m², determine the horizontal force applied by the water to the reducer. (The density of water is 1000 kg/m³.)

15.132   Water enters the horizontal bend in the pipe at a velocity of 4.8 m/s. The entrance and exit gage pressures are 23 kPa and 32 kPa, respectively. Find the horizontal force applied by the water to the bend in the pipe. (The mass density of water is 1000 kg/m³.)

15.133   The chain of mass 3.2 kg/m is raised at the constant speed of 1.5 m/s. Determine the tension in the chain at *A*.

15.134   As the snowblower is being pushed forward at 1 m/s, it removes snow at the rate of 2 kg/s. The discharge velocity relative to the snowblower is 10 m/s in the direction shown. Determine the horizontal force *P* required to push the snowblower forward. Neglect rolling resistance.

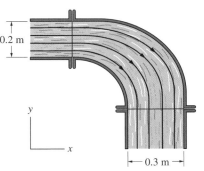

Fig. P15.132

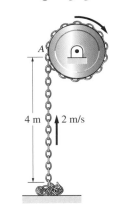

Fig. P15.133

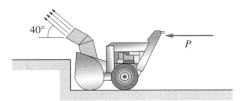

Fig. P15.134

15.135   The diameter of the re-entrant orifice in a water tank is 0.2 m. At section *a–a* the water pressure is 40 kN/m² and the water velocity is negligible. If the velocity of the exit stream is 8 m/s, determine the diameter *d* of the stream. (The density of water is 1000 kg/m³.)

15.136   The takeoff of a jet airplane can be assisted by directing the engine exhaust downward with a deflector. The jet engine shown consumes air at the rate of 80 kg/s and fuel at 1.6 kg/s. The air-fuel mixture is expelled at 660 m/s. If two of these engines are attached to an 8000-kg airplane, what is the smallest deflector

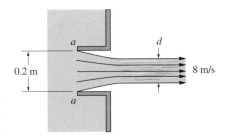

Fig. P15.135

angle θ that would result in instant takeoff? What is the corresponding forward acceleration of the airplane?

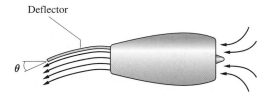

Fig. P15.136

15.137   The 400-kg rocket is in powered flight at an altitude where the gravitational acceleration is $g = 8.5\,\text{m/s}^2$. The engine is consuming fuel at the rate of 16 kg/s, expelling the gases at 600 m/s relative to the rocket. Calculate the angle θ for which the acceleration of the rocket will be horizontal.

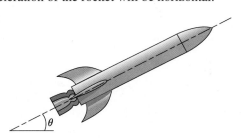

Fig. P15.137

15.138   The 80-kg rocket is connected to a ground control station by a guide wire that uncoils as the rocket ascends. The mass of the wire is 0.005 kg/m and a constant 20-N tension is maintained in the wire at A. At the instant shown, the rocket is unpowered at an elevation of 5 km and flying upward at 250 m/s. Determine the acceleration of the rocket at this instant.

15.139   The force P pulls the chain out of a pile with the constant velocity $\dot{x} = 1\,\text{m/s}$. Determine P if the chain weighs 2 N/m. Neglect friction.

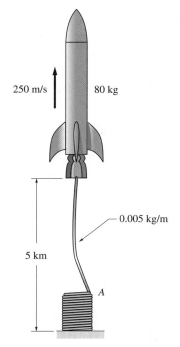

Fig. P15.138

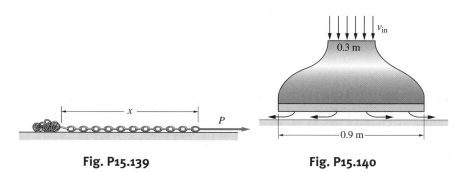

Fig. P15.139   **Fig. P15.140**

15.140   The 5-kg model of a hovercraft takes in air at the speed $v_{\text{in}} = 30\,\text{m/s}$ and exhausts it horizontally from under the skirt. Compute the average pressure under the skirt. (Use $1.2\,\text{kg/m}^3$ for the density of air.)

15.141 A vertical chute discharges coal onto a conveyor at the rate of 60 kg/s. If the speed of the conveyor belt is constant at 3 m/s, find the power required to drive the conveyor. Neglect friction.

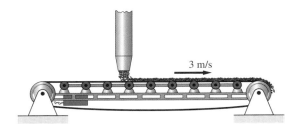

Fig. P15.141

15.142 Each nozzle of the turbine discharges water at the rate of 2 kg/s with the velocity $u = 6$ m/s relative to the nozzle. Find (a) the power generated by the turbine in terms of the constant nozzle velocity v; and (b) the maximum power and the corresponding value of v. Neglect the velocity with which the water enters the nozzles.

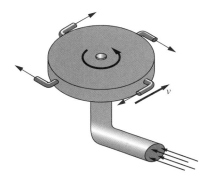

Fig. P15.142

Review of Equations

Relative motion between two particles

$$\mathbf{r}_B = \mathbf{r}_A + \mathbf{r}_{B/A} \quad \mathbf{v}_B = \mathbf{v}_A + \mathbf{v}_{B/A} \quad \mathbf{a}_B = \mathbf{a}_A + \mathbf{a}_{B/A}$$

Motion of the mass center of a particle system

$$\bar{\mathbf{v}} = \frac{d\bar{\mathbf{r}}}{dt} = \frac{1}{m}\sum_{i=1}^{n} m_i\mathbf{v}_i \qquad \bar{\mathbf{a}} = \frac{d\bar{\mathbf{v}}}{dt} = \frac{d^2\bar{\mathbf{r}}}{dt^2} = \frac{1}{m}\sum_{i=1}^{n} m_i\mathbf{a}_i$$

Equation of motion of the mass center

$$\Sigma \mathbf{F} = m\bar{\mathbf{a}}$$

Principle of work and kinetic energy

$$(U_{1-2})_{\text{ext}} + (U_{1-2})_{\text{int}} = T_2 - T_1$$

Conservation of mechanical energy

$$V_1 + T_1 = V_2 + T_2$$

Force-momentum & moment-angular momentum relationships

$$\Sigma \mathbf{F} = \dot{\mathbf{p}} \qquad \Sigma \mathbf{M}_A = \dot{\mathbf{h}}_A$$

$\mathbf{p} = m\bar{\mathbf{v}} = $ linear momentum of the system
$\mathbf{h}_A = $ angular momentum of the system about A

Impulse-momentum equations

$$\mathbf{L}_{1-2} = \mathbf{p}_2 - \mathbf{p}_1$$
$$(\mathbf{A}_A)_{1-2} = (\mathbf{h}_A)_2 - (\mathbf{h}_A)_1 \quad (A \text{ is a fixed point or mass center})$$

$\mathbf{L} = $ linear impulse of external forces
$\mathbf{A}_A = $ angular impulse of external forces about A

Coefficient of restitution of elastic impact

$$e = \frac{\text{velocity of separation}}{\text{velocity of approach}}$$

Review Problems

15.143   Neglecting friction, determine the acceleration of each block when the 12-N horizontal force is applied to block *A*.

15.144   The projectiles *A* and *B* are launched simultaneously at $t = 0$ with the velocities shown. Assuming both trajectories lie in the same vertical plane, determine (a) the relative position vector $\mathbf{r}_{B/A}$ as a function of t; and (b) the smallest distance between the projectiles. Neglect air resistance.

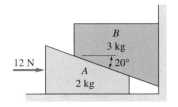

Fig. P15.143

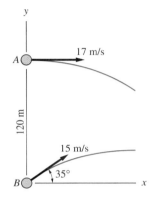

Fig. P15.144

15.145   The 400-kg car drives onto the stationary 3200-kg flatbed railroad car at 25 km/h and brakes to a stop relative to the flatbed car. Determine the final velocity of the cars.

Fig. P15.145

15.146   The coal cars *A* and *B* collide with the velocities shown without becoming coupled together. Knowing that the velocity of *B* after the collision is 5.2 km/h, find the velocity of *A*.

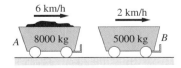

Fig. P15.146

15.147   Blocks *A* and *B* are connected by a cable that runs around the five pulleys. Find the velocity of *B* if *A* is moving downward at 2 m/s.

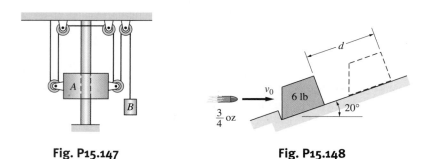

Fig. P15.147   **Fig. P15.148**

15.148   The 20 g bullet strikes the stationary 2-kg block with the horizontal velocity v_0 and becomes embedded. If the maximum displacement of the block after the impact is $d = 0.5$ m, determine v_0. Neglect friction and assume that the block does not leave the inclined surface.

15.149   The airspeed of the helicopter in level flight is 125 km/h, directed east. If the ground speed is 160 km/h in the direction 25° north of east, determine the magnitude and direction of the wind velocity.

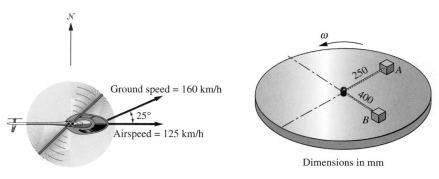

Fig. P15.149   **Fig. P15.150**

Fig. P15.151

15.150   The disk, which has a small peg mounted at its center, is rotating at the angular speed ω. The blocks *A* and *B*, each of mass 2 kg, are connected with a rope that passes around the peg. Initially, the blocks are at rest relative to the disk. If the angular speed of the disk is gradually increased, determine the value of ω at which the blocks start sliding on the disk. The coefficient of static friction under each block is 0.25, and friction at the peg is negligible.

15.151   The three masses are suspended from the cable-and-pulley system shown. Determine the acceleration of each mass in terms of the gravitational acceleration *g*. Neglect the masses of the pulleys.

15.152   The 620-kg rocket *A* is in unpowered flight near the surface of the earth when it breaks into two pieces—the 185-kg nose cone *B* and the 435-kg thruster C. The figure shows the velocities of the two parts 60 s after the breakup (note that A, B, and C are in the same vertical plane). Determine the velocity (magnitude and direction) of the rocket just before the breakup.

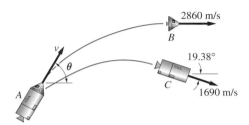

Fig. P15.152

15.153   Two identical billiard balls of radius R are on a horizontal table. After ball A hits stationary ball B with the velocity of 8 m/s, the speed of B is 5.52 m/s. Determine (a) the coefficient of restitution between the balls; and (b) the speed of ball A after the impact. Neglect friction.

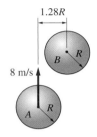

Fig. P15.153

15.154   A small piece of red-hot iron is placed on the anvil and struck with the hammer moving at the velocity v_0. The masses of the hammer and the anvil are m_h and m_a, respectively, and the coefficient of restitution between the hammer and hot iron is e. If the anvil sits on an elastic base (as indicated by the springs in the figure), derive the expression for the impulse delivered by the hammer to the hot metal.

15.155   The rope connecting the slider A and the mass B passes over two small pulleys, one of which is attached to A. Mass B has a constant velocity v_0, directed downward. At the instant when $y_A = b$, determine (a) the velocity of A; and (b) the acceleration of A.

Fig. P15.154

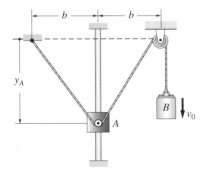

Fig. P15.155

15.156   The two disks A and B lie on a horizontal surface. Disk A is propelled into B, which is initially stationary, with the velocity shown. If $e = 0.75$, calculate the velocity of each disk after the impact. Neglect friction.

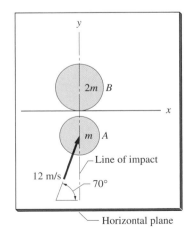

Fig. P15.156

15.157 The identical blocks A and B of mass m each are at rest on a frictionless, horizontal surface. The spring of stiffness k connecting the blocks is undeformed. If block B is given an initial velocity v_0 to the right, determine (a) the relationship between $v_{B/A}$ and $x_{B/A}$ for the ensuing motion; and (b) the maximum value of $x_{B/A}$.

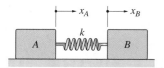

Fig. P15.157

15.158 The 0.2-kg artillery shell A is fired from the gun with the muzzle velocity of 600 m/s. The recoil of the 5-kg barrel B in its mount C is limited by the spring of stiffness 20×10^3 N/m. Determine the maximum deformation of the spring after the shell has been fired. Neglect friction.

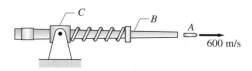

Fig. P15.158

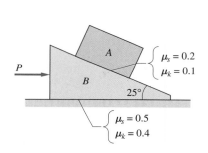

Fig. P15.159

15.159 Determine the largest force P that can be applied to the 2-kg block B without causing the 1-kg block A to slide up relative to B. The static and kinetic coefficients of friction between the contact surfaces are shown in the figure.

15.160 The cars A and B are traveling with the velocities shown when they collide. Assuming the impact is plastic, determine (a) the common velocity of the cars just after the impact; and (b) the percentage of mechanical energy absorbed by the impact.

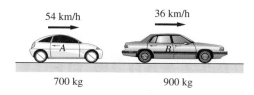

Fig. P15.160

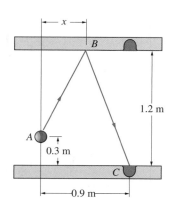

Fig. P15.161

15.161 The billiard ball A is to be banked off the rail at B so that it will enter the pocket C. The coefficient of restitution for the impact between the ball and the rail is 0.85. Neglecting friction, determine distance x that determines the location of B.

16

Planar Kinematics of Rigid Bodies

Rolling without slipping is a commonly encountered problem of kinematics. This chapter contains several problems that analyze rolling, such as Sample Problem 16.4, Prob.16.20 and 16.23. (© Spe/Dreamstime.com)

16.1   Introduction

A body is said to be *rigid* if the distance between any two points in the body remains constant. In other words, a rigid body does not deform. Of course, the rigid-body concept is an idealization, because all bodies deform to some extent when subjected to forces. But if the deformation is sufficiently small ("small" usually means negligible when compared to the dimensions of the body), the assumption of rigidity is justified.

This chapter is concerned only with the kinematics of *plane motion* of rigid bodies. A body undergoes plane motion if all points in the body remain a constant distance from a fixed reference plane, called the *plane of motion* of the body.

In other words, all points in the body move in planes that are parallel to the plane of motion. There are three categories of plane motion:

- *Translation* is the special case in which the body moves without rotation; that is, any line in the body remains parallel to its initial position, as shown in Fig. 16.1(a).* Because all points in the body have the same displacement, the motion of one point determines the motion of the entire body.

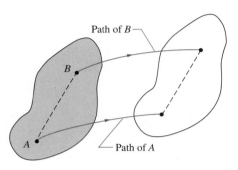

Translation

Fig. 16.1(a)

- *Rotation about a fixed axis* is the special case in which a line in the body, called the *axis of rotation,* is fixed in space. Consequently, each point not on the axis of rotation moves in a circle about the axis, as illustrated in Fig. 16.1(b) (the axis of rotation at *O* is perpendicular to the plane of the paper).
- *General plane motion* is the superposition of translation and rotation. The rolling disk in Fig. 16.1(c) is an example of such motion: the disk is translating and rotating simultaneously.

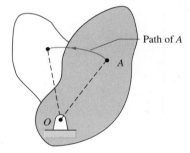

Rotation about a fixed axis

Fig. 16.1(b)

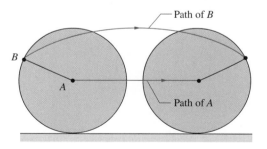

General plane motion

Fig. 16.1(c)

*When illustrating plane motion, it is often convenient to show a representative cross section of the body that is parallel to the plane of motion.

16.2   *Plane Angular Motion*

In this article, we introduce the angular displacement, angular velocity, and angular acceleration of lines and rigid bodies that move in a plane.

Angular Position   Consider the rigid body $\mathcal{B}$ in Fig. 16.2 that is moving in the xy-plane. Let AB be a line that is embedded in $\mathcal{B}$ and lies in the plane of motion.[*] The angle $\theta(t)$ between AB and a fixed reference line, such as the x-axis, is known as the *angular position coordinate of line AB*.

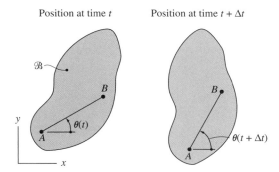

Position at time t Position at time $t + \Delta t$

Fig. 16.2

Angular Displacement   During the time interval Δt, the angular position coordinate of AB changes from $\theta(t)$ to $\theta(t + \Delta t)$, as shown in Fig. 16.2. The angular displacement of line AB during this time interval is defined as

$$\Delta\theta = \theta(t + \Delta t) - \theta(t) \qquad (16.1)$$

We now show that all lines in $\mathcal{B}$ that lie in the plane of motion have the same angular displacement. Figure 16.3 shows two such lines AB and CD, with the

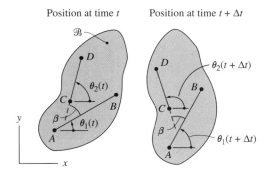

Position at time t Position at time $t + \Delta t$

Fig. 16.3

[*]Here, and in the discussion that follows, the term "plane of motion" also applies to any plane that is parallel to the plane of motion.

angular position coordinates $\theta_1(t)$ and $\theta_2(t)$, respectively. Because the body is rigid, the angle β between the lines does not change between the two positions shown in the figure; that is, $\beta = \theta_2(t) - \theta_1(t) = \theta_2(t + \Delta t) - \theta_1(t + \Delta t)$. It follows that $\theta_1(t + \Delta t) - \theta_1(t) = \theta_2(t + \Delta t) - \theta_2(t)$, or

$$\Delta\theta_1 = \Delta\theta_2$$

Because AB and CD can be chosen arbitrarily in the plane of the motion, we conclude that all lines in the plane of the motion have the same angular displacement. Therefore, $\Delta\theta$ is also called the *angular displacement of body $\mathcal{B}$.*

Angular Velocity   The *angular velocity ω of line AB* is defined as the time derivative of its angular position coordinate θ:

$$\omega = \lim_{\Delta t \to 0} \frac{\Delta\theta}{\Delta t} = \frac{d\theta}{dt} = \dot{\theta} \tag{16.2}$$

Because all lines in the plane of motion have the same angular displacement $\Delta\theta$, they also have the same angular velocity. Therefore, ω is also called the *angular velocity of body $\mathcal{B}$.* Common units for angular velocity are rad/s and rev/min.

Angular Acceleration   The *angular acceleration α of line AB* is defined to be the time derivative of its angular velocity:

$$\alpha = \frac{d\omega}{dt} = \dot{\omega} \quad \text{or} \quad \alpha = \frac{d^2\theta}{dt^2} = \ddot{\theta} \tag{16.3}$$

Because the angular velocity ω is the same for all lines in the plane of motion, α is also called the *angular acceleration of body $\mathcal{B}$.* The units of angular acceleration are usually taken to be rad/s^2.

The time t in Eq. (16.3) can be eliminated as an explicit variable by using the chain rule for differentiation:

$$\alpha = \frac{d\omega}{dt} = \frac{d\omega}{d\theta} \frac{d\theta}{dt} = \omega \frac{d\omega}{d\theta} \tag{16.4}$$

- Observe that positive directions of angular velocity and angular acceleration are the same as the assumed positive direction of θ.
- Equations (16.2)–(16.4) are analogous to Eqs. (12.9) and (12.10) for rectilinear motion of a particle: $v = \dot{x}$, $a = \dot{v} = \ddot{x}$, and $\alpha = v(dv/dx)$. Therefore, the mathematical methods presented in Art. 12.4 can also be used to analyze plane angular rotation.

Vector Representation of Angular Motion   Because $\Delta\theta$, ω, and α possess magnitude and direction, it is sometimes convenient to represent them as vectors using the right-hand rule. Assuming that we have a right-handed coordinate system (the positive z-axis in Fig. 16.2 would point out of the paper), we can write

$$\Delta\boldsymbol{\theta} = \Delta\theta\mathbf{k} \qquad \boldsymbol{\omega} = \omega\mathbf{k} = \dot{\theta}\mathbf{k} \qquad \boldsymbol{\alpha} = \alpha\mathbf{k} = \dot{\omega}\mathbf{k} = \ddot{\theta}\mathbf{k} \qquad (16.5)$$

Comment on Three-Dimensional Motion   In planar motion, angular displacement is a vector quantity because it has magnitude ($\Delta\theta$), has direction (perpendicular to the plane of motion), and obeys the parallelogram law for addition (summing collinear vectors is a special case of parallelogram addition). It follows that angular velocity and angular acceleration are also vectors. However, if the motion is three-dimensional, angular displacements do not obey the parallelogram law for addition. Therefore, angular displacements, in general, are *not* vectors.

As an illustration of the nonvector character of angular displacements, consider a book that is placed in the initial position shown in Fig. 16.4(a) and given the angular rotations $\Delta\theta_x = 90°$ (rotation about the x-axis) and $\Delta\theta_y = 90°$ (rotation about the y-axis). Figure 16.4(b) shows the final position of the book if the rotations are performed in the order $\Delta\theta_x$ followed by $\Delta\theta_y$. Figure 16.4(c) shows the result if the rotations are performed in the order $\Delta\theta_y$ followed by $\Delta\theta_x$. From these figures, we see that the final orientation of the book depends on the order

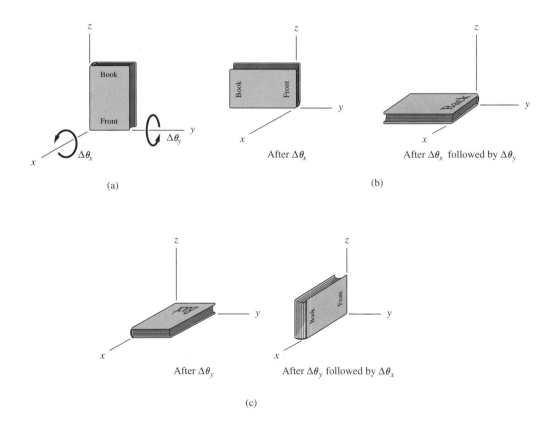

Fig. 16.4

in which the rotations are performed. We conclude that angular displacements generally are not vectors because they do not obey the commutative property of vector addition. However, it can be shown[*] that differential (infinitesimal) angular displacements are vectors, even in three-dimensional motion. For this reason, angular velocity and angular acceleration are *always* vector quantities.

16.3   *Rotation about a Fixed Axis*

Rotation about a fixed axis is the special case of plane motion in which one line in the body, called the *axis of rotation*, is fixed in space. Figure 16.5 shows a rigid body that is rotating about an axis. We let B be a point in the body that is a distance R from the axis. Because the body is rigid, the path of B is a circle of radius R that lies in a plane perpendicular to the axis of rotation. The center O of the circle lies on the axis, and the angular position coordinate of the radial line OB is denoted by θ. Since OB lies in the plane of motion, the angular velocity and angular acceleration of the body are, according to Eqs. (16.2) and (16.3),

$$\omega = \dot{\theta} \qquad \alpha = \dot{\omega} = \ddot{\theta} \tag{16.6}$$

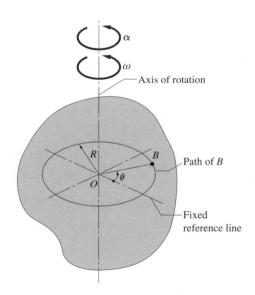

Fig. 16.5

Kinematics of a Point in the Body   The motion of a particle along a circular path of radius R was previously discussed using path coordinates in Art. 13.2. With θ representing the angular velocity of the radial line to the particle,

[*]See *Principles of Dynamics*, 2e, Donald T. Greenwood, Prentice Hall, 1988, p. 350.

the velocity (v) and the normal (a_n) and tangential (a_t) components of acceleration of the particle were found to be

$$v = R\dot{\theta} \qquad \text{(13.10, repeated)}$$

$$a_n = R\dot{\theta}^2 \qquad a_t = R\ddot{\theta} \qquad \text{(13.11, repeated)}$$

Substituting for $\dot{\theta}$ and $\ddot{\theta}$ from Eq. (16.6), the velocity and acceleration components of point B become

$$v = R\omega$$
$$a_n = R\omega^2 = \frac{v^2}{R} = v\omega \qquad \text{(16.7a)}$$
$$a_t = R\alpha$$

The velocity and acceleration vectors of point B lie in the plane of motion (the plane of the circle), as shown in Fig. 16.6(a). The velocity v and the tangential component of acceleration a_t are tangent to the circle, pointing in the direction of increasing θ. The normal component of acceleration a_n is directed toward the center of the circular path.

Vector Representation of Kinematics   It is sometimes convenient to compute the velocity and acceleration of a point using vector algebra.

In Fig. 16.6(b), we let $\mathbf{r}$ be the position vector of B relative to an arbitrary reference point A that lies on the axis of rotation. The angle between $\mathbf{r}$ and the axis is denoted by β. We will show that the following vector equations give the magnitude and direction for the velocity and acceleration of point B:

$$\mathbf{v} = \boldsymbol{\omega} \times \mathbf{r}$$
$$\mathbf{a}_n = \boldsymbol{\omega} \times \mathbf{v} = \boldsymbol{\omega} \times (\boldsymbol{\omega} \times \mathbf{r}) \qquad \text{(16.7b)}$$
$$\mathbf{a}_t = \boldsymbol{\alpha} \times \mathbf{r}$$

Using the properties of the cross product, the magnitudes of the vectors in Eqs. (16.7b) are

$$v = |\boldsymbol{\omega} \times \mathbf{r}| = \omega r \sin\beta = R\omega$$
$$a_n = |\boldsymbol{\omega} \times \mathbf{v}| = \omega v \sin 90^\circ = \omega v = R\omega^2$$
$$a_t = |\boldsymbol{\alpha} \times \mathbf{r}| = \alpha r \sin\beta = R\alpha$$

These results agree with Eqs. (16.7a). Using the right-hand rule for the cross products, it is not difficult to see that the directions of the vectors in Eqs. (16.7b) are the same as those shown in Fig. 16.6(a).

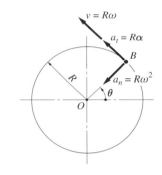

(a)

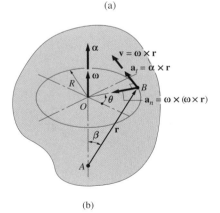

(b)

Fig. 16.6

Sample Problem 16.1

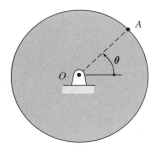

The disk rotates about a fixed axis at O. During the period $t = 0$ to $t = 4\,\text{s}$, the angular position of the line OA in the disk varies as $\theta(t) = t^3 - 12t + 6\,\text{rad}$, where t is in seconds. Determine (1) the angular velocity and the angular acceleration of the disk at the end of the period; (2) the angular displacement of the disk during the period; and (3) the total angle turned through by the disk during the period.

Solution

Part 1

The angular velocity and angular acceleration of the disk are

$$\omega = \dot{\theta} = 3t^2 - 12\,\text{rad/s} \qquad \alpha = \dot{\omega} = 6t\,\text{rad/s}^2$$

When $t = 4\,\text{s}$, we have

$$\omega = 3(4)^2 - 12 = 36\,\text{rad/s (CCW)} \qquad\qquad \textit{Answer}$$

$$\alpha = 6(4) = 24\,\text{rad/s}^2\,\text{(CCW)} \qquad\qquad \textit{Answer}$$

Part 2

The angular positions of the line OA at the beginning and at the end of the period are

$$\theta\big|_{t=0} = 6\,\text{rad}$$

$$\theta\big|_{t=4\,\text{s}} = 4^3 - 12(4) + 6 = 22\,\text{rad}$$

Therefore, the angular displacement of the disk from $t = 0$ to $t = 4$ s is

$$\Delta\theta = \theta\big|_{t=4\,\text{s}} - \theta\big|_{t=0} = 22 - 6 = 16\,\text{rad (CCW)} \qquad \textit{Answer}$$

Part 3

Note that the direction of rotation of the disk changes when $\omega = 0$, that is, when

$$3t^2 - 12 = 0 \qquad t = 2\,\text{s}$$

The angular position of OA at that time is

$$\theta\big|_{t=2\,\text{s}} = 2^3 - 12(2) + 6 = -10\,\text{rad}$$

We conclude that the disk rotates clockwise ($\omega < 0$) between $t = 0$ and $t = 2$ s, its angular displacement being

$$\Delta\theta_1 = \theta\big|_{t=2\,\text{s}} - \theta\big|_{t=0} = -10 - 6 = -16\,\text{rad}$$

Between $t = 2$ s and $t = 4$ s, the rotation is counterclockwise ($\omega > 0$); the corresponding angular displacement is

$$\Delta\theta_2 = \theta\big|_{t=4\,\mathrm{s}} - \theta\big|_{t=2\,\mathrm{s}} = 22 - (-10) = 32 \text{ rad}$$

Therefore, the total angle turned through by the disk from $t = 0$ to $t = 4$ s is

$$|\Delta\theta_1| + |\Delta\theta_2| = 16 + 32 = 48 \text{ rad} \qquad\qquad \textit{Answer}$$

Sample Problem 16.2

Pulley B is being driven by the motorized pulley A that is rotating at $\omega_A = 20$ rad/s. At time $t = 0$, the current in the motor is cut off, and friction in the bearings causes the pulleys to coast to a stop. The angular acceleration of A during the deceleration is $\alpha_A = -2.5t$ rad/s^2, where t is in seconds. Assuming that the drive belt does not slip on the pulleys, determine (1) the angular velocity of B as a function of time; (2) the angular displacement of B during the period of coasting; and (3) the acceleration of point C on the straight portion of the belt as a function of time.

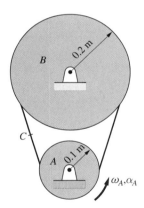

Solution

Part 1

Because the belt does not slip, every point on the belt that is in contact with a pulley has the same velocity as the adjacent point on the pulley. Therefore, the speed of any point on the belt is

$$v = R_A\omega_A = R_B\omega_B \qquad\qquad \text{(a)}$$

so that

$$\omega_B = \frac{R_A}{R_B}\omega_A = \frac{0.10}{0.20}\omega_A = 0.5\omega_A$$

Differentiating with respect to time, we obtain for the angular acceleration of pulley B

$$\alpha_B = 0.5\alpha_A = 0.5(-2.5t) = -1.25t \text{ rad/s}^2$$

Because $\alpha_B = d\omega_B/dt$, we have $d\omega_B = \alpha_B\, dt$, or

$$\omega_B = \int \alpha_B\, dt = \int -1.25t\, dt = -0.625t^2 + C_1$$

The initial condition, $\omega_B = 20$ rad/s when $t = 0$, yields $C_1 = 20$ rad/s. Hence, the angular velocity of pulley B is

$$\omega_B = -0.625t^2 + 20 \text{ rad/s} \qquad\qquad \textit{Answer}$$

Part 2

We let θ_B be the angular position of a line in B measured from a fixed reference line. Recalling that $\omega_B = d\theta_B/dt$, we integrate $d\theta_B = \omega_B dt$ to obtain

$$\theta_B = \int \omega_B\, dt = \int (-0.625t^2 + 20)dt = -0.2083t^3 + 20t + C_2$$

Letting $\theta_B = 0$ when $t = 0$, we have $C_2 = 0$, which gives

$$\theta_B = -0.2083t^3 + 20t \text{ rad}$$

The pulley comes to rest when $\omega_B = -0.625t^2 + 20 = 0$, which yields $t = 5.657$ s. The corresponding angular position of the line in B is

$$\theta_B\big|_{t=5.657 \text{ s}} = -0.2083(5.657)^3 + 20(5.657) = 112.0 \text{ rad}$$

Therefore, the angular displacement of pulley B as it coasts to a stop is

$$\Delta\theta_B = \theta_B\big|_{t=5.657 \text{ s}} - \theta_B\big|_{t=0} = 112.0 - 0 = 112.0 \text{ rad} \qquad \textit{Answer}$$

Because the direction of rotation does not change, the total angle turned through by pulley B during the deceleration is also 112.0 rad.

Part 3

Substituting $R_B = 0.2$ m and $\omega_B = -0.625t^2 + 20$ rad/s into Eq. (a), the speed of point C (which is the same for all points on the belt) is

$$v_C = 0.2(-0.625t^2 + 20) = -0.125t^2 + 4 \text{ m/s} \qquad \text{(b)} \qquad \textit{Answer}$$

Because the path of point C on the belt is a straight line, the acceleration of C is

$$a_C = \dot{v}_C = -0.25t \text{ m/s}^2 \qquad\qquad \textit{Answer}$$

We could obtain the same result by observing that a_C is equal to the *tangential component* of acceleration of a point on the rim of pulley B (pulley A could also be used). Thus $a_C = R_B\alpha_B = 0.2(-1.25t) = -0.25t$ m/s^2.

Observe that the expression for v_C in Eq. (b) is valid for the entire deceleration period ($0 \le t \le 5.657$ s), whereas the answer for a_C applies only for those times when C is not in contact with either pulley. Points on the belt follow circular paths when they are in contact with a pulley.

Sample Problem 16.3

The rigid body is rotating about the y-axis. In the position shown in Fig. (a), the angular velocity and angular acceleration of the body are as specified in the figure. Determine the velocity and acceleration vectors of point A in this position using (1) vector equations; and (2) scalar equations.

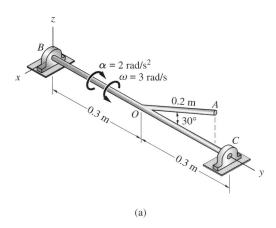

(a)

Solution

Part 1

Using the right-hand rule, the angular velocity and acceleration vectors of the body are

$$\boldsymbol{\omega} = 3\mathbf{j} \text{ rad/s} \qquad \boldsymbol{\alpha} = -2\mathbf{j} \text{ rad/s}^2$$

The position vector of A relative to point O on the axis of rotation is

$$\mathbf{r} = 0.2\cos 30°\mathbf{j} + 0.2\sin 30°\mathbf{k} = 0.1732\mathbf{j} + 0.1\mathbf{k} \text{ m}$$

The velocity and acceleration vectors of point A can now be computed from Eqs. (16.7b):

$$\mathbf{v}_A = \boldsymbol{\omega} \times \mathbf{r} = 3\mathbf{j} \times (0.1732\mathbf{j} + 0.1\mathbf{k}) = 0.3\mathbf{i} \text{ m/s} \qquad \textit{Answer}$$

$$\mathbf{a}_A = \boldsymbol{\omega} \times \mathbf{v} + \boldsymbol{\alpha} \times \mathbf{r} = 3\mathbf{j} \times 0.3\mathbf{i} + (-2\mathbf{j}) \times (0.1732\mathbf{j} + 0.1\mathbf{k})$$

$$= -0.9\mathbf{k} - 0.2\mathbf{i} \text{ m/s}^2 \qquad \textit{Answer}$$

Part 2

Figure (b) shows the plane of motion of point A (looking from C toward B). The path of A is a circle of radius $R = 0.2\sin 38 = 0.1$ m. The magnitudes of the velocity and acceleration components of A are, using Eqs. (16.7a),

$$v_A = R\omega = 0.1(3) = 0.3 \text{ m/s}$$

$$(a_A)_n = R\omega^2 = 0.1(3)^2 = 0.9 \text{ m/s}^2$$

$$(a_A)_t = R\alpha = 0.1(2) = 0.2 \text{ m/s}^2$$

These components are shown in Fig. (b). Their directions were determined as follows:

- $\mathbf{v}_A$ is tangent to the path; its sense is consistent with the counterclockwise direction of ω.
- $(\mathbf{a}_A)_n$ is directed toward the center of the path.
- $(\mathbf{a}_A)_t$ is tangent to the path; its sense is consistent with the clockwise direction of α.

Therefore, the velocity and acceleration vectors of point A are

$$\mathbf{v}_A = 0.3\mathbf{i} \text{ m/s} \qquad \qquad \textit{Answer}$$

$$\mathbf{a}_A = -0.9\mathbf{k} - 0.2\mathbf{i} \text{ m/s}^2 \qquad \qquad \textit{Answer}$$

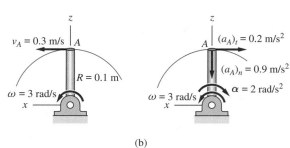

(b)

Problems

16.1   Characterize the motion of bodies A and B of each mechanism shown as (1) translation; (2) rotation about a fixed axis; or (3) general plane motion.

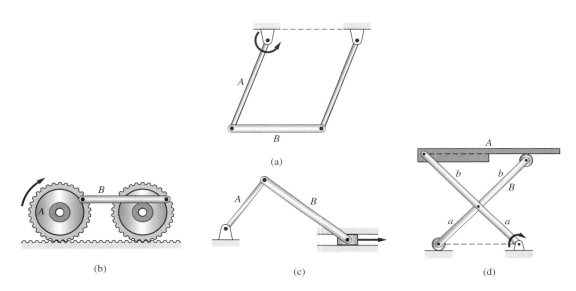

(a)

(b)

(c)

(d)

Fig. P16.1

16.2   When an electric motor is turned on at $t = 0$, its angular acceleration is $\alpha = 10e^{-0.5t}$ rad/s^2, where t is the time in seconds. What is the terminal angular velocity of the motor? How many revolutions are required for the motor to reach half of its terminal angular velocity?

16.3   A rotary engine is running at 8000 rev/min when it runs out of fuel, causing the engine to decelerate at a constant rate. If the angular speed is 4000 rev/min after 3200 revolutions, determine the total time it takes for the engine to stop.

16.4   The angular position of the rod OA varies with time as $\theta = -4t^2 + 24t - 10$, where θ is in radians and t is in seconds. Determine (a) the angular velocity and the angular acceleration of the rod at $t = 4$ s; and (b) the total angle turned through by the rod between $t = 0$ and $t = 4$ s.

16.5   The angular acceleration of rod OA is $\ddot{\theta} = 4 + 6t$ rad/s^2, where t is in seconds. Assuming that the rod was at rest at $t = 0$, calculate its angular displacement between $t = 0$ and the time when the angular speed reaches 24 rad/s.

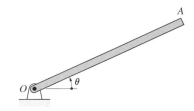

Fig. P16.4–P16.7

16.6   The angular velocity of rod OA is $\omega = 3t^2 - kt$ rad/s, where t is in seconds and k is a constant. When $t = 0$, $\theta = 8$ rad, clockwise; and when $t = 4$ s, $\theta = 16$ rad, clockwise. Determine (a) the constant k; and (b) the total angle that OA turns through between $t = 0$ and $t = 4$ s.

16.7 The angular acceleration α of rod OA is given by $\alpha = k\theta^2$ rad/s^2, where θ is in radians and k is a constant. When $\theta = 0$, $\omega = 2$ rad/s, and when $\theta = 3$ rad, $\omega = 7$ rad/s. Find (a) the constant k; and (b) ω when $\theta = 2$ rad.

16.8 The constant angular acceleration of the rotating disk is $\alpha = 12$ rad/s^2. The angular velocity of the disk is 24 rad/s, clockwise, when $t = 0$. Determine the total angle turned by the disk between $t = 0$ and $t = 4$ s.

16.9 The angular velocity of the rotating disk is $\omega = 4\sqrt{t}$ rad/s, where t is in seconds. Find the angular displacement of the disk for the time interval $t = 0$ to $t = 6$ s.

16.10 The angular acceleration α (rad/s^2) of the rotating disk is related to its angular velocity ω (rad/s) by $\alpha = 4\sqrt{\omega}$. When $t = 0$, the disk is at rest and the angular position coordinate of a line in the disk is $\theta = 8$ rad. Find expressions for the following: (a) $\theta(\omega)$; (b) $\omega(t)$; and (c) $\theta(t)$.

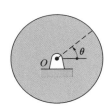

Fig. P16.8–P16.10

16.11 The velocity and acceleration of the belt running between the motor A and the pulley B are $v = 16$ m/s and $a = -9$ m/s^2, respectively. Determine the angular velocities and angular accelerations of pulleys B and C.

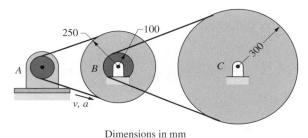

250   100   300

A B C

v, a

Dimensions in mm

Fig. P16.11

16.12 At the instant shown, the speed of the belt connecting the two pulleys is 4 m/s and the magnitude of the acceleration of point A is 250 m/s^2. What is the magnitude of the acceleration of point B at this instant?

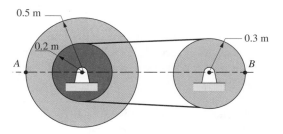

0.5 m   0.3 m

0.2 m

A B

Fig. P16.12

16.13 The rectangular plate rotates in the xy-plane about the corner O. At the instant shown, the acceleration of corner A is $a_A = 60$ m/s^2 in the direction indicated. Determine the acceleration vector of the mid-point C of the plate at this instant.

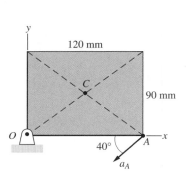

y

120 mm

C

90 mm

O $40°$ A x

a_A

Fig. P16.13

16.14 Just before the two friction wheels are brought into contact, B is rotating clockwise at 18 rad/s and A is stationary. Slipping between the wheels occurs during the first six seconds of contact; during this time the angular speed of each wheel changes uniformly. If the final angular speed of B is 12 rad/s, determine (a) the angular acceleration of A during the period of slipping; and (b) the number of revolutions made by A before it reaches its final speed.

16.15 In the position shown, rod $OABC$ is rotating about the y-axis with the angular velocity $\omega = 2.4$ rad/s and angular acceleration $\alpha = 7.2$ rad/s^2 in the directions shown. For this position, compute the velocity and acceleration vectors of point C using (a) vector equations; and (b) scalar equations.

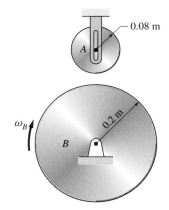

Fig. P16.14

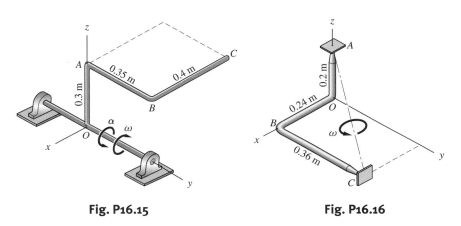

Fig. P16.15 **Fig. P16.16**

16.16 The bent rod is rotating about the axis AC. In the position shown, the angular speed of the rod is $\omega = 2$ rad/s, and it is increasing at the rate of 7 rad/s^2. For this position, determine the velocity and acceleration vectors of point B.

16.17 The bent plate is rotating about the fixed axis OA with the constant angular velocity $\omega = 20$ rad/s in the direction shown. Compute the magnitudes of the velocity and acceleration of point B.

16.18 The bent rod ABC rotates about the axis AC with the constant angular velocity $\omega = 25$ rad/s directed as shown. Determine the velocity and acceleration vectors of point B for the position shown.

Fig. P16.17

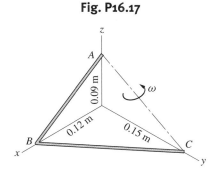

Fig. P16.18

<table>
<tr><td>**16.4**</td><td>*Relative Motion of Two Points in a Rigid Body*</td></tr>
</table>

The kinematics of relative motion of particles (points) was discussed in Art. 15.2. The definition of relative motion between points A and B yielded the following relationships

$$\mathbf{v}_B = \mathbf{v}_A + \mathbf{v}_{B/A} \qquad \text{(15.3, repeated)}$$

$$\mathbf{a}_B = \mathbf{a}_A + \mathbf{a}_{B/A} \qquad \text{(15.4, repeated)}$$

where the subscript B/A denotes the motion of B relative to A. It is useful to recall that $\mathbf{v}_{B/A}$ and $\mathbf{a}_{B/A}$ can be viewed as the velocity and acceleration of B as seen by a *nonrotating* observer attached to point A.

If points A and B belong to the same *translating* (i.e., nonrotating) rigid body, their relative position vector $\mathbf{r}_{B/A}$ is constant, and there is no relative motion ($\mathbf{v}_{B/A} = \mathbf{a}_{B/A} = \mathbf{0}$). Therefore, all points in the translating body have the same velocities and accelerations.

Now consider the body in Fig. 16.7(a) that is undergoing *general plane motion* (simultaneous translation and rotation). The *nonrotating x′y′-axes are attached to point A*. The angular velocity of the body is $\omega = \dot\theta$, and its angular acceleration is $\alpha = \dot\omega = \ddot\theta$, where θ is the angle between $\mathbf{r}_{B/A}$ and the $x′$-axis. As the body moves, the direction of $\mathbf{r}_{B/A}$ changes, but its magnitude $r_{B/A}$ (the distance between A and B) is constant. It follows that the path of B in the *translating x′y′*-coordinate system is a circle of radius $r_{B/A}$ that lies in the plane of motion, as shown in Fig. 16.7(b). Thus, the motion of B relative to A is due only to the rotation of the body about an axis that passes through point A. Therefore, the relative velocity and relative acceleration between two points on a rigid body can be computed using the equations developed in the previous article for rotation about a fixed axis. With the appropriate changes in notation, Eqs. (16.7a) become

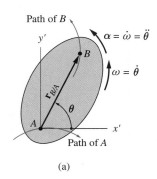

(a)

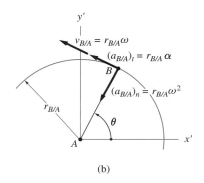

(b)

Fig. 16.7

$$v_{B/A} = r_{B/A}\omega$$
$$(a_{B/A})_n = r_{B/A}\omega^2 = \frac{v_{B/A}^2}{R} = v_{B/A}\omega \qquad \text{(16.8a)}$$
$$(a_{B/A})_t = r_{B/A}\alpha$$

Figure 16.7(b) displays each of the terms in Eqs. (16.8a). This figure is identical to Fig. 16.6(a) for rotation about a fixed axis, except for the notational changes. It is important to understand that the *absolute* angular velocity ($\omega = \dot\theta$) and *absolute* angular acceleration ($\alpha = \dot\omega = \ddot\theta$) are used in the computations. Because the angle θ is measured from the fixed $x′$-direction, all angular measurements are absolute quantities.

The vector forms of Eqs. (16.8a) are

$$\mathbf{v}_{B/A} = \boldsymbol{\omega} \times \mathbf{r}_{B/A}$$
$$(\mathbf{a}_{B/A})_n = \boldsymbol{\omega} \times \mathbf{v}_{B/A} = \boldsymbol{\omega} \times (\boldsymbol{\omega} \times \mathbf{r}_{B/A}) \qquad \text{(16.8b)}$$
$$(\mathbf{a}_{B/A})_t = \boldsymbol{\alpha} \times \mathbf{r}_{B/A}$$

16.5   *Method of Relative Velocity*

There are two methods for analyzing velocities associated with rigid bodies undergoing plane motion. This article discusses the *method of relative velocity*, which utilizes the equation $\mathbf{v}_B = \mathbf{v}_A + \mathbf{v}_{B/A}$ for two points on the same rigid body. The other method, based on *instant centers for velocity*, is described in the next article.

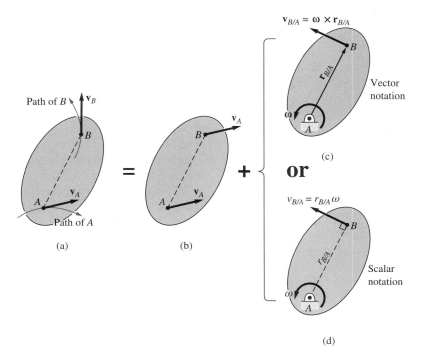

Plane motion   =   Translation   +   Rotation about A

Fig. 16.8

Figure 16.8(a) shows a rigid body that is undergoing general plane motion. If A and B are points in the body, then according to Eqs. (16.8), the velocity of B relative to A is

$$\mathbf{v}_{B/A} = \boldsymbol{\omega} \times \mathbf{r}_{B/A} \qquad (v_{B/A} = r_{B/A}\omega) \qquad (16.9)$$

where $\boldsymbol{\omega}$ is the angular velocity of the body. Substituting Eq. (16.9) into $\mathbf{v}_B = \mathbf{v}_A + \mathbf{v}_{B/A}$, we get

$$\mathbf{v}_B = \mathbf{v}_A + \boldsymbol{\omega} \times \mathbf{r}_{B/A} \qquad (16.10)$$

The physical interpretation of Eq. (16.10) is illustrated in Fig. 16.8. The figure shows that general plane motion is equivalent to the superposition of two simpler motions[*].

1. A rigid-body translation, where the velocity of each point is equal to the velocity of the reference point A ($\mathbf{v}_B = \mathbf{v}_A$), as shown in Fig. 16.8(b).
2. A rigid-body rotation about a fixed axis at A ($\mathbf{v}_{B/A} = \boldsymbol{\omega} \times \mathbf{r}_{B/A}$), illustrated in Figs. 16.8(c) or (d). Showing a pin support at A reinforces the notion that A is considered to be fixed at the instant when the contribution of the rotation is computed.

---

[*]This representation of general plane motion is a two-dimensional form of Chasle's theorem: The most general rigid body displacement is equivalent to a translation of a point in the body plus a rotation about an axis through that point. For a proof of this theorem, see *Principles of Dynamics*, 2e, Donald T. Greenwood, Prentice Hall, 1988, p. 339.

The contribution of the rotation of the body to the velocity of B may be computed using either the vector notation in Fig. 16.8(c) or the scalar notation in Fig. 16.8(d). The choice of notation is a matter of personal preference. When using scalar notation, keep in mind that the direction of the velocity is perpendicular to AB, and its sense is determined by the direction of the angular velocity ω of the body (recall that ω is a property of the body and is independent of the choice of the reference point).

The method of relative velocity consists of writing Eq. (16.10), $\mathbf{v}_B = \mathbf{v}_A + \boldsymbol{\omega} \times \mathbf{r}_{B/A}$, for two points in the *same rigid body*, and then solving for the unknowns. For plane motion, Eq. (16.10) is equivalent to two scalar equations (e.g., the equations that result from equating the horizontal and vertical components of both sides of the vector equation). The number of variables appearing in Eq. (16.10) is five (assuming that $\mathbf{r}_{B/A}$ is known):

$\mathbf{v}_B$: two variables (magnitude and direction, or horizontal and vertical components)

$\mathbf{v}_A$: two variables (magnitude and direction, or horizontal and vertical components)

ω: one variable (magnitude ω of the angular velocity). Note that the direction of vector $\boldsymbol{\omega}$ is known because it is always perpendicular to the plane of motion.

Clearly, Eq. (16.10) cannot be solved unless three of the preceding five variables are known beforehand. Therefore, the choice of A and/or B is restricted to points whose velocities contain less than two unknowns. We refer to these points as a *kinematically important points for velocity.*[*] A common example of a kinematically important point is a fixed point; that is, a point that is pinned to a support. Because the velocity of a fixed point is known to be zero, it contains no unknowns. A second example is a point that travels along a given path. Since the direction of the velocity is known to be tangent to the path, the velocity of such a point contains at most one unknown, namely its magnitude.

The steps in the application of the relative velocity method are as follows:

Step 1: Identify two kinematically important points, say, A and B, on the same rigid body.

Step 2: Write $\mathbf{v}_B = \mathbf{v}_A + \boldsymbol{\omega} \times \mathbf{r}_{B/A}$, identifying the unknown variables (either vector or scalar notation can be used).

Step 3: If the number of unknowns is two, solve the equation.

If the number of unknowns is greater than two, it may still be possible to solve the problem by considering the motion of other kinematically important points.

---

[*]Since we are discussing velocities, we will refer to these points for the time being as simply "kinematically important points".

Rolling Without Slipping   Figure 16.9(a) shows a circular disk of radius R that is rolling on a horizontal surface with the angular velocity ω and angular acceleration α, both clockwise. Observe that the path of the center O is a straight line parallel to the surface. Rolling without slipping occurs if the contact point C on the disk has no velocity, that is, if the disk does not slide along the surface. This case deserves special attention because it occurs in many engineering applications.

Relating the velocities of points O and C, we have

$$\mathbf{v}_O = \mathbf{v}_C + \boldsymbol{\omega} \times \mathbf{r}_{O/C}$$

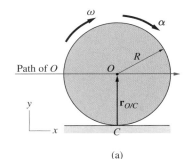

(a)

Substituting $\mathbf{v}_C = \mathbf{0}$, $\boldsymbol{\omega} = -\omega\mathbf{k}$ and $\mathbf{r}_{O/C} = R\mathbf{j}$, we obtain

$$\mathbf{v}_O = -\omega\mathbf{k} \times R\mathbf{j} = R\omega\mathbf{i} \qquad (16.11\text{a})$$

As expected, this result shows that the velocity of the center O is parallel to the surface on which the disk rolls, its magnitude being

$$\boxed{v_O = R\omega} \qquad (16.11\text{b})$$

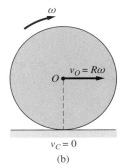

(b)

as shown Fig. 16.9(b).

It is convenient here to also derive the acceleration of O, although this information will not be used until Art. 16.7. The acceleration of O can be obtained by differentiation of Eq. (16.11a). Noting that R and $\mathbf{i}$ are constants, we get

$$\mathbf{a}_O = \dot{\mathbf{v}}_O = R\alpha\mathbf{i} \qquad (16.12\text{a})$$

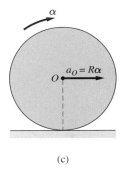

(c)

Fig. 16.9

where $\alpha = \dot{\omega}$ is the angular acceleration of the disk. Thus the acceleration of O is parallel to the horizontal surface, and its magnitude is

$$\boxed{a_O = R\alpha} \qquad (16.12\text{b})$$

as shown in Fig. 16.9(c).

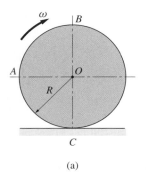

(a)

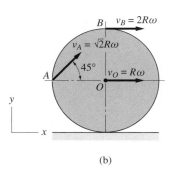

(b)

Sample Problem 16.4

Figure (a) shows a wheel of radius R that is rolling without slipping with the clockwise angular velocity ω. For the position shown, determine the velocity vectors of (1) point A; and (2) point B.

Solution

Introductory Comments

This problem will be solved using both scalar and vector notations. (The results are, of course, identical regardless of which notation is used.) We choose point O (the center of the wheel) as the reference point, because its velocity is known from Eq. (16.11) to be $\mathbf{v}_O = R\omega\mathbf{i}$, as shown in Fig. (b). The reader may find it instructive to repeat the solution using the point of contact C ($\mathbf{v}_C = \mathbf{0}$) as the reference point.

Solution I (using scalar notation)

Part 1

When scalar notation is used to relate the velocities of A and O, $\mathbf{v}_{A/O}$ is computed by assuming that point O is fixed. Therefore, the relative velocity equation becomes

$$\mathbf{v}_A = \mathbf{v}_O + \mathbf{v}_{A/O}$$

from which the velocity of A is found to be

$$\mathbf{v}_A = \sqrt{2}R\omega \quad \angle 45° \qquad \textit{Answer}$$

as shown in Fig. (b).

Part 2

Using scalar notation, the velocities of points B and O are related by

$$\mathbf{v}_B = \mathbf{v}_O + \mathbf{v}_{B/O}$$

which yields

$$\mathbf{v}_B = 2R\omega \rightarrow \qquad \textit{Answer}$$

The velocity of B is also shown in Fig. (b).

Solution II (using vector notation)

Part 1

In vector notation, the relationship between the velocities of A and O is

$$\mathbf{v}_A = \mathbf{v}_O + \boldsymbol{\omega} \times \mathbf{r}_{A/O}$$

Substituting $\mathbf{v}_O = R\omega\mathbf{i}$, $\boldsymbol{\omega} = -\omega\mathbf{k}$, and $\mathbf{r}_{A/O} = -R\mathbf{i}$ [refer to Fig. (c)], we get

$$\mathbf{v}_A = R\omega\mathbf{i} + (-\omega\mathbf{k}) \times (-R\mathbf{i}) = R\omega\mathbf{i} + R\omega\mathbf{j} \qquad \textit{Answer}$$

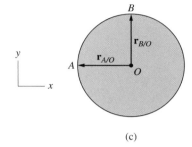

(c)

Part 2

The velocities of B and O are related by

$$\mathbf{v}_B = \mathbf{v}_O + \boldsymbol{\omega} \times \mathbf{r}_{B/O}$$

Substituting $\mathbf{v}_O = R\omega\mathbf{i}$, $\boldsymbol{\omega} = -\omega\mathbf{k}$, and $\mathbf{r}_{B/O} = R\mathbf{j}$ [refer to Fig. (c)] we obtain

$$\mathbf{v}_B = R\omega\mathbf{i} + (-\omega\mathbf{k}) \times (R\mathbf{j}) = 2R\omega\mathbf{i} \qquad \textit{Answer}$$

Sample Problem 16.5

The angular velocity of bar AB in Fig. (a) is 3 rad/s clockwise in the position shown. Determine the angular velocity of bar BC and the velocity of the slider C in this position.

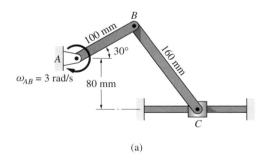

(a)

Solution

Introductory Comments

We note from the geometry in Fig. (b) that $100 \sin 30° + 80 = 160 \sin \beta$, from which we find that $\beta = 54.34°$ for the position shown.

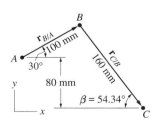

(b)

The following are the kinematically important points:

> A: It is a fixed point.
> C: Its path is a horizontal straight line.
> B: Its path is a circle with center at A; it also connects bars AB and BC.

Because B and C are points in the same rigid bar BC, it seems reasonable to investigate the equation $\mathbf{v}_C = \mathbf{v}_B + \mathbf{v}_{C/B}$ (the equivalent equation $\mathbf{v}_B = \mathbf{v}_C + \mathbf{v}_{B/C}$ could also be used).

Two solutions are presented below—one using scalar notation, the other employing vector notation. The relative position vectors used in the solutions are shown in Fig. (b).

Solution I (using scalar notation)

Using scalar notation, the velocities of B and C are related by

$$\mathbf{v}_C \quad = \quad \mathbf{v}_B \quad + \quad \mathbf{v}_{C/B}$$

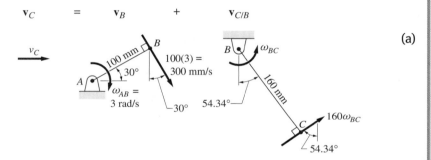

(a)

Comments on Eq. (a):

1. $\mathbf{v}_C$ is assumed to be directed to the right.
2. $\mathbf{v}_B$ is found by recognizing that the path of B is a circle centered at A. Its magnitude is thus $v_B = r_{B/A}\omega_{AB} = 100(3) = 300$ mm/s, directed at right angles to AB. The sense of $\mathbf{v}_B$ is determined by the clockwise direction of ω_{AB}.
3. $\mathbf{v}_{C/B}$ is obtained by considering B to be a fixed point at this instant, which gives $v_{C/B} = r_{C/B}\omega_{BC} = 160\omega_{BC}$. The direction of the vector is perpendicular to BC, with its sense determined by the (assumed) counterclockwise direction of ω_{BC}.

Because there are a total of two unknowns (v_C and ω_{BC}), Eq. (a) can be solved. Equating x- and y-components, we obtain

$$\xrightarrow{+} \quad v_C = 300 \sin 30° + 160\omega_{BC} \sin 54.34° \tag{b}$$

$$+\uparrow \quad 0 = -300 \cos 30° + 160\omega_{BC} \cos 54.34° \tag{c}$$

The solution is

$$v_C = 512 \text{ mm/s} \quad \text{and} \quad \omega_{BC} = 2.785 \text{ rad/s} \qquad \textit{Answer}$$

The positive signs mean that the assumed directions of v_C and ω_{BC} are correct.

Solution II (using vector notation)

The vector equation relating the velocities of points B and C is

$$\mathbf{v}_C = \mathbf{v}_B + \mathbf{v}_{C/B} = \boldsymbol{\omega}_{AB} \times \mathbf{r}_{B/A} + \boldsymbol{\omega}_{BC} \times \mathbf{r}_{C/B} \qquad\qquad \text{(d)}$$

From the given information and inspection of Fig. (b), we obtain

$\mathbf{v}_C = v_C\mathbf{i}$ (assuming the velocity of C to be directed to the right)

$\boldsymbol{\omega}_{AB} = -3\mathbf{k}$ rad/s

$\mathbf{r}_{B/A} = 100 \cos 30°\mathbf{i} + 100 \sin 30°\mathbf{j} = 86.6\mathbf{i} + 50\mathbf{j}$ mm

$\boldsymbol{\omega}_{BC} = \omega_{BC}\mathbf{k}$ (assuming the direction of ω_{BC} to be counterclockwise)

$\mathbf{r}_{C/B} = 160 \cos 54.34°\mathbf{i} - 160 \sin 54.34°\mathbf{j} = 93.28\mathbf{i} - 130\mathbf{j}$ mm

Inspection of the above expressions reveals that there are only two unknowns: v_C and ω_{BC}. Therefore, Eq. (d) can be solved, because it is equivalent to two scalar equations.

Substituting the above expressions into Eq. (d) yields

$$v_C\mathbf{i} = (-3\mathbf{k}) \times (86.6\mathbf{i} + 50\mathbf{j}) + (\omega_{BC}\mathbf{k}) \times (93.28\mathbf{i} - 130\mathbf{j})$$
$$= -259.8\mathbf{j} + 150\mathbf{i} + 93.28\omega_{BC}\mathbf{j} + 130\omega_{BC}\mathbf{i}$$

Equating the coefficients of $\mathbf{i}$ and $\mathbf{j}$ yields

$$v_C = 150 + 130\omega_{BC}$$
$$0 = -259.8 + 93.28\omega_{BC}$$

Solving the two equations, we find that

$$v_C = 512 \text{ mm/s} \quad \text{and} \quad \omega_{BC} = 2.785 \text{ rad/s} \qquad \textit{Answer}$$

The positive signs indicate that the assumed directions of $\mathbf{v}_C$ and ω_{BC} are correct.

Sample Problem **16.6**

In the position shown in Fig. (a), the angular velocity of bar AB is 2 rad/s clockwise. Calculate the angular velocities of bars BC and CD for this position.

Solution

Introductory Comments

A mechanism of the type shown in Fig. (a) is called a *four-bar linkage*. (The ground joining the supports at A and D is considered to be the fourth bar.)

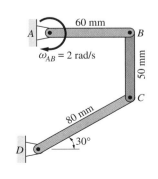

(a)

From Fig. (a) we observe that the following are the kinematically important points:

A: It is a fixed point.
B: Its path is a circle with center at A; it also connects bars AB and BC.
D: It is a fixed point.
C: Its path is a circle centered at D; it also connects bars BC and CD.

Because B and C belong to the same rigid body, we are led to consider the equation $\mathbf{v}_C = \mathbf{v}_B + \mathbf{v}_{C/B}$ (the equivalent equation $\mathbf{v}_B = \mathbf{v}_C + \mathbf{v}_{B/C}$ could also be used). Two solutions are presented—one using scalar notation, the other using vector notation. The relative position vectors used in the solution are shown in Fig. (b).

Solution I (using scalar notation)

The equation relating the velocities of B and C is

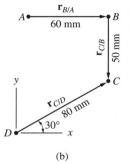

$$\mathbf{v}_C \qquad = \qquad \mathbf{v}_B \qquad + \qquad \mathbf{v}_{C/B} \qquad \text{(a)}$$

(b)

Comments on Eq. (a):

1. $\mathbf{v}_C$ is found by recognizing that the path of C is a circle centered at D. Therefore, its magnitude is $v_C = r_{C/D}\omega_{CD} = 80\omega_{CD}$, its direction being perpendicular to CD. The sense of $\mathbf{v}_C$ is found from the assumed (clockwise) direction of ω_{CD}.
2. $\mathbf{v}_B$ is found by noting that the path of B is a circle centered at A. Its magnitude is $v_B = r_{B/A}\omega_{AB} = 60(2) = 120$ mm/s. The direction of $\mathbf{v}_B$ is perpendicular to AB, and its sense is found from the given clockwise direction of ω_{AB}.
3. $\mathbf{v}_{C/B}$ is constructed by considering B to be a fixed point at this instant, which gives $v_{C/B} = r_{C/B}\omega_{BC} = 50\omega_{BC}$. The direction of $\mathbf{v}_{C/B}$ is perpendicular to BC, and its sense is found from the assumed (counterclockwise) direction of ω_{BC}.

Equation (a) contains a total of two unknowns, ω_{CD} and ω_{BC}, which can be found by solving the two equivalent scalar equations. Equating x- and y-components of Eq. (a), we obtain

$$\xrightarrow{+} \qquad 80\omega_{CD}\cos 60° = 0 + 50\omega_{BC} \qquad \text{(b)}$$

$$+\uparrow \qquad -80\omega_{CD}\sin 60° = -120 + 0 \qquad \text{(c)}$$

Solving these equations, we get

$$\omega_{CD} = 1.732 \text{ rad/s} \quad \text{and} \quad \omega_{BC} = 1.386 \text{ rad/s} \qquad \textit{Answer}$$

The positive signs indicate that the assumed directions of ω_{CD} and ω_{BC} are correct.

Solution II (using vector notation)

The velocities of points B and C are related by

$$\mathbf{v}_C = \mathbf{v}_B + \mathbf{v}_{C/B}$$

$$\boldsymbol{\omega}_{CD} \times \mathbf{r}_{C/D} = \boldsymbol{\omega}_{AB} \times \mathbf{r}_{B/A} + \boldsymbol{\omega}_{BC} \times \mathbf{r}_{C/B} \qquad (d)$$

Using the given information and the vectors shown in Fig. (b), we obtain

$\boldsymbol{\omega}_{CD} = -\omega_{CD}\mathbf{k}$ (the direction of ω_{CD} is assumed to be clockwise)

$\mathbf{r}_{C/D} = 80 \cos 30°\mathbf{i} + 80 \sin 30°\mathbf{j} = 69.28\mathbf{i} + 40.00\mathbf{j}$ mm

$\boldsymbol{\omega}_{AB} = -2\mathbf{k}$ rad/s

$\mathbf{r}_{B/A} = 60\mathbf{i}$ mm

$\boldsymbol{\omega}_{BC} = \omega_{BC}\mathbf{k}$ (the direction of ω_{BC} is assumed to be counterclockwise)

$\mathbf{r}_{C/B} = -50\mathbf{j}$ mm

We see that the above expressions contain only two unknowns: ω_{CD} and ω_{BC}. Therefore, Eq. (d), being equivalent to two scalar equations, can be solved for these unknowns.

Substituting the above expressions into Eq. (d) yields

$$(-\omega_{CD}\mathbf{k}) \times (69.28\mathbf{i} + 40.00\mathbf{j}) = (-2\mathbf{k}) \times (60\mathbf{i}) + (\omega_{BC}\mathbf{k}) \times (-50\mathbf{j})$$

which becomes

$$-69.28\omega_{CD}\mathbf{j} + 40.00\omega_{CD}\mathbf{i} = -120\mathbf{j} + 50\omega_{BC}\mathbf{i}$$

Equating the coefficients of $\mathbf{i}$ and $\mathbf{j}$ yields

$$40.00\omega_{CD} = 50\omega_{BC}$$
$$-69.28\omega_{CD} = -120$$

The solution of these two equations is

$$\omega_{CD} = 1.732 \text{ rad/s} \quad \text{and} \quad \omega_{BC} = 1.386 \text{ rad/s} \qquad \textit{Answer}$$

The positive answers indicate that the assumed directions of ω_{CD} and ω_{BC} are correct.

Problems

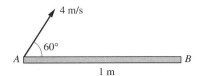

Fig. P16.19

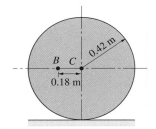

Fig. P16.20

16.19 At a certain instant, the velocity of end A of the bar AB is 4 m/s in the direction shown. Knowing that the magnitude of the velocity of end B is 3 m/s, determine the angular velocity of bar AB.

16.20 The wheel rolls without slipping. In the position shown, the vertical component of the velocity of point B is 4 m/s directed upward. For this position, calculate the angular velocity of the wheel and the velocity of its center C.

16.21 The disk rolls without slipping with the constant angular velocity ω. For the position shown, find the angular velocity of link AB and the velocity of slider A.

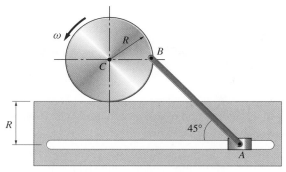

Fig. P16.21

16.22 The pinion gear meshes with the two racks. If the racks are moving with the velocities shown, determine the angular velocity of the gear and the velocity of its center C.

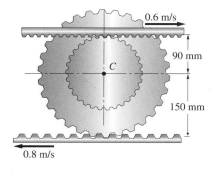

Fig. P16.22

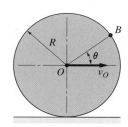

Fig. P16.23

16.23 The wheel rolls without slipping to the right with constant angular velocity. The velocity of the center of the wheel is v_O. Determine the speed of point B on the rim as a function of its angular position θ.

16.24 The arm joining the two friction wheels rotates with the constant angular velocity ω_0. Assuming that wheel A is stationary and that there is no slipping between the wheels, determine the angular velocity of wheel B.

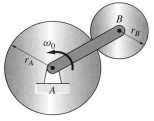

Fig. P16.24, P16.25

16.25   Solve Prob. 16.24 if wheel A is rotating clockwise with the angular velocity $\omega_A = 2\omega_0$.

16.26   Gear A of the planetary gear train is rotating clockwise at $\omega_A = 8$ rad/s. Calculate the angular velocities of gear B and the arm AB. Note that the outermost gear C is stationary.

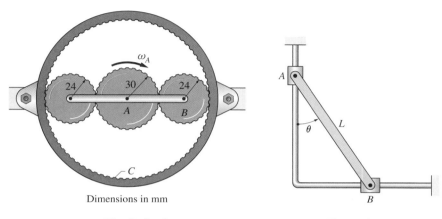

Dimensions in mm

Fig. P16.26

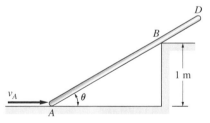

Fig. P16.27

16.27   The bar AB is rotating counterclockwise with the constant angular speed ω_0. (a) Find the velocities of ends A and B as functions of θ. (b) Differentiate the results of part (a) to determine the accelerations of A and B in terms of θ.

16.28   End A of bar AD is pushed to the right with the constant velocity $v_A = 0.6$ m/s. Determine the angular velocity of AD as a function of θ.

16.29   The angular speed of link AB in the position shown is 2.8 rad/s clockwise. Compute the angular speeds of links BC and CD in this position.

Fig. P16.28

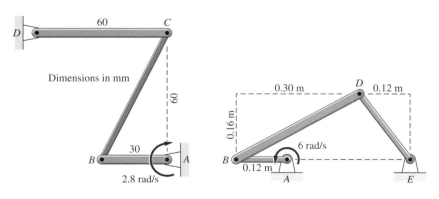

Fig. P16.29

Fig. P16.30

16.30   The link AB of the mechanism rotates with the constant angular speed of 6 rad/s counterclockwise. Calculate the angular velocities of links BD and DE in the position shown.

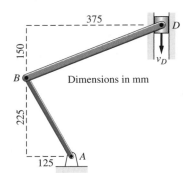

Fig. P16.31

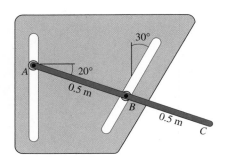

Fig. P16.33

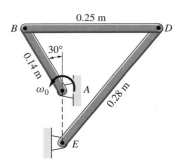

Fig. P16.34

16.31 When the mechanism is in the position shown, the velocity of slider D is $v_D = 1.25$ m/s. Determine the angular velocities of bars AB and BD at this instant.

16.32 When the linkage is in the position shown, bar AB is rotating counterclockwise at 16 rad/s. Determine the velocity of the sliding collar C in this position.

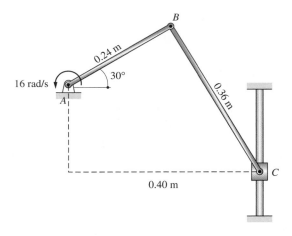

Fig. P16.32

16.33 At the instant shown, end A of the bar ABC has a downward velocity of 2 m/s. Find the angular velocity of the bar and the speed of end C at this instant.

16.34 Bar AB is rotating counterclockwise with the constant angular velocity $\omega_0 = 30$ rad/s. Find the angular velocities of bars BD and DE in the position shown.

16.35 The wheel is rolling without slipping. Its center has a constant velocity of 0.6 m/s to the left. Compute the angular velocity of bar BD and the velocity of end D when $\theta = 0$.

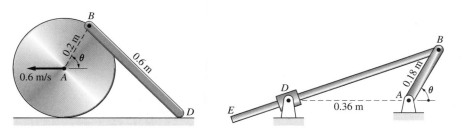

Fig. P16.35　　　　　**Fig. P16.36**

16.36 Crank AB rotates with a constant counterclockwise angular velocity of 16 rad/s. Calculate the angular velocity of bar BE when $\theta = 60°$.

16.37   The hydraulic cylinder raises pin B at the constant rate of 30 mm/s. Determine the speed of end D of the bar AD at the instant shown.

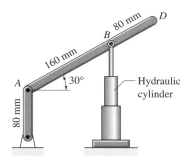

Fig. P16.37

16.38   In the position shown, the speeds of corners A and B of the right triangular plate are $v_A = 3$ m/s and $v_B = 2.4$ m/s, directed as shown. Find (a) the angle α; and (b) the speed of corner D.

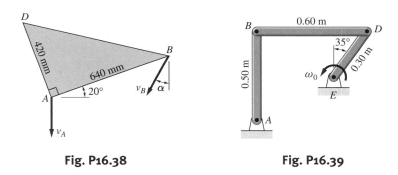

Fig. P16.38   **Fig. P16.39**

16.39   Bar DE is rotating counterclockwise with the constant angular velocity $\omega_0 = 5$ rad/s. Find the angular velocities of bars AB and BD in the position shown.

16.6   *Instant Center for Velocities*

The instant center for velocities of a body undergoing plane motion is defined to be the point that has zero velocity at the instant under consideration.[*] This point may be either in a body or outside the body (in the "body extended"). It is often convenient to use the instant center of the body in computing the velocities of points in the body.

---

[*]Three "centers" are sometimes used in the kinematic analysis of plane motion: the instant center of rotation for virtual motion (see Art. 10.6), the instant center for velocities, and the instant center for accelerations. Each of these points is called simply the instant center when it is clear from the context which center is being used. The discussion of instant center for velocities presented here parallels the discussion of instant center of rotation for virtual motion in Art. 10.6.

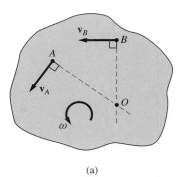

(a)

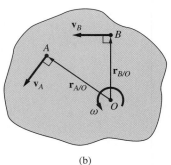

(b)

Fig. 16.10

Figure 16.10(a) shows a rigid body that is undergoing plane motion. It is assumed that the velocity vectors for points A and B are not parallel, and that the angular velocity ω of the body is counterclockwise. To locate the instant center for velocities, we construct a line through A that is perpendicular to $\mathbf{v}_A$, and a line through B that is perpendicular to $\mathbf{v}_B$. These two lines will intersect at a point labeled O in the figure. If point O does not lie in the body, we simply imagine that the body is enlarged to include it, the expanded body being called the *body extended*.

We now show that point O is the instant center—that is, that the velocity of O is zero. Because A, B, and O are points in the same rigid body (or body extended), we can write the following relative velocity equations:

$$\mathbf{v}_O = \mathbf{v}_A - \boldsymbol{\omega} \times \mathbf{r}_{A/O} \qquad \text{(a)}$$

$$\mathbf{v}_O = \mathbf{v}_B - \boldsymbol{\omega} \times \mathbf{r}_{B/O} \qquad \text{(b)}$$

where $\mathbf{r}_{A/O}$ and $\mathbf{r}_{B/O}$ are the relative position vectors shown in Fig. 16.10(b). Because both ω and $\mathbf{r}_{A/O}$ are perpendicular to $\mathbf{v}_A$, it follows that $\boldsymbol{\omega} \times \mathbf{r}_{A/O}$ in Eq. (a) is parallel to $\mathbf{v}_A$. This means that $\mathbf{v}_O$ is parallel to $\mathbf{v}_A$. Similarly, it can be shown from Eq. (b) that $\mathbf{v}_O$ is parallel to $\mathbf{v}_B$. Because a nonzero vector cannot be parallel to two different directions simultaneously, we conclude that $\mathbf{v}_O = \mathbf{0}$. Therefore, point O is indeed the instant center for velocities.

In general, the instant center for velocities is not a fixed point. Therefore, the acceleration of the instant center for velocities is not necessarily zero. However, if a point of a body is fixed, that point is obviously the instant center for both velocities and accelerations.

If O is the instant center, Eq. (a) reduces to

$$\mathbf{v}_A = \boldsymbol{\omega} \times \mathbf{r}_{A/O} \qquad (v_A = r_{A/O}\omega) \qquad \text{(16.13)}$$

This equation has the same form as the first of Eqs. (16.8) for rotation about a fixed axis through O. We must reiterate, however, that for general plane motion the instant center is not a fixed point. Therefore, the analogy with rotation about a fixed axis is valid only for velocities at a given *instant* of time (hence the term "instant center"). The analogy does not apply to accelerations.

The following rules, which follow directly from Eq. (16.13), apply when the instant center is used.

1. The velocity of any point in the body is perpendicular to the line drawn from the point to the instant center.
2. The magnitude of the velocity of any point of the body is proportional to the distance of the point from the instant center.
3. The sense of the velocity vector of any point must be consistent with the sense of the angular velocity of the body.

The construction shown in Fig. 16.10(a) for locating the instant center is obviously valid only if $\mathbf{v}_A$ and $\mathbf{v}_B$ are not parallel. Figure 16.11 illustrates the methods for locating the instant center when $\mathbf{v}_A$ and $\mathbf{v}_B$ are parallel.

In Fig. 16.11(a), the perpendicular lines to the velocity vectors are parallel, which means that the instant center is an infinite distance from A and B. This leads to the conclusion that $\omega = 0$; that is, the body is translating with $\mathbf{v}_A = \mathbf{v}_B$.

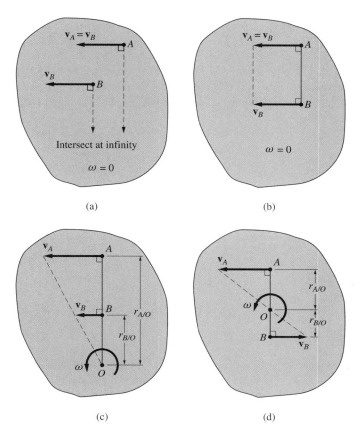

Fig. 16.11

In Fig. 16.11 parts (b) through (d), the lines joining A and B are perpendicular to both velocity vectors, and the usual construction for locating the instant center does not work. In the case where $\mathbf{v}_A = \mathbf{v}_B$, as in Fig. 16.11(b), the instant center is at infinity, so that $\omega = 0$ (i.e., the body is translating). If the magnitudes of the velocities are not equal, the instant center is located using the constructions in Fig. 16.11 parts (c) and (d), which are simply applications of rules 1 through 3 just stated.

The question arises whether the simplification of velocity analysis based on the instant center is worth the labor it takes to find the instant center in the first place. A general rule of thumb is to begin by marking the instant center on a sketch of the body. If the distances between the kinematically important points on the body (or body extended) and the instant center are fairly easy to calculate, the analysis is well suited for using the instant center. However, if a prohibitive amount of trigonometry is required to compute the location of the instant center, the relative velocity method of the preceding article would be the preferred method of solution. Of course, there are many problems that require the same amount of labor using either method, in which case personal preference would dictate which method to use.

Sample Problem 16.7

When the mechanism in Fig. (a) is in the position shown, the angular velocity of bar AB is $\omega_{AB} = 3$ rad/s clockwise. Using instant centers for velocities, calculate the angular velocity of bar BC and the velocity of slider C for this position. (This problem was solved previously as Sample Problem 16.5 by using the method of relative velocity.)

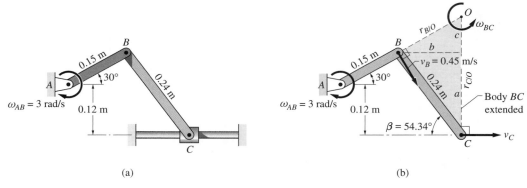

(a)   (b)

Solution

We must first locate the instant centers of the two rigid bodies AB and BC. Because A is a fixed point, it is obviously the instant center of bar AB. The instant center of BC, labeled O in Fig. (b), is located at the point of intersection of the lines that are perpendicular to the velocity vectors of B and C. Note that $\mathbf{v}_B$ is perpendicular to AB, and $\mathbf{v}_C$ is horizontal. Therefore, O is located at the intersection of line AB and the vertical line that passes through C.

From the geometry in Fig. (b), we note that $.15 \sin 30° + 0.12 = 0.24 \sin \beta$, from which we find that $\beta = 54.34°$. Therefore, the distances a, b, and c are

$$a = 0.24 \sin 54.34° = 0.195 \text{ m}$$

$$b = 0.24 \cos 54.34° = 0.140 \text{ m}$$

$$c = b \tan 30° = 0.140 \tan 30° = 0.080 \text{ m}$$

The distances to B and C from O are

$$r_{B/O} = c/\sin 30° = \frac{0.080}{\sin 30°} = 0.16 \text{ m}$$

$$r_{C/O} = a + c = 0.195 + 0.16 = 0.355 \text{ m}$$

Now that the instant center for each bar has been found, we can compute the required velocities.

Considering the motion of AB (its instant center is at A), we find that $v_B = r_{B/A}\omega_{AB} = 0.15(3) = 0.45$ m/s, directed as shown in Fig. (b). Analyzing the motion of BC (its instant center is at O) yields

$$\omega_{BC} = \frac{v_B}{r_{B/O}} = \frac{0.45}{0.16} = 2.81 \text{ rad/s}$$

$$\omega_{BC} = 2.81 \text{ rad/s } \circlearrowleft \qquad\qquad\qquad \textit{Answer}$$

and

$$v_C = r_{C/O}\omega_{BC} = 0.355(2.81) = 0.99 \text{ m/s}$$

$$\mathbf{v}_C = 0.99 \text{ m/s} \rightarrow \qquad\qquad \textit{Answer}$$

Sample Problem   *16.8*

When the linkage in Fig. (a) is in the position shown, the angular velocity of bar
AB is $\omega_{AB} = 2$ rad/s clockwise. For this position, determine the angular velocities
of bars BC and CD and the velocity of C using the instant centers for velocities.
(This problem was solved previously as Sample Problem 16.6 using the method
of relative velocity.)

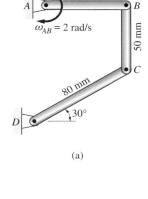

(a)

Solution

Because A and D are fixed points, they are the instant centers for bars AB and CD,
respectively. The instant center for bar BC, labeled O in Fig. (b), is located at the
point of intersection of the lines that are perpendicular to the velocity vectors of
B and C. Because $\mathbf{v}_B$ and $\mathbf{v}_C$ are perpendicular to AB and CD, respectively, the
instant center is at the intersection of these two lines. Note that body BC must be
"extended" to include point O.

(b)

The distances to B and C from O, found from the triangle OBC, are

$$r_{B/O} = 50/\tan 30° = 86.60 \text{ mm}$$

$$r_{C/O} = 50/\sin 30° = 100 \text{ mm}$$

The instant centers A, O, and D can now be used to compute the required angular
velocities directly from Fig. (b).

Considering the motion of AB (its instant center is at A), we find that $v_B = r_{B/A}\omega_{AB} = 60(2) = 120$ mm/s, directed as shown in Fig. (b). Analyzing the motion of BC (its instant center is at O) yields

$$\omega_{BC} = \frac{v_B}{r_{B/O}} = \frac{120}{86.60} = 1.386 \text{ rad/s}$$

$$\omega_{BC} = 1.386 \text{ rad/s} \circlearrowleft \qquad\qquad \textit{Answer}$$

and

$$v_C = r_{C/O}\omega_{BC} = 100(1.386) = 138.6 \text{ mm/s}$$

$$\mathbf{v}_C = 138.6 \text{ mm/s} \qquad\qquad \textit{Answer}$$

Because C is also a point on bar CD (its instant center is at D), the angular velocity of bar CD is

$$\omega_{CD} = \frac{v_C}{r_{C/D}} = \frac{138.6}{80} = 1.733 \text{ rad/s}$$

$$\omega_{CD} = 1.733 \text{ rad/s} \circlearrowleft \qquad\qquad \textit{Answer}$$

Sample Problem 16.9

The wheel in Fig. (a) rolls without slipping with the constant clockwise angular velocity $\omega_0 = 1.6$ rad/s. Calculate the angular velocity of bar AB and the velocity of the slider B when the mechanism is in the position shown. Use the instant centers for velocities.

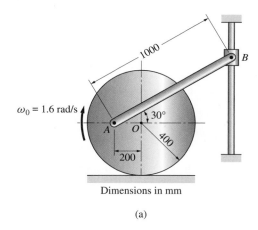

(a)

Solution

The velocity vectors, distances, and points required to solve this problem are shown in Fig. (b).

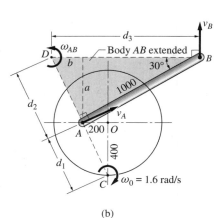

(b)

Because the wheel rolls without slipping, its instant center is at the point of contact C. Therefore, $\mathbf{v}_A$ is perpendicular to the line AC. Note that the slope of AC equals $\overline{OC}/\overline{AO} = 400/200 = 2$.

The instant center of bar AB, labeled D in Fig. (b), is located at the intersection of the lines that are perpendicular to $\mathbf{v}_A$ and $\mathbf{v}_B$. Because $\mathbf{v}_A$ is perpendicular to AC, D also lies on AC. The line BD, which is drawn perpendicular to $\mathbf{v}_B$, is horizontal because $\mathbf{v}_B$ is vertical (it is not necessary to know here that the sense of $\mathbf{v}_B$ is upward).

Referring to Fig. (b), the distances of interest are computed as follows.

$$d_1 = \sqrt{\overline{AO}^2 + \overline{OC}^2} = \sqrt{200^2 + 400^2} = 447.2 \text{ mm}$$

$$a = 1000 \sin 30° = 500 \text{ mm}$$

$$b = a/2 = 250 \text{ mm} \quad \text{(the slope of the line } AC \text{ is 2)}$$

$$d_2 = \sqrt{a^2 + b^2} = \sqrt{500^2 + 250^2} = 559.0 \text{ mm}$$

$$d_3 = 1000 \cos 30° + b = 1000 \cos 30° + 250 = 1116 \text{ mm}$$

Using the distances d_1, d_2, and d_3 and $\omega_0 = 1.6$ rad/s, we find that

$$v_A = d_1 \omega_0 = 447.2(1.6) = 715.5 \text{ mm/s}$$

$$\mathbf{v}_A = 716 \text{ mm/s} \qquad\qquad \textit{Answer}$$

$$\omega_{AB} = \frac{v_A}{d_2} = \frac{715.5}{559.0} = 1.280 \text{ rad/s}$$

$$\boldsymbol{\omega}_{AB} = 1.280 \text{ rad/s} \circlearrowleft \qquad\qquad \textit{Answer}$$

$$v_B = d_3 \omega_{AB} = 1116(1.280) = 1428 \text{ mm/s}$$

$$\mathbf{v}_B = 1428 \text{ mm/s} \uparrow \qquad\qquad \textit{Answer}$$

The sense of $\mathbf{v}_A$ was found by considering that the angular velocity of the wheel is clockwise and its instant center is at C. The counterclockwise direction of ω_{AB} was deduced by inspection of the sense of $\mathbf{v}_A$ and the location of the instant center D. The direction of ω_{AB} and the location of B relative to D determine that the sense of $\mathbf{v}_B$ is upward.

It is frequently convenient to show the instant centers of more than one body on the same sketch, as is done in Fig. (b). However, one must be careful to use the proper instant center when discussing the velocity of a particular point. For example, a common error when referring to Fig. (b) would be to write $v_C = (d_1 + d_2)\omega_{AB}$, which is incorrect, because D is the instant center of bar AB, which does not include point C.

Problems

Note: The following problems are to be solved using instant centers for velocities.

16.40   The end of the cord that is wrapped around the hub of the wheel is pulled to the right with the velocity $v_0 = 0.7$ m/s. Find the angular velocity of the wheel, assuming no slipping.

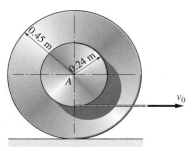

Fig. P16.40

16.41   The wheel rolls without slipping with the angular velocity $\omega = 8$ rad/s. Determine the coordinates of a point B on the wheel for which the velocity vector is $\mathbf{v}_B = -2.4\mathbf{i} + 0.7\mathbf{j}$ m/s.

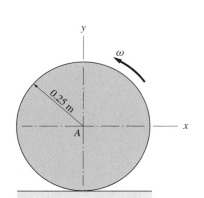

Fig. P16.41

16.42   The unbalanced wheel rolls and slips along the horizontal plane. At the instant shown, the angular velocity ω of the wheel and velocity v_O of its center are as indicated. Find the magnitude and direction of the velocity of G at this instant.

16.43   A 500-mm diameter wheel rolls and slips on a horizontal plane. The angular velocity of the wheel is $\omega = 12$ rad/s (counterclockwise), and the velocity of the center of the wheel is 1.8 m/s to the left. (a) Find the instant center for velocities of the wheel. (b) Calculate the velocity of the point on the wheel that is in contact with the plane.

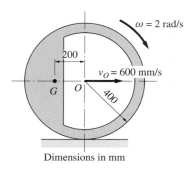

Dimensions in mm

Fig. P16.42

16.44   Determine the coordinates of the instant center for velocities of the bar AB in (a) and (b).

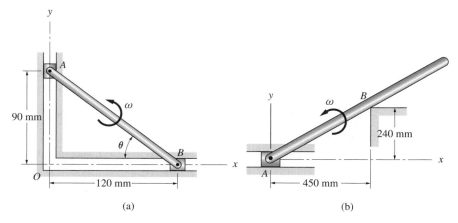

(a)

(b)

Fig. P16.44

16.45 Find the coordinates of the instant center for velocities of bar *AB* in (a) and (b).

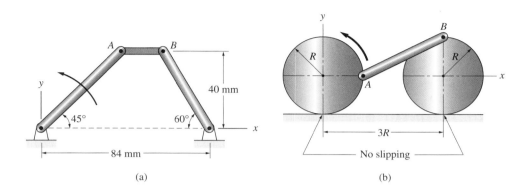

(a)   (b)

Fig. P16.45

16.46 The arm connected between the centers of gears *A* and *B* is rotating counterclockwise with the angular velocity of 4.8 rad/s. At the same time, *A* is rotating at 24 rad/s, also counterclockwise. Determine the angular velocity of *B*.

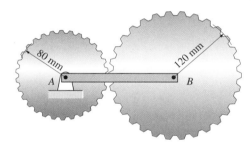

Fig. P16.46

16.47 The pinion gear meshes with the two racks. If the racks are moving with the velocities shown, determine the angular velocity of the gear and the velocity of its center *C*. (Note: This problem was solved as Prob. 16.22 by the method of relative velocity.)

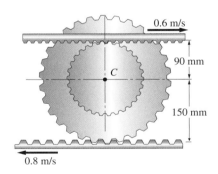

Fig. P16.47

16.48 Bar *AB* is rotating counterclockwise at the constant angular velocity of 6 rad/s. Determine the angular velocity of bar *CD* when the mechanism is in the position shown.

16.49 Sketch the locus of the instant center of velocities of bar *AB* in Fig. P16.44(a) as θ varies from 0° to 90°. (This curve is called a *space centrode*.)

16.50 The 3-m wooden plank is tumbling as it falls in the vertical plane. When the plank is in a horizontal position, the velocities of ends *A* and *B* are as shown in the figure. For this position, determine the location of the instant

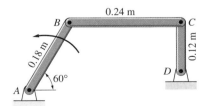

Fig. P16.48

center for velocities, the angular velocity of the plank, and the velocity of the midpoint G.

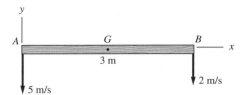

Fig. P16.50

16.51 For the triangular plate undergoing plane motion, $\mathbf{v}_A$ and the direction of $\mathbf{v}_B$ are known. Calculate the angular speed of the plate and the speeds of corners B and C.

16.52 At the instant shown, the angular velocity of the cylinder, which is rolling without slipping, is 2 rad/s, counterclockwise. Find the velocity of end B of the rod that is pinned to the cylinder at A.

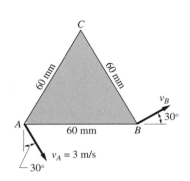

Fig. P16.51

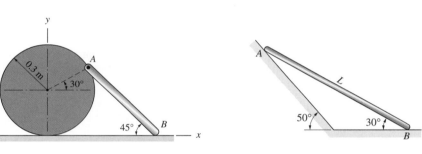

Fig. P16.52   **Fig. P16.53**

16.53 When bar AB is in the position shown, end B is sliding to the right with a velocity of 0.8 m/s. Determine the velocity of end A in this position.

16.54 Slider C of the mechanism has a constant downward velocity of 0.8 m/s. Determine the angular velocity of crank AB when it is in the position shown.

16.55 Bar BC of the linkage slides in the collar D. If bar AB is rotating clockwise with the constant angular velocity of 12 rad/s, determine the angular velocity of BC when it is in the horizontal position shown.

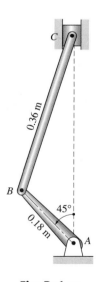

Fig. P16.54

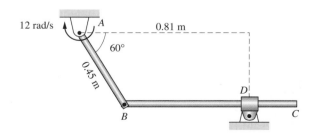

Fig. P16.55

16.56   Bar *BC* of the linkage slides in the collar *D*. If bar *AB* is rotating clockwise with the constant angular velocity of 12 rad/s, determine the angular velocity of bar *BC* in the position shown.

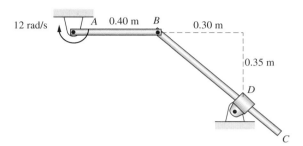

Fig. P16.56

16.57   When the mechanism is in the position shown, the angular velocity of bar *AB* is 72 rad/s, clockwise. For this position, compute the angular velocity of the plate *BCD* and the velocity of corner *D*.

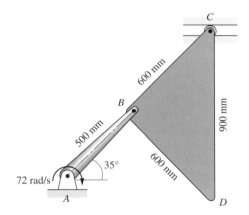

Fig. P16.57

16.58   The crank *AB* of the mechanism rotates counterclockwise at 8 rad/s. Calculate the velocities of sliders *C* and *D* at the instant shown.

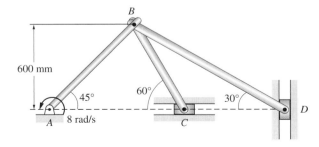

Fig. P16.58

16.59 Bar *AB* of the mechanism rotates clockwise with the angular velocity ω_0. Compute the angular velocities of bars *BD* and *DE* for the position shown.

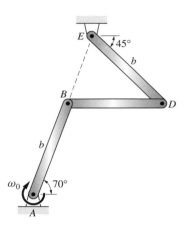

Fig. P16.59

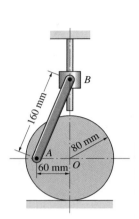

Fig. P16.60

16.60 When the mechanism is in the position shown, the velocity of the center *O* of the disk is 0.4 m/s to the right. Assuming that the disk rolls without slipping, calculate the velocity of the collar *B* in this position.

16.7 Method of Relative Acceleration

In Arts. 16.5 and 16.6, we analyzed the velocities of points in a rigid body undergoing plane motion. Two methods were presented: the method of relative velocity and instant centers for velocities. In this article, we introduce the *method of relative acceleration*, which employs the equation $\mathbf{a}_B = \mathbf{a}_A + \mathbf{a}_{B/A}$ for two points in the same rigid body.

Figure 16.12(a) shows a rigid body that is undergoing general plane motion. The angular velocity and angular acceleration vectors of the body are $\boldsymbol{\omega}$ and $\boldsymbol{\alpha}$, respectively. Letting *A* and *B* be two points in the body, the acceleration of *B* with respect to *A* is, according to Eqs. (16.8),

$$\mathbf{a}_{B/A} = (\mathbf{a}_{B/A})_n + (\mathbf{a}_{B/A})_t \qquad (16.14\text{a})$$

where the normal and tangential components of the relative acceleration are

$$(\mathbf{a}_{B/A})_n = \boldsymbol{\omega} \times (\boldsymbol{\omega} \times \mathbf{r}_{B/A}) \qquad [(a_{B/A})_n = r_{B/A}\omega^2] \qquad (16.14\text{b})$$

$$(\mathbf{a}_{B/A})_t = \boldsymbol{\alpha} \times \mathbf{r}_{B/A} \qquad [(a_{B/A})_t = r_{B/A}\alpha] \qquad (16.14\text{c})$$

Substituting Eqs. (16.14) into $\mathbf{a}_B = \mathbf{a}_A + \mathbf{a}_{B/A}$ gives

$$\mathbf{a}_B = \mathbf{a}_A + \boldsymbol{\omega} \times (\boldsymbol{\omega} \times \mathbf{r}_{B/A}) + \boldsymbol{\alpha} \times \mathbf{r}_{B/A} \qquad (16.15)$$

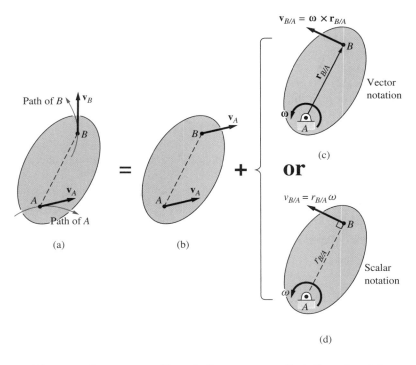

Plane motion   =   Translation   +   Rotation about A

Fig. 16.12

The physical interpretation of Eq. (16.15) is illustrated in Fig. 16.12. The figure shows that general plane motion is equivalent to the superposition of two simpler motions (this equivalence also formed the basis for relative velocity analysis):

1. A rigid-body translation, where the acceleration of each point equals the acceleration of the reference point A ($\mathbf{a}_B = \mathbf{a}_A$), as shown in Fig. 16.12(b).
2. A rigid-body rotation about a fixed axis at A [$\mathbf{a}_{B/A} = (\mathbf{a}_{B/A})_n + (\mathbf{a}_{B/A})_t$] is illustrated in Figs. 16.12 (c) or (d). Showing a pin support at A reinforces the notion that A is considered to be fixed at the instant when the motion is studied.

The contribution of the rotation of the body to the acceleration of B may be computed using either the vector notation in Fig. 16.12(c) or the scalar notation in Fig. 16.12(d). The choice of notation is a matter of personal preference. When using scalar notation, remember that (1) $(\mathbf{a}_{B/A})_n$ is always directed toward the reference point A, and (2) $(\mathbf{a}_{B/A})_t$ is perpendicular to AB, and its sense is determined by the direction of the angular acceleration α of the body (recall that α is a property of the body and is independent of the choice of reference point).

The method of relative acceleration consists of writing the equation $\mathbf{a}_B = \mathbf{a}_A + \boldsymbol{\omega} \times (\boldsymbol{\omega} \times \mathbf{r}_{B/A}) + \boldsymbol{\alpha} \times \mathbf{r}_{B/A}$ for two points in the *same rigid body,* and then solving it for the unknowns. For plane motion, Eq. (16.15) is equivalent to two scalar equations (e.g., the equations that result from equating the horizontal and vertical components of both sides of the vector equation). The number of variables appearing in Eq. (16.15) is six (assuming that $\mathbf{r}_{B/A}$ is known):

$\mathbf{a}_B$: two variables (magnitude and direction, or horizontal and vertical components)

$\mathbf{a}_A$: two variables (magnitude and direction, or horizontal and vertical components)

$\boldsymbol{\omega}$: one variable (magnitude ω of the angular velocity)

$\boldsymbol{\alpha}$: one variable (magnitude α of the angular acceleration)

Solution of Eq. (16.15) is possible only if four of the preceding six variables are known beforehand. The angular velocity ω of the body usually can be found by the velocity analysis described in previous articles, leaving us with the need to know three additional variables (this situation is similar to what we encountered in the relative velocity analysis). It follows that we must restrict the choice of A and/or B to points whose accelerations contain less than two unknowns—points that we call the *kinematically important points for acceleration.*

The steps in the application of the relative acceleration method are:

Step 1: If the angular velocity ω of the body is unknown, find it by the relative velocity analysis described in the previous articles.

Step 2: Identify two kinematically important points (say, A and B) on the same rigid body.

Step 3: Write $\mathbf{a}_B = \mathbf{a}_A + \boldsymbol{\omega} \times (\boldsymbol{\omega} \times \mathbf{r}_{B/A}) + \boldsymbol{\alpha} \times \mathbf{r}_{B/A}$, identifying the unknown variables (either vector or scalar notation can be used).

Step 4: If the number of unknowns is two, solve the equation.

If the number of unknowns is greater than two, it may still be possible to solve the problem by considering the motion of other kinematically important points.

The point in the body that has zero acceleration is called the instant center for accelerations. In general, the instant center for velocities and the instant center for accelerations are not the same point. It can be shown that the two centers coincide only if the angular velocity of the body is zero, or if the body is rotating about a fixed axis. In principle, the instant center for accelerations can be found for any body undergoing plane motion. However, the difficulty in locating this point usually outweighs the advantages gained by its use. For this reason, the instant center for accelerations is not used in this text.

Sample Problem   **16.10**

The wheel of radius R shown in Fig. (a) is rolling without slipping. At the instant shown, its angular velocity and angular acceleration are ω and α, both clockwise. Determine the acceleration vectors of (1) point C, the point of contact on the wheel; and (2) point A.

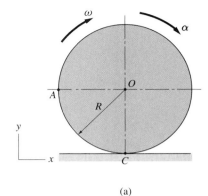

(a)

Solution

Introductory Comments

This problem will be solved using both scalar and vector notations. We choose point O (the center of the wheel) as the reference point, because its acceleration is known from Eq. (16.12a) to be $\mathbf{a}_O = R\alpha\mathbf{i}$, as shown in Fig. (b).

Solution I (using scalar notation)

Part 1

When scalar notation is used to relate the accelerations of C and O, $\mathbf{a}_{C/O}$ is computed by assuming that point O is fixed. Therefore, we have

$$\mathbf{a}_C \;=\; \mathbf{a}_O \;+\; \mathbf{a}_{C/O}$$

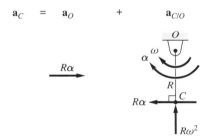

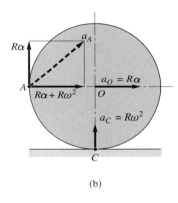

(b)

which gives

$$\mathbf{a}_C = R\omega^2 \uparrow \qquad\qquad \textit{Answer}$$

This result is shown in Fig. (b). Note that although C is the instant center for velocities, its acceleration is not zero. This makes sense, because the velocity of C changes its direction from up to down as it passes over the contact point with the ground. Hence, the acceleration of C at the instant of contact is nonzero and directed upward.

Part 2

The accelerations of A and O are related by

$$\mathbf{a}_A \;=\; \mathbf{a}_O \;+\; \mathbf{a}_{A/O}$$

which yields

$$\mathbf{a}_A = \begin{array}{c} R\alpha \\ \uparrow \\ \longrightarrow R(\alpha + \omega^2) \end{array}$$

Answer

The acceleration vector of A is also shown in Fig. (b).

Solution II (using vector notation)

Part 1

Using vector notation, the relationship between the accelerations of C and O is

$$\mathbf{a}_C = \mathbf{a}_O + \boldsymbol{\alpha} \times \mathbf{r}_{C/O} + \boldsymbol{\omega} \times (\boldsymbol{\omega} \times \mathbf{r}_{C/O})$$

Substituting $\mathbf{r}_{C/O} = -R\mathbf{j}$ [see Fig. (c)], $\boldsymbol{\omega} = -\omega\mathbf{k}$, and $\boldsymbol{\alpha} = -\alpha\mathbf{k}$, we get

$$\begin{aligned}
\mathbf{a}_C &= R\alpha\mathbf{i} + (-\alpha\mathbf{k}) \times (-R\mathbf{j}) + (-\omega\mathbf{k}) \times [(-\omega\mathbf{k}) \times (-R\mathbf{j})] \\
&= R\alpha\mathbf{i} - R\alpha\mathbf{i} + (-\omega\mathbf{k}) \times (-R\omega\mathbf{i}) \\
&= R\omega^2\mathbf{j}
\end{aligned}$$

Answer

Part 2

The relationship between the accelerations of A and O becomes, on substituting $\mathbf{r}_{A/O} = -R\mathbf{i}$ [see Fig. (c)],

$$\begin{aligned}
\mathbf{a}_A &= \mathbf{a}_O + \boldsymbol{\alpha} \times \mathbf{r}_{A/O} + \boldsymbol{\omega} \times (\boldsymbol{\omega} \times \mathbf{r}_{A/O}) \\
&= R\alpha\mathbf{i} + (-\alpha\mathbf{k}) \times (-R\mathbf{i}) + (-\omega\mathbf{k}) \times [(-\omega\mathbf{k}) \times (-R\mathbf{i})] \\
&= R\alpha\mathbf{i} + R\alpha\mathbf{j} + (-\omega\mathbf{k}) \times (R\omega\mathbf{j}) = R\alpha\mathbf{i} + R\alpha\mathbf{j} + R\omega^2\mathbf{i} \\
&= R(\alpha + \omega^2)\mathbf{i} + R\alpha\mathbf{j}
\end{aligned}$$

Answer

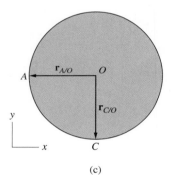

(c)

Sample Problem 16.11

Bar AB of the mechanism shown in Fig. (a) is rotating clockwise with a constant angular velocity of 3 rad/s. Determine the angular acceleration of bar BC and the acceleration of the slider C at the instant when bar AB makes an angle of 30° with the horizontal, as shown.

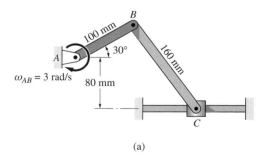

(a)

Solution

Introductory Comments

We will solve this problem using scalar notation (Solution I) and vector notation (Solution II). Because the angular velocity of bar BC is not given, it must be calculated before the accelerations can be found. We assume that this has already been done, the result being $\omega_{BC} = 2.785$ rad/s counterclockwise.[*] Furthermore, we assume that the value of angle β shown in Fig. (b) has been computed by trigonometry.

Clearly, B and C are the kinematically important points on bar BC: The acceleration of B can be computed from the prescribed motion of bar AB, and the path of point C is known. Therefore, the problem can be solved by relating the accelerations of points B and C.

We assume that the acceleration of C is directed to the right and that the angular acceleration of BC is counterclockwise. The angular acceleration of AB is given as zero.

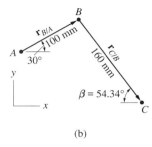

(b)

Solution I (using scalar notation)

The acceleration of C is related to the acceleration of B by

$$\mathbf{a}_C \quad = \quad \mathbf{a}_B \quad + \quad \mathbf{a}_{C/B} \qquad \text{(a)}$$

Inspection of Eq. (a) reveals that it contains two unknowns: a_C and α_{BC}. Equating the horizontal and vertical components, we obtain the following two scalar equations.

$$\xrightarrow{+} \quad a_C = -900\cos 30° + 160\alpha_{BC}\sin 54.34° - 1241\cos 54.34° \qquad \text{(b)}$$

$$+\uparrow \quad 0 = -900\sin 30° + 160\alpha_{BC}\cos 54.34° + 1241\sin 54.34° \qquad \text{(c)}$$

the solution of which is $a_C = -2280$ mm/s^2, and $\alpha_{BC} = -5.99$ rad/s^2. Therefore,

$$\mathbf{a}_C = 2.28 \text{ m/s}^2 \leftarrow \qquad \alpha_{BC} = 5.99 \text{ rad/s}^2 \, \circlearrowleft \qquad \textit{Answer}$$

Note that if ω_{BC} had not been determined previously, Eqs. (b) and (c) would contain ω_{BC} as a third unknown, making the equations unsolvable.

---

[*]See the solution of either Sample Problem 16.5 (method of relative velocity), or Sample Problem 16.7 (instant centers).

Solution II (using vector notation)

The relative position vectors shown in Fig. (b) are

$$\mathbf{r}_{B/A} = 100\cos 30°\mathbf{i} + 100\sin 30°\mathbf{j}$$
$$= 86.6\mathbf{i} + 50\mathbf{j} \text{ mm.} \tag{d}$$

and

$$\mathbf{r}_{C/B} = 160\cos 54.34°\mathbf{i} - 160\sin 54.34°\mathbf{j}$$
$$= 93.28\mathbf{i} - 130\mathbf{j} \text{ mm.} \tag{e}$$

The acceleration equation that solves the problem is

$$\mathbf{a}_C = \mathbf{a}_B + \mathbf{a}_{C/B} \tag{f}$$

Because the direction of $\mathbf{a}_C$ is horizontal, we have

$$\mathbf{a}_C = a_C\mathbf{i} \tag{g}$$

The acceleration of B in Eq. (f) can be determined by noting that B moves on a circular path centered at A. Therefore,

$$\mathbf{a}_B = \boldsymbol{\alpha}_{AB} \times \mathbf{r}_{B/A} + \boldsymbol{\omega}_{AB} \times (\boldsymbol{\omega}_{AB} \times \mathbf{r}_{B/A})$$

Substituting $\boldsymbol{\alpha}_{AB} = \mathbf{0}$, $\boldsymbol{\omega}_{AB} = -3\mathbf{k}$ rad/s, and $\mathbf{r}_{B/A}$ from Eq. (d), we get

$$\mathbf{a}_B = \mathbf{0} + (-3\mathbf{k}) \times [(-3\mathbf{k}) \times (86.6\mathbf{i} + 50\mathbf{j})]$$
$$= (-3\mathbf{k}) \times (-259.8\mathbf{j} + 150\mathbf{i})$$
$$= -779.4\mathbf{i} - 450\mathbf{j} \text{ mm/s}^2 \tag{h}$$

According to Eq. (16.15), the acceleration of C relative to B is

$$\mathbf{a}_{C/B} = \boldsymbol{\alpha}_{BC} \times \mathbf{r}_{C/B} + \boldsymbol{\omega}_{BC} \times (\boldsymbol{\omega}_{BC} \times \mathbf{r}_{C/B})$$

Substituting $\boldsymbol{\alpha}_{BC} = \alpha_{BC}\mathbf{k}$, $\boldsymbol{\omega}_{BC} = 2.785\mathbf{k}$ rad/s, and $\mathbf{r}_{C/B}$ from Eq. (e), we get

$$\mathbf{a}_{C/B} = (\alpha_{BC}\mathbf{k}) \times (93.28\mathbf{i} - 130\mathbf{j})$$
$$+ (2.785\mathbf{k}) \times [(2.785\mathbf{k}) \times (93.28\mathbf{i} - 130\mathbf{j})]$$
$$= 93.28\alpha_{BC}\mathbf{j} + 130\alpha_{BC}\mathbf{i}$$
$$+ (2.785\mathbf{k}) \times (259.8\mathbf{j} + 362\mathbf{i})$$
$$= 93.28\alpha_{BC}\mathbf{j} + 130\alpha_{BC}\mathbf{i} - 723.6\mathbf{i} + 1008.2\mathbf{j} \text{ mm/s}^2 \tag{i}$$

Substituting Eqs. (g), (h), and (i) into Eq. (f) and equating coefficients of $\mathbf{i}$ and $\mathbf{j}$, we obtain the following two scalar equations.

$$a_C = -779.4 + 130\alpha_{BC} - 723.6 \tag{j}$$
$$0 = -450 + 93.28\alpha_{BC} + 1008.2 \tag{k}$$

Solving Eqs. (j) and (k) simultaneously gives $a_C = -2280 \text{ mm/s}^2$, and $\alpha_{BC} = -5.99 \text{ rad/s}^2$. Therefore,

$$\mathbf{a}_C = -2.28\mathbf{i} \text{ m/s}^2 \quad \text{and} \quad \boldsymbol{\alpha}_{BC} = -5.99\mathbf{k} \text{ rad/s}^2 \qquad \textit{Answer}$$

Sample Problem 16.12

When the linkage in Fig. (a) is in the position shown, bar AB is rotating with angular velocity $\omega_{AB} = 2.4 \text{ rad/s}$ and angular acceleration $\alpha_{AB} = 1.5 \text{ rad/s}^2$, both counterclockwise. Determine the angular accelerations of bars BC and CD for this position.

Solution

Preliminary Calculations

This problem will be solved using scalar notation (Solution I) and vector notation (Solution II).

Inspection of the linkage in Fig. (a) reveals that A, B, C, and D are the kinematically important points: A and D are fixed, and the paths of B and C (which are points on the same rigid bar BC) are known to be circles centered at A and D, respectively. Figure (c) shows the relative position vectors between the kinematically important points. The angle between BC and the horizontal was found to be $\theta = \sin^{-1}(40/95) = 24.90°$.

Before the angular accelerations can be found, the angular velocities of bars BC and CD must be known. These velocities can be determined using either the relative velocity method or the instant centers for velocities, with the latter being more convenient for this problem.

Because both $\mathbf{v}_B$ and $\mathbf{v}_C$ are horizontal, as shown in Fig. (b), the instant center for bar BC is at infinity. Therefore, BC is translating at this instant; that is, $\omega_{BC} = 0$.

The magnitude of $\mathbf{v}_B$ is $v_B = r_{B/A}\omega_{AB} = 80(2.4) = 192.0 \text{ mm/s}$, the sense being to the right because ω_{AB} is directed counterclockwise. As the bar BC is translating, it follows that $v_C = v_B = 192.0 \text{ mm/s}$ (all points of a translating body possess the same velocities), also directed to the right. Therefore $\omega_{CD} = v_C/r_{C/D} = 192.0/120 = 1.6 \text{ rad/s}$ with a counterclockwise direction, as shown in Fig. (b). Summarizing these results in vector notation, we have

$$\boldsymbol{\omega}_{AB} = 2.4\mathbf{k} \text{ rad/s} \qquad \boldsymbol{\omega}_{BC} = \mathbf{0} \qquad \boldsymbol{\omega}_{CD} = 1.6\mathbf{k} \text{ rad/s} \qquad \text{(a)}$$

Assuming that α_{BC} and α_{CD} are both counterclockwise, the angular accelerations of the bars are

$$\boldsymbol{\alpha}_{AB} = 1.5\mathbf{k} \text{ rad/s}^2 \qquad \boldsymbol{\alpha}_{BC} = \alpha_{BC}\mathbf{k} \text{ rad/s}^2 \qquad \boldsymbol{\alpha}_{CD} = \alpha_{CD}\mathbf{k} \text{ rad/s}^2 \qquad \text{(b)}$$

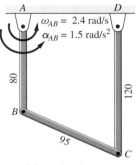

$\omega_{AB} = 2.4 \text{ rad/s}$
$\alpha_{AB} = 1.5 \text{ rad/s}^2$

Dimensions in mm

(a)

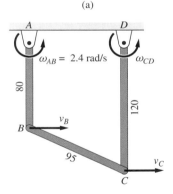

$\omega_{AB} = 2.4 \text{ rad/s}$

(b)

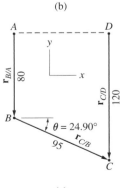

$\theta = 24.90°$

(c)

Solution I (using scalar notation)

The accelerations of points B and C are related by $\mathbf{a}_C = \mathbf{a}_B + \mathbf{a}_{C/B}$. The expression for the relative acceleration $\mathbf{a}_{C/B}$ is obtained by imagining that point B is fixed at the instant of concern. The accelerations $\mathbf{a}_B$ and $\mathbf{a}_C$ are derived from the fact that bars AB and CD rotate about the fixed points A and D, respectively. Therefore, the relationship between the accelerations becomes

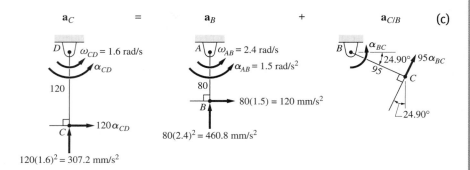

(c)

Inspection of this equation reveals that there are two unknowns: α_{BC} and α_{CD}, which can be found by equating horizontal and vertical components of Eq. (c):

$$\xrightarrow{+} \quad 120\alpha_{CD} = 120 + 95\alpha_{BC}\sin 24.90° \tag{d}$$

$$+\uparrow \quad 307.2 = 460.8 + 95\alpha_{BC}\cos 24.90° \tag{e}$$

Solving Eqs. (d) and (e) gives $\alpha_{BC} = -1.783$ rad/s^2 and $\alpha_{CD} = 0.406$ rad/s^2; that is,

$$\alpha_{BC} = 1.783 \text{ rad/s}^2 \circlearrowright \qquad \alpha_{CD} = 0.406 \text{ rad/s}^2 \circlearrowright \qquad \textit{Answer}$$

Solution II (using vector notation)

The relative position vectors shown in Fig. (c) can be written in vector form as

$$\left. \begin{aligned} \mathbf{r}_{B/A} &= -80\mathbf{j} \text{ mm} \\[2mm] \mathbf{r}_{C/B} &= 95\cos 24.90°\mathbf{i} - 95\sin 24.90°\mathbf{j} \\[1mm] &= 86.17\mathbf{i} - 40.00\mathbf{j} \text{ mm} \\[2mm] \mathbf{r}_{C/D} &= -120\mathbf{j} \text{ mm} \end{aligned} \right\} \tag{f}$$

The relationship between the accelerations of C and B is

$$\mathbf{a}_C = \mathbf{a}_B + \mathbf{a}_{C/B} \tag{g}$$

Because the path of C is a circle centered at D, we have

$$\mathbf{a}_C = \boldsymbol{\alpha}_{CD} \times \mathbf{r}_{C/D} + \boldsymbol{\omega}_{CD} \times (\boldsymbol{\omega}_{CD} \times \mathbf{r}_{C/D}) \tag{h}$$

Substituting the vectors in Eqs. (a), (b), and (f), the result is

$$\mathbf{a}_C = (\alpha_{CD}\mathbf{k}) \times (-120\mathbf{j}) + (1.6\mathbf{k}) \times [(1.6\mathbf{k}) \times (-120\mathbf{j})]$$

$$= 120\alpha_{CD}\mathbf{i} + (1.6\mathbf{k}) \times (192.0\mathbf{i})$$

$$= 120\alpha_{CD}\mathbf{i} + 307.2\mathbf{j} \text{ mm/s}^2 \tag{i}$$

Noting that B moves on a circular path centered at A, we conclude that

$$\mathbf{a}_B = \boldsymbol{\alpha}_{AB} \times \mathbf{r}_{B/A} + \boldsymbol{\omega}_{AB} \times (\boldsymbol{\omega}_{AB} \times \mathbf{r}_{B/A})$$

$$= (1.5\mathbf{k}) \times (-80\mathbf{j}) + (2.4\mathbf{k}) \times [(2.4\mathbf{k}) \times (-80\mathbf{j})]$$

$$= 120\mathbf{i} + (2.4\mathbf{k}) \times (192.0\mathbf{i})$$

$$= 120\mathbf{i} + 460.8\mathbf{j} \text{ mm/s}^2 \tag{j}$$

From Eqs. (16.13), we obtain

$$\mathbf{a}_{C/B} = \boldsymbol{\alpha}_{BC} \times \mathbf{r}_{C/B} + \boldsymbol{\omega}_{BC} \times (\boldsymbol{\omega}_{BC} \times \mathbf{r}_{C/B})$$

$$= (\alpha_{BC}\mathbf{k}) \times (86.17\mathbf{i} - 40.00\mathbf{j}) + \mathbf{0}$$

$$= 86.17\alpha_{BC}\mathbf{j} + 40.00\alpha_{BC}\mathbf{i} \text{ mm/s}^2 \tag{k}$$

Substituting Eqs. (i)–(k) into Eq. (g), and equating the coefficients of $\mathbf{i}$ and $\mathbf{j}$, yields the following two scalar equations.

$$120\alpha_{CD} = 120 + 40.00\alpha_{BC} \tag{l}$$

$$307.2 = 460.8 + 86.17\alpha_{BC} \tag{m}$$

Solving Eqs. (l) and (m) gives $\alpha_{BC} = -1.783$ rad/s^2 and $\alpha_{CD} = 0.406$ rad/s^2, or

$$\boldsymbol{\alpha}_{BC} = -1.783\mathbf{k} \text{ rad/s}^2 \quad \text{and} \quad \boldsymbol{\alpha}_{CD} = 0.406\mathbf{k} \text{ rad/s}^2 \qquad \textit{Answer}$$

Problems

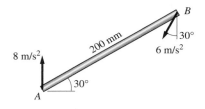
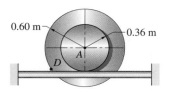

Fig. P16.61

16.61   At a given instant, the endpoints of the bar *AB* have the accelerations shown. Determine the angular velocity and angular acceleration of the bar at this instant.

16.62   The wheel rolls on its 0.36 m-radius hub without slipping. The angular velocity of the wheel is 3 rad/s. Determine the acceleration of point *D* on the rim of the wheel if the angular acceleration of the wheel is (a) 6.75 rad/s² clockwise; and (b) 6.75 rad/s² counterclockwise.

Fig. P16.62

16.63   A string is wrapped around the hub of the spool. A pull at the end of the string causes the spool to roll on the horizontal plane without slipping. At a certain instant, the angular velocity and angular acceleration of the spool are as shown in the figure. For this instant, find (a) the acceleration of point *D* on the spool; (b) the acceleration of point *B*; and (c) the acceleration a_0 of the end of the string.

16.64   A string is wrapped around the hub of the spool. A pull at the end of the string causes the spool to roll and slip on the horizontal plane. At a certain instant, the angular velocity and angular acceleration of the spool are as shown in the figure, while the velocity and acceleration of the end of the string are $v_0 = 1$ m/s and $a_0 = 2$ m/s², respectively. For this instant, find the acceleration of (a) point *D* on the spool; (b) point *A*; and (c) point *B*.

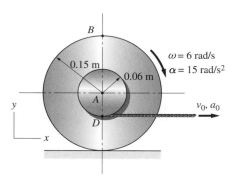

Fig. P16.63, P16.64

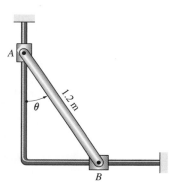

Fig. P16.65

16.65   When $\theta = 30°$, the angular velocity of the bar is 2 rad/s counterclockwise, and the acceleration of slider *B* is 8 m/s², directed to the right. Calculate the acceleration of slider *A* at this instant.

16.66   When the rod *AB* is in the horizontal position shown, the velocity and acceleration of collar *A* are $v_A = 2$ m/s and $a_A = 6$ m/s², directed as shown. Calculate the acceleration of collar *B* and the angular acceleration of the rod in this position.

Fig. P16.66

16.67   The crank *AB* is rotating clockwise with the constant angular velocity of 20 rad/s. Determine the acceleration of piston *C* when $\theta = 90°$.

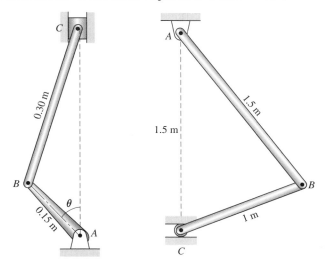

Fig. P16.67   **Fig. P16.68**

16.68   In the position shown, the angular velocity and angular acceleration of the bar *AB* are $\omega_{AB} = 3$ rad/s CW and $\alpha_{AB} = 12$ rad/s² CCW. Calculate the acceleration of roller *C* in this position.

16.69   When the mechanism is in the position shown, bar *AB* is rotating with the angular velocity ω and angular acceleration α, both counterclockwise. Determine the angular acceleration of bar *BC* and the acceleration of roller *C* in this position.

Fig. P16.69

16.70   Rod *AB* of the mechanism is sliding to the right with a constant velocity of 4 m/s. Determine the acceleration of roller *C* in the position shown.

16.71   When the mechanism is in the position shown, the velocity of the sliding collar is $v_A = 2$ m/s, and it is increasing at the rate of 1.2 m/s². For this position, calculate the angular accelerations of bars *AB* and *BC*.

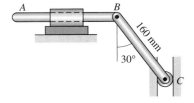

Fig. P16.70

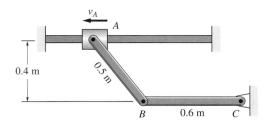

Fig. P16.71

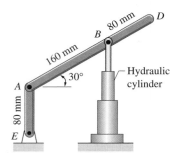

Fig. P16.72

16.72 As the hydraulic cylinder elongates, it raises pin *B* of the mechanism. When the system is in the position shown, the velocity of pin *B* is 40 mm/s upward, and it is increasing at the rate of 80 mm/s². For this instant, determine the angular accelerations of bars *AD* and *AE*.

16.73 Bar *AB* is rotating clockwise with the constant angular velocity of 20 rad/s. For the position shown, determine the angular accelerations of bars *BD* and *DE*.

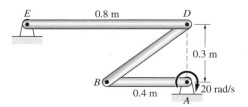

Fig. P16.73

16.74 The wheel rolls without slipping with the constant clockwise angular velocity of 0.8 rad/s, as end *B* of bar *AB* slides on the ground. Calculate the acceleration of *B* in the position shown.

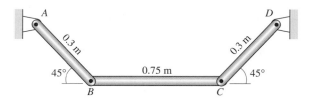

Fig. P16.74

16.75 Bar *BC* of the mechanism rotates clockwise with the constant angular velocity of 24 rad/s. Determine the angular accelerations of bars *AB* and *CD* in the position shown.

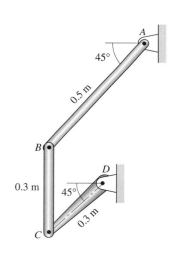

Fig. P16.76

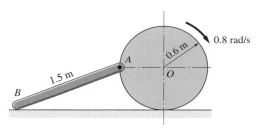

Fig. P16.75

16.76 In the position shown, the angular velocity and angular acceleration of bar *CD* are 6 rad/s and 20 rad/s², respectively, both counterclockwise. Compute the angular accelerations of bars *AB* and *BC* in this position.

16.77 Bar *AB* of the mechanism rotates with the constant angular velocity of 3 rad/s counterclockwise. For the position shown, calculate the angular accelerations of bars *BD* and *DE*.

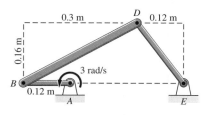

Fig. P16.77

16.78   The wheel rolls without slipping on the horizontal surface. In the position shown, the angular velocity of the wheel is 4 rad/s counterclockwise, and its angular acceleration is 5 rad/s² clockwise. Find the angular acceleration of rod AB and the acceleration of slider B in this position.

16.79   The disk is rotating counterclockwise with the constant angular speed of 2 rad/s. For the position shown, find the angular accelerations of bars AB and BD.

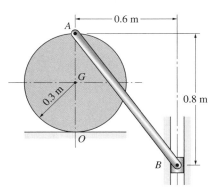

Fig. P16.78

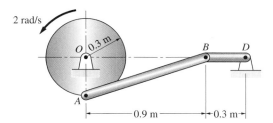

Fig. P16.79

16.80   The arm joining the friction wheels A and B is rotating with the angular velocity $\omega = 5$ rad/s and the angular acceleration $\alpha = 12.5$ rad/s², both counterclockwise. Assuming that wheel A is stationary and that there is no slipping, determine the magnitude of the acceleration of the point on the rim of B that is in contact with A.

16.81   When the mechanism is in the position shown, the angular velocity of the gear is 2 rad/s clockwise, and its angular acceleration is 4 rad/s² counterclockwise. Determine the angular accelerations of bars AB and BD in this position.

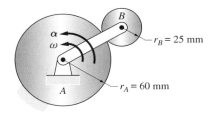

Fig. P16.80

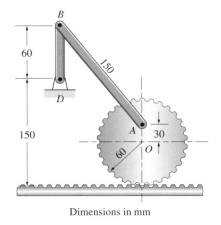

Fig. P16.81

16.82   Bar AB of the mechanism rotates with the constant angular velocity 1.2 rad/s clockwise. For the position shown, (a) verify that the angular velocities of the other two bars are $\omega_{BD} = 1.358$ rad/s counterclockwise and $\omega_{DE} = 1.131$ rad/s clockwise; and (b) determine the acceleration vector of point D.

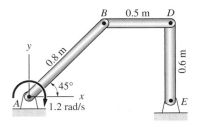

Fig. P16.82

16.8 Absolute and Relative Derivatives of Vectors

a. Introduction

Up to this point, our kinematic analysis of rigid bodies used the formulas for relative motion between points in the *same body*. The coordinate system used was allowed to translate but not rotate. However, there is a class of problems associated with sliding connections in which the point of interest does not lie in a body, but its path *relative to a body* is known. For problems of this type, it is convenient to describe the motion of the point in a reference frame that is embedded in the body. Such a coordinate system may rotate as well as translate.

The utility of a rotating coordinate system may be seen in the following example. Consider the motion of collar *B* sliding along the rotating bar *OA* in Fig. 16.13. We introduce two coordinate systems: the fixed *xy*-axes (with base vectors **i** and **j**), and the rotating *x'y'*-axes (with base vectors **i'** and **j'**), which are embedded in the bar. The *absolute* path of *B* (measured relative to the *xy*-axes), will, in general, be complicated. However, the *relative* path of *B*—that is, the path in the *x'y'*-coordinate system—is known: It is a straight line along the *x'*-axis. Letting $v_{B/OA}$ and $a_{B/OA}$ denote the speed and the magnitude of the acceleration of *B* relative to the bar, the corresponding relative velocity and relative acceleration vectors are simply $\mathbf{v}_{B/OA} = v_{B/OA}\mathbf{i'}$ and $\mathbf{a}_{B/OA} = a_{B/OA}\mathbf{i'}$.

This example shows that in some cases the description of motion is greatly simplified by using a rotating coordinate system. In Art. 16.9, we show how to determine the absolute motion of a point if its relative motion is given in a rotating coordinate system. This article is devoted to the derivation of formulas needed for this computation.

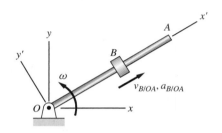

Fig. 16.13

b. Absolute and relative time derivatives of vectors

Consider the rigid body $\mathcal{B}$ that moves in the *xy*-plane, as shown in Fig. 16.14. The figure displays two reference frames:

- The *xyz* reference frame is fixed, the unit vectors along the axes being denoted by **i**, **j**, and **k**. A vector **V** can be expressed in terms of the fixed axes as

$$\mathbf{V} = V_x\mathbf{i} + V_y\mathbf{j} + V_z\mathbf{k} \tag{16.16}$$

The time derivative of this vector, also known as the *absolute derivative*, is

$$\dot{\mathbf{V}} = \frac{d\mathbf{V}}{dt} = \dot{V}_x\mathbf{i} + \dot{V}_y\mathbf{j} + \dot{V}_z\mathbf{k} \tag{16.17}$$

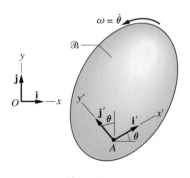

Fig. 16.14

- The *x'y'z'* axes are embedded in the body, the unit vectors being **i'**, **j'**, and **k'** = **k**. The expression for **V** in terms of the embedded coordinates is

$$\mathbf{V} = V_{x'}\mathbf{i'} + V_{y'}\mathbf{j'} + V_{z'}\mathbf{k'} \tag{16.18}$$

The *relative derivative* (relative to the body $\mathcal{B}$) of **V** is defined as

$$\left(\frac{d\mathbf{V}}{dt}\right)_{/\mathcal{B}} = \dot{V}_{x'}\mathbf{i'} + \dot{V}_{y'}\mathbf{j'} + \dot{V}_{z'}\mathbf{k'} \tag{16.19}$$

Equation (16.19) represents the rate of change of $\mathbf{V}$ with respect to the $x'y'z'$ reference frame; that is, the rate of change of $\mathbf{V}$ as seen by an observer attached to the body $\mathcal{B}$. The absolute derivative of the vector in Eq. (16.18) is

$$\frac{d\mathbf{V}}{dt} = \dot{V}_{x'}\mathbf{i}' + \dot{V}_{y'}\mathbf{j}' + \dot{V}_{z'}\mathbf{k}' + V_{x'}\frac{d\mathbf{i}'}{dt} + V_{y'}\frac{d\mathbf{j}'}{dt} + V_{z'}\frac{d\mathbf{k}'}{dt}$$

$$= \left(\frac{d\mathbf{V}}{dt}\right)_{/\mathcal{B}} + V_{x'}\frac{d\mathbf{i}'}{dt} + V_{y'}\frac{d\mathbf{j}'}{dt} + V_{z'}\frac{d\mathbf{k}'}{dt} \qquad (16.20)$$

Since the $x'y'z'$ axes rotate with the body, the time derivatives of the base vectors $\mathbf{i}'$ and $\mathbf{j}'$ are generally not zero (translation of $\mathcal{B}$ does not affect the derivatives). However, for plane motion $d\mathbf{k}'/dt = \mathbf{0}$.

c. Derivatives of embedded unit vectors

We now derive the expressions for the absolute derivatives of $\mathbf{i}'$ and $\mathbf{j}'$. Referring to Fig. 16.14, we can write

$$\mathbf{i}' = \cos\theta\,\mathbf{i} + \sin\theta\,\mathbf{j} \quad \mathbf{j}' = -\sin\theta\,\mathbf{i} + \cos\theta\,\mathbf{j}$$

Therefore, the absolute derivatives of the embedded unit vectors are

$$\frac{d\mathbf{i}'}{dt} = (-\sin\theta\,\mathbf{i} + \cos\theta\,\mathbf{j})\dot{\theta} = \omega\mathbf{j}' \quad \frac{d\mathbf{j}'}{dt} = (-\cos\theta\,\mathbf{i} - \sin\theta\,\mathbf{j})\dot{\theta} = -\omega\mathbf{i}' \quad (16.21)$$

where $\omega = \dot{\theta}$ is the angular velocity of $\mathcal{B}$. Introducing the angular velocity vector $\boldsymbol{\omega} = \omega\mathbf{k}$, the absolute derivatives also can be written as

$$\frac{d\mathbf{i}'}{dt} = \boldsymbol{\omega} \times \mathbf{i}' \qquad \frac{d\mathbf{j}'}{dt} = \boldsymbol{\omega} \times \mathbf{j}' \qquad \frac{d\mathbf{k}'}{dt} = \mathbf{0} \qquad (16.22)$$

d. Relationships between absolute and relative derivatives

Substituting Eqs. (16.22) into Eq. (16.20) we get

$$\frac{d\mathbf{V}}{dt} = \left(\frac{d\mathbf{V}}{dt}\right)_{/\mathcal{B}} + V_{x'}(\boldsymbol{\omega} \times \mathbf{i}') + V_{y'}(\boldsymbol{\omega} \times \mathbf{j}') \qquad (16.23)$$

It can be verified readily that the last equation is equivalent to

$$\frac{d\mathbf{V}}{dt} = \left(\frac{d\mathbf{V}}{dt}\right)_{/\mathcal{B}} + \boldsymbol{\omega} \times \mathbf{V} \qquad (16.24)$$

Equation (16.24) is convenient for evaluating the absolute derivative of a vector when its relative derivative is known.

The expression for the second absolute derivative of $\mathbf{V}$ can be obtained by differentiating Eq. (16.24):

$$\frac{d^2\mathbf{V}}{dt^2} = \frac{d}{dt}\left(\frac{d\mathbf{V}}{dt}\right) = \frac{d}{dt}\left[\left(\frac{d\mathbf{V}}{dt}\right)_{/\mathscr{B}} + \boldsymbol{\omega} \times \mathbf{V}\right]$$

$$= \left(\frac{d^2\mathbf{V}}{dt^2}\right)_{/\mathscr{B}} + \boldsymbol{\omega} \times \left(\frac{d\mathbf{V}}{dt}\right)_{/\mathscr{B}} + \boldsymbol{\omega} \times \mathbf{V} + \boldsymbol{\omega} \times \left[\left(\frac{d\mathbf{V}}{dt}\right)_{/\mathscr{B}} + \boldsymbol{\omega} \times \mathbf{V}\right]$$

After simplifying, this yields

$$\frac{d^2\mathbf{V}}{dt^2} = \left(\frac{d^2\mathbf{V}}{dt^2}\right)_{/\mathscr{B}} + \boldsymbol{\omega} \times \mathbf{V} + \boldsymbol{\omega} \times (\boldsymbol{\omega} \times \mathbf{V}) + 2\boldsymbol{\omega} \times \left(\frac{d\mathbf{V}}{dt}\right)_{/\mathscr{B}} \qquad \text{(16.25)}$$

e. Special case: vector embedded in rotating reference frame

If the vector $\mathbf{V}$ is embedded in the body $\mathscr{B}$, its components $V_{x'}$, $V_{y'}$, and $V_{z'}$ remain constant, so that $(d\mathbf{V}/dt)_{/\mathscr{B}} = \mathbf{0}$. Consequently, Eqs. (16.24) and (16.25) become

$$\frac{d\mathbf{V}}{dt} = \boldsymbol{\omega} \times \mathbf{V} \qquad \text{(16.26)}$$

$$\frac{d^2\mathbf{V}}{dt^2} = \dot{\boldsymbol{\omega}} \times \mathbf{V} + \boldsymbol{\omega} \times (\boldsymbol{\omega} \times \mathbf{V}) \qquad \text{(16.27)}$$

f. Note on general motion

Up to now we have assumed plane motion. If the motion is not plane, it can be shown that[*]

$$\frac{d\mathbf{i}'}{dt} = \boldsymbol{\omega} \times \mathbf{i}' \qquad \frac{d\mathbf{j}'}{dt} = \boldsymbol{\omega} \times \mathbf{j}' \qquad \frac{d\mathbf{k}'}{dt} = \boldsymbol{\omega} \times \mathbf{k}' \qquad \text{(16.28)}$$

Note that the first two equations are identical to those in Eqs. (16.22). As a consequence, the absolute derivative of $\mathbf{V}$ in Eq. (16.20) becomes

$$\frac{d\mathbf{V}}{dt} = \left(\frac{d\mathbf{V}}{dt}\right)_{/\mathscr{B}} + V_{x'}(\boldsymbol{\omega} \times \mathbf{i}') + V_{y'}(\boldsymbol{\omega} \times \mathbf{j}') + V_{z'}(\boldsymbol{\omega} \times \mathbf{k}')$$

which can be written in exactly the same form as Eq. (16.24):

$$\frac{d\mathbf{V}}{dt} = \left(\frac{d\mathbf{V}}{dt}\right)_{/\mathscr{B}} + \boldsymbol{\omega} \times \mathbf{V}$$

Therefore, we conclude that Eqs. (16.24)–(16.27) are not restricted to plane motion.

---

[*]See *Principles of Dynamics*, 2e, Donald T Greenwood, Prentice Hall, 1988, p.33.

Motion Relative to a Rotating Reference Frame

We now apply the formulas relating the absolute and relative derivatives of an arbitrary vector **V**, derived in the previous article, to the velocity and acceleration vectors of a point. As shown in Fig. 16.15, we let $\mathscr{B}$ be a rigid body that is undergoing plane motion in the xy-plane, where the xyz-coordinate system is fixed. The $x'y'z'$-coordinate axes are embedded in the body and, therefore, rotate with the angular velocity $\boldsymbol{\omega} = \omega\mathbf{k}$ of the body. The following three points will be involved in our discussion:

- A is the origin of the embedded $x'y'z'$-coordinate system.
- P is a point or particle that moves independently of $\mathscr{B}$.
- P' (not shown) is a point *embedded* in $\mathscr{B}$ that is coincident with P at the instant shown.

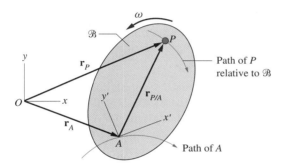

Fig. 16.15

From Fig. 16.15, we have $\mathbf{r}_P = \mathbf{r}_A + \mathbf{r}_{P/A}$, which upon differentiation with respect to time yields

$$\mathbf{v}_P = \mathbf{v}_A + \mathbf{v}_{P/A} \qquad (16.29)$$

where

$$\mathbf{v}_{P/A} = \frac{d\mathbf{r}_{P/A}}{dt} \qquad (16.30)$$

The velocity in Eq. (16.30) is measured relative to the fixed xyz-coordinate system. In many situations, however, the velocity of P relative to the moving body $\mathscr{B}$ is easier to describe. The relationship between these two velocities can be found by replacing **V** by $\mathbf{r}_{P/A}$ in Eq. (16.24) of the previous article, which gives

$$\frac{d\mathbf{r}_{P/A}}{dt} = \left(\frac{d\mathbf{r}_{P/A}}{dt}\right)_{/\mathscr{B}} + \boldsymbol{\omega} \times \mathbf{r}_{P/A} \qquad (16.31)$$

where the subscript "$/\mathscr{B}$" denotes quantities that are measured relative to the body $\mathscr{B}$. Introducing the notation

$$\mathbf{v}_{P/\mathscr{B}} = \left(\frac{d\mathbf{r}_{P/A}}{dt}\right)_{/\mathscr{B}} \qquad (16.32)$$

for the velocity of P relative to $\mathcal{B}$, and substituting Eq. (16.31) into Eq. (16.29), we get

$$\mathbf{v}_P = \mathbf{v}_A + \boldsymbol{\omega} \times \mathbf{r}_{P/A} + \mathbf{v}_{P/\mathcal{B}} \tag{16.33}$$

Because P' and A are embedded in the body $\mathcal{B}$, the velocity of P' relative to A is given by

$$\mathbf{v}_{P'/A} = \boldsymbol{\omega} \times \mathbf{r}_{P/A} \tag{16.34}$$

Therefore, Eq. (16.33) can be written in the form

$\mathbf{v}_P$	$=$	$\mathbf{v}_A$	$+$	$\mathbf{v}_{P'/A}$	$+$	$\mathbf{v}_{P/\mathcal{B}}$	
|		|		|		|	
velocity of P		velocity of origin of $x'y'z'$-axes, which are embedded in body $\mathcal{B}$		velocity of P' *relative* to A (P' is embedded in body $\mathcal{B}$ and coincident with P at this instant)		velocity of P relative to body $\mathcal{B}$	(16.35)

Figure 16.16 displays the terms that appear in Eq. (16.35) (the angular velocity ω is assumed to be counterclockwise). Observe that the motion is represented as a translation and a rotation of the body, plus the velocity of P relative to the body. The velocity $\mathbf{v}_P$ is tangent to the absolute path of P, whereas $\mathbf{v}_{P/\mathcal{B}}$ is tangent to its relative path. The rotation term in Fig. 16.16 is shown as the vector cross product $\boldsymbol{\omega} \times \mathbf{r}_{P/A}$, but this could be replaced by the scalar notation in Fig. 16.8(d).

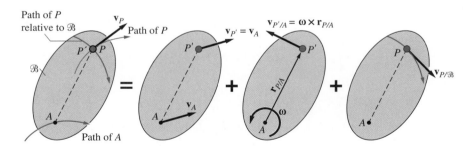

Fig. 16.16

The acceleration of P is found by differentiating Eq. (16.29):

$$\mathbf{a}_P = \mathbf{a}_A + \mathbf{a}_{P/A} \tag{16.36}$$

where the relative acceleration is

$$\mathbf{a}_{P/A} = \frac{d\mathbf{v}_{P/A}}{dt} = \frac{d^2\mathbf{r}_{P/A}}{dt^2} \tag{16.37}$$

Substituting $\mathbf{r}_{P/A}$ for $\mathbf{V}$ in Eq. (16.25), we see that Eq. (16.37) may be written as

$$\mathbf{a}_{P/A} = \left(\frac{d^2\mathbf{r}_{P/A}}{dt^2}\right)_{/\mathcal{B}} + \dot{\boldsymbol{\omega}} \times \mathbf{r}_{P/A} + \boldsymbol{\omega} \times (\boldsymbol{\omega} \times \mathbf{r}_{P/A}) + 2\boldsymbol{\omega} \times \mathbf{v}_{P/\mathcal{B}} \qquad (16.38)$$

The first term of Eq. (16.38) is the acceleration of P relative to body $\mathcal{B}$:

$$\mathbf{a}_{P/\mathcal{B}} = \left(\frac{d^2\mathbf{r}_{P/A}}{dt^2}\right)_{/\mathcal{B}} \qquad (16.39)$$

The next two terms in Eq. (16.38) represent the acceleration of P' relative to A, namely

$$\mathbf{a}_{P'/A} = \dot{\boldsymbol{\omega}} \times \mathbf{r}_{P/A} + \boldsymbol{\omega} \times (\boldsymbol{\omega} \times \mathbf{r}_{P/A}) \qquad (16.40)$$

The last term in Eq. (16.38), called the *Coriolis acceleration* (named after the French mathematician G. G. Coriolis), will be denoted by

$$\mathbf{a}_C = 2\boldsymbol{\omega} \times \mathbf{v}_{P/\mathcal{B}} \qquad (16.41)$$

Note that the Coriolis acceleration represents an interaction between the angular velocity of the body and the velocity of P relative to the body.

Substituting Eqs. (16.38) through (16.41) into Eq. (16.36), the acceleration of P becomes

$$\mathbf{a}_P = \mathbf{a}_A + \left[\dot{\boldsymbol{\omega}} \times \mathbf{r}_{P/A} + \boldsymbol{\omega} \times (\boldsymbol{\omega} \times \mathbf{r}_{P/A})\right] + \mathbf{a}_{P/\mathcal{B}} + 2\boldsymbol{\omega} \times \mathbf{v}_{P/\mathcal{B}} \qquad (16.42)$$

A physically more meaningful form of this equation is

$\mathbf{a}_P$	$=$	$\mathbf{a}_A$	$+$	$\mathbf{a}_{P'/A}$	$+$	$\mathbf{a}_{P/\mathcal{B}}$	$+$	$\mathbf{a}_C$
|		|		|		|		|
acceleration of P		acceleration of origin of $x'y'z'$-axes, which are embedded in body $\mathcal{B}$		acceleration of P' relative to A (P' is embedded in body $\mathcal{B}$ and coincident with P at this instant)		acceleration of P relative to body $\mathcal{B}$		Coriolis acceleration

$$(16.43)$$

Figure 16.17 illustrates the terms that appear in Eq. (16.42) (the angular velocity ω and the angular acceleration $\dot{\omega}$ are assumed to be counterclockwise). Note that the motion is represented as a translation and rotation of the body plus the acceleration of P relative to the body plus the Coriolis acceleration. The rotation terms in Fig. 16.17 are shown using their vector representations, $\dot{\boldsymbol{\omega}} \times \mathbf{r}_{P/A}$ and $\boldsymbol{\omega} \times (\boldsymbol{\omega} \times \mathbf{r}_{P/A})$; they can be replaced by the scalar notations shown in Fig. 16.12(d). As seen in Fig. 16.17, the Coriolis acceleration $\mathbf{a}_C$ is perpendicular to both $\mathbf{v}_{P/\mathcal{B}}$ and $\boldsymbol{\omega}$. When the scalar representation is used, the magnitude of $\mathbf{a}_C$ is $2\omega v_{P/\mathcal{B}}$, and its direction can be determined by fixing the tail of $\mathbf{v}_{P/\mathcal{B}}$ and rotating this vector through $90°$ in the direction of ω.

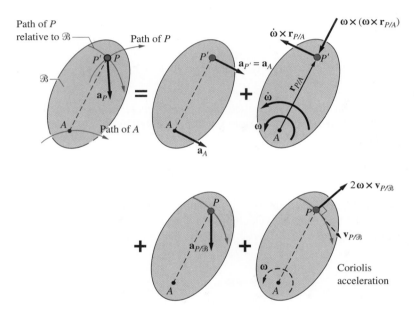

Fig. 16.17

Sample Problem   16.13

As shown in Fig. (a), the collar P slides from A toward B along a semicircular rod AB of radius 200 mm. The rod rotates about the pin at A, and the speed of P relative to the rod is constant at 120 mm/s. When the system is in the position shown, the angular velocity and angular acceleration of the rod are $\omega_{AB} = 0.8$ rad/s counterclockwise and $\alpha_{AB} = 0.5$ rad/s^2 clockwise. For this position, determine the velocity and acceleration vectors of P.

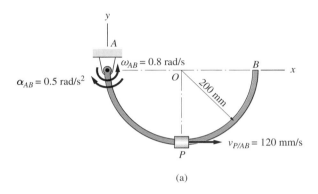

(a)

Solution

Preliminary Comments

This problem will be solved using scalar notation (Solution I) and vector notation (Solution II). In both solutions we employ point P', identified as the point on AB that coincides with P at the instant of concern. The relative position vectors required in the solutions are shown in Fig. (b).

Note that (1) the absolute path of P' is a circle that is centered at the fixed point A, and (2) the path of P relative to AB is a circle that is centered at point O.

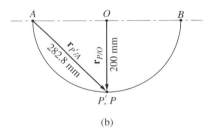

(b)

Solution I (using scalar notation)

Letting body $\mathcal{B}$ in Eq. (16.35) be the rod AB, the velocity of P becomes

$$\mathbf{v}_P = \underset{\mathbf{0}}{\mathbf{v}_A} \quad + \quad \mathbf{v}_{P'/A} \quad + \quad \mathbf{v}_{P/AB} \qquad \text{(a)}$$

from which we find that

$$(v_P)_x = 226.2 \sin 45° + 120 = 280 \text{ mm/s}$$
$$(v_P)_y = 226.2 \cos 45° = 160 \text{ mm/s}$$

or

$$\mathbf{v}_P =$$

Answer

Using Eq. (16.43), the acceleration of P is

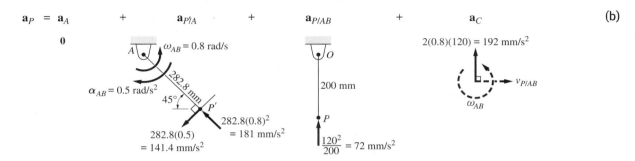

$$\mathbf{a}_P = \mathbf{a}_A \quad + \quad \mathbf{a}_{P'/A} \quad + \quad \mathbf{a}_{P/AB} \quad + \quad \mathbf{a}_C \qquad \text{(b)}$$

Note that in Eq. (b) the direction of the Coriolis acceleration $\mathbf{a}_C$ is found by fixing the tail of $\mathbf{v}_{P/AB}$ and then rotating this vector 90° in the direction of ω_{AB}. Furthermore, observe that $\mathbf{a}_{P/AB}$ contains only the normal component $v^2_{P/AB}/r_{P/O}$ because the magnitude of $\mathbf{v}_{P/AB}$ is constant. Evaluating the components of Eq. (b) gives

$$(a_P)_x = -181\cos 45° - 141.4\sin 45° = -228 \text{ mm/s}^2$$
$$(a_P)_y = 181\sin 45° - 141.4\cos 45° + 72 + 192 = 292 \text{ mm/s}^2$$

or

$$\mathbf{a}_P =$$

Answer

Solution II (using vector notation)

From Eq. (16.35), the velocity of P is

$$\mathbf{v}_P = \mathbf{v}_A + \mathbf{v}_{P'/A} + \mathbf{v}_{P/AB} \qquad \text{(c)}$$

Because A is a fixed point, we have

$$\mathbf{v}_A = \mathbf{0} \qquad \text{(d)}$$

Noting that P', being a point that is embedded in rod AB, travels along a circular path centered at A, we get

$$\mathbf{v}_{P'/A} = \boldsymbol{\omega}_{AB} \times \mathbf{r}_{P'/A} = (0.8\mathbf{k}) \times (200\mathbf{i} - 200\mathbf{j}) = 160\mathbf{j} + 160\mathbf{i} \text{ mm/s} \qquad \text{(e)}$$

The velocity of P relative to bar AB is given as

$$\mathbf{v}_{P/AB} = 120\mathbf{i} \text{ mm/s} \qquad \text{(f)}$$

Substituting Eqs. (d) through (f) into Eq. (c), the velocity of P becomes

$$\mathbf{v}_P = \mathbf{0} + (160\mathbf{j} + 160\mathbf{i}) + (120\mathbf{i})$$
$$= 280\mathbf{i} + 160\mathbf{j} \text{ mm/s} \qquad \textit{Answer}$$

The acceleration of P is, according to Eq. (16.42),

$$\mathbf{a}_P = \mathbf{a}_A + \mathbf{a}_{P'/A} + \mathbf{a}_{P/AB} + \mathbf{a}_C \qquad \text{(g)}$$

Because A is a fixed point, we have

$$\mathbf{a}_A = \mathbf{0} \qquad \text{(h)}$$

Noting that the path of P' is a circle with its center at A, the acceleration of P' relative to A (which is also the absolute acceleration of P' considering that A is a fixed point) is

$$\mathbf{a}_{P'/A} = \boldsymbol{\alpha}_{AB} \times \mathbf{r}_{P'/A} + \boldsymbol{\omega}_{AB} \times (\boldsymbol{\omega}_{AB} \times \mathbf{r}_{P'/A})$$
$$= (-0.5\mathbf{k}) \times (200\mathbf{i} - 200\mathbf{j}) + (0.8\mathbf{k}) \times (160\mathbf{i} + 160\mathbf{j})$$
$$= (-100\mathbf{j} - 100\mathbf{i}) + (128\mathbf{j} - 128\mathbf{i})$$
$$= -228\mathbf{i} + 28\mathbf{j} \text{ mm/s}^2 \qquad \text{(i)}$$

The acceleration of P relative to AB has only a normal component because $v_{P/AB}$ is constant. Because the normal component of the relative acceleration is directed toward the center of curvature of the relative path (i.e., toward point O), we find that

$$\mathbf{a}_{P/AB} = (\mathbf{a}_{P/AB})_n = \frac{v_{P/AB}^2}{r_{P/O}}\mathbf{j} = \frac{(120)^2}{200}\mathbf{j} = 72\mathbf{j} \text{ mm/s}^2 \qquad \text{(j)}$$

The Coriolis acceleration from Eq. (16.41) is

$$\mathbf{a}_C = 2\boldsymbol{\omega}_{AB} \times \mathbf{v}_{P/AB} = 2(0.8\mathbf{k}) \times (120\mathbf{i}) = 192\mathbf{j} \text{ mm/s}^2 \qquad \text{(k)}$$

Substituting Eqs. (h)–(k) into Eq. (g), we obtain

$$\mathbf{a}_P = \mathbf{0} + (-228\mathbf{i} + 28\mathbf{j}) + 72\mathbf{j} + 192\mathbf{j}$$
$$= -228\mathbf{i} + 292\mathbf{j} \text{ mm/s}^2 \qquad \textit{Answer}$$

Sample Problem 16.14

Crank AB of the quick-return mechanism in Fig. (a) rotates counterclockwise with the constant angular velocity $\omega_{AB} = 6$ rad/s. When the mechanism is in the position shown, calculate the velocity and acceleration of the slider B relative to arm DE, and the angular velocity and acceleration of arm DE.

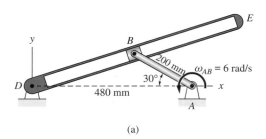

(a)

Solution

Introductory Comments

Note that the path of slider B relative to arm DE is the slot in the arm. Because this relative path is a straight line, both $\mathbf{v}_{B/DE}$ and $\mathbf{a}_{B/DE}$ (the velocity and acceleration of B relative to arm DE) are directed along the slot.

We let B' be the point on DE that is coincident with B at the instant of concern. The velocity and acceleration of B from Eqs. (16.35) and (16.43), respectively, are $\mathbf{v}_B = \mathbf{v}_D + \mathbf{v}_{B'/D} + \mathbf{v}_{B/DE}$ and $\mathbf{a}_B = \mathbf{a}_D + \mathbf{a}_{B'/D} + \mathbf{a}_{B/DE} + \mathbf{a}_C$, where $\mathbf{a}_C$ is the Coriolis acceleration. Noting that D is a fixed point ($\mathbf{v}_D = \mathbf{0}$ and $\mathbf{a}_D = \mathbf{0}$), the velocity and acceleration equations become

$$\mathbf{v}_B = \mathbf{v}_{B'/D} + \mathbf{v}_{B/DE} \qquad \text{(a)}$$

and

$$\mathbf{a}_B = \mathbf{a}_{B'/D} + \mathbf{a}_{B/DE} + \mathbf{a}_C \qquad \text{(b)}$$

These equations will be analyzed using scalar notation (Solution I) and vector notation (Solution II).

Throughout the analyses it must be kept in mind that the paths of B and B' are circles that are centered at A and D, respectively. We assume that ω_{DE} and α_{DE} are counterclockwise and that $\mathbf{v}_{B/DE}$ and $\mathbf{a}_{B/DE}$ are both directed toward D, as indicated in Fig. (b). Figure (c) shows the relative position vectors required for the analysis. The distance $\overline{DB}$ and the angle between DE and the x-axis were determined by trigonometry.

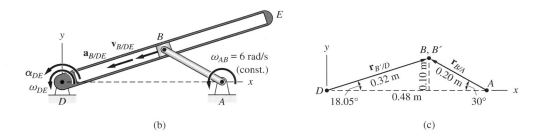

(b)   (c)

The following vectors are involved in Eqs. (a) and (b):

$$\mathbf{r}_{B/A} = 0.20(-\cos 30°\mathbf{i} + \sin 30°\mathbf{j}) = -0.17\mathbf{i} + 0.1\mathbf{j} \text{ m} \tag{c}$$

$$\mathbf{r}_{B'/D} = 0.32(\cos 18.05°\mathbf{i} + \sin 18.05°\mathbf{j}) = 0.30\mathbf{i} + 0.1\mathbf{j} \text{ m} \tag{d}$$

$$\boldsymbol{\omega}_{AB} = 6\mathbf{k} \text{ rad/s} \qquad \boldsymbol{\alpha}_{AB} = \mathbf{0} \tag{e}$$

$$\boldsymbol{\omega}_{DE} = \omega_{DE}\mathbf{k} \text{ rad/s} \qquad \boldsymbol{\alpha}_{DE} = \alpha_{DE}\mathbf{k} \text{ rad/s}^2 \tag{f}$$

$$\mathbf{v}_{B/DE} = v_{B/DE} \; \text{\fbox{$\diagup$}}\!-18.05°$$
$$= v_{B/DE}(-\cos 18.05°\mathbf{i} - \sin 18.05°\mathbf{j}) \text{ m/s} \tag{g}$$

$$\mathbf{a}_{B/DE} = a_{B/DE} \; \text{\fbox{$\diagup$}}\!-18.05°$$
$$= a_{B/DE}(-\cos 18.05°\mathbf{i} - \sin 18.05°\mathbf{j}) \text{ m/s}^2 \tag{h}$$

Solution I (using scalar notation)

Velocity

Equation (a) is

$$\mathbf{v}_B \qquad = \qquad \mathbf{v}_{B'/D} \qquad + \qquad \mathbf{v}_{B/DE} \tag{i}$$

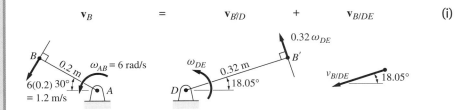

Inspection of Eq. (i) reveals that there are two unknowns, ω_{DE} and $v_{B/DE}$, which can be found by solving the two equivalent scalar equations.

Equating horizontal and vertical components of both sides of Eq. (i) gives

$$\xrightarrow{+} \quad -1.2 \sin 30° = -0.32\omega_{DE} \sin 18.05° - v_{B/DE} \cos 18.05° \qquad (j)$$

$$+\uparrow \quad -1.2 \cos 30° = 0.32\omega_{DE} \cos 18.05° - v_{B/DE} \sin 18.05° \qquad (k)$$

Solving Eqs. (j) and (k) yields $\omega_{DE} = -2.536$ rad/s and $v_{B/DE} = 0.9$ m/s, from which we find that

$$\boldsymbol{\omega}_{DE} = 2.536 \text{ rad/s} \circlearrowleft \quad \mathbf{v}_{B/DE} = 0.9 \text{ m/s} \quad \nearrow\!\!-18.05° \qquad \textit{Answer} \quad (l)$$

Acceleration

Equation (b) is

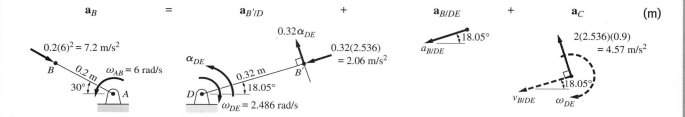

$$\mathbf{a}_B \quad = \quad \mathbf{a}_{B'/D} \quad + \quad \mathbf{a}_{B/DE} \quad + \quad \mathbf{a}_C \qquad (m)$$

Note that in Eq. (m) the magnitude of the Coriolis acceleration is $a_C = 2\omega_{DE}v_{B/DE}$, and its direction is found by fixing the tail of $\mathbf{v}_{B/DE}$ and rotating this vector 90° in the direction of ω_{DE}. The unknowns in Eq. (m) are α_{DE} and $a_{B/DE}$, which can be solved by equating horizontal and vertical components.

$$\xrightarrow{+} \quad 7.2 \cos 30° = -0.32\alpha_{DE} \sin 18.05°$$
$$- 2.06 \cos 18.05° - a_{B/DE} \cos 18.05°$$
$$- 4.57 \sin 18.05° \qquad (n)$$

$$+\uparrow \quad -7.2 \sin 30° = 0.32\alpha_{DE} \cos 18.05°$$
$$- 2.06 \sin 18.05° - a_{B/DE} \sin 18.05°$$
$$+ 4.57 \cos 18.05° \qquad (o)$$

Solving Eqs. (n) and (o) simultaneously gives $\alpha_{DE} = -31.39$ rad/s^2 and $a_{B/DE} = -6.8$ m/s^2. Therefore the results are

$$\alpha_{DE} = 31.39 \text{ rad/s}^2\circlearrowleft \quad \text{and} \quad \mathbf{a}_{B/DE} = 6.8 \text{ m/s}^2 \quad \nearrow\!\!-18.05° \qquad \textit{Answer}$$

Solution II (using vector notation)

Velocity

We begin by computing the three terms appearing in Eq. (a).
  Because B is a point on AB, we have

$$\mathbf{v}_B = \boldsymbol{\omega}_{AB} \times \mathbf{r}_{B/A}$$
$$= (6\mathbf{k}) \times (-0.17\mathbf{i} + 0.1\mathbf{j})$$
$$= -1.02\mathbf{j} - 0.6\mathbf{i} \text{ m/s} \qquad\qquad \text{(p)}$$

Using the fact that B' is a point on DE, we have

$$\mathbf{v}_{B'/D} = \boldsymbol{\omega}_{DE} \times \mathbf{r}_{B'/D}$$
$$= (\omega_{DE}\mathbf{k}) \times (0.3\mathbf{i} + 0.1\mathbf{j})$$
$$= 0.3\omega_{DE}\mathbf{j} - 0.1\omega_{DE}\mathbf{i} \text{ m/s} \qquad\qquad \text{(q)}$$

The relative velocity vector $\mathbf{v}_{B/DE}$ was found previously in Eq. (g):

$$\mathbf{v}_{B/DE} = v_{B/DE}(-\cos 18.05°\mathbf{i} - \sin 18.05°\mathbf{j}) \text{ m/s} \qquad\qquad \text{(r)}$$

  Substituting Eqs. (p)–(r) into Eq. (a) and equating coefficients of $\mathbf{i}$ and $\mathbf{j}$, respectively, we obtain

$$-0.6 = -0.1\omega_{DE} - v_{B/DE}\cos 18.05° \qquad\qquad \text{(s)}$$
$$-1.02 = 0.3\omega_{DE} - v_{B/DE}\sin 18.05° \qquad\qquad \text{(t)}$$

Solving Eqs. (s) and (t) simultaneously gives $\omega_{DE} = -2.536$ rad/s and $v_{B/DE} = 0.9$ m/s, from which we find that

$$\boldsymbol{\omega}_{DE} = -2.536 \text{ rad/s} \qquad\qquad\qquad \textit{Answer}$$

and

$$\mathbf{v}_{B/DE} = 0.9(-\cos 18.05°\mathbf{i} - \sin 18.05°\mathbf{j})$$
$$= -0.86\mathbf{i} - 0.28\mathbf{j} \text{ m/s} \qquad\qquad \textit{Answer}$$

Acceleration

The terms in Eq. (b) will be computed next.
  Because B is a point on AB, the acceleration of B becomes

$$\mathbf{a}_B = \boldsymbol{\omega}_{AB} \times (\boldsymbol{\omega}_{AB} \times \mathbf{r}_{B/A})$$
$$= (6\mathbf{k}) \times (-1.02\mathbf{j} - 0.6\mathbf{i})$$
$$= 6.12\mathbf{i} - 3.6\mathbf{j} \text{ m/s}^2 \qquad\qquad \text{(u)}$$

Because B' is a point on arm DE, we obtain

$$\mathbf{a}_{B'/D} = \boldsymbol{\alpha}_{DE} \times \mathbf{r}_{B'/D} + \boldsymbol{\omega}_{DE} \times (\boldsymbol{\omega}_{DE} \times \mathbf{r}_{B'/D})$$

$$= (\alpha_{DE}\mathbf{k}) \times (0.3\mathbf{i} + 0.1\mathbf{j})$$
$$+ (-2.536\mathbf{k}) \times [(-2.536\mathbf{k}) \times (0.3\mathbf{i} + 0.1\mathbf{j})]$$

$$= 0.3\alpha_{DE}\mathbf{j} - 0.1\alpha_{DE}\mathbf{i}$$
$$+ (-2.536\mathbf{k}) \times (-0.76\mathbf{j} + 0.25\mathbf{i})$$

$$= .30\alpha_{DE}\mathbf{j} - 0.1\alpha_{DE}\mathbf{i} - 1.93\mathbf{i} - 0.63\mathbf{j} \text{ m/s}^2 \qquad \text{(v)}$$

The acceleration vector of B relative to arm DE was given in Eq. (h) to be

$$\mathbf{a}_{B/DE} = a_{B/DE}(-\cos 18.05°\mathbf{i} - \sin 18.05°\mathbf{j}) \text{ m/s}^2 \qquad \text{(w)}$$

From Eq. (16.41), the Coriolis acceleration becomes

$$\mathbf{a}_C = 2\boldsymbol{\omega}_{DE} \times \mathbf{v}_{B/DE}$$

$$= 2(-2.536\mathbf{k}) \times (-0.86\mathbf{i} - 0.28\mathbf{j})$$

$$= 4.36\mathbf{j} - 1.42\mathbf{i} \text{ m/s}^2 \qquad \text{(x)}$$

Substituting Eqs. (u)–(x) into Eq. (b) and equating the coefficients of $\mathbf{i}$ and $\mathbf{j}$, respectively, we obtain

$$6.12 = -0.1\alpha_{DE} - 1.93 - a_{B/DE}\cos 18.05° - 1.42 \qquad \text{(y)}$$

$$-3.6 = 0.3\alpha_{DE} - 0.63 - a_{B/DE}\sin 18.05° + 4.36 \qquad \text{(z)}$$

Solving Eqs. (y) and (z) simultaneously yields $\alpha_{DE} = -31.39$ rad/s^2 and $a_{B/DE} = -6.8$ m/s^2. Therefore, the results written in vector notation are

$$\boldsymbol{\alpha}_{DE} = -31.39\mathbf{k} \text{ rad/s}^2 \qquad \textit{Answer}$$

and

$$\mathbf{a}_{B/DE} = 6.8(\cos 18.05°\mathbf{i} + \sin 18.05°\mathbf{j})$$

$$= 6.47\mathbf{i} + 2.11\mathbf{j} \text{ m/s}^2 \qquad \textit{Answer}$$

Problems

16.83 Rod *OB* rotates counterclockwise with the constant angular speed of 30 rev/min. At the same time, collar *A* is sliding toward *B* with the constant speed 1 m/s relative to the rod. Using a rotating reference frame attached to *OB*, calculate the acceleration of the collar when $R = 0.2$ m and $\theta = 0$. (This problem could also be solved using polar coordinates—see Prob. 13.33).

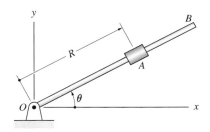

Fig. P16.83

16.84 The triangular frames $\mathscr{B}$ in Figs. (a) and (b) rotate about *A* with a constant angular velocity of 2 rad/s. At the same time, the slider *P* is moving to the right at the constant speed of 0.2 m/s relative to the frame. Determine the acceleration of *P* in the positions shown.

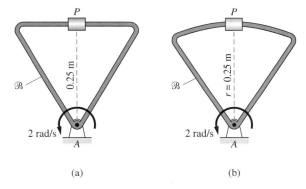

(a)   (b)

Fig. P16.84

16.85 Rod *OAB* is rotating counterclockwise with the constant angular velocity $\omega = 5$ rad/s. In the position shown, collar *P* is sliding toward *A* with the speed of 0.8 m/s, increasing at the rate of 8 m/s², both measured relative to the rod. Determine the acceleration of *P* in this position.

16.86 In the position shown, the slotted plate $\mathscr{B}$ is rotating about pin *A* with the angular velocity $\omega = 3$ rad/s CCW and the angular acceleration $\alpha = 6$ rad/s² CW. The slider *P* moves along the slot at the constant speed of 0.7 m/s relative to the plate, in the direction indicated. Compute the velocity and acceleration vectors of *P* at this instant.

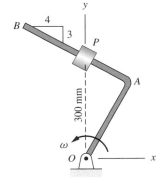

Fig. P16.85

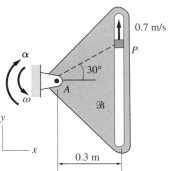

Fig. P16.86

16.87   The telescoping rod *AB* is being extended at the constant rate of 2 m/s as it rotates about *A* with the constant counterclockwise angular velocity ω. At the instant shown, the velocity vector of end *B* is directed as indicated. Determine ω and the acceleration vector of *B* at this instant.

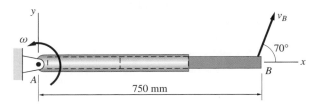

Fig. P16.87

16.88   Collar *P* slides along the semicircular guide rod. A pin attached to the collar engages the slot in the rotating arm *AB*. When $\theta = 45°$, the angular velocity and angular acceleration of *AB* are $\omega = 4$ rad/s and $\alpha = 12$ rad/s^2 with both counterclockwise. Determine the speed and the acceleration vector of the collar *P* at this instant.

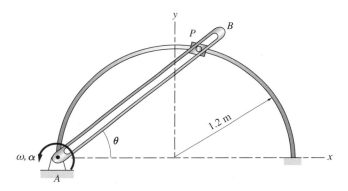

Fig. P16.88

16.89   The pin *P*, attached to the sliding rod *PD*, engages a slot in the rotating arm *AB*. Rod *PD* is sliding to the left with the constant velocity of 1.2 m/s. Determine the angular velocity and angular acceleration of *AB* when $\theta = 60°$.

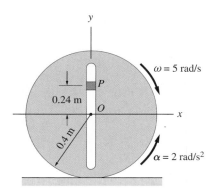

Fig. P16.90

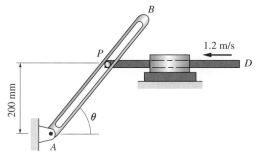

Fig. P16.89

16.90   The slotted disk rolls without slipping. In the position shown, the angular velocity of the disk is 5 rad/s clockwise, and the angular acceleration is 2 rad/s^2

counterclockwise. In the same position, the velocity and acceleration of slider P relative to the wheel are 2 m/s and 10 m/s^2, respectively, both directed downward. Find the acceleration of P in this position.

*16.91   End A of bar ABD is being pushed to the right with the constant speed $v_A = 1$ m/s. When $\theta = 30°$, determine (a) the angular velocity of the bar; and (b) the components of the acceleration vector of point B that are parallel and perpendicular to the bar.

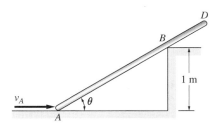

Fig. P16.91

16.92   The collar C is pushed along the horizontal bar by the pin P that slides in the slotted arm AB. The arm is rotating counterclockwise with the constant angular velocity $\omega = 4$ rad/s. In the position shown, determine (a) velocity of P relative to AB; and (b) acceleration of P relative to AB.

16.93   Crank AD rotates with the constant clockwise angular velocity of 8 rad/s. For the position shown, determine the angular speed of rod BE and the velocity of slider D relative to BE.

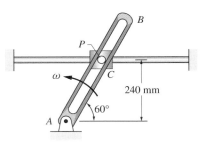

Fig. P16.92

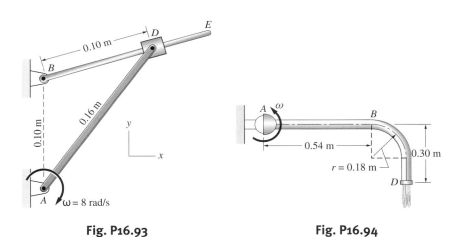

Fig. P16.93   **Fig. P16.94**

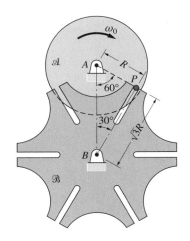

Fig. P16.95

16.94   Water entering the curved pipe at A is discharged at D. The pipe is rotating about A at the constant angular velocity $\omega = 10$ rad/s, and the water has a constant speed of 6 m/s relative to the pipe. Determine the acceleration of the water (a) just after it enters the bend at B; and (b) just before it is discharged at D.

16.95   The figure shows a mechanism, called the Geneva stop, which converts the constant angular velocity of disk $\mathcal{A}$ into stop-and-go motion of the slotted disk $\mathcal{B}$. In the position shown, the pin P, which is attached to $\mathcal{A}$, is just entering a slot in disk $\mathcal{B}$. Compute the angular acceleration of $\mathcal{B}$ for this position. (Note that the angular velocity of $\mathcal{B}$ is zero at this instant.)

16.96   Arm AB is rotating counterclockwise with the constant angular speed of 4 rad/s. At the same time, the disk is rotating clockwise with the angular speed 8 rad/s relative to AB. Determine the acceleration of point P on the rim of the disk by (a) considering AB as a rotating reference frame; and (b) using the relative acceleration method of Art. 16.7.

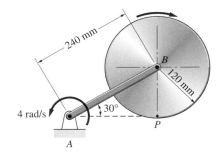

Fig. P16.96

16.97   The disk is rolling without slipping on the horizontal surface with the constant counterclockwise angular velocity of 2 rad/s. The pin *P*, attached to the rim of the disk, engages the slot in the rotating arm *AB*. Calculate the angular velocity and angular acceleration of *AB* when the system is in the position shown.

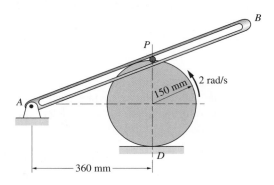

Fig. P16.97

16.98   Rod *AB* of the mechanism rotates at the constant angular speed 8 rad/s clockwise. For the position shown, calculate the angular velocity of rod *BE*.

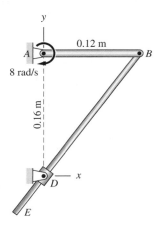

Fig. P16.98

*16.10   *Method of Constraints*

In Art. 15.3, we showed how equations of constraint and their time derivatives can be used to solve problems of particle kinematics. This technique is equally applicable to kinematics of rigid bodies. Let us begin by extending the terminology introduced in Art. 15.3 to rigid bodies:

- *Kinematic constraints:* geometric restrictions imposed on the motion of points in bodies.
- *Equations of constraint:* mathematical expressions that describe the kinematic constraints in terms of position coordinates.

- *Kinematically independent coordinates:* position coordinates that are not subject to kinematic constraints.
- *Number of degrees of freedom:* the number of kinematically independent coordinates that are required to completely describe the configuration of a body or a system of bodies.

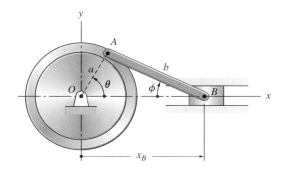

Fig. 16.18

The term "position coordinate" may refer to the coordinate of a point or to the angular position coordinate of a line. As an example, consider the system shown in Fig. 16.18. The figure displays three position coordinates:

1. The angle θ is the angular position coordinate of the line OA on the disk.
2. The angle ϕ is the angular position coordinate of the connecting rod AB.
3. The distance x_B is the rectilinear position coordinate of the slider B.

Because point A is common to the disk and the connecting rod, the x- and y-coordinates of A on the disk are equal to the coordinates of A on the connecting rod; that is,

$$a \cos \theta = x_B - b \cos \phi \qquad \text{(a)}$$

$$a \sin \theta = b \sin \phi \qquad \text{(b)}$$

Equations (a) and (b) are the equations of constraint for the system. Note that due to the connection at A, the *system* has only a single degree of freedom; that is, only one of the coordinates (θ, ϕ, or x_B) is kinematically independent. For example, if θ is given, the equations of constraint determine ϕ and x_B.

The relationships between the velocities is obtained by differentiating the equations of constraint with respect to time:

$$-a \sin \theta \cdot \dot{\theta} = \dot{x}_B + b \sin \phi \cdot \dot{\phi} \qquad \text{(c)}$$

$$a \cos \theta \cdot \dot{\theta} = b \cos \phi \cdot \dot{\phi} \qquad \text{(d)}$$

where $\dot{\theta}$ and $\dot{\phi}$ are the angular velocities of the disk and the connecting rod, respectively, and $\dot{x}_B$ is the speed of point B. If the accelerations are needed, Eqs. (c) and (d) can be differentiated again with respect to time.

Sample Problem 16.15

The angular position of bar AB is controlled by the sliding rod CD. If the constant velocity of CD is v_0 in the direction shown, determine the angular velocity and the angular acceleration of AB as functions of the angle θ.

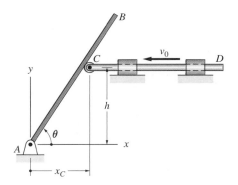

Solution

The figure shows two position coordinates: the angular position coordinate θ of bar AB and the coordinate x_C of point C in the sliding rod. However, because C is always in contact with bar AB (see the figure), the two coordinates are subject to the constraint

$$x_C = h \cot \theta \qquad \text{(a)}$$

Hence the system has only one degree of freedom.

Differentiation of Eq. (a) with respect to time yields

$$\dot{x}_C = -h\dot{\theta} \, \text{cosec}^2\theta$$

Substituting $\dot{x}_C = -v_0$ and solving for $\dot{\theta}$, we obtain for the angular velocity of bar AB

$$\omega = \dot{\theta} = \frac{v_0}{h} \sin^2 \theta \qquad \qquad \textit{Answer}$$

The angular acceleration of AB is obtained by differentiating its angular velocity:

$$\alpha = \ddot{\theta} = \frac{v_0}{h}(2 \sin \theta \cos \theta)\dot{\theta}$$

Substitution for $\dot{\theta}$ yields

$$\alpha = \frac{v_0}{h}(2 \sin \theta \cos \theta) \left(\frac{v_0}{h} \sin^2 \theta \right) = \frac{2v_0^2}{h^2} \sin^3 \theta \cos \theta \qquad \textit{Answer}$$

Sample Problem 16.16

Bar AB is rotating clockwise with the constant angular speed of 3 rad/s. When $\theta = 30°$, determine (1) the angular velocity and angular acceleration of bar BC; and (2) the velocity and acceleration of sliding collar C. (Note: This problem appeared previously as Sample Problems 16.5, 16.7, and 16.11.)

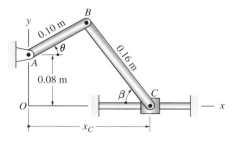

Solution

Preliminary Calculations

The positions of bars AB and BC are defined by the three coordinates (θ, β, and x_C) shown in the figure. However, the mechanism has a single degree of freedom, because there are two equations of constraints due to the pin connection at B. Equating the coordinates of point B in bar AB and in bar BC, we obtain

$$0.10 \cos \theta = x_C - 0.16 \cos \beta \tag{a}$$

$$0.08 + 0.10 \sin \theta = 0.16 \sin \beta \tag{b}$$

When $\theta = 30°$, Eq. (b) yields

$$\beta = \sin^{-1} \frac{0.08 + 0.10 \sin 30}{0.16} = 54.34°$$

Part 1

Differentiating Eq. (b) with respect to time, we get

$$0.10 \cos \theta \cdot \dot{\theta} = 0.16 \cos \beta \cdot \dot{\beta} \tag{c}$$

or

$$\dot{\beta} = \frac{5 \cos \theta}{8 \cos \beta} \dot{\theta}$$

Substituting $\theta = 30°$, $\beta = 54.34°$, and $\dot{\theta} = -3$ rad/s (minus indicates that θ and $\dot{\theta}$ have opposite sense), we arrive at

$$\dot{\beta} = \frac{5 \cos 30°}{8 \cos 54.34°} (-3) = -2.785 \text{ rad/s}$$

Hence the angular velocity of bar BC is

$$\omega_{BC} = 2.79 \text{ rad/s (CCW)} \qquad \qquad \textit{Answer}$$

Differentiation of Eq. (c) with respect to time yields

$$0.10(\cos\theta \cdot \ddot{\theta} - \sin\theta \cdot \dot{\theta}^2) = 0.16(\cos\beta \cdot \ddot{\beta} - \sin\beta \cdot \dot{\beta}^2)$$

Therefore,

$$\ddot{\beta} = \frac{0.10(\cos\theta \cdot \ddot{\theta} - \sin\theta \cdot \dot{\theta}^2) + 0.16\sin\beta \cdot \dot{\beta}^2}{0.16\cos\beta}$$

At $\theta = 30°$, we have

$$\ddot{\beta} = \frac{0.10[0 - \sin 30°(-3)^2] + 0.16\sin 54.34°(-2.785)^2}{0.16\cos 54.34°} = 5.985 \text{ rad/s}^2$$

Thus the angular acceleration of BC is

$$\alpha_{BC} = 5.99 \text{ rad/s}^2 \text{ (CW)} \qquad \qquad \textit{Answer}$$

Part 2

Upon differentiation with respect to time, Eq. (a) becomes

$$-0.10\sin\theta \cdot \dot{\theta} = \dot{x}_C + 0.16\sin\beta \cdot \dot{\beta}$$

Solving for $\dot{x}_C$ yields

$$\dot{x}_C = -(0.10\sin\theta \cdot \dot{\theta} + 0.16\sin\beta \cdot \dot{\beta}) \qquad \qquad \text{(d)}$$

At $\theta = 30°$, this becomes

$$\dot{x}_C = -[(0.10\sin 30°)(-3) + (0.16\sin 54.34°)(-2.785)] = 0.512 \text{ m/s}$$

Therefore, the velocity of C is

$$v_C = 0.512 \text{ m/s} \rightarrow \qquad \qquad \textit{Answer}$$

The acceleration of C is obtained by differentiating Eq. (d):

$$\ddot{x}_C = -0.10(\sin\theta \cdot \ddot{\theta} + \cos\theta \cdot \dot{\theta}^2) - 0.16(\sin\beta \cdot \ddot{\beta} + \cos\beta \cdot \dot{\beta}^2)$$

which yields, at $\theta = 30°$,

$$\ddot{x}_C = -0.10[0 + \cos 30°(-3)^2] - 0.16[\sin 54.34°(5.985) + \cos 54.34°(-2.785)^2]$$
$$= -2.28 \text{ m/s}^2$$

Therefore, the acceleration of C is

$$a_C = 2.28 \text{ m/s}^2 \leftarrow \qquad \qquad \textit{Answer}$$

Problems

Note: The following problems are to be solved by differentiating the equation of constraint.

16.99   When the crank *AB* of the scotch yoke is in the position $\theta = 40°$, its angular velocity is 8 rad/s, and its angular acceleration is 140 rad/s², both clockwise. Calculate the velocity and acceleration of the sliding rod *D* in this position.

16.100   Collar *B* is sliding to the right at the constant speed of 1.4 m/s. When bar *AB* is in the position $\theta = 20°$, determine (a) the velocity of collar *A*; and (b) the acceleration of *A*. The length of the bar is $L = 1.8$ m.

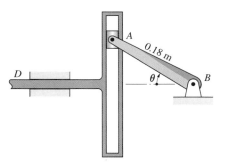

Fig. P16.99

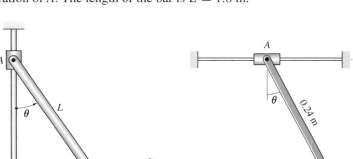

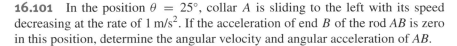

Fig. P16.100   **Fig. P16.101**

16.101   In the position $\theta = 25°$, collar *A* is sliding to the left with its speed decreasing at the rate of 1 m/s². If the acceleration of end *B* of the rod *AB* is zero in this position, determine the angular velocity and angular acceleration of *AB*.

16.102   When $\theta = 60°$, rod *AB* of the mechanism is sliding to the left at the speed of 1.2 m/s. Find the angular velocity of bar *CD* in this position.

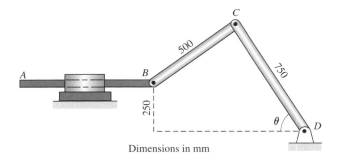

Dimensions in mm

Fig. P16.102

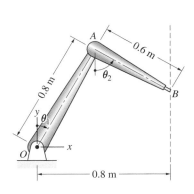

16.103   The link *OA* of the robot's arm is rotating clockwise with the constant angular velocity $\dot{\theta}_1 = 0.8$ rad/s. At the same time, end *B* of arm *AB* is tracing the vertical line $x = 0.8$ m. Determine the angular velocity and acceleration of link *AB* when $\theta_1 = 30°$. Assume that $\theta_2 < 90°$.

Fig. P16.103

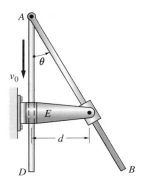

Fig. P16.104

16.104   The rod *AD* is sliding in the fixed collar *E* at the constant speed v_0. Find (a) the angular velocity; and (b) the angular acceleration of bar *AB* as functions of the angle θ.

16.105   The rod *AD* is sliding in the fixed collar *E* at the constant speed v_0. Determine (a) the angular velocity; and (b) the angular acceleration of bar *AB* as functions of the angle θ. Note that the axis of the disk is attached to the collar *E*.

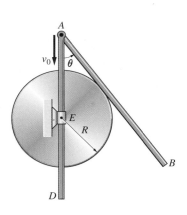

Fig. P16.105

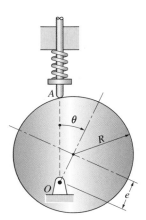

Fig. P16.106, P16.107

16.106   The circular cam has a radius *R* and an eccentricity $e = R$. The follower *A* is kept in contact with the surface of the cam by a compression spring. Assuming that the cam starts from rest at $\theta = 0$ and accelerates at the constant rate $\ddot{\theta} = \alpha_0$, find the acceleration of the follower as a function of θ.

16.107   The radius of the circular cam is $R = 100$ mm and its eccentricity is $e = 60$ mm. If the angular speed of the cam is 1000 rev/min, calculate the velocity of the follower *A* when $\theta = 60°$.

*16.108   The free end of the rope attached to bar *AB* is being pulled down at the rate of 1 m/s. Find the angular velocity of *AB* when $\theta = 20°$.

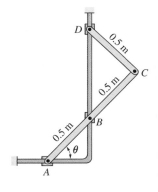

Fig. P16.109

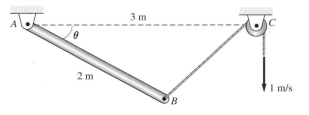

Fig. P16.108

16.109   At the instant when $\theta = 30°$, collar *D* is sliding upward with a constant velocity of 3 m/s. Determine the angular velocity and angular acceleration of bar *ABC* at this instant.

16.110   Rod BC slides in the pivoted sleeve D as bar AB is rotating at the constant angular velocity $\dot{\theta}_1 = 12$ rad/s. Determine the angular velocity of rod BC in the *two positions* where $\theta_2 = 30°$.

16.111   Collar C slides on the horizontal guide rod with the constant velocity v_0. The rod CD is free to slide in sleeve B, which is rigidly attached to bar AB. Determine the angular velocity and angular acceleration of bar AB in terms of v_0, b, and θ. (Hint: $\omega_{AB} = \omega_{CD}$.)

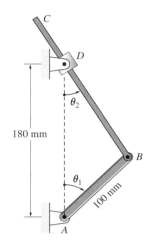

Fig. P16.110

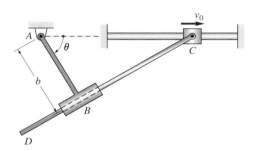

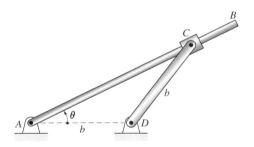

Fig. P16.111

16.112   The gear of radius R rolls on the horizontal rack. Pin G in the center of the gear engages a slot in the arm AB, which rotates at the constant angular velocity $\omega = \dot{\theta}$. Determine the angular acceleration of the gear when $\theta = 50°$.

16.113   For the mechanism shown, determine the speed and the magnitude of the acceleration of collar C in terms of b, θ, $\dot{\theta}$, and $\ddot{\theta}$.

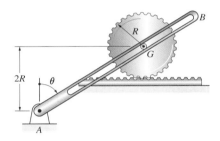

Fig. P16.112

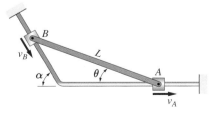

Fig. P16.113

16.114   If the velocity of slider A is constant, derive the expressions for (a) the velocity; and (b) the acceleration of slider B in terms of the angle θ.

Fig. P16.114

Review of Equations

Plane angular motion of a rigid body

$$\omega = \frac{d\theta}{dt} \quad \alpha = \frac{d\omega}{dt} = \frac{d^2\theta}{dt^2} = \omega\frac{d\omega}{d\theta}$$

Rotation about a fixed axis

The velocity and acceleration components of a point in a body are:

$$v = R\omega \qquad a_n = R\omega^2 = \frac{v^2}{R} = v\omega \qquad a_t = R\alpha$$

$$\mathbf{v} = \boldsymbol{\omega} \times \mathbf{r} \quad \mathbf{a}_n = \boldsymbol{\omega} \times \mathbf{v} = \boldsymbol{\omega} \times (\boldsymbol{\omega} \times \mathbf{r}) \quad \mathbf{a}_t = \boldsymbol{\alpha} \times \mathbf{r}$$

R = radial distance from the axis
$\mathbf{r}$ = vector from any point on the axis to the point in the body

Relative motion

Points A and B are in the same rigid body. The relative velocity and acceleration are:

$$v_{B/A} = r_{B/A}\omega \qquad (a_{B/A})_n = r_{B/A}\omega^2 = \frac{v_{B/A}^2}{r_{B/A}} \qquad (a_{B/A})_t = r_{B/A}\alpha$$

$$\mathbf{v}_{B/A} = \boldsymbol{\omega} \times \mathbf{r}_{B/A} \quad (\mathbf{a}_{B/A})_n = \boldsymbol{\omega} \times \mathbf{v}_{B/A} = \boldsymbol{\omega} \times (\boldsymbol{\omega} \times \mathbf{r}_{B/A})$$

$$(\mathbf{a}_{B/A})_t = \boldsymbol{\alpha} \times \mathbf{r}_{B/A}$$

Rolling without slipping

The velocity and acceleration of the center O of a disk are:

$$v_o = R\omega \quad a_o = R\alpha$$

R = radius of the disk

Motion relative to a rotating reference frame

Reference frame is embedded in body $\mathcal{B}$ that rotates with angular velocity $\boldsymbol{\omega}$ and angular acceleration $\dot{\boldsymbol{\omega}}$. The velocity and acceleration of point P (not necessarily in the body) are

$$\mathbf{v}_P = \mathbf{v}_A + \boldsymbol{\omega} \times \mathbf{r}_{P/A} + \mathbf{v}_{P/\mathcal{B}}$$

$$\mathbf{a}_P = \mathbf{a}_A + \left[\dot{\boldsymbol{\omega}} \times \mathbf{r}_{P/A} + \boldsymbol{\omega} \times (\boldsymbol{\omega} \times \mathbf{r}_{P/A})\right] + \mathbf{a}_{P/\mathcal{B}} + 2\boldsymbol{\omega} \times \mathbf{v}_{P/\mathcal{B}}$$

Review Problems

16.115   In the position shown, velocities of corners A and B of the plate are $\mathbf{v}_A = 2.38\mathbf{j} - 1.0\mathbf{k}$ m/s and $\mathbf{v}_B = u\mathbf{i} + 1.73\mathbf{k}$ m/s, where u is an unknown. Knowing that the plate is rotating at a constant angular velocity about an axis that passes through O, determine (a) the angular velocity of the plate; and (b) the acceleration of corner A.

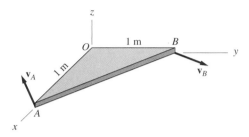

Fig. P16.115

16.116   The angular acceleration of the body undergoing plane motion is $\alpha = 6t^2 + k$ rad/s^2, where t is in seconds and k is a constant. When $t = 0$, the angular position coordinate of the body is $\theta = -4$ rad, and its angular velocity is $\omega = 6$ rad/s. When $t = 2$ s, the angular position coordinate is $\theta = -8$ rad. Determine the angular acceleration of the body when $t = 3$ s.

16.117   Points A and B are fixed in the disk that is rotating about the pin at O. At the instant shown, the acceleration of A is $\mathbf{a}_A = -6.28\mathbf{i} - 0.72\mathbf{j}$ m/s^2. Determine the acceleration vector of B at this instant.

16.118   When $\theta = 30°$, the angular velocity and angular acceleration of the arm AB are $\dot{\theta} = 2$ rad/s and $\ddot{\theta} = -1.5$ rad/s^2. Compute the velocity and acceleration vectors of point C for this position.

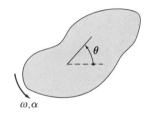

Fig. P16.116

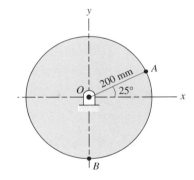

Fig. P16.117

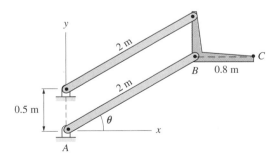

Fig. P16.118

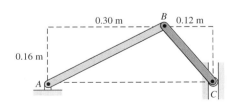

Fig. P16.119

16.119   The roller C is moving up the slot with the constant speed of 1.2 m/s. Find the angular velocities of bars AB and BC in the position shown.

16.120 The bent rod *ABC* rotates about the axis *AB*. In the position shown, the angular velocity and acceleration of the rod are $\omega = 6$ rad/s and $\dot{\omega} = -25$ rad/s². Determine the velocity and acceleration of end *C* in this position.

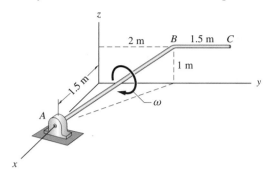

Fig. P16.120

16.121 Knowing that roller *D* is moving to the left at the constant speed of 2 m/s, find the velocity of roller *A* in the position shown.

16.122 The disk is rotating clockwise with the constant angular velocity of 2 rad/s. Determine the angular accelerations of bars *AB* and *BC* in the position shown.

16.123 Bar *AB* of the mechanism rotates with constant angular velocity of 4.5 rad/s. Determine the angular acceleration of bar *BC* and the acceleration of slider *C* in the position shown.

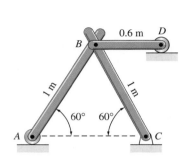

Fig. P16.121

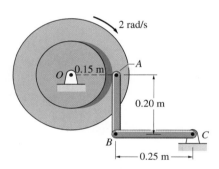

Fig. P16.122

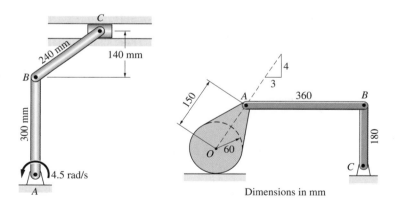

Fig. P16.123   **Fig. P16.124**

16.124 The rocker *OA* rolls without slipping on the horizontal surface. In the position shown, the angular velocity of the rocker is 7 rad/s clockwise. Determine the angular velocity of link *BC* in this position.

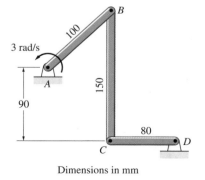

Dimensions in mm

Fig. P16.125

16.125 Bar *AB* is rotating counterclockwise with a constant angular velocity of 3 rad/s. (a) Verify that in the position shown, the angular velocities of bars *BC* and *CD* are 1.2 rad/s (CCW) and 3 rad/s (CW), respectively. (b) Determine the angular accelerations of bars *BC* and *CD* in the same position.

16.126 Bar *AB* is rotating clockwise with the constant angular velocity of 72 rad/s. When the mechanism is the position shown, determine the angular velocity of the plate *BCD* and the magnitude of the velocity of corner *D*.

16.127 In the position shown, the bent rod *AB* is rotating about *A* with the angular velocity $\omega = 6$ rad/s and the angular acceleration $\dot\omega = 18$ rad/s². At the same time, collar *D* is sliding on the rod with the velocity $v = 1$ m/s and acceleration $\dot v = -1.5$ m/s² (measured relative to the rod). Determine the acceleration of collar *D* in this position.

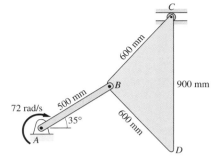

Fig. P16.126

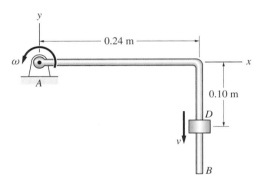

Fig. P16.127

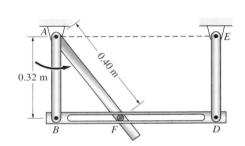

Fig. P16.128

16.128 The pin *F*, which is attached to the rod *AF*, engages a slot in bar *BD* of the parallelogram linkage *ABDE*. Bar *AB* of the linkage has a constant angular velocity of 15 rad/s counterclockwise. For the position shown, determine the angular acceleration of rod *AF* and the acceleration of pin *F* relative to bar *BD*.

16.129 Bar *AB* of the linkage rotates with the constant angular velocity of 10 rad/s. For the position shown, determine (a) the angular velocities of bars *BC* and *CD*; and (b) the angular accelerations of bars *BC* and *CD*.

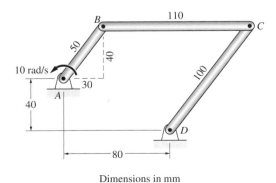

Dimensions in mm

Fig. P16.129

16.130   In the position shown, the angular velocity and angular acceleration of the rod *AB* are 6 rad/s and 8 rad/s², respectively, both counterclockwise. For this position, calculate (a) the velocity of collar *A*; and (b) the acceleration of collar *A*.

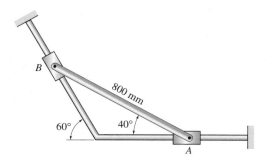

Fig. P16.130

16.131   The curved, slender bar *OC* rotates about *O*. At the instant shown, the angular velocity of *OC* is 2 rad/s and its angular acceleration is zero. Find the angular acceleration of bar *AB* at this instant.

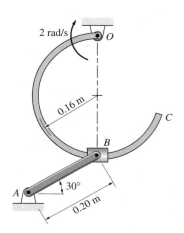

Fig. P16.131

17

Planar Kinetics of Rigid Bodies: Force-Mass-Acceleration Method

If the acceleration is sufficiently large, the front wheel of a motorcycle will lift off the ground, resulting in a "wheelie." This problem is investigated in Prob. 17.18. (© iStockphoto.com/Bernhard Weber)

17.1 Introduction

This chapter presents the force-mass-acceleration (FMA) method of kinetic analysis for rigid bodies in plane motion. The equations of motion, which are the basis of the method, can be obtained from the results for particle systems in Chapter 15 by viewing a rigid body as a collection of particles where the distances between the particles remain constant. The resulting equations are known as *Euler's laws of motion*. The first law governs the motion of the mass center of the body; it is identical to the equation of motion of the mass center of a particle system. The second law, which governs the rotational motion of the body, is the moment–angular momentum relationship for a system of particles.

When this law is specialized for a rigid body, it gives rise to the concept of *mass moment of inertia,* which is the topic of the next article. The subsequent articles derive the expression for the angular momentum of a rigid body and discuss the equations of motion and their application using free-body and mass-acceleration diagrams. At the conclusion of the chapter we show how the history of motion can be determined by integrating the equations of motion.

17.2   *Mass Moment of Inertia; Composite Bodies*

In this article we introduce the mass moment of inertia of a body about an axis. A comprehensive discussion of mass moment of inertia, including its computation by integration, is contained in Appendix F.

a.   *Mass moment of inertia*

Figure 17.1 shows a body of mass m that occupies the region $\mathcal{V}$; r is the perpendicular distance from the a-axis to the differential mass dm of the body. The mass moment of inertia of the body about the a-axis is defined as

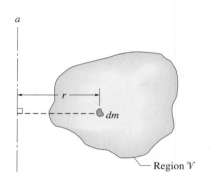

Fig. 17.1

$$I_a = \int_{\mathcal{V}} r^2\, dm \tag{17.1}$$

It will be seen shortly that this integral is a measure of the ability of the body to resist a change in its angular motion about the a-axis, just as the mass of the body is a measure of its ability to resist a change in its translational motion.

From its definition we see that mass moment of inertia is a positive quantity with the dimension $[ML^2]$. In SI units, I_a is measured in kg $\cdot$ m^2.

b.   *Radius of gyration*

The radius of gyration k_a of the body about the a-axis is defined as

$$I_a = mk_a^2 \quad \text{or} \quad k_a = \sqrt{\frac{I_a}{m}} \tag{17.2}$$

Although the unit of radius of gyration is length (e.g., feet, meters), it is not a distance that can be measured physically. Instead, its value can be found only by computation using Eq. (17.2). The radius of gyration allows us to compare the rotational resistances of bodies that have the same mass.

c.   *Parallel-axis theorem*

Consider the two parallel axes shown in Fig. 17.2. The location of the *a*-axis is arbitrary. We call the other axis, which passes through the mass center *G* of the body, the *central a-axis.*[*]

With *d* being the distance between the two axes, the parallel-axis theorem states that

$$I_a = \bar{I}_a + md^2 \qquad (17.3)$$

where *m* is the mass of the body, I_a is the moment of inertia of the body about the *a*-axis, and $\bar{I}_a$ is its moment of inertia about the central *a*-axis. Note that if the moment of inertia about a central axis is known, the parallel-axis theorem can be used to calculate the moment of inertia about any parallel axis without resorting to integration. Table 17.1 lists the moments of inertia about central axes for a few homogeneous bodies.

Caution   The parallel-axis theorem applies only if one of the axes is a *central* axis.

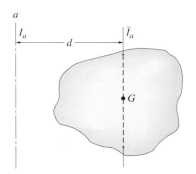

Fig. 17.2

Proof

To prove the parallel-axis theorem, consider the body of mass *m* occupying the region $\mathcal{V}$ that is shown in Fig. 17.3. The origin *O* of the *xyz*-axes is located arbitrarily, but the $x'y'z'$-axes are central axes that are parallel to the *xyz*-axes. The coordinates of the mass center *G* relative to the *xyz*-axes are denoted by $\bar{x}$, $\bar{y}$, and $\bar{z}$. Let *dm* be a differential mass element of the body that is located at *P*. Because the perpendicular distance from the *z*-axis to *P* is $r = (x^2 + y^2)^{1/2}$, the moment of inertia of the body about the *z*-axis is

$$I_z = \int_{\mathcal{V}} r^2 \, dm = \int_{\mathcal{V}} (x^2 + y^2) \, dm \qquad (a)$$

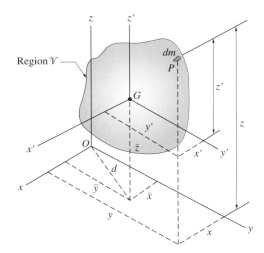

Fig. 17.3

---

[*]We will refer to any axis that passes through the mass center as a *central axis*. The *central a-axis* is the central axis that is parallel to the *a*-axis.

Slender rod (A = cross-sectional area)	Cylinder
$m = AL\rho$	$m = \pi R^2 h\rho$
$\bar{I}_x = \bar{I}_y = \frac{1}{12}mL^2 \quad \bar{I}_z \approx 0$	$\bar{I}_x = \bar{I}_y = \frac{1}{12}m(3R^2 + h^2) \quad \bar{I}_z = \frac{1}{2}mR^2$
Thin ring (A = cross-sectional area)	Sphere
$m = 2\pi AR\rho$	$m = \frac{4\pi}{3}R^3\rho$
$\bar{I}_x = \bar{I}_y = \frac{1}{2}mR^2 \quad \bar{I}_z = mR^2$	$\bar{I}_x = \bar{I}_y = \bar{I}_z = \frac{2}{5}mR^2$
Rectangular prism	Cone
$m = abc\rho$	$m = \frac{\pi}{3}R^2 h\rho$
$\bar{I}_x = \frac{1}{12}m(b^2 + c^2)$ $\bar{I}_y = \frac{1}{12}m(c^2 + a^2)$ $\bar{I}_z = \frac{1}{12}m(a^2 + b^2)$	$\bar{I}_x = \bar{I}_y = \frac{3}{80}m(4R^2 + h^2)$ $\bar{I}_z = \frac{3}{10}mR^2$

Table 17.1 Mass Moments of Inertia of Homogeneous Bodies
(ρ = mass density)

Substituting $x = x' + \bar{x}$ and $y = y' + \bar{y}$ yields

$$I_z = \int_V \left[(x' + \bar{x})^2 + (y' + \bar{y})^2 \right] dm \qquad \text{(b)}$$

Expanding and rearranging terms, we obtain

$$I_z = \int_V (x'^2 + y'^2)\, dm + \int_V (\bar{x}^2 + \bar{y}^2)\, dm + 2\bar{x} \int_V x'\, dm + 2\bar{y} \int_V y'\, dm \qquad \text{(c)}$$

Consider now each of the integrals that appear in this equation. The first integral is equal to $I_{z'}$, the moment of inertia about the z'-axis. Because z' is a central axis (passes through G), this term may be written as $\bar{I}_z$. Letting $d = (\bar{x}^2 + \bar{y}^2)^{1/2}$, the distance between the z- and z'-axes, the second integral in Eq. (c) equals md^2. The last two integrals in Eq. (c) vanish, because $\int_V x'\, dm = 0$ and $\int_V y'\, dm = 0$ when the y'- and x'-axes are central axes. Therefore, Eq. (c) becomes

$$I_z = \bar{I}_z + md^2 \qquad \text{(d)}$$

Because the z-axis can be chosen arbitrarily, comparison of Eqs. (17.3) and (d) shows that the parallel-axis theorem has been proved.

d.   *Method of composite bodies*

From Eq. (17.1) we see that the computation of the mass moment of inertia requires that an integration be performed over the body. The various integration techniques are discussed in Appendix F. Here we will consider only the *method of composite bodies*, a method that follows directly from the property of definite integrals:

The integral of a sum is equal to the sum of the integrals.[*]

Using this property, it can be shown that if a body is divided into composite parts, the moment of inertia of the body about a given axis equals the sum of the moments of inertia of its parts about that axis. The following sample problems illustrate the application of this method.

---

[*]This property also formed the basis for the method of composite shapes discussed in Chapter 8.

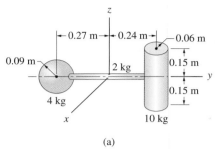

(a)

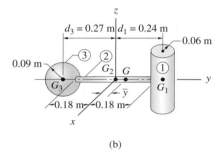

(b)

Sample Problem   17.1

The assembly in Fig. (a) is composed of three homogeneous bodies: the 10-kg cylinder, the 2-kg slender rod, and the 4-kg sphere. For this assembly, calculate (1) I_x, the mass moment of inertia about the x-axis; and (2) $\bar{I}_x$ and $\bar{k}_x$, the mass moment of inertia and radius of gyration about the central x-axis of the assembly.

Solution

The mass centers of the cylinder (G_1), the rod (G_2), and the sphere (G_3) are shown in Fig. (b). By symmetry, the mass center of the assembly (G) lies on the y-axis, with its coordinate $\bar{y}$ to be determined.

Part 1

Cylinder   Using Table 17.1, the moment of inertia of the cylinder about its own central x-axis is

$$(\bar{I}_x)_1 = \frac{1}{12}m_1(3R^2 + h^2) = \frac{1}{12}(10)\left[3(0.06)^2 + (0.3)^2\right]$$

$$= 0.084 \text{ kg} \cdot \text{m}^2$$

Utilizing the parallel-axis theorem, the moment of inertia of the cylinder about the x-axis becomes

$$(I_x)_1 = (\bar{I}_x)_1 + m_1 d_1^2$$

$$= 0.084 + 10(0.24)^2 = 0.66 \text{ kg} \cdot \text{m}^2$$

Slender Rod   Because G_2 coincides with the origin of the xyz-axes, the moment of inertia of the rod about the x-axis is obtained directly from Table 17.1.

$$(I_x)_2 = (\bar{I}_x)_2 = \frac{1}{12}mL^2$$

$$= \frac{1}{12}(2)(0.36)^2 = 0.0216 \text{ kg} \cdot \text{m}^2$$

Sphere   According to Table 17.1, the moment of inertia of the sphere about its own central x-axis is

$$(\bar{I}_x)_3 = \frac{2}{5}mR^2 = \frac{2}{5}(4)(0.09)^2 = 0.01296 \text{ kg} \cdot \text{m}^2$$

Using the parallel-axis theorem, the moment of inertia about the x-axis is given by

$$(I_x)_3 = (\bar{I}_x)_3 + m_3 d_3^2$$

$$= 0.01296 + 4(0.27)^2 = 0.30456 \text{ kg} \cdot \text{m}^2$$

Assembly The moment of inertia of the assembly about an axis equals the sum of the moments of inertia of its parts about that axis. Therefore, adding the values that were found above, we get

$$I_x = (I_x)_1 + (I_x)_2 + (I_x)_3$$

$$= 0.66 + 0.0216 + 0.30456$$

$$= 0.98616 \text{ kg} \cdot \text{m}^2 \qquad\qquad \textit{Answer}$$

Part 2

Referring to Fig. (b), the y-coordinate of G is

$$\bar{y} = \frac{\Sigma_i m_i y_i}{\Sigma_i m_i} = \frac{10(0.24) + 2(0) - 4(0.27)}{10 + 2 + 4}$$

$$= \frac{1.32}{16} = 0.0825 \text{ m}$$

Because $\bar{y}$ is the distance between the x-axis and the central x-axis of the assembly, the moment of inertia of the assembly about the latter axis is found from the parallel-axis theorem:

$$\bar{I}_x = I_x - m\bar{y}^2 = 0.98616 - 16(0.0825)^2$$

$$= 0.877 \text{ kg} \cdot \text{m}^2 \qquad\qquad \textit{Answer}$$

The corresponding radius of gyration is

$$\bar{k}_x = \sqrt{\frac{\bar{I}_x}{m}} = \sqrt{\frac{0.877}{16}} = 0.234 \text{ m} \qquad\qquad \textit{Answer}$$

Alternative Method for Computing $\bar{I}_x$

In the preceding solution, we first computed I_x for the assembly by summing the values of I_x for each part. Then $\bar{I}_x$ for the assembly was found by applying the parallel-axis theorem. An alternative method for computing $\bar{I}_x$ for the assembly is to first compute the moments of inertia for each part about the central x-axis of the assembly, and then sum these values. Using this approach, we obtain

$$\bar{I}_x = \left[(\bar{I}_x)_1 + m_1(d_1 - \bar{y})^2\right] + \left[(\bar{I}_x)_2 + m_2\bar{y}^2\right] + \left[(\bar{I}_x)_3 + m(d_3 + \bar{y})^2\right]$$

In this equation, note that $\bar{I}_x$ is the moment of inertia for the entire assembly about the central x-axis of the assembly (axis passing through G), whereas $(\bar{I}_x)_1$,

$(\bar{I}_x)_2$, and $(\bar{I}_x)_3$ are the moments of inertia of the parts about their own central axes. The distances between G and the mass center of each part, namely $(d_1 - \bar{y})$, $\bar{y}$, and $(d_3 + \bar{y})$, are found from Fig. (b). Substituting the numerical values, we obtain

$$I_x = \left[.084 + 10(0.24 - 0.0825)^2\right]$$
$$+ \left[0.0216 + 2(0.0825)^2\right] + \left[0.01296 + 4(0.27 + 0.0825)^2\right]$$
$$= 0.332 + 0.0352 + 0.5099 = 0.877 \text{ kg} \cdot \text{m}^2 \qquad \textit{Answer}$$

which agrees with the result found previously.

Sample Problem 17.2

The 290-kg machine part in Fig. (a) is made by drilling an off-center, 160-mm diameter hole through a homogeneous, 400-mm cylinder of length 350 mm. Determine (1) I_z (the mass moment of inertia of the part about the z-axis); and (2) $\bar{k}_z$ (the radius of gyration of the part about its central z-axis).

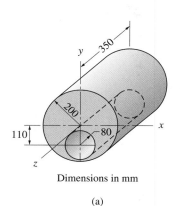

Dimensions in mm

(a)

Solution

The machine part in Fig. (a) can be considered to be the difference between the homogeneous cylinders A and B shown in Figs. (b) and (c), respectively. The mass density ρ of the machine part is

$$\rho = \frac{m}{\pi(R_A^2 - R_B^2)h} = \frac{290}{\pi(0.20^2 - 0.08^2)(0.35)} = 7849 \text{ kg/m}^3$$

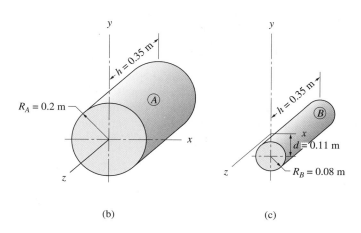

(b)   (c)

Consequently, the masses of cylinders A and B are

$$m_A = \rho \pi R_A^2 h = (7849)\pi(0.20)^2(0.35) = 345.2 \text{ kg}$$

$$m_B = \rho \pi R_B^2 h = (7849)\pi(0.08)^2(0.35) = 55.2 \text{ kg}$$

As a check on our computations, we note that $m_A - m_B = m$, as expected.

Part 1

From Table 17.1, the moment of inertia of cylinder A about the z-axis, which coincides with its central z-axis, is

$$(I_z)_A = (\bar{I}_z)_A = \frac{1}{2}m_A R_A^2 = \frac{1}{2}(345.2)(0.20)^2 = 6.904 \text{ kg} \cdot \text{m}^2$$

The moment of inertia of cylinder B about its central z-axis is

$$(\bar{I}_z)_B = \frac{1}{2}m_B R_B^2 = \frac{1}{2}(55.2)(0.080)^2 = 0.177 \text{ kg} \cdot \text{m}^2$$

Because the distance between the z-axis and the central z-axis of B is $d = 0.11$ m, the moment of inertia of B about the z-axis is found from the parallel-axis theorem:

$$(I_z)_B = (\bar{I}_z)_B + m_B d^2 = 0.177 + (55.2)(0.11)^2 = 0.845 \text{ kg} \cdot \text{m}^2$$

Therefore, the moment of inertia of the machine part about the z-axis is

$$I_z = (I_z)_A - (I_z)_B = 6.904 - 0.845 = 6.059 \text{ kg} \cdot \text{m}^2 \qquad \textit{Answer}$$

Part 2

By symmetry, the x- and z-coordinates of the mass center of the machine part are $\bar{x} = 0$ and $\bar{z} = -0.175$ m. The y-coordinate is given by

$$\bar{y} = \frac{m_A \bar{y}_A - m_B \bar{y}_B}{m} = \frac{345.2(0) - 55.2(-0.11)}{290} = 0.020\,94 \text{ m}$$

The moment of inertia of the machine part about its central z-axis can now be found from the parallel-axis theorem:

$$\bar{I}_z = I_z - m\bar{y}^2 = 6.059 - 290(0.020\,94)^2 = 5.932 \text{ kg} \cdot \text{m}^2$$

The corresponding radius of gyration is

$$\bar{k}_z = \sqrt{\frac{\bar{I}_z}{m}} = \sqrt{\frac{5.932}{290}} = 0.1430 \text{ m} \qquad \textit{Answer}$$

Problems

Fig. P17.1

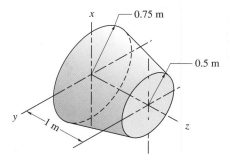

Fig. P17.2

17.1   The homogeneous body of total mass m consists of a cylinder with hemispherical ends. Calculate the moment of inertia of the body about the z-axis in terms of R and m.

17.2   Determine the moment of inertia of the truncated cone about the z-axis. The cone is made of wood that density $10 \, \text{kg/m}^3$.

17.3   A hole of radius R_2 is drilled at the center of the cylinder of radius R_1 and length h. Show that the mass moment of inertia of the resulting body about the z-axis is $I_z = m(R_1^2 + R_2^2)/2$, where m is the mass of the body.

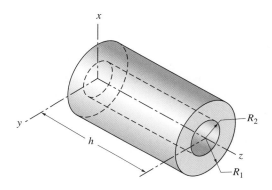

Fig. P17.3

17.4   The inertial properties of a three-stage rocket are shown in the figure. Note that $\bar{k}_i$ is the radius of gyration for the ith stage about the axis that is parallel to the x-axis and passes through the center of gravity G_i of the stage. Find $\bar{z}$ and $\bar{I}_x$ of the rocket.

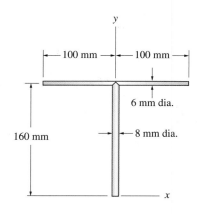

Fig. P17.5

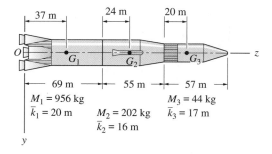

Fig. P17.4

17.5   Two steel rods of different diameters are welded together as shown. Locate the mass center of the assembly and compute $\bar{I}_z$. For steel, $\rho = 7850 \, \text{kg/m}^3$.

17.6 The equilateral triangle is formed by connecting three identical slender rods. If the total mass of the triangle is m, compute its mass moment of inertia about the z-axis.

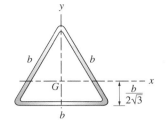

Fig. P17.6

17.7 Three thin plates, each of thickness t, are welded together as shown. Knowing the total mass of the assembly is m, compute its mass moment of inertia about the z-axis. Assume $t \ll b$.

17.8 Calculate I_x and I_z for the cast aluminum machine part. The mass density of aluminum is 2650 kg/m^3.

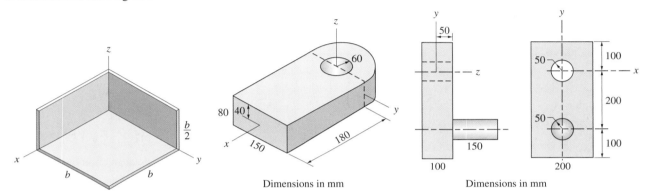

Fig. P17.7	**Fig. P17.8**	**Fig. P17.9**

17.9 The machine part is made of steel with density $\rho = 7850 \text{ kg/m}^3$. Compute I_z, the mass moment of inertia of the part about the z-axis.

17.10 Calculate I_z and $\bar{I}_z$ for the bent slender rod that mass 1 kg.

17.11 The solid body consists of a steel cylinder and a copper cone. The mass density of copper is 1.10 times the mass density of steel. Locate the mass center of the body and compute $\bar{k}_x$.

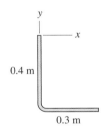

Fig. P17.10

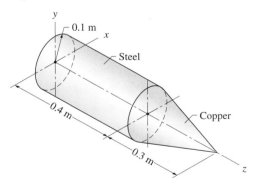

Fig. P17.11

17.12 Referring to Table 17.1, $\bar{I}_x$ for a cylinder can be approximated by $\bar{I}_x$ for a slender rod if the radius R of the cylinder is sufficiently small compared with its length h. Determine the largest ratio R/h for which the relative error for this approximation does not exceed 3 percent.

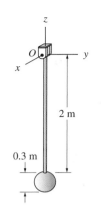

Fig. P17.13

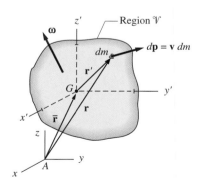

Fig. P17.16

17.13 (a) Compute I_x for the pendulum, which consists of a 4-kg sphere attached to a 1-kg slender rod. (b) Determine the relative error in I_x if the mass of the rod is neglected and the sphere is approximated as a particle.

17.14 The moments of inertia of the 120-kg helicopter blade about the vertical axes passing through O and C are known from experiments to be 408.5 kg · m² and 145.5 kg · m², respectively. Determine the location of the mass center G and the moment of inertia about the vertical axis passing through G.

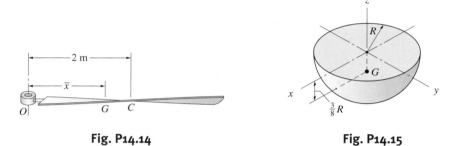

Fig. P14.14   **Fig. P14.15**

17.15 Using the properties of a sphere in Table 17.1, derive the expression for $\bar{I}_x$ of the homogeneous hemisphere of mass m.

*17.16 If the wall thickness t of the hollow sphere of mass m is sufficiently small, its moment of inertia can be approximated by $I_x = (2/3)mR^2$. Derive this result using the properties of a solid sphere in Table 17.1. (Hint: For $t \ll R$, the binomial series yields the following approximation: $R_o^n - R_i^n \approx nR^{n-1}t$.)

17.3	*Angular Momentum of a Rigid Body*

The angular momentum of a body plays a major role in the equations of motion that will be introduced in the next article. Here we derive the angular momentum for three-dimensional motion and then specialize the results for plane motion.

The starting point of our derivation is the angular momentum of a particle of mass m about a point A, which was defined in Art. 14.7 as the moment of its linear momentum about that point:

$$\mathbf{h}_A = \mathbf{r} \times (m\mathbf{v}) \qquad \text{(14.45 repeated)}$$

where $\mathbf{v}$ is the velocity of the particle and $\mathbf{r}$ represents its position vector relative to A. By viewing the body as a collection of an infinite number of particles (differential elements), we can compute its angular momentum by adding (integrating) the angular momenta of the elements.

a. *General motion*

Angular Momentum about the Mass Center Consider a rigid body of mass m that occupies the region $\mathcal{V}$, as shown in Fig. 17.4. The $x'y'z'$ coordinate system

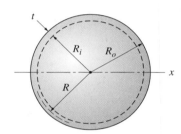

Fig. 17.4

has its origin at the mass center G of the body. The linear momentum of a typical differential element of mass dm moving with the velocity $\mathbf{v}$ is $d\mathbf{p} = \mathbf{v}\,dm$, as shown in the figure. The angular momentum of this element about G is

$$d\mathbf{h}_G = \mathbf{r}' \times (\mathbf{v}\,dm) \tag{a}$$

where $\mathbf{r}' = x'\mathbf{i} + y'\mathbf{j} + z'\mathbf{k}$ is the position vector of the element relative to G. Because dm and G are in the same rigid body, their velocities are related by

$$\mathbf{v} = \bar{\mathbf{v}} + \boldsymbol{\omega} \times \mathbf{r}' \tag{b}$$

where $\bar{\mathbf{v}}$ is the velocity of G, and $\boldsymbol{\omega}$ is the angular velocity of the body. Substituting Eq. (b) in Eq. (a), we get

$$d\mathbf{h}_G = \mathbf{r}' \times (\bar{\mathbf{v}} + \boldsymbol{\omega} \times \mathbf{r}')\,dm$$

Therefore, the angular momentum of the body is

$$\mathbf{h}_G = \int_{\mathscr{V}} d\mathbf{h}_G = \int_{\mathscr{V}} \mathbf{r}' \times (\bar{\mathbf{v}} + \boldsymbol{\omega} \times \mathbf{r}')\,dm$$

Noting that $\int_{\mathscr{V}} \mathbf{r}' \times \bar{\mathbf{v}}\,dm = \left(\int_{\mathscr{V}} \mathbf{r}'\,dm\right) \times \bar{\mathbf{v}} = \mathbf{0}$ (according to the definition of mass center), the angular momentum of the body about G becomes

$$\mathbf{h}_G = \int_{\mathscr{V}} \mathbf{r}' \times (\boldsymbol{\omega} \times \mathbf{r}')dm \tag{17.4}$$

Angular Momentum about an Arbitrary Point   The angular momentum of the body about an arbitrary point A is

$$\mathbf{h}_A = \int_{\mathscr{V}} \mathbf{r} \times (\mathbf{v}\,dm) \tag{c}$$

where $\mathbf{r}$ is the position vector of the element measured from A, as shown in Fig. 17.4. Letting $\bar{\mathbf{r}}$ be the position vector of G relative to A, we can write $\mathbf{r} = \mathbf{r}' + \bar{\mathbf{r}}$, which upon substitution in Eq. (c) yields

$$\mathbf{h}_A = \int_{\mathscr{V}} \mathbf{r}' \times (\mathbf{v}\,dm) + \bar{\mathbf{r}} \times \int_{\mathscr{V}} \mathbf{v}\,dm$$

Recognizing that the first integral is $\mathbf{h}_G$ and substituting Eq. (b) into the second integral, we get

$$\mathbf{h}_A = \mathbf{h}_G + \bar{\mathbf{r}} \times \int_{\mathscr{V}} (\bar{\mathbf{v}} + \boldsymbol{\omega} \times \mathbf{r}')\,dm = \mathbf{h}_G + \bar{\mathbf{r}} \times \bar{\mathbf{v}} \int_{\mathscr{V}} dm + \bar{\mathbf{r}} \times \left(\boldsymbol{\omega} \times \int_{\mathscr{V}} \mathbf{r}'dm\right)$$

Because $\int_{\mathscr{V}} dm = m$ and $\int_{\mathscr{V}} \mathbf{r}'dm = 0$,

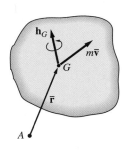

Fig. 17.5

we are left with

$$\mathbf{h}_A = \mathbf{h}_G + \bar{\mathbf{r}} \times (m\bar{\mathbf{v}}) \tag{17.5}$$

A physical interpretation of Eq. (17.5) is shown in Fig. 17.5. The figure, called the *momentum diagram*, is a sketch of the body that displays (1) the linear momentum vector $m\bar{\mathbf{v}}$ of the body acting at the mass center G, and (2) the angular momentum $\mathbf{h}_G$ of the body about G, represented as a couple. According to Eq. (17.5), the angular momentum of the body about A is the vector sum of $\mathbf{h}_G$ and the moment of $m\bar{\mathbf{v}}$ about A (this operation is analogous to computing the resultant moment of a force-couple system about a point).

b.   Plane motion

Angular Momentum about the Mass Center   If the plane of motion is parallel to the xy-plane, we have $\boldsymbol{\omega} = \omega\mathbf{k}$. It follows that $\boldsymbol{\omega} \times \mathbf{r}' = \omega\mathbf{k} \times (x'\mathbf{i} + y'\mathbf{j} + z'\mathbf{k}) = \omega(-y'\mathbf{i} + x'\mathbf{j})$ and

$$\mathbf{r}' \times (\boldsymbol{\omega} \times \mathbf{r}') = \omega \begin{vmatrix} \mathbf{i} & \mathbf{j} & \mathbf{k} \\ x' & y' & z' \\ -y' & x' & 0 \end{vmatrix} = \omega\left[-x'z'\mathbf{i} - y'z'\mathbf{j} + (x'^2 + y'^2)\mathbf{k}\right]$$

Consequently, the angular momentum about G in Eq. (17.4) becomes

$$\mathbf{h}_G = \omega\left[-\mathbf{i}\int_{\mathcal{V}} x'z'\,dm - \mathbf{j}\int_{\mathcal{V}} y'z'\,dm + \mathbf{k}\int_{\mathcal{V}} (x'^2 + y'^2)\,dm\right]$$

The first two integrals, called the *products of inertia*, are discussed in Chapter 19. In this chapter, we assume that the body is *symmetric* about the $x'y'$-plane, in which case the products of inertia vanish.[*] This conclusion follows from the observation that for every element located at (x', y', z') there is a corresponding element at $(x', y', -z')$ so that their combined contribution to each of the first two integrals is zero. The last integral is, according to Eq. (17.1), the moment of inertia of the body about the z'-axis, which we denote by $\bar{I}$. Therefore, the angular momentum of a symmetric body about its mass center reduces to $\mathbf{h}_G = \bar{I}\omega\mathbf{k}$, or

$$h_G = \bar{I}\omega \tag{17.6}$$

Angular Momentum about an Arbitrary Point   The momentum diagram for plane motion is shown in Fig. 17.6(a). The figure represents the cross-section of the body that contains the mass center G and is parallel to the plane of motion. Assuming that point A lies on the same cross section, the resultant moment of the momenta about A is

$$h_A = \bar{I}\omega + m\bar{v}d \quad (A: \text{arbitrary point}) \tag{17.7}$$

where d is the "moment arm" of the linear momentum with respect to A, as indicated in the figure.

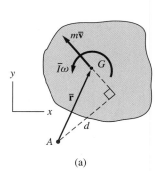

(a)

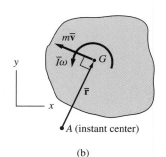

(b)

Fig. 17.6

---

[*]This assumption is overly restrictive. As explained in Chapter 19, the products of inertia also vanish when the z'-axis is a principal axis of inertia of the body.

Angular Momentum about the Instant Center   If point A is the instant center for velocities (this includes the case where the body rotates about a fixed point A), the momentum diagram takes the form shown in Fig. 17.6(b). Note that now $\bar{v} = \bar{r}\omega$ and $m\bar{v}$ is perpendicular to $\bar{r}$. The resultant moment of momenta about A is $h_A = \bar{I}\omega + m\bar{r}^2\omega$. Recognizing that $\bar{I} + m\bar{r}^2 = I_A$ by the parallel-axis theorem, we can write the angular momentum about A as

$$h_A = I_A\omega \quad (A\text{: instant center}) \tag{17.8}$$

17.4   *Equations of Motion*

a.   *Introductory comments*

In this article we investigate the equations of motion for a rigid body, such as the body shown in Fig. 17.7. We consider the body to be a collection of particles and assume that the internal forces between the particles occur in equal and opposite, collinear pairs. By implication, the equations of motion for a system of particles derived in Chapter 15, namely

$$\Sigma\mathbf{F} = m\bar{\mathbf{a}} \tag{15.19, repeated}$$

$$\Sigma\mathbf{M}_A = \dot{\mathbf{h}}_A + \mathbf{v}_A \times (m\bar{\mathbf{v}}) \tag{15.34, repeated}$$

are thus also applicable to a rigid body.[*] In these equations, m is the mass of the body, $\Sigma\mathbf{F}$ is the resultant external force acting on the body, and $\Sigma\mathbf{M}_A$ represents the resultant moment of the external forces about point A.

As indicated in Fig. 17.7, the reference point A is not necessarily a fixed point or a point in the body. Therefore, we must be careful not to confuse absolute velocities and accelerations (which refer to the inertial reference frame) with velocities and accelerations measured relative to A. For example, $\bar{\mathbf{v}}$ denotes the absolute velocity of the mass center G, whereas $\bar{\mathbf{r}}$ is the position vector of G relative to A. Consequently, $d\bar{\mathbf{r}}/dt = \mathbf{v}_{G/A} = \bar{\mathbf{v}} - \mathbf{v}_A$.

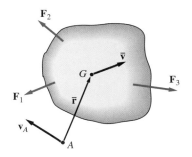

Fig. 17.7

b.   *General motion*

The first equation of motion, Eq. (15.19),

$$\Sigma\mathbf{F} = m\bar{\mathbf{a}} \tag{17.9}$$

which relates the external forces to the acceleration $\bar{\mathbf{a}}$ of the mass center, can be used for a rigid body without modification. However, the second equation of motion, Eq. (15.34), can be made more useful for a rigid body if we incorporate the results of the previous article.

---

[*]Although the particle model is not entirely accurate for a rigid body, it does yield the correct equations of motion. A more rigorous derivation based on the concept of stress is beyond the scope of this text. Historically, the equations of motion for a rigid body were not derived from particle mechanics, but were postulated outright. Presumably, the inspiration came from the analysis of a system of particles.

The angular momentum of a rigid body about an arbitrary point A was found to be

$$\mathbf{h}_A = \mathbf{h}_G + \bar{\mathbf{r}} \times (m\bar{\mathbf{v}}) \qquad \text{(17.5, repeated)}$$

where $\mathbf{h}_G$ is the angular momentum about the mass center G. Differentiating with respect to time, we get

$$\dot{\mathbf{h}}_A = \dot{\mathbf{h}}_G + \frac{d\bar{\mathbf{r}}}{dt} \times (m\bar{\mathbf{v}}) + \bar{\mathbf{r}} \times (m\bar{\mathbf{a}}) \qquad \text{(a)}$$

With the substitution $d\bar{\mathbf{r}}/dt = \bar{\mathbf{v}} - \mathbf{v}_A$, the second term on the right side of Eq. (a) is

$$\frac{d\bar{\mathbf{r}}}{dt} \times (m\bar{\mathbf{v}}) = (\bar{\mathbf{v}} - \mathbf{v}_A) \times (m\bar{\mathbf{v}}) = -\mathbf{v}_A \times (m\bar{\mathbf{v}})$$

Therefore, Eq. (a) becomes

$$\dot{\mathbf{h}}_A = \dot{\mathbf{h}}_G - \mathbf{v}_A \times (m\bar{\mathbf{v}}) + \bar{\mathbf{r}} \times (m\bar{\mathbf{a}}) \qquad \text{(b)}$$

which upon substitution in Eq. (15.34) yields for the second equation of motion

$$\Sigma \mathbf{M}_A = \dot{\mathbf{h}}_G + \bar{\mathbf{r}} \times (m\bar{\mathbf{a}}) \qquad \text{(17.10)}$$

The physical interpretation of the equations of motion, Eqs. (17.9) and (17.10), is shown in Fig. 17.8. The free-body diagram (FBD) displays the external forces $\mathbf{F}_1$, $\mathbf{F}_2$, ... acting on the body. The mass-acceleration diagram (MAD) shows the *inertia vector* $m\bar{\mathbf{a}}$ acting at the mass center G, and the *inertia couple* $\dot{\mathbf{h}}_G$. The equal sign between the diagrams implies that the force systems in the FBD and the MAD are *equivalent*; that is, the two force systems have the same resultant force and the same resultant moment about any point. It is easily verified that the two conditions of equivalence $(\Sigma \mathbf{F})_{\text{FBD}} = (\Sigma \mathbf{F})_{\text{MAD}}$ and $(\Sigma \mathbf{M}_A)_{\text{FBD}} = (\Sigma \mathbf{M}_A)_{\text{MAD}}$ reproduce the two equations of motion.

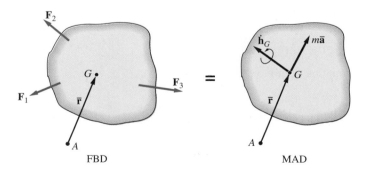

Fig. 17.8

c. Plane motion

The FBD and the MAD for a body in plane motion in the xy-plane are shown in Fig. 17.9. The diagrams actually display the cross section of the body that is parallel to the plane motion and contains the mass center G of the body. To obtain

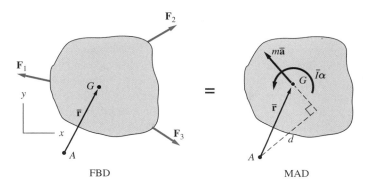

Fig. 17.9

a two-dimensional formulation of plane motion, we assume that all the external forces $\mathbf{F}_1$, $\mathbf{F}_2$, ... lie in the plane of the cross section, and that the cross section is a plane of symmetry of the body.

In plane motion, the angular momentum of the body about G is given in Eq. (17.6) as $h_G = \bar{I}\omega$, where ω is the angular velocity of the body, and $\bar{I}$ is the moment of inertia about G. Hence the inertia couple shown in the MAD is $\dot{h}_G = \bar{I}\alpha$, where $\alpha = \dot{\omega}$ is the angular acceleration of the body. From the equivalence of the force systems on the two diagrams in Fig. 17.9, we get for the equations of motion

$$\Sigma \mathbf{F} = m\bar{\mathbf{a}} \qquad (17.11)$$

$$\Sigma M_A = \bar{I}\alpha + m\bar{a}d \qquad (17.12)$$

where d is the moment arm of the inertia vector shown in the figure.

Moment Equation about the Mass Center   If A coincides with the mass center, then $d = 0$ and Eq. (17.12) becomes

$$\Sigma M_G = \bar{I}\alpha \qquad (17.13)$$

Moment Equation for Rotation about a Fixed Point   If the body is rotating about a fixed point A (i.e., A is fixed in the body as well as in the inertial reference frame), then we can use Eq. (17.8): $h_A = I_A\omega$. Differentiating with respect to time (note that I_A is constant because A is fixed in the body), we get $\dot{h}_A = I_A\alpha$. Therefore, for rotation about a fixed point Eq. (15.34) takes the form

$$\Sigma M_A = I_A\alpha \quad (A: \text{fixed in the body and in space}) \qquad (17.14)$$

17.5   *Force-Mass-Acceleration Method: Plane Motion*

a.   *General plane motion*

The *force-mass-acceleration* (FMA) *method* of kinetic analysis is based on the equations of motion for a rigid body derived in the previous article. In this article, our discussion will be limited to plane motion; three-dimensional problems will be introduced in Chapter 19.

There are three independent scalar equations of plane motion. One example of an independent set of equations is $\Sigma F_x = m\bar{a}_x$, $\Sigma F_y = m\bar{a}_y$ and the moment equation $\Sigma M_A = \bar{I}\alpha + m\bar{a}d$. However, because choices for the coordinate directions and the moment center are arbitrary, the number of available (but not necessarily independent) equations of motion is infinite. The restrictions on the equations that guarantee their independence are identical to those used in coplanar equilibrium in statics. For example, if three moment equations are used, the moment centers must not be collinear.

It is usually convenient to obtain the equations of motion from the FBD and the MAD. The procedure is essentially the same as used for particles in Art. 12.3:

Step 1: Draw the FBD of the body showing all external forces and couples.
Step 2: If there are kinematic constraints imposed on the motion, use kinematics to determine the relationship between $\bar{a}$, ω, and α.
Step 3: Draw the MAD of the body showing the inertia vector $m\bar{a}$ acting at the mass center and the inertia couple $\bar{I}\alpha$, using the results of Step 2.
Step 4: Derive three independent equations of motion from the equivalence of the FBD and the MAD.

If the number of unknowns that appear in the FBD and the MAD is three, then they can be determined by solving the equations of motion. In some problems, however, additional information must be obtained from the history of motion. For example, if ω appears in the MAD, it may be necessary to obtain its value at the instant of concern by integration: $\omega = \int \alpha \, dt$.

One advantage of the FBD-MAD technique is that the FBD displays the unknown forces, and the MAD displays the unknown accelerations. Consequently, one is less likely to attempt to derive and solve the equations of motion before the unknown variables have been correctly identified.

b.   Translation

Figure 17.10 shows the MAD for a rigid body that is translating. Because $\alpha = 0$, the MAD reduces simply to the inertia vector passing through G. It can be seen from this diagram that the resultant moment is zero about any point that lies on the same line as the inertia vector. For any other point, the resultant moment is equal to the moment of the inertia vector.

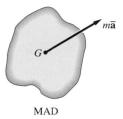

MAD

Fig. 17.10

c. Rotation about a fixed axis

We distinguish between two types of rotation about a fixed axis—central and noncentral rotation.[*]

In *central rotation*, the fixed axis passes through the mass center G. Because $\bar{\mathbf{a}} = \mathbf{0}$, the MAD reduces to the inertia couple, as shown in Fig. 17.11(a). In this case the inertia vector is zero and the resultant moment about every point is equal to $\bar{I}\alpha$.

In *noncentral rotation*, the axis of rotation passes through a fixed point A that is not the mass center. The MAD for this case is shown in Fig. 17.11(b), where the components of $\bar{\mathbf{a}}$ have been determined from the fact that the path of G is a circle centered at A. Since A is a fixed point, the special case of the moment equation, $\Sigma M_A = I_A \alpha$, could be used. However, the same moment equation is obtained by equating the resultant moments about A for the FBD and the MAD:

$$\stackrel{\curvearrowleft}{+} \quad \Sigma M_A = \bar{I}\alpha + (m\bar{r}\alpha)\bar{r} = (\bar{I} + m\bar{r}^2)\alpha = I_A \alpha$$

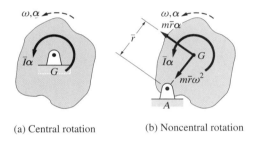

(a) Central rotation   (b) Noncentral rotation

Fig. 17.11

d. Systems of connected rigid bodies

When analyzing the motion of a system of connected rigid bodies, an FBD and an MAD can be drawn for each component body, because each component must satisfy a separate set of the three equations of motion. Therefore, a system containing N rigid bodies must satisfy $3N$ independent equations of motion.

Frequently it is convenient to use the FBD and MAD of the assembly, because then the internal forces do not appear in the FBD. For this reason, the FBD and the MAD for the entire system is often a good choice for beginning the analysis.

For systems of connected bodies, kinematic constraints require that special attention be paid to the MAD. Figure 17.12 shows the MAD for a system consisting of a slender bar AB that is pinned at B to a disk C (the MAD for the system is

---

[*]The types of rotation are sometimes identified as *centroidal* and *noncentroidal* rotation. We avoid using the term "centroid" to eliminate the confusion that often exists between centroids, which are properties of geometric shapes, and mass centers, which are properties associated with mass.

the superposition of the MADs for the respective bodies). In this figure the properties (masses, accelerations, moments of inertia, and mass centers) of the bar and the disk are identified by the subscripts 1 and 2, respectively. Note that α_1 does not necessarily equal α_2 because of the pin at B. However, $\bar{\mathbf{a}}_1$, $\bar{\mathbf{a}}_2$, α_1, and α_2 are related kinematically by the equation of constraint $\mathbf{a}_B = \bar{\mathbf{a}}_1 + \mathbf{a}_{B/G_1} = \bar{\mathbf{a}}_2 + \mathbf{a}_{B/G_2}$, which expresses the fact that B is a point on both bodies.

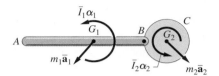

Fig. 17.12

Figure 17.13(a) shows the MAD when the bar and disk are rigidly connected at B. In this case, $\alpha_1 = \alpha_2 = \alpha$. Furthermore, $\bar{\mathbf{a}}_1$ and $\bar{\mathbf{a}}_2$ are related by $\bar{\mathbf{a}}_2 = \bar{\mathbf{a}}_1 + \mathbf{a}_{G_2/G_1}$ because G_1 and G_2 are on the same rigid body.

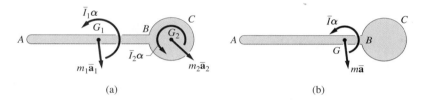

(a)   (b)

Fig. 17.13

Figure 17.13(b) shows an alternate form of the MAD for the rigid body in Fig. 17.13(a). Here, $m\bar{\mathbf{a}}$ and $\bar{I}\alpha$ are the inertia vector and inertia couple, respectively, for the entire body and G is its mass center. Whether one employs the MAD in Fig. 17.13 part (a) or (b) is a matter of personal preference, because the analysis using either diagram involves about the same amount of computation.

Sample Problem 17.3

The mass center of the 10-kg sliding door in Fig. (a) is located at G. The door is supported on the horizontal rail by sliders at A and B. The coefficient of static as well as kinetic friction is 0.4 at A and 0.3 at B. The door was at rest before the horizontal force $P = 60\,N$ was applied. (1) Find the maximum value of h for which the door will slide to the right without tipping and the corresponding acceleration of the door. (2) If $h = 1.5$ m, find all forces acting on the door, and calculate its acceleration.

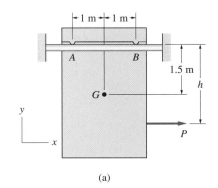

(a)

Solution

Part 1

The free-body and mass-acceleration diagrams for the door, shown in Fig. (b), are described in detail as follows.

Free-body diagram The FBD contains the following forces: the $10(9.8) = 98\,N$ weight acting at G, the applied force P, the normal force N_A, and the friction force $F_A = \mu_A N_A = 0.4 N_A$. Because the door starts from rest, its velocity will be directed to the right (i.e., in the same direction as the acceleration), which means that F_A is directed to the left. There is no normal force and, therefore, no friction force at B, because the problem statement implies that the door is sliding to the right in a state of impending tipping about A.

Mass-acceleration diagram The MAD contains only the inertia vector of magnitude $m\bar{a}$ acting at G. There is no inertia couple because the door is translating ($\alpha = 0$).

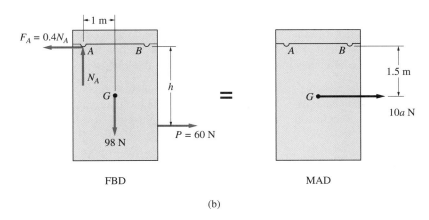

FBD   MAD

(b)

Inspection of Fig. (b) reveals that the unknowns are N_A and h on the FBD, and $\bar{a}$ on the MAD. These three unknowns can be computed by deriving and solving any three independent equations of motion. When employing the FBD-MAD technique, remember that (1) the resultant force on the FBD can be equated to the inertia vector $m\bar{a}$ on the MAD, and (2) the resultant moment about any point on the FBD can be equated to the resultant moment about the same point on the MAD.

Equating the x- and y-components of the forces on the FBD to the corresponding components of the inertia vector, we obtain

$$\Sigma F_y = m\bar{a}_y \quad +\uparrow \quad N_A - 98 = 0$$
$$N_A = 98 \text{ N}$$

and

$$\Sigma F_x = m\bar{a}_x \quad \xrightarrow{+} \quad 60 - 0.4N_A = 10a$$
$$60 - 0.4(98) = 10a$$

which gives

$$\bar{a} = 2.08 \text{ m/s}^2 \qquad\qquad \textit{Answer}$$

The third independent equation is a moment equation about any point. If we choose point A as the moment center, the resultant moment on the FBD is equated to the resultant moment on the MAD:

$$(\Sigma M_A)_{\text{FBD}} = (\Sigma M_A)_{\text{MAD}} \quad \overset{+}{\curvearrowright} \quad 60h - 10(9.8)(1.5) = 10\bar{a}(1.5)$$

With $\bar{a} = 2.08 \text{ m/s}^2$, we get

$$h = 2.97 \text{ m} \qquad\qquad \textit{Answer}$$

Part 2

The FBD and MAD for $h = 1.5$ m are shown in Fig. (c). The details of these diagrams are as follows.

Free-body diagram   The FBD contains the 98-N weight, the applied force $P = 60$ N passing through G, the normal forces N_A and N_B, and the friction forces F_A and F_B. From the solution of Part 1 we know that the door will be sliding to the right while maintaining contact at both A and B. The friction forces, determined by the kinetic coefficients of friction, are directed to the left—that is, opposite to the motion.

Mass-acceleration diagram   Because the door is sliding to the right without rotating, the MAD contains only the inertia vector acting through G.

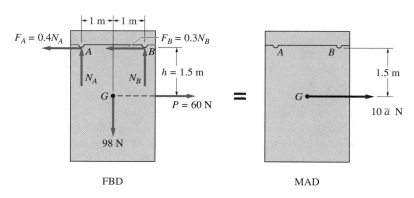

FBD   MAD

(c)

The number of unknowns in the FBD and MAD is three: N_A, N_B, and $\bar{a}$, which can be found from any three independent equations of motion. One such set of equations is given below; their validity can be determined by referring to the FBD and MAD in Fig. (c).

$$\Sigma F_y = m\bar{a}_y \quad +\uparrow \quad N_A + N_B - 98 = 0 \tag{a}$$

$$\Sigma M_G = 0 \quad \overset{+}{\curvearrowright} \quad N_B(1) + 0.3N_B(1.5) - N_A(1) + 0.4N_A(1.5) = 0 \tag{b}$$

$$\Sigma F_x = m\bar{a}_x \quad \overset{+}{\longrightarrow} \quad 60 - 0.4N_A - 0.3N_B = 10\,\bar{a} \tag{c}$$

Solving Eqs. (a), (b), and (c) gives

$$N_A = 76.85 \text{ N} \qquad N_B = 21.2 \text{ N} \qquad \bar{a} = 2.29 \text{ m/s}^2 \qquad \textit{Answer}$$

The positive values of N_A and N_B confirm that the door will not tip.

Sample Problem **17.4**

The homogeneous bar in Fig. (a) has mass m and length L. The bar, which is free to rotate in the vertical plane about a pin at O, is released from rest in the position $\theta = 0$. Find the angular acceleration α when $\theta = 60°$.

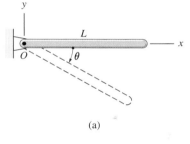

(a)

Solution

Figure (b) shows the FBD and the MAD of the bar when $\theta = 60°$. The FBD contains the weight W of the bar, acting at its mass center G (located at the midpoint of the bar) and the components of the pin reaction at O. In the MAD, the inertia couple $\bar{I}\alpha$ was drawn assuming that α is clockwise, and using $\bar{I} = mL^2/12$ from Table 17.1. The components of the inertia vector $m\mathbf{a}$ were found by noting that the path of G is a circle centered at O. Therefore, the normal and tangential components of $\bar{\mathbf{a}}$ are $\bar{a}_n = (L/2)\omega^2$ and $\bar{a}_t = (L/2)\alpha$. The direction of $\bar{a}_n$ is toward O, regardless of the direction of ω. The direction of $\bar{a}_t$ is consistent with the assumed direction of α.

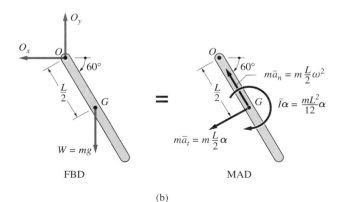

(b)

We note that there are a total of four unknowns in Fig. (b): O_x, O_y, α, and ω. Because there are only three independent equations of motion, we will not be able to determine all unknowns using only the FBD and the MAD. The reason for this is that ω depends on the history of motion: $\omega = \int \alpha \, dt + C$. Therefore, the equations of motion at a specific position of the bar will not determine the angular velocity in that position. However, inspection of the FBD and MAD reveals that it is possible to determine the angular acceleration α, because it is the only unknown that appears in the moment equation when O is used as the moment center. Referring to the diagrams in Fig. (b), this moment equation is

$$(\Sigma M_O)_{\text{FBD}} = (\Sigma M_O)_{\text{MAD}}$$

$$\circlearrowright_{+} \quad mg\frac{L}{2}\cos 60° = \frac{mL^2}{12}\alpha + \left(m\frac{L}{2}\alpha\right)\frac{L}{2} = \frac{mL^2}{3}\alpha \quad \text{(a)}$$

from which we find that

$$\alpha = \frac{3g}{2L}\cos 60° = 0.750\frac{g}{L} \qquad \textit{Answer} \quad \text{(b)}$$

Because the acceleration of O is zero, the above moment equation could have also been obtained from Eq. (17.14): $\Sigma M_O = I_O \alpha$, where

$$I_O = \bar{I} + md^2 = \frac{mL^2}{12} + m\left(\frac{L}{2}\right)^2 = \frac{mL^2}{3} \qquad \text{(c)}$$

from the parallel-axis theorem. We now see that $\Sigma M_O = I_O \alpha$ will yield an equation that is identical to Eq. (a).

Sample Problem   **17.5**

The body shown in Fig. (a) consists of the homogeneous slender bar 1 that is rigidly connected to the homogeneous sphere 2. The body is rotating in the vertical plane about the pin at O. When the body is in the position where $\theta = 30°$, its angular velocity is $\omega = 1.2$ rad/s clockwise. At this instant, determine the angular acceleration α and the magnitude of the pin reaction at O.

$L = 800$ mm
$\theta = 30°$
$\omega = 1.2$ rad/s
$m_1 = 30$ kg
$m_2 = 80$ kg
$R = 200$ mm

(a)

Solution

The FBD and MAD of the body in the position $\theta = 30°$ are shown in Fig. (b). In these diagrams, the bar and the sphere are treated as separate entities, each with its own inertia couple and inertia vector. (An equivalent form of the MAD would be obtained by showing the inertia couple and inertia vector for the assembly; refer to Fig. 17.13.) Details of the diagrams are described in the following.

Free-body diagram   The forces O_n and O_t are the components of the pin reaction relative to the n- and t-axes shown in the figure. The weights W_1 and W_2 of the bar and sphere, respectively, act at their mass centers G_1 and G_2. The distances

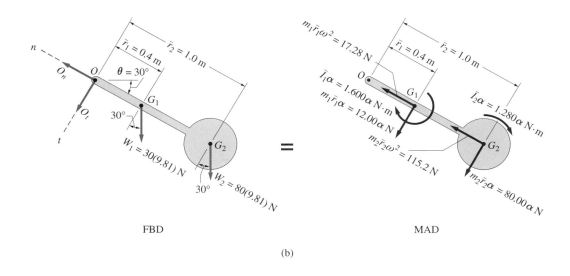

FBD   =   MAD

(b)

$\bar{r}_1 = 0.4$ m and $\bar{r}_2 = 1.0$ m, measured from O to the mass centers, are deduced from the dimensions in Fig. (a).

Mass-acceleration diagram In the MAD, we assume that the angular acceleration α, measured in rad/s^2, is clockwise. Using the fact that the body rotates about the fixed point O, kinematic analysis enables us to express the accelerations of G_1 and G_2 in terms of α and ω of the body. The inertia terms that appear in the MAD have been computed in the following manner.
For the slender bar:

$$\bar{I}_1\alpha = \frac{m_1 L^2}{12}\alpha = \frac{30(0.8)^2}{12}\alpha = 1.600\alpha \text{ N} \cdot \text{m}$$
$$m_1\bar{r}_1\omega^2 = 30(0.4)(1.2)^2 = 17.28 \text{ N}$$
$$m_1\bar{r}_1\alpha = 30(0.4)\alpha = 12.00\alpha \text{ N}$$

For the sphere:

$$\bar{I}_2\alpha = \frac{2}{5}m_2 R^2\alpha = \frac{2}{5}(80)(0.2)^2\alpha = 1.280\alpha \text{ N} \cdot \text{m}$$
$$m_2\bar{r}_2\omega^2 = 80(1.0)(1.2)^2 = 115.2 \text{ N}$$
$$m_2\bar{r}_2\alpha = 80(1.0)\alpha = 80.00\alpha \text{ N}$$

In the MAD, the directions of the tangential components of the inertia vectors (those containing α) are consistent with the assumed clockwise direction of α. The normal components of the inertia vectors (those containing ω^2) are directed toward the center of rotation O, regardless of the direction of ω.

From Fig. (b) we see that there are two unknowns in the FBD (O_n and O_t) and one unknown (α) in the MAD. Therefore, all that remains is to derive and solve the three independent equations of motion for the unknowns.

Equating moments about O on the FBD and the MAD in Fig. (b), we obtain

$$(\Sigma M_O)_{\text{FBD}} = (\Sigma M_O)_{\text{MAD}}$$

$\overset{+}{\curvearrowright}$ $\quad 30(9.8)(0.4)\cos 30° + 80(9.8)(1.0)\cos 30°$
$$= 1.600\alpha + (12.00\alpha)(0.4) + 1.280\alpha + (80.00\alpha)(1.0)$$
$$= 87.68\alpha$$

from which we find that
$$\alpha = 8.905 \text{ rad/s}^2 \qquad\qquad \textit{Answer}$$

Because the acceleration of point O is zero, this result could also have been derived using the special case $\Sigma M_O = I_O\alpha$.

Using $\alpha = 8.914 \text{ rad/s}^2$ and referring to Fig. (b), the force equations in the t and n directions give

$$\Sigma F_t = m\bar{a}_t$$

$\overset{+}{\nearrow}$ $\quad O_t + 30(9.8)\cos 30° + 80(9.8)\cos 30° = 12.00(8.905) + 80.00(8.905)$
$$O_t = -114.3 \text{ N}$$

and

$$\Sigma F_n = m\bar{a}_n$$

$\overset{+}{\searrow}$ $\quad O_n - 30(9.8)\sin 30° - 80(9.8)\sin 30° = 17.28 + 115.2$
$$O_n = 671.5 \text{ N}$$

Therefore, the magnitude of the pin reaction at O is

$$O = \sqrt{O_t^2 + O_n^2} = \sqrt{(-114.3)^2 + (671.5)^2} = 681.2 \text{ N} \qquad \textit{Answer}$$

Sample Problem 17.6

The cable connected to block B in Fig. (a) is wound tightly around disk A, which is free to rotate about the axle at its mass center G. The masses of A and B are 60 kg and 20 kg, respectively, and $\bar{k} = 400$ mm for the disk. Determine the angular acceleration of A and the tension in the cable.

Solution

The free-body and mass-acceleration diagrams of the system are shown in Fig. (b). The FBD contains the weights $W_A = 60(9.8) = 588$ N and $W_B = 20(9.8) = 196$ N together with the unknown pin reactions at G. The tension in the cable, being an internal force, does appear on this FBD.

The MAD displays the inertia couple of the disk and the inertia vector of the block. There is no inertia vector of the disk because its mass center G is stationary. The angular acceleration α of the disk is assumed to be directed clockwise. The corresponding inertia couple of the disk is

$$\bar{I}\alpha = m\bar{k}^2\alpha = 60(0.4)^2\alpha = 9.600\alpha \text{ N} \cdot \text{m}$$

$R = 500$ mm

$m_A = 60$ kg
$\bar{k} = 400$ mm

G

A

$B \quad m_B = 20$ kg

(a)

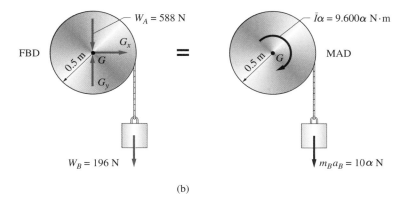

$W_A = 588$ N

$\bar{I}\alpha = 9.600\alpha$ N·m

G_x

0.5 m

G

$=$

0.5 m

G

MAD

G_y

$W_B = 196$ N

$m_B a_B = 10\alpha$ N

(b)

which also is directed clockwise. Because the cable does not slip on the disk, the acceleration of the block is $a_B = R\alpha$, which results in the inertia vector

$$m_B a_B = m_B(R\alpha) = 20(0.5\alpha) = 10\alpha \text{ N} \quad \downarrow$$

There are three unknowns on the FBD and the MAD: two components of the pin reaction at G and the angular acceleration α of the disk. Since the number of independent equations available from the FBD and MAD is also three, all the unknowns can be computed.

The angular acceleration α can be found by equating the resultant moments about G in the FBD and the MAD.

$$(\Sigma M_G)_{FBD} = (\Sigma M_G)_{MAD} \quad \circlearrowright + \quad 196(0.5) = 9.600\alpha + 10\alpha(0.5)$$

$$\alpha = 6.712 \text{ rad/s}^2 \qquad \textit{Answer}$$

To find the tension in the cable, we analyze the block separately (the disk could also be used). The FBD and MAD for the block are shown in Fig. (c), where T is the cable tension. Summing forces in the y-direction yields

$$\Sigma F_y = ma_y \quad +\downarrow \quad 196 - T = 10\alpha = 10(6.712)$$

$$T = 129.0 \text{ N} \qquad \textit{Answer}$$

T

FBD

$=$

MAD

$W_B = 196$ N

$m_B a_B = 10\alpha$ N

(c)

Sample Problem   17.7

The 40-kg unbalanced wheel in Fig. (a) is rolling without slipping under the action of a counterclockwise couple $C_0 = 20$ N $\cdot$ m. When the wheel is in the position shown, its angular velocity is $\omega = 2$ rad/s, clockwise. For this position, calculate the angular acceleration α and the forces exerted on the wheel at C by the rough horizontal plane. The radius of gyration of the wheel about its mass center G is $\bar{k} = 200$ mm.

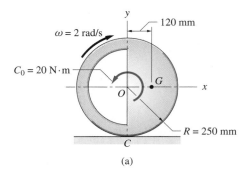

(a)

Solution

The free-body and mass-acceleration diagrams for the wheel, shown in Fig. (b), were constructed as follows.

Free-body diagram   The FBD consists of the applied couple C_0, the weight $W = 40(9.8) = 392$ N, and the normal and friction forces that act at the contact point C, denoted by N_C and F_C, respectively. Observe that F_C has been assumed to be directed to the right.

Mass-acceleration diagram   In the MAD of Fig. (b) the angular acceleration α, measured in rad/s^2, has been assumed to be clockwise. The corresponding inertia couple shown on this diagram is

$$\bar{I}\alpha = m\bar{k}^2\alpha = 40(0.200)^2\alpha = 1.600\alpha \text{ N} \cdot \text{m}$$

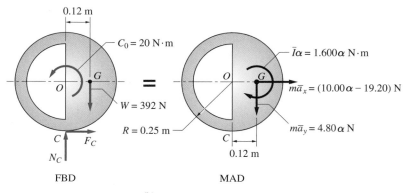

FBD   MAD

(b)

Because the wheel does not slip, the acceleration of its center O is $a_O = R\alpha = 0.250\alpha$ m/s^2, directed to the right. Applying the relative acceleration equation between G and O, we obtain (the units of each term are m/s^2)

$$\bar{\mathbf{a}} = \mathbf{a}_G = \mathbf{a}_O + \mathbf{a}_{G/O}$$

from which we find $\bar{a}_x = 0.250\alpha - 0.480$ m/s^2 and $\bar{a}_y = 0.120\alpha$ m/s^2. Multiplying these results by $m = 40$ kg, the components of the inertia vector become $m\bar{a}_x = (10.00\alpha - 19.20)$ N, directed to the right, and $m\bar{a}_y = 4.80\alpha$ N, directed downward.

The FBD and MAD in Fig. (b) now contain only three unknowns: N_C, F_C, and α, which can be found using any three independent equations of motion.

Because N_C and F_C act at C, it is convenient to use that point as a moment center, the corresponding moment equation being

$$(\Sigma M_C)_{\text{FBD}} = (\Sigma M_C)_{\text{MAD}}$$

$\overset{+}{\curvearrowleft}$ $\quad -20 + 392(0.120) = 1.600\alpha + 0.250(10.00\alpha - 19.20) + 0.120(4.80\alpha)$

The solution to this equation is

$$\alpha = 6.820 \text{ rad/s}^2 \qquad\qquad \textit{Answer}$$

Because α is positive, its direction is clockwise, as assumed.

The forces at C can now be found from force equations of motion:

$$\Sigma F_x = m\bar{a}_x \quad \overset{+}{\longrightarrow} \quad F_C = 10.00\alpha - 19.20 = 10.00(6.820) - 19.20$$

and

$$\Sigma F_y = m\bar{a}_y \quad +\!\downarrow \quad 392 - N_C = 4.80\alpha = 4.80(6.820)$$

which yield

$$F_C = 49.0 \text{ N} \quad \text{and} \quad N_C = 359.3 \text{ N} \qquad\qquad \textit{Answer}$$

Because each force is positive, it is directed as shown in the FBD.

Sample Problem 17.8

Figure (a) shows a 10-kg homogeneous disk of radius 0.2 m. The disk is at rest before the horizontal force $P = 60$ N is applied to its mass center G. The coefficients of static and kinetic friction for the surfaces in contact are 0.20 and 0.15, respectively. Determine the angular acceleration of the disk and the acceleration of G after the force is applied.

$R = 0.2$ m

$P = 60$ N

$\mu_s = 0.20$
$\mu_k = 0.15$

(a)

Solution

Two motions of the disk are possible: rolling without slipping, and rolling with slipping. We will solve the problem by assuming that the disk rolls without slipping. This assumption will then be checked by comparing the required friction force with its maximum static value.

The free-body diagram (FBD) and the mass-acceleration diagram (MAD) based on the no-slip assumption are shown in Fig. (b). The FBD contains the $10(9.8) = 98$ N weight, the 60 N applied force, the normal force N_A, and the friction force F, assumed acting to the left. The MAD contains the inertia couple and inertia vector, where the angular acceleration α, measured in rad/s^2, has been assumed to be clockwise. The values of $\bar{I}\alpha$ and $m\bar{a}$ were computed as follows.

$$\bar{I}\alpha = \frac{mR^2}{2}\alpha = \frac{10(0.2)^2}{2}\alpha = 0.2\alpha \text{ N} \cdot \text{m}$$

$$m\bar{a} = mR\alpha = 10(0.2)\alpha = 2\alpha \text{ N}$$

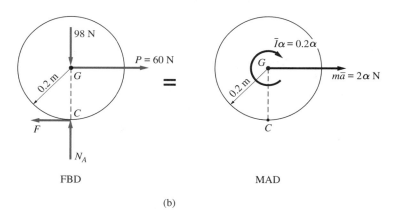

FBD   MAD

(b)

Note that $\bar{a} = R\alpha$ is a valid kinematic equation because the disk is assumed to be rolling without slipping. There are a total of three unknowns in the FBD and MAD: F, N, and α, which can be computed using any three independent equations of motion.

A convenient solution is to first equate the resultant moment about C on the FBD to the resultant moment about C on the MAD and then utilize the force equations of motion.

$$(\Sigma M_C)_{FBD} = (\Sigma M_C)_{MAD} \quad \overset{+}{\curvearrowright} \quad 60(0.2) = 0.2\alpha + 2\alpha(0.2)$$

$$\alpha = 20 \text{ rad/s}^2$$

$$\Sigma F_x = m\bar{a}_x \quad \overset{+}{\longrightarrow} \quad 60 - F = 2\alpha = 2(20)$$

$$F = 20 \text{ N}$$

$$\Sigma F_y = m\bar{a}_y \quad +\uparrow \quad N_A - 98 = 0$$

$$N_A = 98 \text{ N}$$

Because α, F, and N_A are all positive, their directions are as shown in Fig. (b).

Next we note that the maximum possible static friction force is $F_{max} = \mu_s N_A = 0.20(98) = 19.6$. The friction force required for rolling without slipping is, according to our solution, $F = 20$ N. Because $F > F_{max}$, we conclude that the disk does slip, and we must reformulate the problem.

The FBD and MAD for the case where the disk rolls and slips simultaneously are shown in Fig. (c). The friction force F in the FBD has been set equal to its kinetic value, $\mu_k N$. This force must be shown acting to the left in order to oppose slipping. The inertia couple $\bar{I}\alpha$ in the MAD is identical to that used in Fig. (b). However, the important difference here is that the magnitude of the inertia vector is now $m\bar{a} = 10\bar{a}$ N, where $\bar{a}$ is measured in m/s^2. Because the disk is slipping, the kinematic constraint $\bar{a} = R\alpha$ does not apply. Once again we see that there are three unknowns on the FBD and the MAD, except that now the unknowns are N, α, and $\bar{a}$.

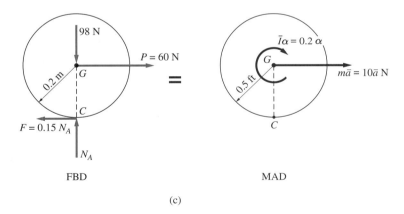

FBD   MAD

(c)

The three unknowns can be calculated as follows (of course, any other three independent equations could also be used):

$$\Sigma F_y = m\bar{a}_y \quad +\uparrow \quad N_A - 98 = 0 \quad N = 98 \text{ N} \qquad \text{(a)}$$
$$\Sigma F_x = m\bar{a}_x \quad \overset{+}{\longrightarrow} \quad 60 - 0.15N_A = 10\bar{a} \qquad \text{(b)}$$
$$\Sigma M_G = \bar{I}\alpha \quad \overset{+}{\curvearrowright} \quad 0.2(0.15N_A) = 0.2\alpha \qquad \text{(c)}$$

Substituting $N = 50$ lb from Eq. (a) into Eqs. (b) and (c) yields

$$\bar{a} = 4.53 \text{ m/s}^2 \quad \text{and} \quad \alpha = 14.7 \text{ rad/s}^2 \qquad \textit{Answer}$$

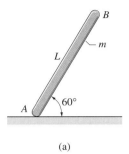

(a)

Sample Problem 17.9

A homogeneous slender bar AB of mass m and length L is released from rest in the position shown in Fig. (a). Determine the acceleration of end A, the reaction at A, and the angular acceleration of the bar immediately after the release. Assume that the horizontal plane is frictionless.

Solution

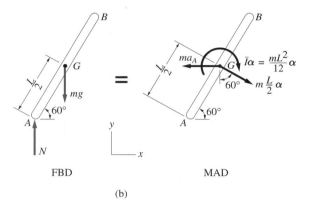

(b)

The free-body diagram (FBD) and mass-acceleration diagram (MAD) of the bar at the instant after the release are shown in Fig. (b). The FBD contains the weight of the bar, mg, and the vertical reaction N. The MAD contains the inertia couple, $\bar{I}\alpha$, and the inertia vector, $m\bar{\mathbf{a}}$. The latter consists of components ma_A and $m(L/2)\alpha$, which were obtained from kinematics as follows:

$$\bar{\mathbf{a}} \; = \; \mathbf{a}_G \; = \; \mathbf{a}_A \; + \; \mathbf{a}_{G/A} \qquad\qquad \text{(a)}$$

In Eq. (a), the senses of $\mathbf{a}_A$ and α were assumed to be to the left and clockwise, respectively. The angular velocity ω is zero because the bar has just been released from rest in the position being considered. Multiplying the right-hand side of Eq. (a) by the mass m and placing the results at G gives the components of the inertia vector shown in the MAD of Fig. (b).

Inspection of Fig. (b) reveals that there are a total of three unknowns: N, a_A, and α. Therefore, the solution can be obtained by deriving and solving any three independent equations of motion.

Equating moments about A on the FBD and MAD in Fig. (b) yields

$$(\Sigma M_A)_{\text{FBD}} = (\Sigma M_A)_{\text{MAD}}$$

$$\underset{+}{\circlearrowright} \quad mg\left(\frac{L}{2}\cos 60°\right) = \frac{mL^2}{12}\alpha + m\frac{L}{2}\alpha\left(\frac{L}{2}\right) - ma_A\left(\frac{L}{2}\sin 60°\right)$$

which, on simplification, becomes

$$a_A = 0.7698L\alpha - 0.5774g \tag{b}$$

Referring again to Fig. (b), the force equation for the horizontal direction becomes

$$\Sigma F_x = m\bar{a}_x \quad \xrightarrow{+} \quad 0 = -ma_A + m\frac{L}{2}\alpha \sin 60°$$

which reduces to

$$a_A = 0.4330L\alpha \tag{c}$$

Solving Eqs. (b) and (c) simultaneously yields

$$a_A = 0.742g \quad \text{and} \quad \alpha = 1.714\frac{g}{L} \qquad \textit{Answer}$$

Using the diagrams in Fig. (b), the force equation for the vertical direction is

$$\Sigma F_y = m\bar{a}_y \quad +\uparrow \quad -mg + N = -m\frac{L}{2}\alpha \cos 60° \tag{d}$$

Substituting the expression for α found above, and solving for N, gives

$$N = 0.572\,mg \qquad \textit{Answer}$$

It must be emphasized that the values obtained for N, α, and a_A are valid only at the instant after the release. Each of these variables will vary throughout the subsequent motion of the bar. However, it is interesting to note that there is never a horizontal force acting on the bar because the plane is frictionless. Therefore, the path followed by the mass center G will be a vertical straight line.

Problems

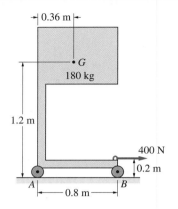

Fig. P17.17

17.17 The 400-N horizontal force is applied to the cabinet that is supported on frictionless casters at A and B. The mass of the cabinet is 180 kg, and G is its mass center. (a) Determine the acceleration of the cabinet assuming that it does not tip. (b) Verify that the cabinet does not tip by computing the reactions at A and B.

17.18 The combined mass center of the motorcycle and the cyclist is located at G. (a) Find the smallest acceleration for which the cyclist can perform a "wheelie"—that is, raise the front wheel off the ground. (b) What minimum coefficient of static friction between the tires and the road is required for the wheelie?

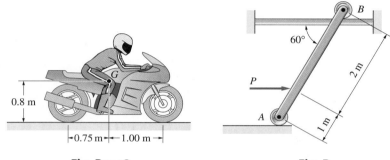

Fig. P17.18 **Fig. P17.19**

17.19 Find the largest force P that would accelerate the 1-kg uniform bar AB to the right without causing the roller at B to lift off the guide rail. Neglect the masses of the rollers.

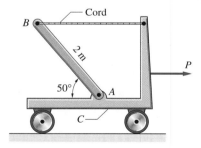

Fig. P17.20

17.20 The uniform 5-kg bar AB is attached to the 7-kg cart C with a pin at A and a horizontal cord at B. If the force in the cord is 40 N, determine the horizontal force P acting on the cart.

17.21 The mass center of the 200-kg trailer is located at G. The trailer is attached to the car by a ball-and-socket joint at C. Determine (a) the maximum acceleration of the car for which the trailer does not pull up on the hitch; and (b) the corresponding horizontal force on the hitch.

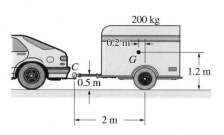

Fig. P17.21

17.22   The 10-kg homogeneous panel, pinned to a frictionless roller A and a light sliding collar B, is acted on by the 20 N horizontal force. Determine the acceleration of the panel and the roller reaction at A, given that the velocity of the panel is 1.8 m/s to the left. Neglect friction.

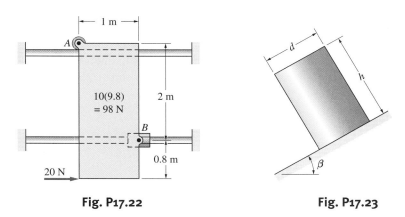

Fig. P17.22   **Fig. P17.23**

17.23   The homogeneous cylinder of mass m slides down the incline of slope angle β. The kinetic coefficient of friction between the cylinder and the incline is μ. Determine the expression for the smallest ratio d/h for which the cylinder will not tip.

17.24   The 20-kg pallet B carrying the 40-kg homogeneous box A rolls freely down the inclined plane. The static and kinetic coefficients of friction between A and B are 0.4 and 0.35, respectively. (a) Show that the box will slide on the pallet, assuming that the dimensions of the box are such that it does not tip. (b) Determine the smallest ratio b/h for which the box will not tip.

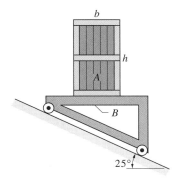

Fig. P17.24

17.25   The 20-kg box rides on a platform that is attached to two parallel arms. The arms are being driven at the constant angular velocity of 2 rad/s CCW. Determine the normal and friction forces exerted on the box by the platform when the system is in the position shown. Assume that the box does not slide relative to the platform.

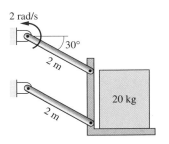

Fig. P17.25

Fig. P17.26

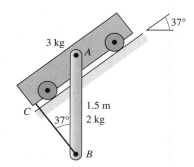

Fig. P17.27

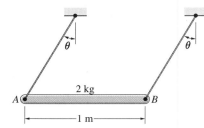

Fig. P17.29

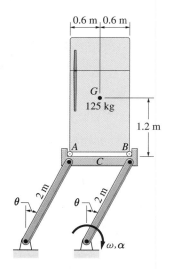

Fig. P17.30

17.26   The uniform bar *AB* is released from rest in the horizontal position. Find the initial acceleration of end *B* of the bar.

17.27   The vertical 2-kg bar *AB* is attached to the 3-kg cart by a pin at *A* and the cord *BC*. Determine the tension in the cord immediately after the assembly is released from rest in the position shown.

17.28   The two identical uniform bars *AC* and *DE*, each of mass *m* and length *L*, are connected by cords at their ends. The assembly hangs in the vertical plane from a pin at *B*. Find the tension in cord *CE* the instant after cord *AD* has been cut.

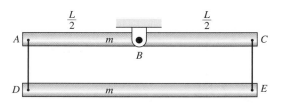

Fig. P17.28

17.29   The 2-kg homogeneous bar *AB* is supported by the parallel strings attached at *A* and *B*. If the bar is released from rest when $\theta = 35°$, calculate the forces in the strings immediately after release.

17.30   The 125-kg refrigerator is being lowered to the ground by the platform *C*, which is controlled by the parallelogram linkage shown. Lips on the platform prevent the refrigerator from rolling on its wheels at *A* and *B*. Prove that the refrigerator will not tip when $\theta = 0$, $\omega = 1.2$ rad/s, and $\alpha = 1.6$ rad/s². (Hint: Assume that the refrigerator does not tip, and find the reactions at *A* and *B*.)

17.31   Determine the angular accelerations of the homogeneous pulleys shown in (a) and (b). The mass moment of inertia for each pulley about its mass center *G* is 4 kg · m².

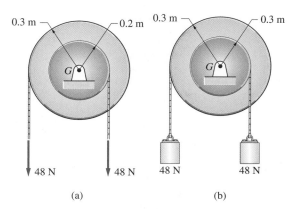

(a)   (b)

Fig. P17.31

17.32   The radius of gyration of the 20-kg pulley about its mass center G is 300 mm. Compute the angular acceleration of the pulley and the tension in the cord AB.

17.33   Determine the maximum possible acceleration of the car if the coefficient of static friction between its tires and the road is 0.8. The car has front-wheel drive, and its mass center is located at G.

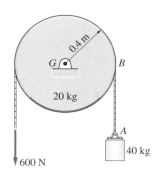

Fig. P17.32

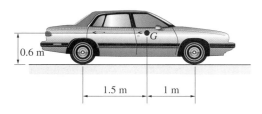

Fig. P17.33

17.34   The two identical 8-kg homogenous cylinders are connected by a cable. Compute the tension in the cable the instant after the cylinders are released from rest in the position shown. Assume that friction allows the cylinders to roll without slipping and neglect the weight of the small pulley.

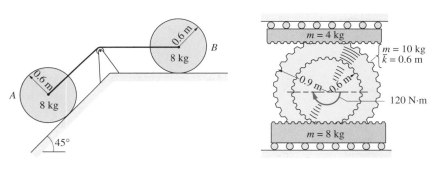

Fig. P17.34

Fig. P17.35

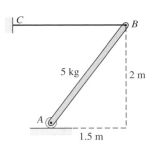

Fig. P17.36

17.35   The compound gear of mass 10 kg engages the two sliding racks. If the 120 N · m couple is applied to the gear, determine the acceleration of the center of the gear.

17.36   The homogeneous, 5-kg bar AB is released from rest in the position shown. Compute the angular acceleration of the bar and the tension in the cord BC immediately after the release.

17.37   Gears A and B, of masses 4 kg and 10 kg, respectively, are rotating about their mass centers. The radius of gyration about the axis of rotation is 100 mm for A and 300 mm for B. A constant couple $C_0 = 0.75$ N · m acts on gear A. Neglecting friction, compute the angular acceleration of each gear and the tangential contact force between the gears at C.

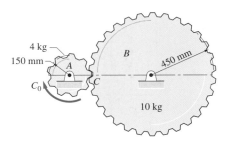

Fig. P17.37

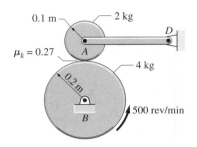

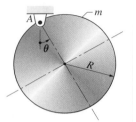

Fig. P17.38

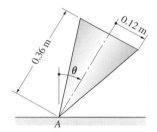

Fig. P17.39

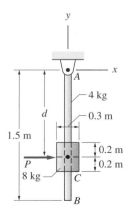

Fig. P17.40

17.38 Uniform disks *A* and *B*, having masses of 2 kg and 4 kg, respectively, can rotate freely about their mass centers. The kinetic coefficient of friction between the disks is 0.27. Disk *B* is spinning at 500 rev/min counterclockwise when it is placed in contact with the stationary disk *A*. Calculate the angular acceleration of each disk during the time that slipping occurs between the disks. Neglect the mass of bar *AD*.

17.39 The uniform disk of radius *R* and mass *m* is free to rotate about the pin at *A*. Determine the magnitude of the pin reaction at *A* immediately after the disk is released from rest when $\theta = 90°$.

17.40 The solid steel cone of density 7850 kg/m³ is released from rest when $\theta = 22°$. Assume that there is sufficient friction at *A* to prevent slipping. Determine the angular acceleration of the cone and the normal and friction forces at *A* immediately after release.

17.41 The uniform 2-kg slender bar *AB* is mounted on a vertical shaft at *C*. A constant couple of 9 N-m is applied to the bar. Calculate the angular acceleration of the bar and the magnitude of the horizontal reaction at *C* at the instant when the angular velocity of the bar is 6 rad/s.

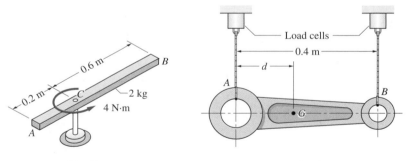

Fig. P17.41   **Fig. P17.42**

17.42 To determine the inertial properties of the connecting rod *AB*, it is suspended from two wires, one of which is subsequently cut. Load cells are used to measure the force in each wire. When the rod is hanging in the position shown, the forces in the wires are measured to be 6.80 N at *A* and 5.20 N at *B*. Immediately after the wire at *B* is cut, the force in the wire at *A* is reduced to 3.6 N. Compute (a) the distance *d* that locates the mass center *G*; and (b) the radius of gyration of the rod about *G*.

17.43 The homogeneous 8-kg collar *C* is fastened to the uniform 4-kg rod *AB*. The mass moment of inertia of *C* about its center is $\bar{I}_z = 0.105$ kg · m². The system is at rest in the position shown when the horizontal force *P* is applied through the mass center of the collar. Compute the distance *d* for which the pin reaction at *A* would not change immediately after *P* is applied.

Fig. P17.43

17.44   The uniform 40-kg bar *AB* is attached to the frame with a pin at *A* while end *B* is resting against the frame. The assembly rotates about a vertical axis at *C*. At the instant shown, the angular velocity and angular acceleration of the assembly are 2 rad/s and 6 rad/s², both clockwise. Determine the magnitude of the pin reaction at *A* at this instant.

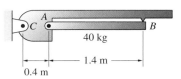

Fig. P17.44

17.45   The 1.8-kg uniform bar rotates in the vertical plane about the pin at *O*. When the bar is in the position shown, its angular velocity is 4 rad/s, clockwise. For this position, find (a) the angular acceleration of the bar; and (b) the magnitude of the pin reaction at *O*.

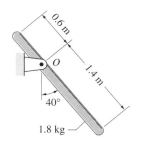

Fig. P17.45

17.46   The axle of the 4-kg homogeneous disk is mounted on the end of bar *AB*, which rotates freely in the vertical plane about the pin at *B*. The cable *BC* maintains the disk in a fixed position relative to the bar. Find the cable tension immediately after the assembly is released from rest in the position shown. Neglect the mass of the bar *AB*.

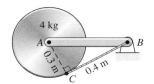

Fig. P17.46

17.47   A rope is wrapped around the uniform 50-kg pulley *B* and attached to the 20-kg block *A*. If the system is released from rest, find (a) the initial acceleration of *A*; and (b) the velocity of *A* after it has moved 1 m down the incline. Neglect friction.

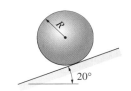

Fig. P17.49

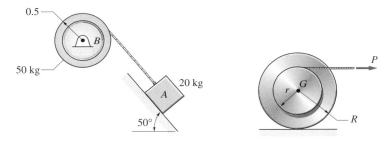

Fig. P17.47   **Fig. P17.48**

17.48   (a) Show that the spool of mass *m* and radius *R* can roll without slipping on the frictionless horizontal surface if the radius of the hub is $r = \bar{I}/(mR)$, where $\bar{I}$ is the mass moment of inertia of the spool about *G*. (b) If the horizontal surface is not frictionless, what is the direction of the friction force acting on the spool if (i) $r > \bar{I}/(mR)$; and (ii) $r < \bar{I}/(mR)$?

17.49   The homogeneous sphere of mass *M* and radius *R* is released from rest and moves down the rough, inclined plane. Calculate the acceleration of the center of the sphere if the coefficient of static friction is insufficient to prevent slipping. The coefficient of kinetic friction is 0.075.

17.50   The mass moment of inertia of the 60-kg spool is $\bar{I} = 1.35 \text{ kg} \cdot \text{m}^2$. The static and kinetic coefficients of friction between the spool and the ground are 0.30 and 0.27, respectively. A cable wound around the hub of the spool is pulled with the constant horizontal force $P = 200$ N. Find the acceleration of the center of the spool.

17.51   Solve Prob. 17.50 if $P = 450$ N.

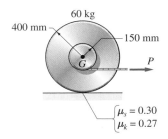

Fig. P17.50, P17.51

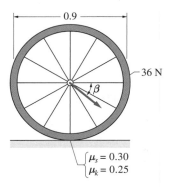

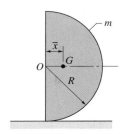

Fig. P17.52, P17.53

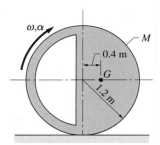

Fig. P17.54

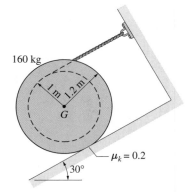

Fig. P17.55

17.52 The rim of the wheel weighs 36 N; the weights of the spokes and hub may be neglected. An 36 N force, inclined at the angle β to the horizontal, is applied to the center of the wheel. The static and kinetic coefficients of friction between the wheel and the ground are 0.30 and 0.25, respectively. If $\beta = 0$, (a) show that the wheel slips on the ground; and (b) find the angular acceleration of the wheel and the acceleration of its center.

17.53 For the wheel described in Prob. 17.52, find the smallest angle β for which the wheel will roll without slipping, and determine the corresponding angular acceleration.

17.54 The homogeneous semicylinder of mass m and radius R is released from rest in the position shown. Assuming no slipping, determine (a) the initial angular acceleration of the semicylinder; and (b) the smallest static coefficient of friction that is consistent with the no-slip condition. (Note: The mass center of the semicylinder is located at $\bar{x} = 4R/3\pi$; and $I_O = mR^2/2$.)

17.55 The radius of gyration of the eccentric disk of mass M about its mass center G is 0.4 m. In the position shown, the angular acceleration of the disk is 3.0 rad/s². Assuming rolling without slipping, find the angular velocity of the disk for this position.

17.56 If the uniform 2-kg bar is released from rest in the position shown, determine the initial angular acceleration of the bar. Neglect friction.

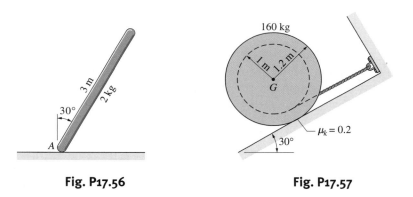

Fig. P17.56   **Fig. P17.57**

17.57 The radius of gyration of the 160-kg spool about its mass center G is 0.5 m. The cable that is wrapped tightly around the inner radius of the spool is attached to a rigid support as shown. If the spool is moving down the rough plane, determine its angular acceleration and the tension in the cable.

17.58 Repeat Prob. 17.57 if the cable unwinds from the top of the hub as shown.

Fig. P17.58

17.59   The 3.6-kg homogeneous bar *AB* is pinned to the 2-kg slider at *A*. The system was at rest in the position $\theta = 0$ before the 12-N force was applied to the collar. Neglecting friction, compute the acceleration of the collar and the angular acceleration of the bar immediately after the 12-N force is applied.

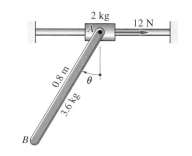

Fig. P17.59

17.60   The 8-kg uniform slender bar was at rest on a frictionless horizontal plane before the application of the force $F = 16$ N. For the instant immediately after *F* was applied, determine (a) the acceleration of end *A*; and (b) the *x*-coordinate of the point on the bar that has zero acceleration.

17.61   The uniform bar *AB* of mass *m* and length *L* is released from rest in the position shown. If the inclined plane is frictionless, calculate the initial acceleration of end *A*.

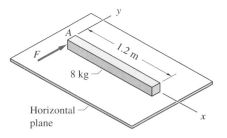

Fig. P17.60

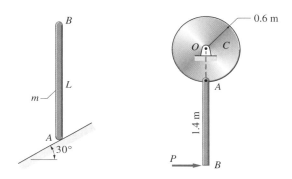

Fig. P17.61   Fig. P17.62

17.62   The uniform 30-kg bar *AB* hangs from a pin attached to the 50-kg homogeneous disk *C*. The system is at rest when the horizontal force $P = 200$ N is applied. Find the angular accelerations of the disk and the bar immediately after *P* is applied.

17.63   The mass moment of inertia of the 128-kg disk about its mass center *G* is 20 kg · m². The axle of the disk is supported by the lower half of a split bearing. The rope wrapped around the periphery of the disk is pulled horizontally with speed v_0. Find the largest value of v_0 for which the axle of the disk will stay in the bearing. Neglect friction.

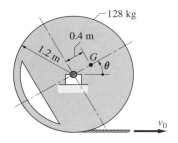

Fig. P17.63

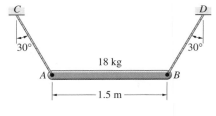

Fig. P17.64

17.64   The 18-kg homogeneous bar *AB* is at rest in the position shown when rope *BD* is cut. Determine the initial values of (a) the angular acceleration of the bar; and (b) the acceleration of end *B*.

17.65   The uniform rod of mass *m* and length *L* is supported by rollers at *A* and *B*. Find the acceleration of roller *B* immediately after the rod is released in the position shown. Neglect the masses of the rollers.

Fig. P17.65

*17.66   The pin *B* attached to the end of the uniform 0.3-kg crank *AB* slides in a vertical slot in the 0.45-kg slider *CD*. A constant counterclockwise angular velocity of 2000 rev/min is maintained by the couple C_A. Determine C_A as a function of the crank angle θ, and use this expression to show that the gravitational forces are negligible compared with the inertial forces. Neglect friction.

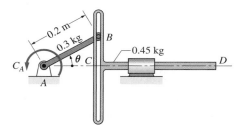

Fig. P17.66

*17.6   *Differential Equations of Motion*

The preceding article discussed the derivation of the equations of motion for a rigid body using the FBD-MAD method. The accompanying problems were restricted to the computation of the forces and accelerations at the instant when the body was in a *specified position*. In this article we consider the more practical problem of determining the motion of a rigid body as a function of time and/or position. As noted in previous chapters, the determination of motion involves two steps: the equations of motion must first be derived, and then they must be integrated (solved).

The equations of motion are obtained by applying the FBD-MAD method to an *arbitrary position* of the body. The forces appearing in the resulting equations of motion fall into two categories: applied loads and constraint forces. The applied loads are usually given as functions of time, position, or velocity. On the other hand, the constraint forces, such as pin reactions, are unknowns. Because the equations of motion are integrable when they contain only kinematic variables (the generalized coordinates and their derivatives) as the unknowns, the

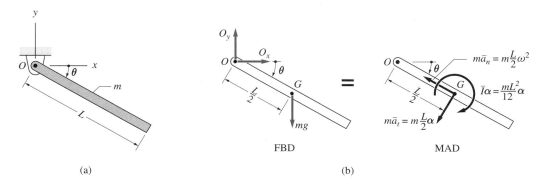

Fig. 17.14

constraint forces must be eliminated from the equations of motion. The equations that remain after the elimination procedure are called the *differential equations of motion*.

As an illustration, consider the homogeneous bar of mass m and length L shown in Fig. 17.14(a). We are to determine the resulting motion when the bar is released from rest at $\theta = 0$. The FBD and MAD of the bar for an arbitrary value of θ are shown in Fig. 17.14(b). From these diagrams, we obtain the following three independent equations of motion:

$$\Sigma F_x = m\bar{a}_x \quad \xrightarrow{+} \quad O_x = -\frac{mL}{2}\omega^2\cos\theta - \frac{mL}{2}\alpha\sin\theta \tag{a}$$

$$\Sigma F_y = m\bar{a}_y \quad +\uparrow \quad O_y - mg = \frac{mL}{2}\omega^2\sin\theta - \frac{mL}{2}\alpha\cos\theta \tag{b}$$

$$\Sigma M_G = \bar{I}\alpha \quad \overset{+}{\curvearrowright} \quad O_x\frac{L}{2}\sin\theta + O_y\frac{L}{2}\cos\theta = \frac{mL^2}{12}\alpha \tag{c}$$

These equations of motion contain the components of the unknown pin reaction at O in addition to the kinematic variables θ, ω, and α. Substituting the expressions for O_x and O_y, obtainable from Eqs. (a) and (b), into Eq. (c) yields the differential equation of motion

$$\alpha = \ddot{\theta} = (3g/2L)\cos\theta \tag{d}$$

Equation (d) could also be obtained directly from the special case of the moment equation of motion: $\Sigma M_O = I_O\alpha$.

The following conclusions can now be drawn.

1. If a problem is solvable, the number of independent equations of motion equals the number of degrees of freedom (DOFs) plus the number of unknown forces. In the preceding illustration, the number of independent equations of motion was three, the body had one DOF (θ being the generalized coordinate), and there were two unknown components of the pin reaction at O.

2. Because an equation of motion is required to eliminate each of the unknown constraint forces, the number of differential equations of motion is equal to the number of DOFs. In the foregoing illustration, two equations, Eqs. (a) and (b), were used to eliminate O_x and O_y, resulting in the single differential equation of motion, Eq. (d).

The differential equations of motion can be solved analytically only in a few special cases, making numerical integration the primary method of solution. However, in this article we consider both analytical and numerical solutions. In either case, the solution of the differential equations of motion (note that they are second-order equations) requires the knowledge of two *initial conditions* for every DOF: the values of the generalized coordinates and the generalized velocities at some instant of time (usually at $t = 0$).

Sometimes it is possible to obtain a partial solution to the differential equations of motion in closed form, whereas the complete solution would require numerical integration. For example, to obtain the expression for ω versus θ for the bar shown in Fig. 17.14, we substitute α from Eq. (d) into $\alpha\, d\theta = \omega\, d\omega$, obtaining

$$\omega\, d\omega = (3g/2L) \cos\theta\, d\theta \tag{e}$$

Integrating both sides of this equation, and applying the initial condition $\omega = 0$ when $\theta = 0$, the relationship between ω and θ becomes

$$\omega = \sqrt{(3g/2L) \sin\theta} \tag{f}$$

The relationships between the kinematic variables and *time*, however, cannot be determined this easily. To find θ versus time, for example, we substitute ω from Eq. (f) into $\omega = d\theta/dt$. Solving for dt, we find

$$dt = d\theta/\sqrt{(3g/2L) \sin\theta} \tag{g}$$

The right-hand side of Eq. (g) cannot be integrated in a closed form. Therefore, we see that even relatively simple problems may require numerical integration for the complete determination of the motion.

It should be mentioned that after the values of ω and α have been found as functions of θ, Eqs. (a) and (b) can be used to find O_x and O_y, also as functions of θ. Similarly, if θ, ω, and α have been found as functions of time, Eqs. (a) and (b) will determine O_x and O_y as functions of time.

Sample Problem 17.10

The 180 kg uniform plate shown in Fig. (a) rotates in the vertical plane about a pin at A. The plate is released from rest when $\theta = 0$. (1) Show that the differential equation of motion for the plate is $\alpha = 0.588(4 \cos \theta - 3 \sin \theta) \text{ rad/s}^2$. (2) Integrate the differential equation of motion analytically to obtain the angular velocity of the plate as a function of θ. (3) Find the maximum value of θ.

Solution

Part 1

The mass of the plate is $m = 180$ kg, and the moment of inertia about its mass center G is (see Table 17.1)

$$\bar{I} = \frac{1}{12}m(b^2 + c^2) = \frac{1}{12}(180)(4^2 + 3^2) = 375 \text{ kg} \cdot \text{m}^2$$

Figure (b) shows the free-body diagram (FBD) and mass-acceleration diagram (MAD) for an arbitrary position of the plate. The FBD contains A_n and A_t, the unknown components of the pin reactions at A, and the $180(9.8) = 1764$ N weight acting at G. The MAD consists of the inertia couple $\bar{I}\alpha$ and the components of the inertia vector $m\bar{a}$ acting at G. Because the path of G is a circle of radius $r = 2.5$ m centered at A, the normal component of $\bar{a}$ is $\bar{a}_n = r\omega^2 = 2.5\omega^2 \text{ ft/s}^2$, and its tangential component is $\bar{a}_t = r\alpha = 2.5\alpha \text{ ft/s}^2$. Observe that the angular acceleration α is assumed to be clockwise in the MAD. The units for ω and α are rad/s and rad/s^2, respectively.

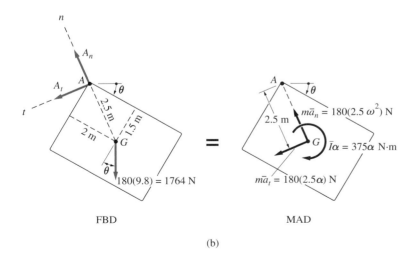

FBD   MAD

(b)

Because the angle θ completely determines the position of the plate, the plate has a single degree of freedom. Therefore, there is only one differential equation of motion. The most convenient method of deriving this equation is by equating moments about A in the FBD and MAD, thereby obtaining

$$(\Sigma M_A)_{\text{FBD}} = (\Sigma M_A)_{\text{MAD}}$$

$$\overset{+}{\circlearrowleft} \quad (1764\cos\theta)(2) - (1764\sin\theta)(3) = 375\alpha + 180(2.5\alpha)(2.5)$$

which reduces to

$$\alpha = 0.588(4\cos\theta - 3\sin\theta) \text{ rad/s}^2 \qquad \textit{Answer} \quad \text{(a)}$$

The identical result could be obtained by using the special case of the moment equation: $\Sigma M_A = I_A\alpha$.

Part 2

To find the angular velocity as a function of θ, we substitute α from Eq. (a) into $\omega\,d\omega = \alpha\,d\theta$, which yields

$$\omega\,d\omega = 0.588(4\cos\theta - 3\sin\theta)\,d\theta$$

The result of integrating this equation is

$$\frac{\omega^2}{2} = 0.588(4\sin\theta + 3\cos\theta) + C$$

The constant of integration C is evaluated by applying the initial condition $\omega = 0$ when $\theta = 0$, which gives $C = -3(0.588)$ rad^2/s^2. Therefore, the above equation becomes

$$\frac{\omega^2}{2} = 0.588(4\sin\theta + 3\cos\theta - 3)$$

from which the angular velocity is found to be

$$\omega = \pm 1.08\sqrt{4\sin\theta + 3\cos\theta - 3} \text{ rad/s} \qquad \textit{Answer} \quad \text{(b)}$$

Part 3

The maximum value of θ occurs when $\omega = 0$. According to Eq. (b), this value is the nonzero root of the equation $4\sin\theta + 3\cos\theta - 3 = 0$. The numerical solution of this equation yields

$$\theta_{\max} = 106.3° \qquad \textit{Answer}$$

Only one equation of motion was used to determine the motion of the plate. The remaining two independent equations of motion—for example, $\Sigma F_n = m\bar{a}_n$

and $\Sigma F_t = m\bar{a}_t$—could now be utilized to find the components of the pin reaction, A_n and A_t, in terms of θ, ω, and α. By substituting for α from Eq. (a) and ω from Eq. (b), we would then obtain A_n and A_t as functions of θ only.

Sample Problem 17.11

For the plate described in Sample Problem 17.10, (1) solve the differential equation of motion numerically from the time of release until θ reaches its maximum value for the first time; and (2) use the numerical solution to determine the maximum value of θ and the time when it first occurs.

Solution

Part 1

The differential equation of motion given in Sample Problem 17.10 is $\alpha = 0.9660(4\cos\theta - 3\sin\theta)$ rad/s^2. The equivalent first-order equations are

$$\dot{\theta} = \omega \qquad \dot{\omega} = 0.588(4\cos\theta - 3\sin\theta) \text{ rad/s}^2 \qquad \text{(a)}$$

with the initial conditions $\theta = \omega = 0$ at $t = 0$ (the time of release). Letting

$$x_1 = \theta \quad x_2 = \omega$$

we obtain the following MATLAB program for the solution of the differential equations in Eq. (a):

```
function example17_11
x0 =[0 0];
time =[0:0.05:1.6];
[t,x] = ode45 (@f,time,x0);
axes('fontsize',14)
plot(t,x(:,1),'linewidth',1.5)
grid on
xlabel('t (s)'); ylabel('theta (rad)')
printSol(t,x)

    function dxdt = f(t,x)
    dxdt = [x(2)
            0.588*(4*cos(x(1))-3*sin(x(1)))];
    end
end
```

The period of integration is from $t = 0$ to the time where θ reaches its maximum value for the first time (where $\omega = 0$). By trial and error, this time was found to be approximately 1.5 s. The plot of θ versus time is shown in the following figure.

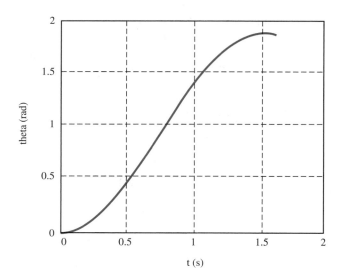

Part 2

To determine the time when θ is at its maximum, we must examine the numerical output in the vicinity where ω changes sign. The two lines of printout that span $\omega = 0$ are shown below.

```
    t        x1        x2
 1.9000   1.8531    0.0842
 1.9500   1.8543   -0.0334
```

Letting t_1 be the time when $\omega = 0$, linear interpolation yields

$$\frac{-0.0334 - 0.0842}{1.95 - 1.90} = \frac{0 - 0.0842}{t_1 - 1.90}$$

The solution is

$$t_1 = 1.936 \, \text{s} \qquad \qquad \textit{Answer}$$

From the printout we deduce by inspection that

$$\theta_{\text{max}} = 1.854 \, \text{rad} = 106.2° \qquad \qquad \textit{Answer}$$

Problems

17.67   The uniform bar AB of mass m, which is supported by two wires, is released from rest when $\theta = \theta_0$. (a) Show that the differential equation of motion is $\alpha = -(g/L)\sin\theta$. (b) Integrate the differential equation of motion analytically to find ω as a function of θ. (c) Determine the force in each wire as a function of θ.

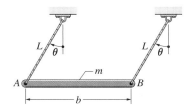

Fig. P17.67

17.68   The uniform pole AB of mass m is hoisted from rest in the horizontal ($\theta = 90°$) position by a cable passing over the pulley C. The force acting at the end of the cable has constant magnitude $P = mg/\sqrt{2}$ (it is equal to the smallest force necessary to initiate the motion). (a) Show that the differential equation of motion is $\alpha = 3g/(2L)[\sin\theta - \sqrt{2}\cos(\theta/2)]$. (b) Integrate the differential equation of motion analytically to obtain ω in terms of θ. (c) What is the angular velocity of the pole when it reaches the vertical position?

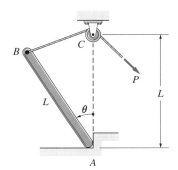

Fig. P17.68

17.69   The dimensions of the homogeneous 40-kg box are $h = 0.8$ m and $b = 0.6$ m. The box is at rest on the lubricated horizontal surface, then the constant force $P = 220$ N is applied at time $t = 0$. The hydrodynamic drag force of the lubricant during the ensuing motion is $F_D = 20v$ N, where v is the speed of the box in meters/second. (a) Show that the differential equation of motion of the box is $a = 5.5 - 0.5v$ m/s². (b) Find the time when the box tips over.

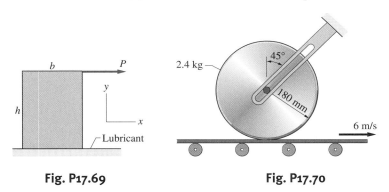

Fig. P17.69	**Fig. P17.70**

17.70   The axle of the 2.4-kg homogeneous disk fits into an inclined slot. The disk is initially at rest and is then lowered onto the conveyor belt that is being driven at the constant speed of 6 m/s. The kinetic coefficient of friction between the disk and the belt is 0.15. Calculate (a) the angular acceleration of the disk during the time that it slips on the belt; and (b) the number of revolutions made by the disk before it reaches its final angular speed.

17.71   The thin rim of the wheel weighs 3-kg; the weights of the spokes and hub may be neglected. Before the homogeneous 1-kg bar AB was lowered into the position shown, the wheel was rotating freely at 400 rev/min clockwise. If the kinetic coefficient of friction at B is 0.75, determine (a) the angular acceleration of the wheel; and (b) the time it takes the wheel to stop.

17.72   Repeat Prob. 17.71 for the case where the wheel was initially rotating counterclockwise at 400 rev/min.

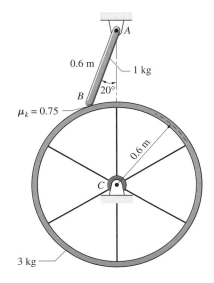

Fig. P17.71, P17.72

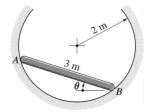

Fig. P17.73

17.73   The axle of the 1-kg uniform disk A can slide in the vertical slot, whereas the axle of the 1.5-kg uniform disk B is fixed. The coefficient of kinetic friction between the disks is 0.4. Disk A is rotating clockwise at 200 rev/min when it is lowered onto the stationary disk B. Determine (a) the angular acceleration of each disk during the time that slipping occurs; and (b) the final angular speed of each disk.

17.74   The homogeneous bar AB of mass m and length L is released from rest in the position $\theta = \theta_0$. (a) Show that the differential equation of motion is $\alpha = -(3g/2L) \sin \theta$. (b) Integrate the differential equations analytically to find ω as a function of θ. (c) Find the maximum value of the vertical component of the pin reaction at A and the value of θ at which it occurs.

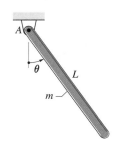

Fig. P17.74

17.75   After the 3-kg uniform bar AB is released from rest in the position $\theta = 35°$, it slides inside the frictionless cylindrical surface. (a) Show that the differential equation of motion is $\alpha = -4.9 \sin \theta$ rad/s². (b) Integrate the differential equation analytically to determine ω in terms of θ. (c) Find the normal reaction at A as a function of θ.

17.76   The radius of gyration of the 9-kg pulley about its mass center G is 0.30 m. The 1.125-kg weight hangs from a cord that is wrapped around the pulley. (a) Derive the differential equation of motion for the pulley. (b) If the pulley is given an initial counterclockwise angular velocity of 10 rad/s when $\theta = 0$, find θ when the system comes to rest.

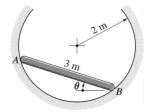

Fig. P17.75

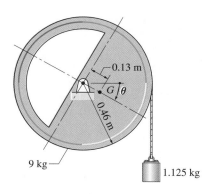

Fig. P17.76

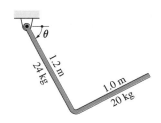

Fig. P17.77

17.77   The L-shaped bar is released from rest when $\theta = 0$. (a) Show that the differential equation of motion is $\alpha = 8.017 \cos \theta + 2.088 \sin \theta$ rad/s². (b) Integrate the equation of motion analytically to obtain the expression for ω in terms of θ. (c) Determine the maximum value of ω and the corresponding value of θ. (d) Find the maximum value of θ.

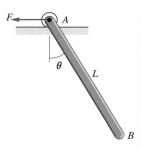

Fig. P17.78

17.78   The uniform bar AB of mass m is at rest in the position $\theta = 0$ when the force F imposes the constant acceleration $a_A = g$ on end A. (a) Neglecting the mass of the roller, show that the differential equation of motion for AB is $\alpha = (3g/2L)(\cos \theta - \sin \theta)$. (b) Integrate the differential equation of motion analytically to find the angular velocity ω as a function of θ. (c) Determine the maximum value of θ and the corresponding value of F.

17.79   The T-shaped assembly of total mass m was made by welding together two identical rods each of length L. If the assembly is released from rest when $\theta = 0$, determine its angular velocity ω and angular acceleration α of the assembly as functions of θ.

17.80   The homogeneous 6-kg block is pinned to the parallel links AB and CD of negligible weight. The assembly is initially at rest in the $\theta = 0$ position when the constant 60-N · m couple is applied to AB. (a) Show that the differential equation of motion is $\alpha = 10 - 9.8 \cos \theta$ rad/s². (b) Integrate the differential equation of motion analytically to obtain ω as a function of θ.

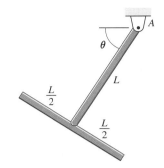

Fig. P17.79

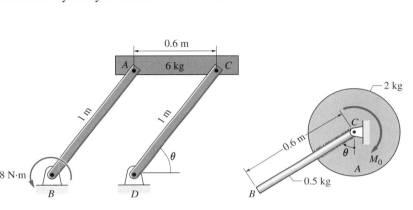

Fig. P17.80

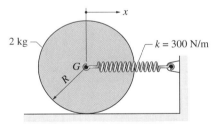

Fig. P17.81

17.81   The 0.5-kg uniform rod BC is welded to the 2-kg drum A. The radius of gyration of the drum about its mass center C is 0.1 m. The constant couple M_0 lb · ft is applied when the system is at rest at $\theta = 0$. (a) Show that the differential equation of motion for the system is $\alpha = 12.5M_0 - 18.375 \sin \theta$ rad/s². (b) Find the value of M_0 for which the bar reaches the position where both its angular velocity and angular acceleration are zero. Compute θ in that position.

17.82   The 2-kg uniform disk of radius R rolls without slipping on the rough, horizontal surface. The stiffness of the spring attached to the center G of the disk is 300 N/m. The disk is released from rest in the position shown with the spring stretched by 75 mm. (a) Determine the acceleration and velocity of G in terms of its coordinate x. (b) Compute the maximum velocity of G and the corresponding value of x. (c) Find the smallest static coefficient of friction that would prevent slipping.

Fig. P17.82

17.83   The thin 0.5-kg hoop is launched on a horizontal surface with the velocity $v_0 = 4$ m/s and the angular velocity $\omega_0 = 12$ rad/s, both directed as shown in the figure. The kinetic coefficient of friction between the hoop and the surface is 0.25. Calculate (a) the angular acceleration of the hoop and the acceleration of the mass center G during the period of slipping; (b) the time elapsed before slipping stops; and (c) the final velocity of G.

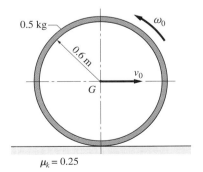

Fig. P17.83

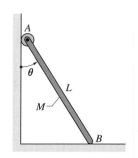

Fig. P17.84

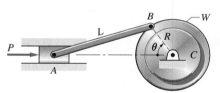

Fig. P17.85, P17.86

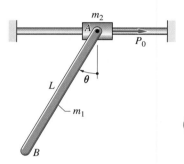

Fig. P17.87

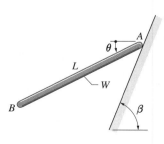

Fig. P17.88

*17.84 The uniform ladder AB of length $L = 3$ m and mass $M = 25$ kg is released from rest when $\theta = 30°$. Friction between the ladder and the ground is negligible. (a) Derive the expressions for the angular velocity and angular acceleration of the ladder, assuming that end A remains in contact with the vertical wall. (b) Determine the expression for the contact force at A as a function of θ. (c) At what value of θ will end A lose contact with the wall?

17.85 Bar AB of mass m_1 and length L is pinned to the sliding collar of mass m_2. The system is at rest with $\theta = 0$ when the constant horizontal force P_0 is applied to the collar. (a) Assuming that friction is negligible, show that the differential equation of motion for bar AB is

$$\alpha = \frac{2P_0 \cos\theta - 2g(m_1 + m_2)\sin\theta - m_1 L\omega^2 \sin\theta \cos\theta}{(4L/3)(m_1 + m_2) + m_1 L \cos^2\theta}$$

(b) Use numerical integration to find the maximum value of θ during the first two seconds of motion. Use the following data: $m_1 = 3.6$ kg, $m_2 = 2.0$ kg, $L = 0.8$ m, and $P_0 = 12$ N. (c) Plot θ versus time for the period of integration.

17.86 When the angular position θ of a body is small enough, its equations of motion can be simplified by the approximations $\sin\theta \approx \theta$ and $\cos\theta \approx 1$. Applying these approximations to the differential equation of motion in Prob. 17.85, re-solve parts (b) and (c). Compare the maximum value of θ with 24.6°, the value obtained if the small angle approximations are not used.

17.87 The steam engine consists of the balanced flywheel C pinned to the connecting rod AB. The rod, in turn, is pinned to the double-action piston A. The steam pressure is regulated so that the force P exerted on the piston varies with the angle θ as $P = P_0 \sin\theta$, where P_0 is a constant. The weight of the flywheel is W and its radius of gyration about its mass center is $\bar{k}$; the weights of the piston and the connecting rod may be neglected. (a) Show that the differential equation of motion for the flywheel is

$$\alpha = \frac{gP_0 R \sin\theta}{W\bar{k}^2}\left(\sin\theta + \frac{\cos\theta}{\sqrt{(L/R)^2 - \sin^2\theta}}\right)$$

(b) If the flywheel starts from rest when $\theta = 90°$, determine by numerical integration the time required for it to reach a speed of 10 rad/s. Use the following data: $W = 810$ N, $\bar{k} = 0.2$ m, $R = 0.225$ m, $L = 0.45$ m, $P_0 = 108$ N. (c) Plot the angular speed versus time for the period of integration.

17.88 The uniform rod AB is released from rest when $\theta = 0$ with end A in contact with the frictionless, inclined surface. (a) Assuming that end A maintains contact with the surface, show that the differential equation governing the angular motion of the rod is

$$\alpha = \frac{(2g/L)(\cos\theta - \sin\beta \sin\phi) - \omega^2 \sin\phi \cos\phi}{(4/3) - \sin^2\phi} \quad \text{rad/s}^2$$

where $\phi = \beta - \theta$. (b) Solve the differential equation of motion in part (a) numerically from the time of release until end B makes contact with the inclined surface.

Use this solution to determine ω at the time of contact. Use the data $L = 2.4$ m, $W = 30$ lb, $\beta = 60°$.

17.89   The small collar C of mass m_2 slides with negligible friction on rod AB of mass m_1 and length L. The system is released from rest when $\theta = \theta_0$ and $r = L$. (a) Show that the differential equations of motion of the system are

$$\ddot{r} = -g\sin\theta + r\dot{\theta}^2$$

$$\ddot{\theta} = -\frac{3}{2}\frac{g\cos\theta[L(m_1/m_2) + 2r] + 4r\dot{r}\dot{\theta}}{L^2(m_1/m_2) + 3r^2}$$

(b) Integrate the equations numerically from the time of release until the collar leaves the rod at B and plot r versus θ. Use $\theta_0 = 60°$, $m_1/m_2 = 2$, $L = 0.45$ m.

17.90   The small collar D of mass m slides on the rod AB. A light spring of stiffness k joins the collar to end A of the rod. The spring is undeformed when the collar is in the position $r = r_0$. The rod and the frame ACB rotate freely about the vertical axis OC, their combined moment of inertia about that axis being I. A shaker (not shown) drives the base E in such a manner that the position of E is given by $y(t) = a\sin pt$, where a and p are constants. (a) Show that the differential equations of motion for the system are

$$\ddot{r} = -\frac{k}{m}(r - r_0) + r\dot{\theta}^2 + ap^2\sin pt\sin\theta$$

$$\ddot{\theta} = \frac{r}{(I/m) + r^2}(ap^2\sin pt\cos\theta - 2\dot{r}\dot{\theta})$$

(b) Solve the differential equations of motion numerically from $t = 0$ to 3 s and plot θ versus t. Assume that the motion begins at $t = 0$ with the system at rest in the position $\theta = 0$ and $r = r_0$. From the solution, determine the direction of rotation (as viewed from above). Use the following data: $m = 0.125$ kg, $k = 3.125$ N/m, $r_0 = 50$ mm, $I = 312.5 \times 10^{-6}$ kg · m^2, $a = 10$ mm, and $p = 10$ rad/s.

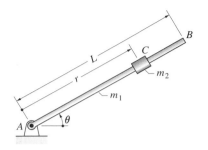

Fig. P17.89

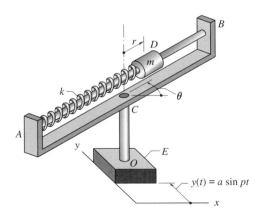

Fig. P17.90

Review of Equations

Mass moment of inertia

The mass moment of inertia of a body about an axis (the a-axis) is

$$I_a = \int_{\mathcal{V}} r^2 \, dm = mk_a^2$$

r = perpendicular distance from the axis to the differential mass dm
$\mathcal{V}$ = volume occupied by the body of mass m
k_a = radius of gyration of the body about the a-axis

Parallel-axis theorem

$$I_a = \bar{I}_a + md^2$$

$\bar{I}_a$ = mass moment of inertia about the centroidal axis that is parallel to a-axis
d = distance between the centroidal axis and the a-axes

Angular momentum of a rigid body in plane motion

$h_G = \bar{I}\omega$ (about the mass center G)
$h_A = I_A\omega$ (about the instant center A for velocities)
$h_A = \bar{I}\omega + m\bar{v}d$ (about an arbitrary point A)

d = moment arm of the momentum vector $m\bar{\mathbf{v}}$ about A

Equations of plane motion

$\Sigma\mathbf{F} = m\bar{\mathbf{a}}$

$\Sigma M_G = \bar{I}\alpha$ (G is the mass center)

$\Sigma M_A = I_A\alpha$ (point A is fixed in the body and in space)

$\Sigma M_A = \bar{I}\alpha + m\bar{a}d$ (A is an arbitrary point)

d = moment arm of the inertia vector $m\bar{\mathbf{a}}$ about A

Review Problems

17.91   The figure shown is made of uniform, thin wire; its mass is m. Determine (a) the x-coordinate of the mass center; (b) the moment of inertia about the z-axis; and (c) the moment of inertia about the $\bar{z}$-axis.

17.92   For the homogeneous bracket shown, determine (a) the x-coordinate of the mass center; (b) the radius of gyration about the z-axis; and (c) the radius of gyration about the $\bar{z}$-axis (the axis that passes through the mass center).

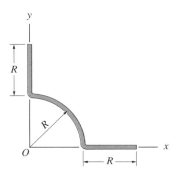

Fig. P17.91

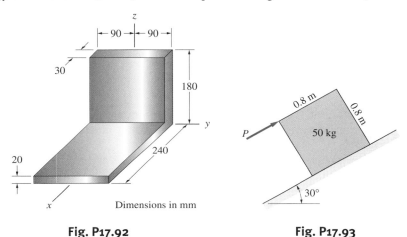

Fig. P17.92   **Fig. P17.93**

17.93   The 50-kg block is pushed up the 30° incline with the constant force P. The coefficient of kinetic friction between the block and the incline is 0.2. Determine the largest force P that can be applied without tipping the block.

17.94   The 6-kg uniform bar AB is swinging freely from two rods of negligible weight. In the position $\theta = 30°$, the tension in rod AC is measured to be 20 N. Determine the magnitudes of the acceleration and velocity vectors of the mass center of the bar AB in the same position.

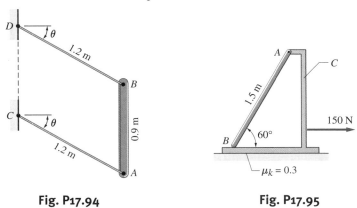

Fig. P17.94   **Fig. P17.95**

17.95   The uniform bar AB of weight W is attached to the 12-kg frame C with a pin at A. The kinetic coefficient of friction between the frame and the horizontal surface is 0.3. Determine the smallest value of W for which the bar remains in contact with the frame at B after the 150 N force is applied to the frame.

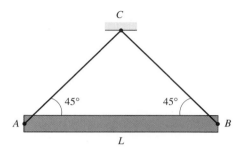

Fig. P17.96

17.96 The uniform beam AB of mass m and length L is suspended by two cables. Determine the tension in the cable BC immediately after cable AC breaks.

17.97 The homogeneous 6-kg bar AB is pinned to the disk at A. A rope connects end B of the bar to the center C of the disk. Calculate the tension in the rope when the disk is rotating in the horizontal plane with the constant angular velocity $\omega = 10$ rad/s.

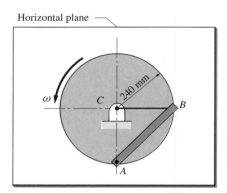

Fig. P17.97

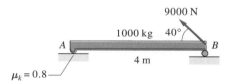

Fig. P17.98

17.98 The uniform 1000-kg beam is resting on a rough surface at A and a roller support at B before the 9000 N force is applied. Determine the angular acceleration of the beam and the acceleration of end A immediately after the force is applied. Assume that end A slides on the rough surface. The kinetic coefficient of friction at A is 0.8.

Fig. P17.99

17.99 The identical blocks A and B, each of mass m, are connected by a rod of length L and negligible mass. Block A is free to slide on the frictionless, inclined surface. If the system is released from rest in the position shown, determine expressions for the acceleration of block A and the axial force in the rod immediately after the release.

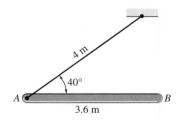

Fig. P17.100

17.100 The uniform 50-kg bar AB is supported by a cable at A. Determine the angular acceleration of the bar and the tension in the cable immediately after the bar is released from rest in the position shown.

17.101   The T-bar consists of two identical rods, each of mass m and length L. Determine the pin reaction at A immediately after the bar is released from rest in the position shown.

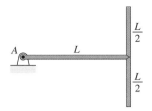

Fig. P17.101

17.102   The uniform 50-kg log AB is released from rest at an unknown value of θ. Knowing that the vertical reaction at A is 250 N when $\theta = 40°$, determine the angular speed of the log in this position.

17.103   The homogeneous box of mass m is released from rest in the position shown. Determine the initial acceleration of the small roller at A.

17.104   The spool mass 150 kg, and its radius of gyration with respect to its mass center G is 1 m. A cable is wound tightly around the hub of the spool with one end attached to the wall at A. The 500 N force causes the spool to slip at its point of contact with the horizontal plane. Determine the acceleration of G and the tension in the cable. The coefficient of kinetic friction between the spool and the plane is 0.4.

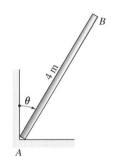

Fig. P17.102

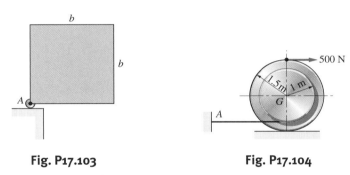

Fig. P17.103   **Fig. P17.104**

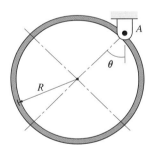

Fig. P17.105

17.105   The uniform thin ring of mass m and radius R is released from rest when $\theta = 90°$. (a) Show that the differential equation of motion is $\alpha = (-g/2R)\sin\theta$. (b) Integrate the expression for α analytically to obtain the relationship between the angular velocity ω and θ. (c) Derive the expression for the magnitude of the pin reaction at A as a function of θ.

17.106   The homogeneous slender 20-kg bar is moving in the vertical plane. The ends of the bar are pinned to sliders A and B of negligible weight. Determine the largest horizontal force P that will slide the bar to the right without lifting slider B from the surface. Neglect friction.

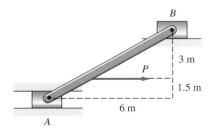

Fig. P17.106

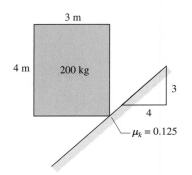

Fig. P17.107

17.107   The uniform 200-kg box is released from rest in the position shown. Calculate the initial acceleration of corner A, assuming it slips on the rough inclined surface.

17.108   The uniform bar AB of mass m and length L is at rest in the position $\theta = 0$ when the constant vertical force P is applied at A. Assuming $P > mg/2$, determine the following as functions of θ: (a) the angular acceleration and angular velocity of the bar; and (b) the horizontal component of the pin reaction at B.

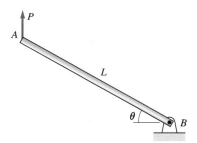

Fig. P17.108

17.109   The homogeneous disk of mass m is stationary when it is lowered onto a conveyor belt that is moving at the constant speed of 5 m/s. If the kinetic coefficient of friction between the disk and the belt is 0.25, determine the time when the disk reaches its final angular speed?

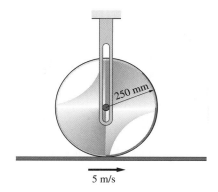

Fig. P17.109

18

Planar Kinetics of Rigid Bodies: Work-Energy and Impulse-Momentum Methods

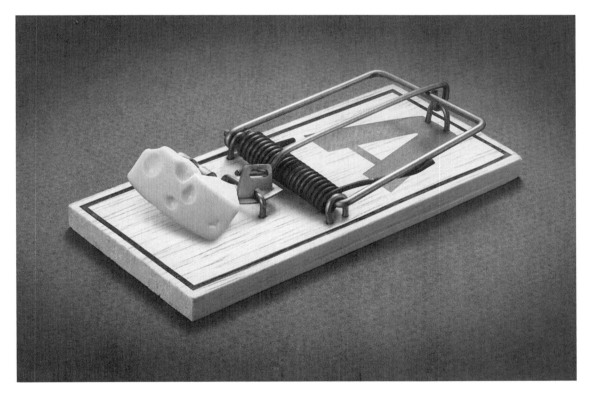

| 18.1 | Introduction |

This chapter continues the kinetic analysis of plane motion of rigid bodies that was begun in the previous chapter.

Part A of the chapter extends the work-energy method for a system of particles (see Chapter 15) to rigid bodies. The work-energy method, when applied to rigid-body motion, relates the work done by the applied forces and *couples* to the change in the kinetic energy of the body. Therefore, our presentation of this method is preceded by discussions of the work done by a couple and the kinetic energy of a rigid body. As is the case for particle motion, the work-energy method is convenient for finding the change in the speed of the body as it moves between two spatial positions.

In Part B, the impulse-momentum method for systems of particles (see Chapter 15) is extended to rigid bodies. This method relates the linear and angular impulses of the applied forces and couples to the changes in the body's linear and *angular momenta*. The computation of angular momentum makes extensive use of momentum diagrams, which were introduced in the previous chapter.

As discussed in Chapter 15, the conservation of momentum (linear and/or angular) is one of the more useful concepts in dynamics. It is invaluable in rigid body impact, which is also included in Part B. As pointed out before, impact problems can be analyzed only by impulse-momentum techniques, because impact forces are generally unknown, and energy is not conserved during impact.

PART A: *Work-Energy Method*

18.2   *Work and Power of a Couple*

a.   *Work*

In applying the work-energy method to rigid body motion, it is frequently necessary to calculate the work done by a couple. The work of a couple can be derived with the assistance of two tools already at our disposal: the definition of work of a force and rigid body kinematics.

Figure 18.1 shows a couple applied to a rigid body. The couple is represented by two parallel, but oppositely directed, forces of magnitude F, acting at points A and B. The corresponding couple-vector is

$$\mathbf{C} = \mathbf{r}_{B/A} \times \mathbf{F} \tag{a}$$

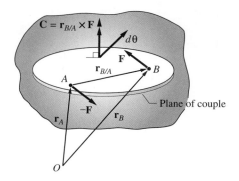

Fig. 18.1

The work done by the couple is calculated by summing the work of the two forces that constitute the couple. Therefore, the work of the couple during an infinitesimal displacement of the body is

$$dU = -\mathbf{F} \cdot d\mathbf{r}_A + \mathbf{F} \cdot d\mathbf{r}_B = \mathbf{F} \cdot (d\mathbf{r}_B - d\mathbf{r}_A) = \mathbf{F} \cdot d\mathbf{r}_{B/A} \tag{b}$$

where $d\mathbf{r}_A$ and $d\mathbf{r}_B$ are the infinitesimal displacements of points A and B, respectively. According to rigid body kinematics, the velocity of B relative to A is

$$\mathbf{v}_{B/A} = \boldsymbol{\omega} \times \mathbf{r}_{B/A} \quad \text{or} \quad \frac{d\mathbf{r}_{B/A}}{dt} = \frac{d\boldsymbol{\theta}}{dt} \times \mathbf{r}_{B/A}$$

where $\boldsymbol{\omega}$ is the angular velocity vector and $d\boldsymbol{\theta}$ represents the infinitesimal rotation of the body. Multiplying both sides of the second equation by dt, we obtain

$$d\mathbf{r}_{B/A} = d\boldsymbol{\theta} \times \mathbf{r}_{B/A}$$

which, on substitution into Eq. (b), yields

$$dU = \mathbf{F} \cdot (d\boldsymbol{\theta} \times \mathbf{r}_{B/A}) = (\mathbf{r}_{B/A} \times \mathbf{F}) \cdot d\boldsymbol{\theta}$$

Using Eq. (a), we finally get

$$dU = \mathbf{C} \cdot d\boldsymbol{\theta} \tag{18.1}$$

From Eq. (18.1) we see that the work of a couple depends on the rotation of the body and is independent of the translation. This conclusion was anticipated; it is obvious that during rigid body translation the work of $\mathbf{F}$ is canceled by the work of $-\mathbf{F}$, because the displacements of A and B are equal.

For the special case of *plane motion,* $\mathbf{C}$ and $d\boldsymbol{\theta}$ are parallel, both being perpendicular to the plane of the motion. Consequently, $dU = C\,d\theta$, and the work during a finite displacement of the body is

$$U_{1-2} = \int_{\theta_1}^{\theta_2} C\,d\theta \tag{18.2}$$

where θ_1 and θ_2 are the initial and final angular positions of the body, measured from a convenient reference line. If the magnitude of the couple remains constant during the plane motion, its work becomes

$$U_{1-2} = C(\theta_2 - \theta_1) = C\,\Delta\theta \qquad C\text{: constant} \tag{18.3}$$

where $\Delta\theta = \theta_2 - \theta_1$ is the angular displacement of the body. It must be remembered that Eqs. (18.2) and (18.3) are valid only for plane motion.

In order to determine the correct sign for the work, the directions of C and the rotation must be compared. If C and the rotation are in the same direction, the work is positive; if C and the rotation have opposite directions, the work is negative.

It should also be noted that the angular displacement must be measured in radians for Eqs. (18.1)–(18.3) to be valid.

b. Power

In Art. 14.5, the power P was defined to be the time rate at which work is done:

$$P = \frac{dU}{dt} \tag{14.30, repeated}$$

It was also shown that the power of a force **F** may be expressed as

$$P = \mathbf{F} \cdot \mathbf{v}$$

(14.31, repeated)

where **v** is the velocity vector of the point of application of **F**.

When a couple **C** acts on a rigid body, its power is, according to Eq. (18.1),

$$P = \frac{dU}{dt} = \frac{\mathbf{C} \cdot d\boldsymbol{\theta}}{dt} = \mathbf{C} \cdot \boldsymbol{\omega}$$

(18.4)

where $\omega = d\theta/dt$ is the angular velocity of the body.

For plane motion, where **C** and **ω** are parallel, the power of the couple becomes

$$P = C\omega$$

(18.5)

Observe that in Eq. (18.5) the power will be positive if C and ω are in the same direction, and negative if they are oppositely directed. The SI units for power are watts ($1\,\text{W} = 1\,\text{J/s} = 1\,\text{N} \cdot \text{m/s}$).

We also recall that the efficiency η of a machine was defined in Art. 14.5 as

$$\eta = \frac{\text{output power}}{\text{input power}} \times 100\%$$

(14.32, repeated)

18.3   *Kinetic Energy of a Rigid Body*

a.   *General motion*

The kinetic energy of a body can be obtained by adding the kinetic energies of the constituent particles. Thus the kinetic energy of a body occupying the region $\mathcal{V}$, as shown in Fig. 18.2, is

$$T = \int_{\mathcal{V}} \frac{1}{2} v^2 \, dm = \int_{\mathcal{V}} \frac{1}{2} \mathbf{v} \cdot \mathbf{v} \, dm$$

(a)

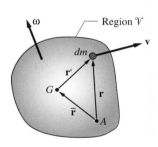

Fig. 18.2

where **v** is the velocity of the differential element (particle) of mass dm. If the body is rigid, the velocities of the differential elements are not independent, but are determined by the velocity of a reference point A (a point in the body) and the angular velocity **ω** of the body:

$$\mathbf{v} = \mathbf{v}_A + \boldsymbol{\omega} \times \mathbf{r}$$

(b)

where **r** is the position vector of dm relative to A. There are two convenient choices for the reference point: the mass center G of the body and the instant center for velocities.

Reference Point: Mass Center   Using the mass center G as the reference point, Eq. (b) becomes

$$\mathbf{v} = \bar{\mathbf{v}} + \boldsymbol{\omega} \times \mathbf{r}'$$

where $\mathbf{r}'$ represents the position vector of dm relative to G, and $\bar{\mathbf{v}} = d\bar{\mathbf{r}}/dt$ is the velocity of G. Substitution in Eq. (a) yields, after expanding the dot product,

$$T = \frac{1}{2} \int_{\mathcal{V}} [\bar{\mathbf{v}} \cdot \bar{\mathbf{v}} + 2\bar{\mathbf{v}} \cdot (\boldsymbol{\omega} \times \mathbf{r}') + (\boldsymbol{\omega} \times \mathbf{r}') \cdot (\boldsymbol{\omega} \times \mathbf{r}')]\, dm \qquad \text{(c)}$$

Applying the identity $\mathbf{A} \cdot (\mathbf{B} \times \mathbf{C}) = \mathbf{B} \cdot (\mathbf{C} \times \mathbf{A})$ to last term in the brackets, we get $(\boldsymbol{\omega} \times \mathbf{r}') \cdot (\boldsymbol{\omega} \times \mathbf{r}') = \boldsymbol{\omega} \cdot [\mathbf{r}' \times (\boldsymbol{\omega} \times \mathbf{r}')]$. Consequently, Eq. (c) can be written as

$$T = \frac{1}{2}\bar{v}^2 \int_{\mathcal{V}} dm + \bar{\mathbf{v}} \cdot \left[\boldsymbol{\omega} \times \int_{\mathcal{V}} \mathbf{r}'\, dm \right] + \frac{1}{2}\boldsymbol{\omega} \cdot \int_{\mathcal{V}} \mathbf{r}' \times (\boldsymbol{\omega} \times \mathbf{r}')\, dm$$

We note that $\int_{\mathcal{V}} dm = m$ is the mass of the body, and $\int_{\mathcal{V}} \mathbf{r}'\, dm = \mathbf{0}$ by definition of the mass center. Also, according to Eq. (17.4), $\int_{\mathcal{V}} \mathbf{r}' \times (\boldsymbol{\omega} \times \mathbf{r}')\, dm = \mathbf{h}_G$ is the angular momentum of the body about G. Therefore, the kinetic energy of a rigid body takes the form

$$T = \frac{1}{2}m\bar{v}^2 + \frac{1}{2}\boldsymbol{\omega} \cdot \mathbf{h}_G \qquad \text{(18.6a)}$$

Reference Point: Instant Center   If A is the instant center for velocities, then $\mathbf{v}_A = \mathbf{0}$ and Eq. (b) becomes $\mathbf{v} = \boldsymbol{\omega} \times \mathbf{r}$. Equation (a) can then be written as

$$T = \frac{1}{2} \int_{\mathcal{V}} \mathbf{v} \cdot (\boldsymbol{\omega} \times \mathbf{r})\, dm$$

Applying the identity $\mathbf{A} \cdot (\mathbf{B} \times \mathbf{C}) = \mathbf{B} \cdot (\mathbf{C} \times \mathbf{A})$ to the integrand, we obtain $\mathbf{v} \cdot (\boldsymbol{\omega} \times \mathbf{r}) = \boldsymbol{\omega} \cdot (\mathbf{r} \times \mathbf{v})$, so that the kinetic energy becomes

$$T = \frac{1}{2}\boldsymbol{\omega} \cdot \left(\int_{\mathcal{V}} \mathbf{r} \times \mathbf{v}\, dm \right)$$

Recognizing that the integral is, by definition, the angular momentum $\mathbf{h}_A$ of the body about A, we have

$$T = \frac{1}{2}\boldsymbol{\omega} \cdot \mathbf{h}_A \quad (A:\text{instant center}) \qquad \text{(18.6b)}$$

b.   *Plane motion*

Reference Point: Mass Center   If the motion takes place in a plane, then both $\boldsymbol{\omega}$ and $\mathbf{h}_G$ are perpendicular to the plane of motion. Moreover, the magnitude of the angular momentum about G is given by Eq. (17.6): $h_G = \overline{I}\omega$, where $\overline{I}$ is the moment of inertia of the body about G (more precisely, about the axis through G that is perpendicular to the plane of motion). Hence Eq. (18.6a) reduces to

$$T = \frac{1}{2}m\overline{v}^2 + \frac{1}{2}\overline{I}\omega^2 \qquad \text{(18.7a)}$$

The first term on the right side is called the *kinetic energy of translation*, and the second term is known as the *kinetic energy of rotation*.

Reference Point: Instant Center   If point A is the instant center for velocities, the magnitude of the angular momentum about A is $h_A = I_A\omega$—see Eq. (17.8). Therefore, Eq. (18.6b) takes the form

$$T = \frac{1}{2}I_A\omega^2 \qquad (A\text{: instant center}) \qquad \text{(18.7b)}$$

This equation is commonly used to compute the kinetic energy of a body that rotates about a fixed axis at A. However, it is not necessary for A to be fixed; it is sufficient for its velocity to be zero at the instant of interest.

Sample Problem   18.1

Figure (a) shows a counterclockwise couple $C(\theta)$ that acts on the uniform 1.5-kg bar AB. Calculate the total work done on the bar as it rotates in the vertical plane about A from $\theta = 0$ to $\theta = 180°$ if (1) $C(\theta) = 4.9 \sin \theta$ N·m; and (2) if $C(\theta)$ varies as shown in Fig. (b).

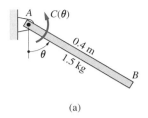

(a)

Solution

The total work done on the bar is the sum of the work done by the weight W and the couple $C(\theta)$. The work done by W can be computed from $(U_{1-2})_W = -W \Delta h$, where Δh is the upward vertical distance moved by the center of gravity of the bar between $\theta = 0$ (position 1) and $\theta = 180°$ (position 2). Because the bar is homogeneous, $\Delta h = 0.4$ m, which yields $(U_{1-2})_W = -1.5(9.8)(0.4) = -5.9$ N · m. The computation of the work done by the couple C is somewhat more complicated, because its magnitude is not constant.

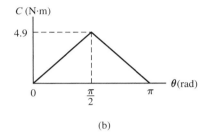

(b)

Part 1

Using Eq. (18.2), the work done by $C(\theta)$ as the bar rotates from $\theta = 0$ to $\theta = 180°$ (π rad) becomes

$$(U_{1-2})_C = \int_0^{\pi} C(\theta)\, d\theta = \int_0^{\pi} 4.9 \sin \theta \, d\theta = [-4.9 \cos \theta]_0^{\pi} = 9.8 \text{ N} \cdot \text{m}$$

Therefore, the total work done on the bar is

$$U_{1-2} = (U_{1-2})_C + (U_{1-2})_W = 9.8 - 5.9 = 3.9 \text{ N} \cdot \text{m} \qquad \textit{Answer}$$

Part 2

Recognizing that the work done by the couple equals the area under the C-θ diagram in Fig. (b), we have

$$(U_{1-2})_C = \frac{1}{2}(4.9)(\pi) = 7.7 \text{ N} \cdot \text{m}$$

and the total work done on the bar becomes

$$U_{1-2} = (U_{1-2})_C + (U_{1-2})_W = 7.7 - 5.9 = 1.8 \text{ N} \cdot \text{m} \qquad \textit{Answer}$$

Sample Problem 18.2

A couple (not shown) causes the eccentric disk of mass M in Fig. (a) to roll without slipping at the constant angular velocity ω. The radius of gyration of the disk about its mass center G is $\bar{k}$. Determine the maximum and minimum kinetic energies of the disk and the corresponding values of the angle θ (the angle between OG and the vertical). Use the following data: $M = 40$ kg, $R = 240$ mm, $e = 50$ mm, $\bar{k} = 160$ mm, and $\omega = 10$ rad/s.

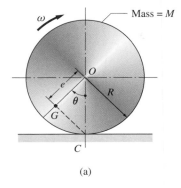

Mass = M

(a)

Solution

Because the disk is rolling without slipping, the point of contact C is its instant center for velocities. Hence the kinetic energy of the disk can be calculated from Eq. (18.7b): $T = I_C\omega^2/2$. Using the parallel-axis theorem and $\bar{I} = M\bar{k}^2$, we obtain

$$T = \frac{1}{2}(\bar{I} + M\,\overline{CG}^2)\omega^2 = \frac{1}{2}M(\bar{k}^2 + \overline{CG}^2)\omega^2$$

We see that T_{max} occurs when the distance $\overline{CG}$ is largest—that is, when the disk is in the position shown in Fig. (b). Conversely, the position of T_{min}, shown in Fig. (c), is where $\overline{CG}$ has its smallest value. Therefore, the maximum and minimum values of the kinetic energy of the disk are

$$T_{max} = \frac{1}{2}(40)[(0.16)^2 + (0.29)^2](10)^2 = 219.4 \text{ J} \quad \text{at } \theta = 180° \qquad \textit{Answer}$$

$$T_{min} = \frac{1}{2}(40)[(0.16)^2 + (0.19)^2](10)^2 = 123.4 \text{ J} \quad \text{at } \theta = 0 \qquad \textit{Answer}$$

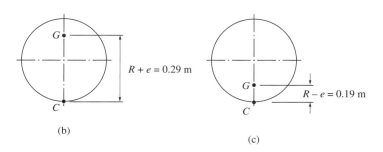

(b)

(c)

Sample Problem 18.3

Bar AB of the mechanism shown in Fig. (a) is rotating counterclockwise at the constant angular velocity of $\omega_{AB} = 2.5$ rad/s. Calculate the total kinetic energy of the mechanism when it is in the position shown. The mass per unit length of each bar is 2 kg/m.

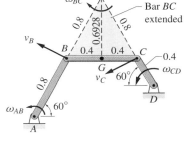

(a)

Solution

We will calculate the kinetic energy of each bar using instant centers for velocities. Before these calculations can be carried out, we must find the angular velocities of bars BC and CD by kinematics, and compute the moment of inertia of each bar about its instant center.

Kinematic Analysis

Because A and D are fixed points, they are the instant centers for velocity for bars AB and CD, respectively. The instant center for BC, labeled O in Fig. (b), is located at the intersection of the lines that are perpendicular to the velocity vectors of B and C. Because $\mathbf{v}_B$ and $\mathbf{v}_C$ are perpendicular to AB and CD, respectively, point O is at the intersection of lines AB and CD. Letting G be the mass center of bar BC, we can calculate the following dimensions shown on triangle OBC:

$$\overline{BO} = \overline{CO} = \frac{\overline{BG}}{\cos 60°} = \frac{0.4}{\cos 60°} = 0.8 \text{ m}$$

$$\overline{GO} = \overline{BG}\tan 60° = 0.4\tan 60° = 0.6928 \text{ m}$$

Dimensions in meters

(b)

The angular velocities of BC and CD can now be found by the following series of computations.

1. From the motion of bar AB (instant center is A):

$$v_B = \overline{AB}\,\omega_{AB} = 0.8(2.5) = 2.0 \text{ m/s}$$

2. From the motion of bar BC (instant center is O):

$$\omega_{BC} = v_B/\overline{BO} = \frac{2.0}{0.8} = 2.50 \text{ rad/s}$$

$$v_C = \overline{OC}\,\omega_{BC} = 0.8(2.5) = 2.0 \text{ m/s}$$

3. From the motion of CD (instant center is D):

$$\omega_{CD} = v_C/\overline{CD} = \frac{2.0}{0.4} = 5.0 \text{ rad/s}$$

Inertial Properties

The masses of the bars are $m_{AB} = m_{BC} = 2(0.8) = 1.6\,\text{kg}$, and $m_{CD} = 2(0.4) = 0.8\,\text{kg}$. The moments of inertia of the bars about their respective instant centers are

$$(I_A)_{AB} = \frac{m_{AB}L_{AB}^2}{3} = \frac{1.6(0.8)^2}{3} = 0.3413 \text{ kg} \cdot \text{m}^2$$

$$(I_O)_{BC} = (\bar{I})_{BC} + m_{BC}\,\overline{GO}^2 = \frac{m_{BC}L_{BC}^2}{12} + m_{BC}\,\overline{GO}^2$$

$$= \frac{1.6(0.8)^2}{12} + 1.6(0.6928)^2 = 0.8533 \text{ kg} \cdot \text{m}^2$$

$$(I_D)_{CD} = \frac{m_{CD}L_{CD}^2}{3} = \frac{0.8(0.4)^2}{3} = 0.04267 \text{ kg} \cdot \text{m}^2$$

Kinetic Energy

The kinetic energy T of the linkage is found by summing the kinetic energies of the three bars. Using Eq. (18.7b), we obtain

$$T = (T)_{AB} + (T)_{BC} + (T)_{CD}$$

$$= \frac{1}{2}(I_A)_{AB}\omega_{AB}^2 + \frac{1}{2}(I_O)_{BC}\omega_{BC}^2 + \frac{1}{2}(I_D)_{CD}\omega_{CD}^2$$

$$= \frac{1}{2}(0.3413)(2.50)^2 + \frac{1}{2}(0.8533)(2.50)^2 + \frac{1}{2}(0.04267)(5.0)^2$$

$$= 1.067 + 2.667 + 0.533 = 4.27 \text{ J} \qquad \textit{Answer}$$

Problems

18.1   The constant 2 N · m couple acts on the homogeneous disk that rolls without slipping down the incline. Determine the displacement of the disk, measured from the position where the spring is undeformed, for which the total work done on the disk is zero.

18.2   The torsion spring at A applies the counterclockwise couple $C(\theta) = -10\theta$ N · m to the uniform bar. The angle θ (in radians) is measured counterclockwise from position 1, where the spring is undeformed. Calculate the work done by the spring on the bar as it rotates from position 2 to position 3.

18.3   The platform carrying the 200-kg crate is supported by two parallel links. The couple $C(\theta) = C_0(1 - 2\theta/\pi)$ acts on one of the links, where θ is the angle (in radians) between the links and the horizontal. Determine the value of C_0 for which the total work done on the system is zero as it moves from $\theta = 0$ to $\theta = 90°$. Neglect the weights of the platform and the links.

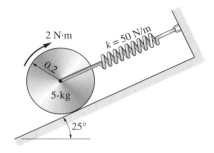

Fig. P18.1

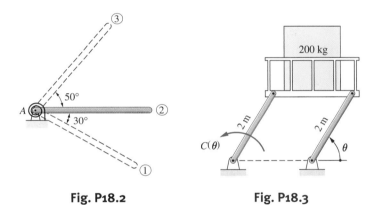

Fig. P18.2   **Fig. P18.3**

18.4   The system, which lies in the vertical plane, consists of the 5-kg homogeneous bars AB and BC and the spring AC. The free length of the spring is 1 m, and its stiffness is 200 N/m. A constant 70-N · m couple acts on bar AB. Determine the work done on the bars as θ changes from 0 to 45°.

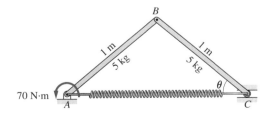

Fig. P18.4

18.5   The system consists of the vertical uniform 8-kg bar AB and a linear spring of stiffness 360 N/m. When the bar is in position 1, the spring is horizontal and undeformed. A constant 30-N · m couple is applied to the bar. Calculate the total work done on the bar as it moves from position 1 to position 2.

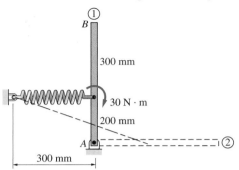

Fig. P18.5

18.6   The spools in Figs. (a) and (b) are rolling without slipping along horizontal rails. In each case, a constant 200-N force is applied to a cord that is wound around the outer radius of the spool. The angular velocity of each spool is 6 rad/s clockwise. Determine the power of the 200-N force in each case.

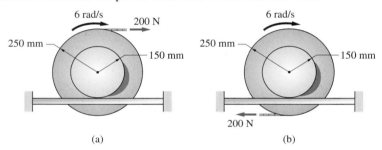

Fig. P18.6

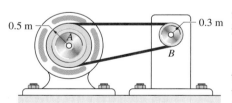

Fig. P18.9, P18.10

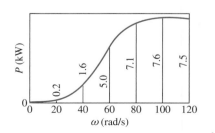

Fig. P18.11

18.7   An electric motor of 78% efficiency consumes 12 kW of power. What is the torque carried by the output shaft when the speed of the motor is (a) 1800 rev/min; and (b) 3600 rev/min?

18.8   A 10-kg flywheel with a central radius of gyration of 0.3 m is driven with constant power and efficiency. If the angular acceleration of the flywheel is 2 rad/s² at 600 rev/min, calculate the angular acceleration at 800 rev/min.

18.9   The machine B is belt-driven by the electric motor A that has an efficiency of 85%. When the motor is running at 25 rad/s, the tension in the upper part of the belt is 90 N greater than in the lower part. Determine the power consumption of the motor in horsepower (hp).

18.10   When the system described in Prob. 18.9 is running at 40 rad/s, the power consumption of the motor is 2.4 hp. Calculate the torque in (a) the output shaft of motor A; and (b) the input shaft of machine B.

18.11   The figure shows a plot of the power output P of a reciprocating engine versus the shaft speed ω. (a) Plot the torque (couple) developed by the engine versus ω. (b) From the plot in part (a), estimate the maximum torque and the corresponding angular speed.

18.12 The torque (couple) M developed by a turbine as a function of its angular speed ω is shown in the figure. (a) Plot the horsepower of the turbine versus ω. (b) Use the plot in part (a) to estimate the maximum power and the corresponding angular velocity.

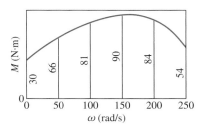

Fig. P18.12

18.13 Each homogeneous slender bar AB has mass m and length L. The bar in Fig. (a) is guided by pins at G and B, which slide in slots, and the bar in Fig. (b) rotates about a pin at C. Calculate the kinetic energy of each bar in terms of its angular speed ω, m, and L.

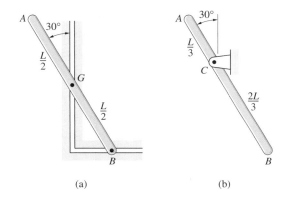

(a)   (b)

Fig. P18.13

18.14 Each of the bodies has mass m and rotates about a pin at O. The body in Fig. (a) is made from a uniform rod, and the body in Fig. (b) is a homogeneous rectangular plate. Determine the kinetic energy of each body in terms of its angular velocity ω, m, and L.

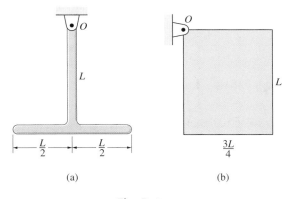

(a)   (b)

Fig. P18.14

18.15 The 60-kg unbalanced wheel rolls without slipping on the horizontal plane. The radius of gyration of the wheel about its mass center G is 240 mm. Calculate the kinetic energy of the wheel in the position shown, where its angular velocity is $\omega = 2$ rad/s.

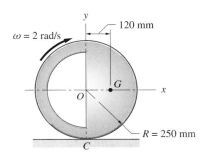

Fig. P18.15

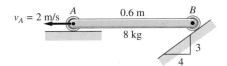

Fig. P18.16

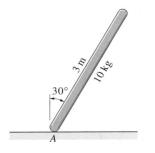

Fig. P18.17

18.16 When the 8-kg uniform bar *AB* is in the position shown, the velocity of end *A* is 2 m/s to the left. Determine the kinetic energy of the bar in this position.

18.17 In the position shown, end *A* of the uniform 10-kg bar is moving to the left with the velocity $v_A = 0.6$ m/s. At the same time, the angular velocity of the bar is $\omega = 0.3$ rad/s, clockwise. Find the kinetic energy of the bar in this position.

18.18 End *A* of the uniform bar *AB* is pinned to the homogeneous disk, and end *B* is connected to the 1-kg slider. In the position shown, the disk is rotating counterclockwise with an angular velocity of 4 rad/s. Compute the kinetic energy of the system in this position.

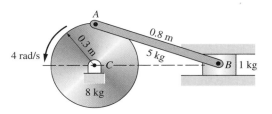

Fig. P18.18

18.19 The mechanism consists of the uniform bars *AB* and *BC*. In the position shown, roller *C* is moving to the right with the velocity $v_C = 2$ m/s. Determine the kinetic energy of the mechanism in this position.

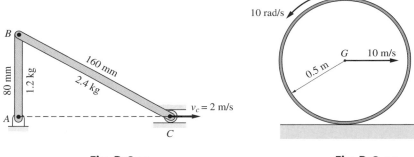

Fig. P18.19   **Fig. P18.20**

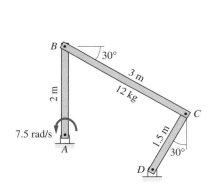

Fig. P18.21

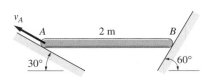

Fig. P18.22

18.20 The 2-kg uniform thin hoop is launched to the right with the velocity of 10 m/s and a backspin of 10 rad/s. Calculate the kinetic energy of the hoop at the instant of launch.

18.21 At the instant shown, bar *AB* of the mechanism is rotating counterclockwise at 7.5 rad/s. Determine the kinetic energy of the 12-kg homogenous bar *BC* at this instant.

18.22 Ends *A* and *B* of the 80-kg uniform bar slide along the inclined planes. In the position shown, the velocity of end *A* is $v_A = 3$ m/s. Determine the kinetic energy of the bar in this position.

18.23 Rod *AB* of the mechanism rotates clockwise with a constant angular velocity of 20 rad/s. Determine the total kinetic energy of the mechanism in the position shown. The mass of each bar is 3 kg/m.

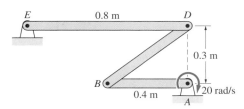

Fig. P18.23

18.24 The radius of gyration of disk *A* about its mass center *G* is $\bar{k} = 0.5$ m. In the position shown, the disk rolls without slipping with the angular velocity $\omega_A = 4$ rad/s. Determine the kinetic energy of the system in this position.

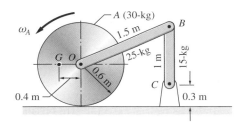

Fig. P18.24

18.4 *Work-Energy Principle and Conservation of Mechanical Energy*

The work-energy principle for a system of particles, derived in Art. 15.5, was

$$(U_{1-2})_{\text{ext}} + (U_{1-2})_{\text{int}} = \Delta T \qquad \text{(15.24, repeated)}$$

where the subscripts 1 and 2 refer to the initial and final positions of the system, respectively, $(U_{1-2})_{\text{ext}}$ is the work done by the external forces (including the weights of the particles), $(U_{1-2})_{\text{int}}$ is the work done by the internal forces, and $\Delta T = T_2 - T_1$ is the change in the kinetic energy of the system.

Considering a rigid body to be made up of particles, Eq. (15.24) can be applied directly to the motion of a rigid body, or a system of connected rigid bodies, as described in the following.

Single Rigid Body   The forces internal to a rigid body hold the body together; that is, they are the constraint forces that impose the condition of rigidity. These internal forces occur in equal and opposite collinear pairs. Because the body is rigid, the distances between the particles do not change, which in turn implies that the distances between the points of application of the internal forces remain constant. Consequently, the internal forces do no work on a rigid body, so that the work-energy principle governing the motion of a single rigid body becomes

$$(U_{1-2})_{\text{ext}} = \Delta T \tag{18.8}$$

System of Connected Rigid Bodies   As explained above, the forces internal to a rigid body are workless. However, the internal connections between rigid bodies may either be workless or capable of doing work. Inextensible strings and pins are examples of workless internal connectors, whereas springs and friction give rise to internal forces that can do work. (Workless internal forces and internal forces that can do work were discussed in Art. 15.5.) Therefore, the work-energy principle for the motion of a system of connected rigid bodies must be used in its general form

$$(U_{1-2})_{\text{ext}} + (U_{1-2})_{\text{int}} = \Delta T \tag{18.9}$$

In this equation, T represents the kinetic energy *of the system,* which equals the sum of the kinetic energies of the constituent bodies.

The principle of conservation of mechanical energy for a system of particles, stated in Art. 15.5, is also applicable to a rigid body (or a system of connected rigid bodies): If all forces, internal as well as external, are conservative, the mechanical energy of the rigid body (or system of connected rigid bodies) is conserved. This principle may be written as

$$V_1 + T_1 = V_2 + T_2 \tag{18.10}$$

where V_1 and V_2 are the initial and final potential energies, and T_1 and T_2 are the initial and final kinetic energies.

Sample Problem 18.4

The uniform 20-kg slender bar AC shown in Fig. (a) rotates in a vertical plane about the pin at B. The ideal spring AD has a spring constant $k = 15$ N/m and an undeformed length $L_0 = 2$ m. When the bar is at rest in the position $\theta = 0$, it is given a small angular displacement and released. Find the angular velocity of the bar when it reaches the horizontal position.

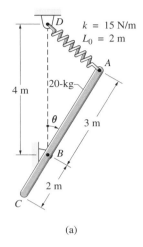

(a)

Solution

This problem is well suited for solution by the work-energy method because it is concerned with the change in velocity that occurs during a change in position. Because the system is conservative, it may be analyzed by the work-energy or conservation of mechanical energy principles. We use the first of these methods; you may wish to try the second method as an exercise.

Figure (b) shows the initial and final positions of the bar, labeled 1 and 2, respectively. Work is done on the bar by its weight and by the linear spring. Noting that the mass center G of the bar moves downward through a distance of 0.5 m between positions 1 and 2, the work of the weight is

$$U_{1-2} = 20(9.8)(0.5) = 98 \text{ N} \cdot \text{m} \qquad \text{(a)}$$

From Fig. (b) we see that the length of the spring is $L_1 = 1$ m in the initial position and $L_2 = \sqrt{3^2 + 4^2} = 5$ m in the final position. Because the unstretched length of the spring is $L_0 = 2$ m, the initial and final deformations of the spring are $\delta_1 = L_1 - L_0 = 1 - 2 = -1$ m and $\delta_2 = L_2 - L_0 = 5 - 2 = 3$ m. (The signs indicate that the spring is in compression in the initial position and in tension in the final position.) The work done by the spring on the bar, therefore, is

$$U_{1-2} = -\frac{1}{2}k(\delta_2^2 - \delta_1^2) = -\frac{1}{2}(15)[3^2 - (-1)^2] = -60 \text{ N} \cdot \text{m} \qquad \text{(b)}$$

Because point B is fixed, the kinetic energy of the bar can be calculated from $T = (1/2)I_B\omega^2$. Using the parallel-axis theorem, we find that

$$I_B = \overline{I} + md^2 = m\left(\frac{L^2}{12} + d^2\right) = 20\left(\frac{5^2}{12} + 0.5^2\right) = 46.67 \text{ kg} \cdot \text{m}^2$$

From Eqs. (a) and (b), and knowing that the bar is released from rest ($T_1 = 0$), the work-energy principle becomes

$$U_{1-2} = T_2 - T_1 = \frac{1}{2}I_B\omega_2^2 - 0$$

$$98 - 60 = \frac{1}{2}(46.67)\omega_2^2$$

which yields

$$\omega_2 = 1.276 \text{ rad/s} \qquad \qquad \textit{Answer}$$

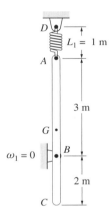

Position ①

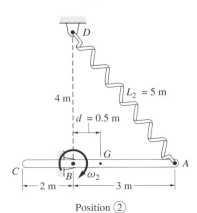

Position ②

(b)

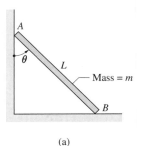

Sample Problem 18.5

Figure (a) shows a homogeneous, slender bar AB of mass m and length L. When the bar was at rest in the position $\theta = 0$, it was displaced slightly and released. Determine the angular velocity and angular acceleration of the bar as functions of the angle θ. Neglect friction and assume that end A does not lose contact with the vertical surface.

Solution

Our solution utilizes the fact that the mechanical energy is conserved (the weight of the bar is the only force that does work as the bar falls). An identical solution could be obtained just as easily from the work-energy principle.

Figure (b) shows the bar in the release ($\theta = 0$) position 1, and in an arbitrary position 2 defined by the angle θ. Using the horizontal plane as the datum for potential energy, the potential energies of the bar in the two positions are

$$V_1 = mg\frac{L}{2} \quad \text{and} \quad V_2 = mg\frac{L}{2}\cos\theta \tag{a}$$

Position ①   Position ②

(b)

The initial kinetic energy is obviously $T_1 = 0$, because the bar is released from rest. The kinetic energy in position 2 can be readily calculated by observing that the point O, shown in Fig. (b), is the instant center for velocities of the bar. (Point O is located where the perpendiculars to the velocity vectors of ends A and B intersect.) Utilizing the parallel-axis theorem, the mass moment of inertia of the bar about O is $I_O = \bar{I} + md^2 = (mL^2/12) + m(L/2)^2 = mL^2/3$. Consequently, the kinetic energies in the two positions are

$$T_1 = 0 \quad \text{and} \quad T_2 = \frac{1}{2}I_O\omega^2 = \frac{1}{2}\frac{mL^2}{3}\omega^2 \tag{b}$$

where ω is the angular velocity of the bar in position 2.

Because mechanical energy is conserved, we have

$$V_1 + T_1 = V_2 + T_2$$

Substituting from Eqs. (a) and (b), we get

$$mg\frac{L}{2} + 0 = mg\frac{L}{2}\cos\theta + \frac{1}{2}\frac{mL^2}{3}\omega^2$$

from which the angular velocity in position 2 is found to be

$$\omega = \left[\frac{3g}{L}(1 - \cos\theta) \right]^{1/2} \qquad \textit{Answer}$$

The angular acceleration α is obtained by taking the time derivative of ω, which yields

$$\alpha = \frac{d\omega}{dt} = \frac{1}{2}\left[\frac{3g}{L}(1 - \cos\theta) \right]^{-(1/2)} \frac{3g}{L}\dot\theta \sin\theta$$

Substituting the previously found expression for $\dot\theta = \omega$, the angular acceleration reduces to

$$\alpha = \frac{3g}{2L}\sin\theta \qquad \textit{Answer}$$

Sample Problem 18.6

Figure (a) shows a slider-crank mechanism that is being driven by a constant clockwise couple $M = 0.5$ N · m. All the components are homogeneous, with the mass and dimensions as indicated. When the mechanism is in the position shown in Fig. (a), the angular velocity of the crank is $\omega_1 = 12$ rad/s clockwise. Determine the angular velocity of the crank after it has rotated 90° from the position shown. Neglect friction and assume that motion is in a vertical plane.

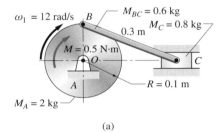

(a)

Solution

The solution of this problem lends itself to a work-energy analysis because it is concerned with the change in velocity between two positions. The problem can be solved by using either the work-energy principle or by noting that mechanical energy is conserved (we chose the latter approach). Regardless of which method is used, it is convenient to analyze the entire mechanism, thereby eliminating the need to consider the work done at the connections (pins B and C). The initial and final positions of the mechanism, labeled 1 and 2, respectively, are shown in Fig. (b).

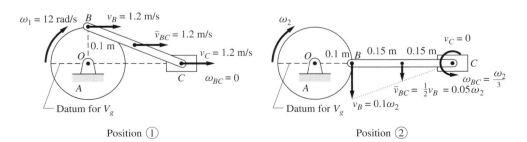

(b)

Kinematic Analysis

The steps used in the kinematic analysis that resulted in the velocities shown in Fig. (b) are

Position 1

1. Because O is a fixed point, $v_B = R\omega_1 = (0.1)(12) = 1.2$ m/s.
2. Recognizing that both $\mathbf{v}_B$ and $\mathbf{v}_C$ are horizontal, we conclude that $\omega_{BC} = 0$; that is, bar BC is translating in this position.
3. Because bar BC is translating, the velocities of all points on the bar are the same. In particular, the velocity of the mass center of BC is $\bar{v}_{BC} = 1.2$ m/s.

Position 2

1. Because O is a fixed point, $v_B = R\omega_2 = 0.1\omega_2$ m/s (directed downward). It was assumed that ω_2 is directed clockwise.
2. Because $\mathbf{v}_B$ is vertical and the path of C is horizontal, we conclude that C is the instant center for bar BC; that is, $v_C = 0$.
3. Because $v_C = 0$, we know that $\omega_{BC} = v_B/L_{BC} = (0.1\omega_2)/0.3 = \omega_2/3$ rad/s (counterclockwise) and that the velocity of the mass center of BC is $\bar{v}_{BC} = v_B/2 = 0.1\omega_2/2 = 0.05\omega_2$.

Potential Energy

The system possesses gravitational potential energy due to the weights of its components; in addition, there is the potential energy of the constant couple.

As indicated in Fig. (b), we choose the horizontal plane passing through OC to be the datum for the gravitational potential energy V_g. In position 1, the mass centers of A and C lie in the datum plane, whereas the mass center of bar BC is 0.05 m above this plane. Therefore, we obtain $(V_g)_1 = W_{BC}h = 0.6(9.8)(0.05) = 0.294$ N $\cdot$ m. In position 2, there is no gravitational potential energy because the mass center of each part lies in the datum plane; that is, $(V_g)_2 = 0$.

The constant couple M is conservative because its work $U_{1-2} = M\Delta\theta$ depends on only the magnitude M and the initial and final angular positions of crank A. The work done by the couple may be expressed as $U_{1-2} = -[(V_M)_2 - (V_M)_1]$, where V_M is the potential energy of the couple. Choosing position 1 to be the datum, we have $(V_M)_2 = (V_M)_1 - U_{1-2} = 0 - (0.5)\frac{\pi}{2} = -0.785$ N $\cdot$ m.

In summary, the initial and final potential energies are

$$V_1 = (V_g)_1 + (V_M)_1 = 0.294 + 0 = 0.294 \text{ N} \cdot \text{m} \tag{a}$$

$$V_2 = (V_g)_2 + (V_M)_2 = 0 + (-0.785) = -0.785 \text{ N} \cdot \text{m} \tag{b}$$

Kinetic Energy

The kinetic energy of the entire mechanism is the sum of the kinetic energies of the three parts:

$$T = \left(\frac{1}{2}\bar{I}\omega^2\right)_A + \left(\frac{1}{2}\bar{I}\omega^2 + \frac{1}{2}m\bar{v}^2\right)_{BC} + \left(\frac{1}{2}mv^2\right)_C \tag{c}$$

The central moments of inertia of the crank A and the arm BC are

$$\overline{I}_A = \left(\frac{mR^2}{2}\right)_A = \frac{1}{2}(2)(0.1)^2 = 0.01 \text{ kg-m}^2$$

$$\overline{I}_{BC} = \left(\frac{mL^2}{12}\right)_{BC} = \frac{1}{12}(0.6)(0.3)^2 = .0045 \text{ kg-m}^2$$

Position 1 Substituting the values of $\overline{I}_A$, $\overline{I}_{BC}$, $\omega_A = \omega_1 = 12 \text{ rad/s}$, $\omega_{BC} = 0$, and $\overline{v}_{BC} = v_C = 1.2 \text{ m/s}$ into Eq. (c), we obtain

$$T_1 = \left[\frac{1}{2}(0.01)(12)^2\right] + \left[0 + \frac{1}{2}(0.6)(1.2)^2\right] + \left[\frac{1}{2}(0.8)(1.2)^2\right]$$

which yields

$$T_1 = 0.72 + 0.432 + 0.576 = 1.728 \text{ N} \cdot \text{m} \qquad \text{(d)}$$

Position 2 Substituting the values of $\overline{I}_A$, $\overline{I}_{BC}$, $\omega_A = \omega_2 \text{ rad/s}$, $\omega_{BC} = \omega_2/3 \text{ rad/s}$, $\overline{v}_{BC} = 0.05\omega_2 \text{ m/s}$, and $v_C = 0$ into Eq. (c), we find

$$T_2 = \left[\frac{1}{2}(0.01)\omega_2^2\right] + \left[\frac{1}{2}(.0045)\left(\frac{\omega_2}{3}\right)^2 + \frac{1}{2}(0.6)(0.05)^2\right] + 0$$

which simplifies to

$$T_2 = \left(6 \times 10^{-3}\right)\omega_2^2 \qquad \text{(e)}$$

Note that the kinetic energy of bar BC in position 2 could also be computed from $(1/2)I_C\omega^2$ because $v_C = 0$.

Conservation of Mechanical Energy

Using the results of Eqs. (a), (b), (d), and (e), the principle of conservation of mechanical energy becomes

$$V_1 + T_1 = V_2 + T_2$$

$$0.294 + 1.728 = -0.785 + \left(6 \times 10^{-3}\right)\omega_2^2$$

from which the angular velocity in the final position is found to be

$$\omega_2 = 21.6 \text{ rad/s} \qquad \qquad \textit{Answer}$$

Problems

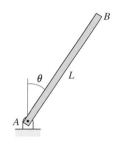

Fig. P18.25

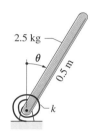

Fig. P18.28

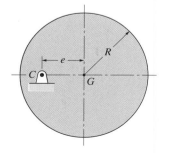

Fig. P18.29

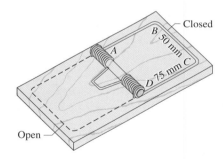

Fig. P18.30

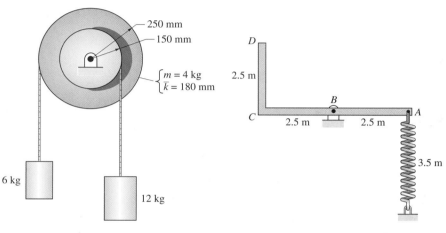

Fig. P18.26   **Fig. P18.27**

18.25 The uniform slender bar *AB* of mass *m* is given a small angular displacement from the position $\theta = 0$ and then released. Determine the angular velocity of the bar as a function of θ.

18.26 The system consisting of two blocks and a compound pulley is released from rest. Find the angular velocity of the pulley after it has rotated 90°.

18.27 The slender L-shaped bar *ABCD* mass 15 kg/m is free to rotate about the pin at *B*. The spring connected to the bar at *A* has a free length of 7 ft, and its stiffness is 180 N/m. If the system is released from rest in the position shown, determine the angular velocity of the bar when *A* is directly above *B*.

18.28 The torsion spring applies the couple $C = -k\theta$ to the 2.5-kg uniform bar. The angular position θ (rad) of the bar is measured from the vertical, and *k* (N · m/rad) is the spring stiffness. If the bar is rotated to the position $\theta = 30°$ and released, determine the smallest value of *k* for which the bar would return to the vertical position.

18.29 Jaw *ABCD* of the mousetrap is made of uniform steel wire weighing 0.4×10^{-4} kg/mm. The torque exerted by the torsion spring on the jaw in the closed position is 0.675 N · m, and it is 1.188 N · m in the open position. If the jaw is released from the open position, determine the speed of segment *BC* when it reaches the closed position. Assume that the torque-angular displacement relationship for the spring is linear.

***18.30** The homogeneous disk of mass *m* and radius *R* is released from rest in the position shown. (a) Derive the expression for the angular velocity of the disk when *CG* is vertical. (b) Find the distance *e* that would maximize the angular velocity found in part (a); and (c) determine this maximum angular velocity.

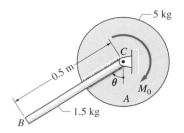

Fig. P18.31

18.31   The 1.5-kg uniform rod BC is welded to drum A, which weighs 10 lb and has a radius of gyration of 0.1 m about C. The system is initially at rest in the position $\theta = 0$ when the constant clockwise couple $M_0 = 2$ N · m is applied. (a) Derive the angular velocity (in rad/s) and angular acceleration (in rad/s^2) of the system as functions of θ. (b) Determine the maximum angular velocity and the value of θ at which it occurs.

18.32   A bicycle chain, 2 m in length and mass 0.75 kg/m, hangs over a sprocket as shown. The sprocket mass 3 kg, and its central radius of gyration is 0.22 m. If the sprocket is given a small clockwise angular displacement from the position shown and then released, calculate its angular velocity after the entire chain has left the sprocket.

18.33   For the system described in Prob. 18.32, what is the angular velocity of the sprocket when end A of the chain has reached point C?

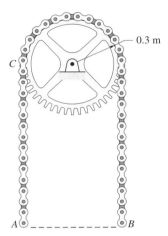

Fig. P18.32, P18.33

18.34   A cylinder of radius R_C and mass m_C, and a sphere of radius R_S and mass m_S are released from rest on a rough surface that is inclined at the angle β with the horizontal. Both bodies are homogeneous and roll without slipping. Determine $\bar{v}_C/\bar{v}_S$ (the ratio of the central velocities) after each body has moved the same distance d down the inclined plane.

18.35   The mass moment of inertia of the 40-kg spool about its mass center G is 1.6 kg · m^2. The spool is at rest on a rough surface when the constant force P is applied to the cable that is wound around its hub. Determine P if the spool is to have an angular speed of 6 rad/s after turning through one revolution. Assume rolling without slipping.

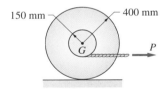

Fig. P18.35

18.36   The end of the string wrapped around the hub of the spool is attached to a fixed support. The spool mass 3-kg, and its radius of gyration about G is 0.11 m. (a) If the spool is released from rest, determine the velocity of G after a 2 m drop. (b) What would the velocity of G be if the spool fell 6 ft from rest without the string being present?

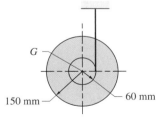

Fig. P18.36

18.37   The uniform rod AB of mass m is pinned to the sliding collar C at its midpoint. If the rod is released from rest when $\theta = \theta_0$, find its angular velocity as a function of θ. Neglect friction and the mass of the collar.

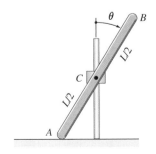

Fig. P18.37

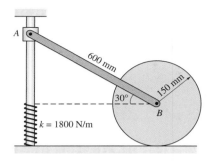

Fig. P18.38

18.38 The uniform bar *AB* of a mass of 4.5 kg connects the small sliding collar *A* to the homogeneous disk of a mass of 6.8 kg. When the system is released from rest in the position shown, the collar slides down the vertical rod and the disk rolls without slipping on the horizontal surface. Determine (a) the velocity of the collar when it is about collide with the spring; and (b) the maximum deflection of the spring after the collision.

18.39 The mass center of the 7-kg unbalanced wheel is located at *G*. The mass moment of inertia of the wheel about its center *O* is $I_O = 0.3$ kg-m². In the position shown, the angular velocity of the wheel is 3 rad/s clockwise and the horizontal spring is undeformed. Find the angular velocity of the wheel after it has rolled clockwise through 180° from this position. Assume that the wheel does not slip.

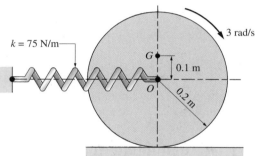

Fig. P18.39

18.40 The movement of the uniform bar *ABC* of mass *m* is controlled by rollers at *A* and *B*. The bar is at rest in the position $\theta = 30°$ when the constant horizontal force *P* is applied at *C*. Determine the angular velocity of the bar as it passes through the vertical ($\theta = 0$) position.

18.41 The eccentric wheel of mass $M = 300$ kg and radius $R = 1.2$ m rolls without slipping. The radius of gyration of the wheel about its mass center *G* is $\bar{k} = 0.5$ m, and its eccentricity is $e = 0.4$ m. Knowing that the angular velocity of the wheel is $\omega = 2.69$ rad/s (clockwise) when $\theta = 0$, calculate (a) the maximum angular velocity; and (b) the minimum angular velocity.

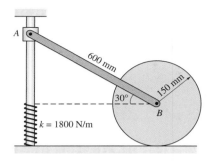

Fig. P18.40

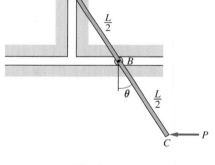

Fig. P18.42

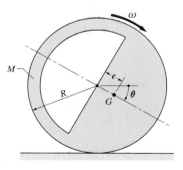

Fig. P18.41

18.42 The 10-kg uniform bar *AB* slides inside a frictionless cylindrical surface. If the bar is released from rest in the position $\theta = 35°$, determine its angular velocity (in rad/s) and angular acceleration (in rad/s²) as functions of θ.

18.43   A bar of negligible mass is pinned to the center of the 100 gm uniform disk. The torsion spring connected between the bar and the disk has a stiffness of 4×10^{-4} N · m/rad. The spring is preloaded by rotating the bar though three clockwise revolutions relative to the disk. The system is then placed on the horizontal surface and released from rest in the position shown. Determine the maximum speed reached by the center of the disk. Assume that the disk rolls without slipping.

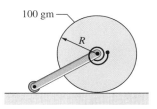

Fig. P18.43

18.44   The 100-kg uniform bar AB is attached to the 300-kg homogeneous disk by a pin at A. The weight of the slider attached at B is negligible. The disk rolls without slipping along the horizontal plane. In the position shown, the kinetic energy of the system is 2400 N · m. After the disk has rolled 180° clockwise, calculate (a) the kinetic energy of the system; and (b) the angular velocity of the disk.

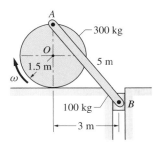

Fig. P18.44

18.45   The linkage shown consists of two identical bars, each of length L and mass m. If the linkage is released from rest at $\theta = 0$, find the angular velocity of each link when $\theta = 90°$. Neglect friction.

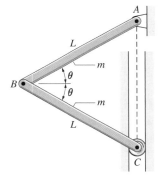

Fig. P18.45

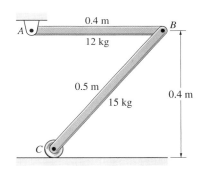

Fig. P18.46

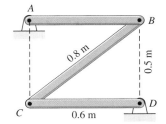

Fig. P18.47

18.46   The mechanism consisting of two uniform rods is released from rest in the position shown. Neglecting friction, find the velocity of roller C when AB reaches the vertical position.

18.47   The three links of the mechanism mass m_0 kg/m. If the mechanism is released from rest in the position shown, determine the angular velocity of link AB when it reaches the vertical position.

18.48   The uniform bar (mass m_2) is pinned to the circumference of the homogeneous disk (mass m_1). The system is displaced slightly from the position shown and released from rest. Find the maximum angular velocity of the disk. Assume rolling without slipping.

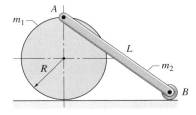

Fig. P18.48

18.49 The three bars of the linkage are uniform and of equal weight. If the linkage is released from rest in the position shown, determine the angular velocity of link *AB* when it reaches the horizontal position.

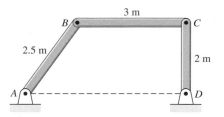

Fig. P18.49

18.50 The homogeneous 30-kg bar *AC* is pinned to the parallel links *AB* and *CD* of negligible weight. The assembly is at rest in the position $\theta = 60°$ when the constant 320-N · m couple is applied to *AB*. Determine the speed of bar *AC* when $\theta = 90°$.

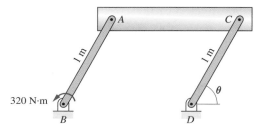

Fig. P18.50

18.51 The homogeneous 80-kg disk *A* is connected by a cable to the 80-kg block *B*. The coefficient of kinetic friction between the block and the inclined surface is 0.4. If the system is released from rest when $x = 0$, determine x when the speed of the block is 5 m/s. Assume that the disk rolls without slipping.

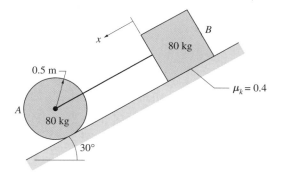

Fig. P18.51

18.52 End *B* of the uniform 2.7-kg bar *AB* is connected to a small roller that moves in a horizontal slot. The other end of the bar is pinned to the homogeneous 1.8-kg disk that rolls without slipping on the vertical surface. The spring attached to *A* has a free length of 0.3 m and a stiffness of 58 N/m. After the system is released from rest in the position shown, the velocity of A is 3 m/s when AB becomes horizontal. Determine the magnitude of the force P that acts on the bar at B.

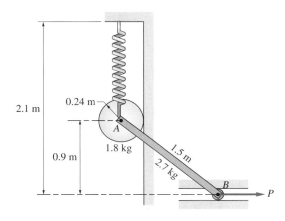

Fig. P18.52

18.53   The 45-kg bucket is suspended from a cable that is wrapped around the periphery of the 180-kg drum. The radius of gyration of the drum about point O is 0.3 m. The system is released from rest in position 1 and moves with negligible friction until the bucket reaches position 2, when a brake is applied to the drum that exerts a constant couple of 778 N · m. (a) Find the height h for which the bucket will land on the ground with zero velocity. (b) Determine the velocity of the bucket in position 2.

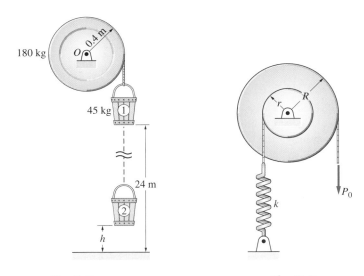

Fig. P18.53   **Fig. P18.54**

*18.54   The moment of inertia of the compound pulley about its center is $\bar{I}$. A rope wound around the inner pulley of radius r is attached to a spring of stiffness k. A constant force P_0 acts on the rope wrapped around the outer pulley, the radius

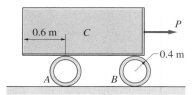

Fig. P18.55

of which is R. If the pulley is at rest with the spring undeformed when P_0 is applied, derive the expression for the maximum angular velocity of the pulley.

18.55 The 700-kg stone slab C is supported by two thin-walled steel cylinders A and B weighing 120-kg each. The system is at rest in the position shown when the constant force $P = 800\,\text{N}$ is applied. Determine the velocity of stone slab C when cylinder A has reached the left corner of the slab. Assume no slipping.

18.56 Gear A, which has a mass of 4 kg and a central radius of gyration of 90 mm, rolls on the fixed gear B. A constant couple M_0 acts on the arm CD, which has negligible mass. Determine the magnitude of the couple, knowing that the angular speed of CD increased from 200 to 320 rev/min, both counterclockwise, as it rotated through six revolutions.

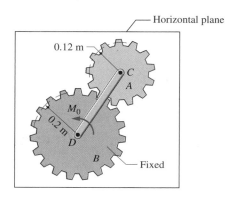

Fig. P18.56

PART B: *Impulse-Momentum Method*

18.5 *Momentum Diagrams*

We derived the angular momentum of a rigid body in Art. 17.3. For plane motion, the angular momentum of the body about its mass center G was found to be

$$h_G = \bar{I}\omega \tag{18.11}$$

where $\bar{I}$ is the mass moment of inertia about G, and ω represents the angular velocity of the body.

We also found that the angular momentum about an arbitrary point A can be obtained from the momentum diagram of the body, shown in Fig. 18.3. The diagram is a sketch of the body that displays the following momenta:

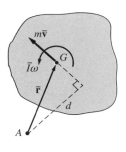

Fig. 18.3

- The linear momentum vector $\mathbf{p} = m\bar{\mathbf{v}}$ of the body acting at G, where m is the mass of the body and $\bar{\mathbf{v}}$ is the velocity of G.
- The angular momentum $h_G = \bar{I}\omega$ of the body, represented as a couple.

The resultant moment of these momenta about A equals the angular momentum of the body about A. For example, if we let d be the "moment arm" of the linear momentum, as shown in Fig. 18.3, the angular momentum about A becomes

$$h_A = \bar{I}\omega + m\bar{v}d \quad (A: \text{arbitrary point}) \tag{18.12}$$

An important special case arises when A is a point in the body (or body extended) that has zero velocity. It was shown in Art. 17.3 that Eq. (18.12) then reduces to

$$h_A = I_A\omega \quad (A: \text{instant center}) \tag{18.13}$$

where I_A is the moment of inertia of the body about A.

It is recommended that a momentum diagram be used to compute the momenta of a rigid body. Momentum diagrams not only provide a convenient pictorial representation of the linear and angular momenta of the body, but they also eliminate the need to memorize formulas such as Eq. (18.12).

To illustrate the use of momentum diagrams, consider the homogeneous slender bar of mass m and length L shown in Fig. 18.4(a). The bar is rotating about the fixed point A with the counterclockwise angular velocity ω at the instant shown. The momentum diagram for the bar, shown in Fig. 18.4(b), consists of the linear momentum $m\bar{v} = m(L/2)\omega$ and the couple $\bar{I}\omega = (mL^2/12)\omega$. The resultant angular momentum (moment of the momentum) about A is

$$\overset{\curvearrowleft}{+} \quad h_A = \frac{mL^2}{12}\omega + \left(m\frac{L}{2}\omega\right)\frac{L}{2} = \frac{mL^2}{3}\omega$$

Because $I_A = mL^2/3$, we see that $h_A = I_A\omega$, as would be expected, because the velocity of A is zero.

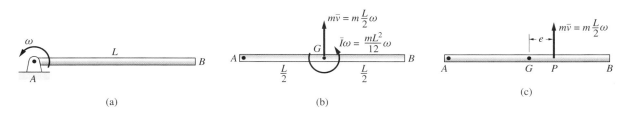

(a)   (b)   (c)

Fig. 18.4

Referring again to Fig. 18.4(b), the angular momentum about B is

$$\circlearrowleft_{+} \quad h_B = \frac{mL^2}{12}\omega - \left(m\frac{L}{2}\omega\right)\frac{L}{2} = -\frac{mL^2}{6}\omega$$

Note that the angular momentum about B is negative—that is, clockwise (recall that the angular velocity of the bar is counterclockwise). Furthermore, observe that h_B is not equal to $I_B\omega = (mL^2/3)\omega$. Because the velocity of B is not zero, Eq. (18.13) is clearly not applicable.

An equivalent form of the momentum diagram is shown in Fig. 18.4(c), where the momenta in Fig. (b) have been reduced to an equivalent single momentum vector $m\bar{\mathbf{v}}$ acting at point P. The distance e that locates P is found from the condition $h_P = 0$; that is, $m\bar{v}e - \bar{I}\omega = 0$. Substituting $\bar{v} = (L/2)\omega$ and $\bar{I} = mL^2/12$, we obtain $e = L/6$. If a body rotates about a fixed axis, a point about which the angular momentum of the body is zero, such as P in Fig. 18.4(c), is called the *center of percussion*.

18.6   *Impulse-Momentum Principles*

a.   *Impulse-momentum relations*

Assuming that a rigid body consists of a large number of particles, our previous discussions of impulse and momentum for particle systems can be applied to rigid-body motion. The equations presented in this article are valid for three-dimensional motion, but we restrict the applications to plane motion.

Repeating the results obtained for particle systems in Art. 14.6, the linear impulse-momentum equation for rigid-body motion is

$$\mathbf{L}_{1-2} = \mathbf{p}_2 - \mathbf{p}_1 = \Delta\mathbf{p} \qquad\qquad (18.14)$$

where $\mathbf{L}_{1-2}$ represents the linear impulse of external forces acting on the body during the time interval from time t_1 to time t_2, and $\Delta\mathbf{p}$ is the change in the linear momentum of the body during the same time interval. Recall that the linear momentum of the body is

$$\mathbf{p} = m\bar{\mathbf{v}} \qquad\qquad (18.15)$$

where m is the mass of the body and $\bar{\mathbf{v}}$ is the velocity of its mass center.

Applying the angular impulse-momentum equation for systems of particles (Art. 15.7) to rigid-body motion, we have

$$(\mathbf{A}_A)_{1-2} = (\mathbf{h}_A)_2 - (\mathbf{h}_A)_1 = \Delta\mathbf{h}_A$$
$$(A: \text{fixed point}^* \text{ or mass center}) \tag{18.16}$$

where $(\mathbf{A}_A)_{1-2}$ is the angular impulse of the external forces about point A for the time interval t_1 to t_2, and $\Delta\mathbf{h}_A$ is the change in the angular momentum of the rigid body about A during the same period.

The preceding results can also be applied to systems of rigid bodies provided that (1) the impulses refer only to forces that are external to the system, (2) the momenta are interpreted as the sum of the momenta of the bodies that constitute the system, and (3) the mass center referred to in Eq. (18.16) is the mass center of the system.

b.   *Conservation of momentum*

As is the case with systems of particles, the linear momentum for a rigid body (or system of rigid bodies) is conserved when the linear impulse of the external forces is zero. Similarly, the angular momentum of a rigid body (or system of rigid bodies) about a point is conserved when the angular impulse of the external forces about that point is zero. The reference point must either be fixed in space, or the mass center, because Eq. (18.16) is valid only if A is restricted to these points. Of course, the best method for determining whether the linear or angular impulse about a point vanishes is to examine the free-body diagram of the body or system of bodies.

---

*Point A is fixed in space; it is not necessarily a point in the body.

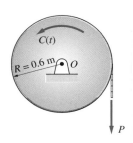

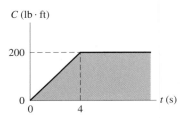

Sample Problem 18.7

The homogeneous 150-kg disk rotates about the fixed axis at O. The disk is acted on by the constant force $P = 320$ N (applied to a rope wound around the disk) and the counterclockwise couple $C(t)$, which varies with time as shown on the diagram. If the disk was at rest at time $t = 0$, determine (1) its angular velocity when $t = 4$ s; and (2) the time when the angular velocity reverses direction.

Solution

The resultant angular impulse about O is obtained by summing the angular impulses of all the forces that act on the disk. The support reactions and the weight of the disk do not contribute to the angular impulse about O (they pass through O). Therefore, we need to consider only the angular impulse of the force P and the couple $C(t)$.

Because P is constant, its angular impulse about O over the time interval $\Delta t = t_2 - t_1$ is $(A_O)_{1-2} = PR\,\Delta t$ (clockwise). However, because $C(t)$ is time-dependent, its angular impulse must be obtained by integration: $(A_O)_{1-2} = \int_{t_1}^{t_2} C(t)\,dt$ (counterclockwise). For the purposes of computation, it is useful to note that the integral represents the *area under the C-t diagram* between t_1 and t_2.

Part 1

The moment of inertia of the disk about its mass center O is

$$\overline{I} = \frac{1}{2}mR^2 = \frac{1}{2}(150)(0.6)^2 = 27 \text{ kg} \cdot \text{m}^2$$

With $t_1 = 0$ and $t_2 = 4$ s, the resultant angular impulse about O becomes

$$(A_O)_{1-2} = PR\,\Delta t - \int_{t_1}^{t_2} C(t)\,dt = (320)(0.6)(4) - \frac{1}{2}(300)(4) = 168 \text{ N} \cdot \text{m} \cdot \text{s}$$

Applying the angular impulse-momentum equation to the disk, we get

$$(A_O)_{1-2} = \Delta h_O = \overline{I}(\omega_2 - \omega_1)$$

$$168 = 27(\omega_2 - 0)$$

which yields

$$\omega_2 = 6.22 \text{ rad/s (CW)} \qquad\qquad \textit{Answer}$$

Part 2

Let t_3 be the time when the angular velocity of the disk becomes zero. Noting that the initial angular velocity is also zero, we conclude there is no change in angular momentum between t_1 and t_3. Consequently, the resultant angular impulse must vanish; that is,

$$(A_O)_{1-3} = PR\,\Delta t - \int_{t_1}^{t_3} C(t)\,dt = 0 \qquad\qquad \text{(a)}$$

In order to evaluate the integral (area under the C-t diagram), we must know whether $t_3 < 4$ s or $t_3 > 4$ s. Noting that $C = 0$ at $t = 0$, we conclude that the initial angular velocity is clockwise, because this is the direction of angular impulse of P. Because the disk still spins clockwise at $t = 4$ s (see the solution of Part 1), we conclude that $t_3 > 4$ s. Therefore, Eq. (a) is

$$\circlearrowleft^{+} \quad (320)(0.6)t_3 - \left[\frac{1}{2}(300)(4) + 300(t_3 - 4)\right] = 0$$

The solution is

$$t_3 = 5.56 \text{ s} \qquad \qquad \textit{Answer}$$

Sample Problem 18.8

The 4-kg uniform disk in Fig. (a) is at rest when the 30-N force is applied to the center of the disk for a period of 1.5 s, after which the force drops to zero. During the 1.5 s period, friction is insufficient to prevent the disk from slipping. Determine the time when slipping stops and the corresponding angular velocity of the disk. The kinetic coefficient of friction between the disk and the horizontal surface is 0.2.

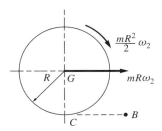

200 mm

30 N

4 kg

$\mu_k = 0.2$

(a)

Solution

Free-Body Diagram (FBD)

Figure (b) shows the FBD of the disk at an arbitrary time t during the period of slipping. Because there is no acceleration in the y-direction, the normal force at the contact point C was obtained from $\Sigma F_y = 0$, yielding $N_C = mg$. Because the disk is slipping, the friction force is $F_C = \mu_k mg$.

Final Momentum Diagram

The final momentum diagram of the disk at the instant ($t = t_2$) when slipping stops is also shown in Fig. (b). This diagram displays the angular momentum about the mass center G: $\bar{I}\omega_2 = (mR^2/2)\omega_2$, and the linear momentum of the disk: $m\bar{v}_2 = m(R\omega_2)$, where $\bar{v}_2 = R\omega_2$ is the kinematic condition for rolling without slipping.

Impulse-Momentum Analysis

We start by applying the angular impulse-momentum equation about the mass center G of the disk, covering the time period $t_1 = 0$ (when the 30-N force is applied) to t_2 (when slipping stops). Referring to Fig. (b), we have:

$$(A_G)_{1-2} = (h_G)_2 - (h_G)_1$$

$$\circlearrowleft^{+} \quad \int_0^{t_2} (\mu_k mg) R \, dt = \frac{mR^2}{2}\omega_2 - 0$$

$$\mu_k g t_2 = \frac{R}{2}\omega_2$$

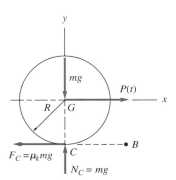

y

mg

$P(t)$

x

R G

$F_C = \mu_k mg$ C B

$N_C = mg$

FBD (during slipping)

$\frac{mR^2}{2}\omega_2$

R G $mR\omega_2$

C B

Final momentum diagram

(b)

447

Therefore,

$$\omega_2 = \frac{2\mu_k g}{R} t_2 = \frac{2(0.2)(9.81)}{0.2} t_2 = 19.62 t_2 \qquad \text{(a)}$$

The linear impulse-momentum equation in the x-direction yields

$$(L_x)_{1-2} = (p_x)_2 - (p_x)_1$$

$$\xrightarrow{+} \int_0^{t_2} P(t)\,dt - \mu_k mg t_2 = mR\omega_2 - 0 \qquad \text{(b)}$$

Noting that $P(t) = 30\,\text{N}$ when $0 < t < 1.5\,\text{s}$ and zero otherwise, we have $\int_0^{t_2} P(t)\,dt = (30)(1.5) = 45\,\text{N}\cdot\text{s}$. Therefore, Eq. (b) becomes

$$45 - (0.2)(4)(9.81)t_2 = (4)(0.2)\omega_2$$

$$\omega_2 = 56.25 - 9.81 t_2 \qquad \text{(c)}$$

The solution of Eqs. (a) and (c) is

$$t_2 = 1.911\,\text{s} \qquad \omega_2 = 37.5\,\text{rad/s} \qquad \textit{Answer}$$

Alternate Solution

We can also apply the angular impulse-momentum equation about any fixed point, such as B shown in Fig. (b). The result is

$$(A_B)_{1-2} = (h_B)_2 - (h_B)_1$$

$$\circlearrowleft + \int_0^{t_2} RP(t)\,dt = \left[\frac{mR^2}{2}\omega_2 + (mR\omega_2)R\right] - 0$$

$$\int_0^{t_2} P(t)\,dt = \frac{3}{2}mR\omega_2$$

which yields

$$\omega_2 = \frac{2}{3mR}\int_0^{t_2} P(t)\,dt = \frac{2}{3(4)(0.2)}(45) = 37.5\,\text{rad/s}$$

as before. The time t_2 when slipping stops can now be calculated from either Eq. (a) or Eq. (c).

Sample Problem 18.9

Figure (a) shows two gears A and B that are supported by pins at D and E. The gears are stationary when the constant counterclockwise couple C_0 is applied at time $t = 0$. Determine the value of C_0 for which the angular velocity of gear A will be 50 rad/s when $t = 5$ s, and determine the corresponding tangential contact force between the gears.

Solution

Figure (b) shows the FBDs and the momentum diagrams at $t = 0$ and $t = 5$ s for each gear. The FBDs contain the weights of the gears, the applied couple C_0, the tangential contact force F, and the support reactions. (Because C_0 is constant, F is also constant.) Note that each momentum diagram contains only the angular momentum $\bar{I}\omega$, because each gear rotates about its mass center; that is, $m\bar{v} = 0$.

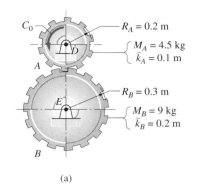

(a)

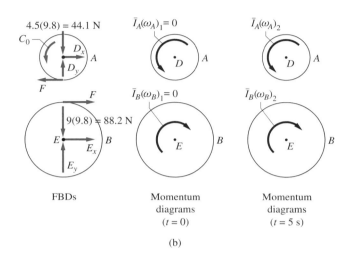

FBDs   Momentum diagrams $(t = 0)$ Momentum diagrams $(t = 5$ s$)$

(b)

The force-mass-acceleration (FMA) method, described in Chapter 17, or the impulse-momentum method may be used with equal facility in solving this problem. The FMA method would be straightforward here, because the forces, and therefore the accelerations, are constant. The impulse-momentum method, which we will employ, is equally convenient, because we are required to calculate the change in velocity that occurs during a given time interval.

The moments of inertia of the gears are $\bar{I}_A = m_A \bar{k}_A^2 = (4.5)(0.1)^2 = 0.045$ kg·m^2 and $\bar{I}_B = m_B \bar{k}_B^2 = 9(0.2)^2 = 0.36$ kg·m^2. Using Fig. (b), the angular impulse-momentum equation for the gear A about point D is

$$(A_D)_{1-2} = \Delta h_D$$

$$\curvearrowleft{+} \quad C_0(\Delta t) - FR_A \Delta t = \bar{I}_A[(\omega_A)_2 - (\omega_A)_1]$$

$$C_0(5) - F(0.2)(5) = 0.045(50 - 0) \qquad \text{(a)}$$

Note that the angular impulses were easy to compute, because C_0 and F are constant.

Observing that the point of contact on each gear has the same velocity, we conclude that $0.2\omega_A = 0.3\omega_B$, which gives $(\omega_B)_2 = (2/3)(\omega_A)_2 = (2/3)(50) = 33.33$ rad/s. Therefore, from Fig. (b), the angular impulse-momentum equation for gear B about point E is

$$(A_E)_{1-2} = \Delta h_E$$

$$\curvearrowright{+} \quad FR_B \Delta t = \bar{I}_B[(\omega_B)_2 - (\omega_B)_1]$$

$$F(0.3)(5) = 0.36(33.33 - 0) \qquad \text{(b)}$$

From Eq. (b) we find that the contact force is

$$F = 7.99 \text{ N} \qquad \textit{Answer}$$

Substituting this value into Eq. (a), we obtain

$$C_0 = 2.048 \text{ N} \cdot \text{m} \qquad \textit{Answer}$$

Sample Problem *18.10*

Figure (a) shows a rod B of mass m_B that is placed inside the tube A of mass m_A. Both bodies are slender, homogeneous, and of length L. Initially, B is held inside A ($a = 0$) by a lightweight cap (not shown) that covers the end of A while the assembly is spinning freely with the angular velocity ω_1 about the z-axis. The cap then falls off, allowing the rod to slide out of the tube. Determine the angular velocity ω_2 of the assembly when the rod is fully out of the tube ($a = L$). Neglect friction.

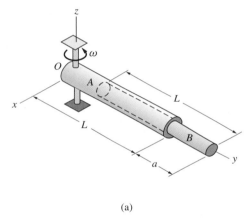

(a)

Solution

Free-Body Diagram (FBD)

Figure (b) shows the FBD of the assembly when rod B extends an arbitrary distance a beyond the end of the tube A. This FBD shows only the forces that act in the xy-plane. The complete FBD would also include the weights of the bodies, a reaction O_z, and a reactive couple C_x acting at O. However, because these have no effect on the motion in the xy-plane, they have been omitted. It should be noted that the contact forces between the tube and rod do not appear on the FBD because they are internal to the system. This is the primary reason why we choose to analyze the motion of the entire assembly, instead of considering each body separately.

Initial Momentum Diagram

Figure (b) also shows the initial momentum diagram, with rod B being entirely inside tube A, before the cap falls off. Note that this diagram includes the angular and linear momentum vectors of A and B. The angular velocities of both bodies are ω_1, so that the velocities of their mass centers are both equal to $(L/2)\omega_1$.

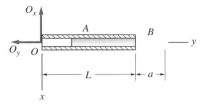

FBD (arbitrary position)

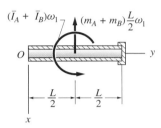

Initial momentum diagram

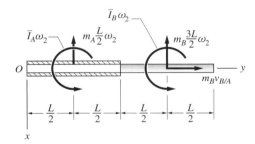

Final momentum diagram

(b)

Final Momentum Diagram

In the final momentum diagram in Fig. (b), rod B is about to leave the tube A with the relative velocity $v_{B/A}$. The angular velocities of the two bodies are both equal to ω_2, giving rise to the angular momenta shown. The linear momentum vector of A follows from the fact that the velocity of its mass center is $(L/2)\omega_2$. The two components of the linear momentum vector of B correspond to the polar components of the velocity of its mass center, namely $v_R = v_{B/A}$ and $v_\theta = (3L/2)\omega_2$.

Impulse-Momentum Analysis

The remainder of the analysis consists of writing and solving the impulse-momentum equations using the three diagrams shown in Fig. (b). Because we are interested only in ω_2, the most convenient solution uses the angular impulse-momentum equation with O as the reference point, which eliminates the unknown reactions O_x and O_y. From the FBD in Fig. (b), we see that $(A_O)_{1-2} = 0$ (O_x and O_y do not contribute to the angular impulse about O). Noting that O is a fixed point, we can use the angular impulse-momentum equation $(A_O)_{1-2} = \Delta h_O$, which in our case reduces to $\Delta h_O = 0$. In other words, angular momentum of the system about O is conserved.

Equating the moments of the momenta about O in the two momentum diagrams shown in Fig. (b), we get

$$(h_O)_1 = (h_O)_2$$

$$\overset{+}{\curvearrowleft} \quad (\bar{I}_A + \bar{I}_B)\omega_1 + (m_A + m_B)\frac{L}{2}\omega_1\left(\frac{L}{2}\right)$$

$$= \left[\bar{I}_A\omega_2 + \left(m_A\frac{L}{2}\omega_2\right)\left(\frac{L}{2}\right)\right] + \left[\bar{I}_B\omega_2 + \left(m_B\frac{3L}{2}\omega_2\right)\left(\frac{3L}{2}\right)\right]$$

Substituting $\bar{I} = mL^2/12$ for each bar and solving for ω_2, we obtain

$$\omega_2 = \frac{m_A + m_B}{m_A + 7m_B}\omega_1 \qquad\qquad \textit{Answer}$$

It should be mentioned that this solution would be valid even if there were friction between the rod and tube—provided, of course, that the coefficient of friction were small enough to permit relative motion between them.

Sample Problem **18.11**

The assembly shown in Fig. (a) consists of an arm AOC, to which are pinned two homogeneous slender rods AB and CD. The assembly rotates about the z-axis in a frictionless bearing at O. An internal mechanism (not shown in the figure) can position and lock the two rods at any angle θ. The moment of inertia of the arm AOC about the z-axis is 1.04 kg · m², and rods AB and CD mass 1.5-kg each. Initially the assembly is rotating freely about the z-axis with the angular velocity $\omega_1 = 10$ rad/s with $\theta = 90°$. Calculate the angular velocity of the assembly when the rods have been moved to the position $\theta = 180°$.

Solution

Figure (b) contains the FBD of the assembly in an arbitrary position, showing only forces that act in the xy-plane—namely O_x and O_y, two components of the bearing reaction at O. Note that this FBD does not change during the motion of the assembly. Also shown in Fig. (b) are the initial ($\theta = 90°$) and final ($\theta = 180°$) momentum diagrams of the assembly. The numerical values shown in the momentum diagrams were computed as described below.

Initial Momentum Diagram

Arm AOC:

$$\bar{I}_z\omega_1 = 1.04(10) = 10.4\,\text{N} \cdot \text{m} \cdot \text{s}$$

(a)

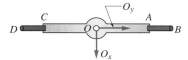

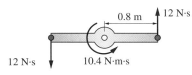

Initial momentum diagram

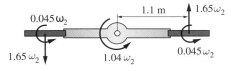

Final momentum diagram

(b)

Rods AB and CD (both rods are perpendicular to the plane of the figure):

$$m\bar{v}_1 = mr_1\omega_1 = 1.5(0.8)(10) = 12\,\text{N}\cdot\text{s}$$

Final Momentum Diagram

Arm AOC:

$$\bar{I}_z\omega_2 = 1.04\omega_2$$

Rods AB and CD (both rods are in the plane of the figure):

$$m\bar{v}_2 = mr_2\omega_2 = 1.5(1.1)\omega_2 = 1.65\omega_2$$

$$\bar{I}_z\omega_2 = \frac{1}{12}mL^2\omega_2 = \frac{1}{12}(1.5)(0.6)^2\omega_2 = 0.045\omega_2$$

Conservation of Angular Momentum

From the FBD in Fig. (b) we see that there is no angular impulse about the z-axis. Hence angular momentum about the z-axis is conserved. Referring to the momentum diagrams in Fig. (b), we obtain

$$(h_z)_1 = (h_z)_2$$

$$\overset{+}{\curvearrowleft} \quad 10.4 + 2[0.8(12)] = 1.04\omega_2 + 2(0.045\omega_2) + 2[1.1(1.65\omega_2)]$$

which yields

$$\omega_2 = 6.22\,\text{rad/s} \qquad\qquad \textit{Answer}$$

Problems

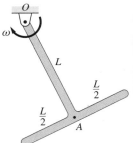

Fig. P18.57

18.57 The T-shaped body is made of two identical uniform rods, each of mass $m/2$. If the body rotates about O with the angular velocity ω, calculate its angular momentum about (a) the mass center of the body; (b) point O; and (c) point A.

18.58 The uniform plate of mass m rotates about its corner O with the angular speed ω. Determine the angular momentum of the plate about (a) the mass center; (b) the corner O; and (c) the corner A.

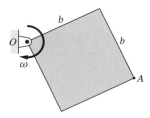

Fig. P18.58

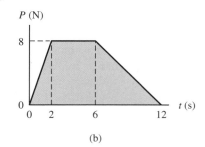

Fig. P18.59

18.59 The uniform slender rods AB and BC are rigidly connected together. If the assembly is free to rotate about the pin at A, find the distance d that locates its center of percussion.

18.60 The uniform disk of mass m and radius R rotates about the pin at C, located a distance d from its center. Determine d so that the center of percussion will be at point A on the rim of the disk.

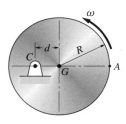

Fig. P18.60

18.61 The radius of gyration about O of the 20-kg pulley in part (a) of the figure is 0.16 m. The end of the rope wound around the pulley is pulled with the force $P(t)$, which varies with time as shown in part (b). Assuming that the pulley starts from rest at $t = 0$, determine its angular velocity at $t = 12$ s.

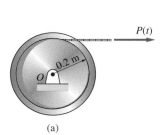

(a)

(b)

Fig. P18.61

18.62 The radius of gyration of the 18-kg flywheel that rotates about O is $\bar{k} = 0.23$ m. The couple acting on the flywheel is $C(t) = 16(1 - e^{-2t})$ lb·ft, where t is the time in seconds. If the flywheel was at rest when $t = 0$, determine its angular speed when $t = 60$ s.

Fig. P18.62

18.63   The block B is suspended from a cable attached to the center of the pulley A. The 75-N vertical force is applied to one end of the cable that wraps around A. If at time $t = 0$ the angular velocity of the pulley is 3 rad/s clockwise, determine the time when the system will come momentarily to rest.

18.64   The torque M acting on the input shaft of a generator varies with time t as $M(t) = M_0 \exp(-t/t_0)$, where $M_0 = 4.6$ N · m and $t_0 = 3.8$ s. If the generator is at rest at $t = 0$, calculate its terminal angular speed. Assume that the generator rotates without resistance, and use $\overline{I} = 0.72$ kg · m^2 for its rotor.

18.65   The uniform disk of mass m and radius R is initially at rest on a frictionless horizontal surface. The constant force of magnitude P is applied at time $t = 0$ to the cord wrapped around the disk. For an arbitrary time t, determine the expression for the velocity of B (the point on the rim where the cord leaves the disk).

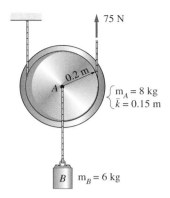

Fig. P18.63

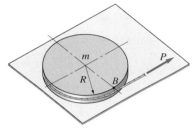

Fig. P18.65

18.66   The disk in part (a) of the figure has a mass of 20 kg and its radius of gyration about O is 160 mm. The disk is spinning freely at $\omega = 400$ rev/min when the force $P(t)$ is applied to the handle of the brake at $t = 0$. The P-t relationship is shown in part (b). Determine the peak value P_0 of the force for which the disk will come to rest at $t = 12$ s. The kinetic coefficient of friction between the disk and brake is 0.3.

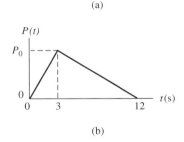

Dimensions in mm

(a)

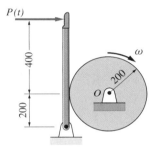

(b)

Fig. P18.66

18.67   The 16-kg homogeneous disk A is attached to the 6-kg uniform rod BC by a bearing at B, the axis of which is vertical. The rod rotates freely about the vertical axis at C. The system is at rest when the constant couple $M_0 = 8$ N · m is applied to the rod. Determine the angular velocity of rod BC after 2 s, assuming that the bearing at B is (a) free to turn; and (b) locked.

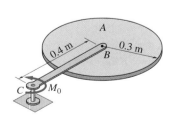

Fig. P18.67

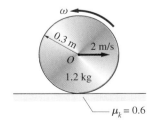

Fig. P18.68

18.68   The 1.2-kg homogeneous cylinder is launched on the horizontal surface with a forward speed of 2 m/s and a backspin of ω rad/s. Determine the required

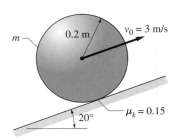

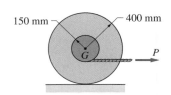

Fig. P18.69

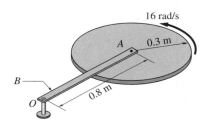

Fig. P18.70

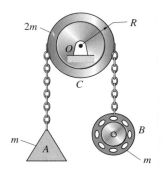

Fig. P18.71

value of ω if the cylinder is to come to a complete stop. The kinetic coefficient of friction between the cylinder and the surface is 0.6.

18.69 The homogeneous, solid ball of mass m is launched on the inclined plane at time $t = 0$ with the forward speed of 3 m/s and no spin. The kinetic coefficient of friction between the ball and the plane is 0.15. Calculate the time when the ball stops slipping on the plane and the angular velocity of the ball at that time.

*18.70 The mass moment of inertia of the 60-kg spool about its mass center G is 1.35 kg · m². The spool is at rest on the horizontal surface when the constant force $P = 200$ N is applied to the cable that is wound around its hub. Calculate the angular velocity of the spool 2 s later, assuming that the static as well as the kinetic coefficient of friction between the spool and the surface is (a) $\mu = 0.2$; and (b) $\mu = 0.1$.

18.71 The 3-kg homogeneous disk A is pinned to the the uniform arm B of mass 2 kg. The assembly is free to rotate about the vertical axis at O. The arm is initially at rest with the disk spinning with an angular velocity of 16 rad/s. If an internal brake (not shown) locks the disk and the arm together, what will be the final angular velocity of the assembly?

18.72 The 0.6 kg uniform rod AB and the 0.1 kg small slider C rotate freely about the vertical axis at A. The angular velocity of the system is 5 rad/s when the cord holding C breaks. Determine the angular velocity of AB when (a) C is just about to leave the rod; and (b) just after C has left the rod.

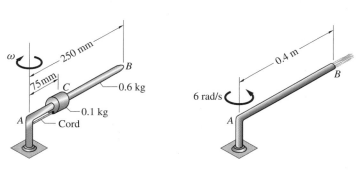

Fig. P18.72 **Fig. P18.73**

*18.73 The pipe AB of inner diameter 10 mm is rotating without friction about the vertical axis at A. Water flows through the pipe with the constant speed 1.8 m/s relative to the pipe. Find the couple that must be applied to the pipe in order to maintain its angular speed at 6 rad/s.

18.74 A chain running over the pulley C supports the weight A and the motorized spool B, both of mass m. The pulley can be modeled as a uniform disk of mass $2m$. The system is at rest when the spool starts reeling in the chain at the speed v_0 relative to the spool. Determine the velocities of A and B in terms of v_0.

Fig. P18.74

18.75   The 7.5 kg stationary disk B is placed onto the 5-kg disk A, which is rotating freely at 3000 rev/min. Assuming that both disks are homogeneous, determine their final angular speed.

18.76   The pendulum in part (a) of the figure consists of the 0.8-kg slender rod and the 3.2-kg homogeneous cylinder. The pendulum is at rest when it receives a short impulse from the force $F(t)$, which varies with time as shown in part (b). Determine the angular speed of the pendulum immediately after the impulse has been received. Neglect the small angular displacement of the pendulum during the impulse.

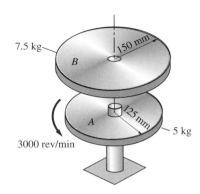

Fig. P18.75

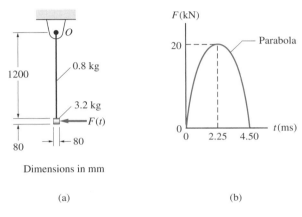

(a)

(b)

Fig. P18.76

18.77   The 15-kg uniform slender rod AB is placed in a smooth collar at O. A pin holds the bar in the position shown as the assembly rotates freely at 12 rad/s about the vertical axis at O. If the pin is removed, allowing the rod to slide in the collar, determine the final angular velocity of the assembly. Note that sliding is limited by the small collars at the ends of the rod.

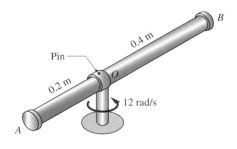

Fig. P18.77

18.78   The 0.15-kg slider at A is attached to the thin, uniform 1-kg ring with a pin. The assembly is rotating in the horizontal plane about O with the angular speed $\omega_1 = 40$ rad/s when the pin holding the slider falls out. The slider then moves along the rotating ring, ending up at B. Determine the final angular speed of the assembly.

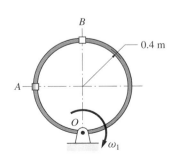

Fig. P18.78

*18.79   The combined mass of the rider and the bicycle without its wheels is 80-kg. Each wheel mass 3 kg, which is due primarily to the weights of the rim and the tire. If the bicycle starts from rest, determine its speed after 10 s, assuming that the chain provides a constant couple of 20 N · m on the rear wheel and that the wheels do not slip on the ground.

Fig. P18.79

18.80   The skater is spinning about the z-axis with her arms outstretched as shown. Neglecting friction, determine the percentage increase in the skater's angular speed after her arms are lowered to her sides. The arms may be modeled as slender, nonuniform rods hinged at the shoulders. The mass of the torso and each arm are 45 kg and 2.8 kg, respectively. For the torso, $\bar{k}_z = 0.14$ m, and $\bar{k}_z = 0.26$ m for each arm in the horizontal position shown.

18.81   Bar AB of negligible weight carries two identical 10-kg uniform, thin disks. The angle ϕ between each disk and the bar can be varied slowly by an internal mechanism (not shown). The entire assembly is free to rotate about the z-axis. If the angular velocity of the assembly about the z-axis is ω_0 when $\phi = 0$, determine ϕ for which the angular velocity is $\omega_0/2$.

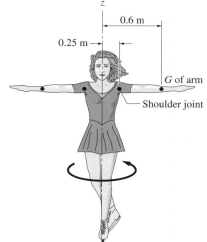

Fig. P18.80

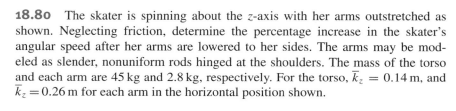

Fig. P18.81

18.82   The two thin square plates, each mass 6 kg, are attached to ends of the 4-kg uniform bar AB. An internal mechanism can rotate the plates simultaneously about the axis of bar AB. When $\theta = 0$, the assembly is rotating freely about the vertical axis at O with the angular velocity $\omega = 12$ rad/s. Determine the angular velocity after the plates have been rotated to the position $\theta = 90°$.

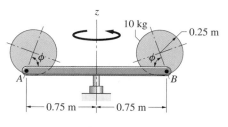

Fig. P18.82

18.7 *Rigid-Body Impact*

The impact of systems of particles was discussed in Arts. 15.8–15.10, where we introduced a simplified analysis of elastic impact that utilized the coefficient of restitution, an experimental constant. Rigid-body impact is a more complex problem that depends on the geometries of the impacting bodies and their surface characteristics, as well as their relative velocities. Any attempt to extend the concept of a constant coefficient of restitution to rigid-body impact greatly oversimplifies the real problem and renders the results meaningless. Therefore, we will not consider problems of rigid-body impact that require the use of an experimental constant analogous to the coefficient of restitution.

A useful simplification arises in the analysis of rigid-body impact when the motion is assumed to be impulsive, meaning that the duration of the impact is negligible—see Art. 15.9. As we have stated before, the expression "angular impulse equals change in angular momentum" is, in general, valid only about the mass center or a fixed point. However, for impulsive motion, "angular impulse equals change in angular momentum" is valid about *all* points. The reason for this simplification is that by assuming the time of impact to be infinitesimal, we are neglecting all displacements during the impact. Consequently, all points are, in effect, fixed during the impact.

The general steps in the analysis of rigid-body impact problems parallel those given for particle impact in Art. 15.9.

Step 1: Draw the FBD of the impacting bodies and/or of the system of impacting bodies. Identify the impulsive forces—use a special symbol, such as a caret (^), to label each impulsive force. (It is advisable to redraw the FBD, showing only the impulsive forces.)

Step 2: Draw the momentum diagrams for the bodies at the instant immediately before impact.

Step 3: Draw the momentum diagrams for the bodies at the instant immediately after impact.

Step 4: Using the diagrams drawn in Steps 1 through 3, derive and solve the appropriate impulse-momentum equations for the individual bodies and/or the system of bodies.

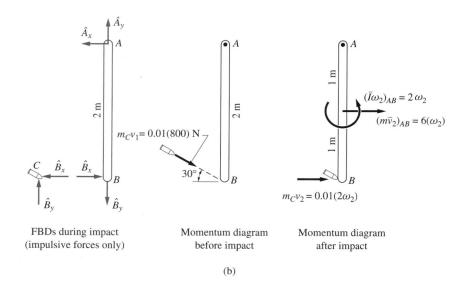

Sample Problem 18.12

Figure (a) shows a 0.01 kg bullet C that is fired at end B of the 6-kg homogeneous slender bar AB. The bar is initially at rest, and the initial velocity of the bullet is $v_1 = 800$ m/s, directed as shown. Assuming that the bullet becomes embedded in the bar, calculate (1) the angular velocity ω_2 of the bar immediately after the impact; (2) the impulse exerted on the bar at A during the impact; and (3) the percentage loss of energy as a result of the impact. Neglect the duration of the impact.

Solution

Introductory Comments

Because the duration of impact is negligible, the motion is impulsive, with the bar and bullet occupying essentially the same positions before, during, and after the impact.

The FBDs for the bullet and bar during the impact are shown in Fig. (b), with only the impulsive forces (denoted with carets) shown. Because the weights of C and AB are finite forces, they will not enter into the impact analysis and consequently are omitted from the FBD. Observe that the FBD contains $\hat{B}_x$ and $\hat{B}_y$, the components of the impulsive contact force at B, and $\hat{A}_x$ and $\hat{A}_y$, the components of the impulsive pin reaction at A.

| FBDs during impact (impulsive forces only) | Momentum diagram before impact | Momentum diagram after impact |

(b)

Figure (b) also contains the momentum diagrams immediately before and after the impact. The momentum diagram before impact contains only the initial linear momentum of the bullet C.

The momentum diagram after the impact contains both the final momentum of the bullet and the final linear and angular momentum of the bar AB (the angular velocity ω_2 is assumed to be measured in rad/s). Note that the kinematic relationships used in the diagram, $v_2 = 2\omega_2$ and $(\bar{v}_2)_{AB} = \omega_2$, follow from the fact that A is a fixed point. Also, the central moment of inertia of AB is $\bar{I}_{AB} = mL^2/12 = \frac{1}{12}(6)(2)^2 = 2\,\text{kg} \cdot \text{m}^2$, which gives $(\bar{I}\omega_2)_{AB} = (2)\omega_2\,\text{N} \cdot \text{m} \cdot \text{s}$, as indicated in the diagram.

Inspection of Fig. (b) reveals that there are a total of five unknowns: the impulses of $\hat{A}_x$, $\hat{A}_y$, $\hat{B}_x$, and $\hat{B}_y$, and the angular velocity ω_2. There are also a total of five independent impulse-momentum equations: two for the bullet C and three for the bar AB. Therefore, all five unknowns can be determined from the five independent equations. However, because we are required to find only three of the unknowns (ω_2 and the impulses of $\hat{A}_x$ and $\hat{A}_y$), it will not be necessary to use all the equations.

Part 1

The most efficient means of computing ω_2 is to consider the system consisting of both the bullet and the rod, as opposed to considering each of them separately. For the system, $\hat{B}_x$ and $\hat{B}_y$ are internal forces; consequently, $\hat{A}_x$ and $\hat{A}_y$ are the only external forces. Therefore, angular impulse acting on the system about A is zero, which leads us to conclude that angular momentum about A is conserved. Referring to the momentum diagrams in Fig. (b), we obtain for the system

$$(h_A)_1 = (h_A)_2$$

$\overset{+}{\curvearrowright}$ $0.01(800\cos 30°)(2) = [2\omega_2 + 6(\omega_2)(1)] + [0.01(2\omega_2)(2)]$

Solving for ω_2 yields

$$\omega_2 = 1.72\,\text{rad/s} \qquad\qquad \textit{Answer}$$

Part 2

As mentioned previously, $\hat{A}_x$ and $\hat{A}_y$ are the only external forces that act on the system during the impact, because $\hat{B}_x$ and $\hat{B}_y$ are internal forces. Referring to Fig. (b), the x-component of the linear impulse-momentum equation for the system is

$$(L_x)_{1-2} = (p_x)_2 - (p_x)_1$$

$\overset{+}{\longrightarrow}$ $-\int \hat{A}_x\,dt = [6(\omega_2) + 0.01(2\omega_2)] - [0.01(800\cos 30°)]$

Substituting $\omega_2 = 1.72\,\text{rad/s}$ from the solution to Part 1 gives

$$\int \hat{A}_x\,dt = -3.43\,\text{N} \cdot \text{s} \qquad\qquad \text{(a)}$$

The negative sign means, of course, that the direction of the impulse of $\hat{A}_x$ is opposite to the direction of $\hat{A}_x$ assumed in the FBD.

The y-component of the linear impulse-momentum equation for the system is

$$(L_y)_{1-2} = (p_y)_2 - (p_y)_1$$

$\overset{+}{\uparrow}$ $\int \hat{A}_y\,dt = 0 - [-0.01(800\sin 30)]$

or

$$\int \hat{A}_y \, dt = 4 \text{ N} \cdot \text{s} \qquad \text{(b)}$$

From the results given in Eqs. (a) and (b), the resultant impulse acting on the bar at A during impact is

4

$\int \hat{A} \, dt = 5.27 \text{ N} \cdot \text{s}$

3.43

Answer

Part 3

The kinetic energy of the system before impact is

$$T_1 = \frac{1}{2} m_C v_1^2 = \frac{1}{2}(0.01)(800)^2 = 3200 \text{ N} \cdot \text{m}$$

After impact, the kinetic energy is

$$T_2 = \frac{1}{2} m_C v_2^2 + \left[\frac{1}{2} \overline{I} \omega_2^2 + \frac{1}{2} m \overline{v}_2^2 \right]_{AB}$$

which, on substitution of the numerical values, becomes

$$T_2 = \frac{1}{2}(0.01)(2 \times 1.72)^2 + \left[\frac{1}{2}(2)(1.72)^2 + \frac{1}{2}(6)(1.72)^2 \right]$$

$$= 11.89 \text{ N} \cdot \text{m}$$

Therefore, the percentage loss of energy during the impact is

$$\frac{T_1 - T_2}{T_1} \times 100\% = \frac{3200 - 11.89}{3200} = 99.6\% \qquad \textit{Answer}$$

Alternate Method of Computing $\int \hat{A}_x \, dt$

The impulse of reaction $\hat{A}_x$ could also be computed from an angular impulse-momentum equation. Referring to Fig. (b) and considering either the entire system or only bar AB, we have

$$(A_B)_{1-2} = (h_B)_2 - (h_B)_1$$

$$\overset{+}{\curvearrowleft} \quad 2 \int \hat{A}_x \, dt = [2\omega_2 - 6(\omega_2)(1)] - 0$$

which yields

$$\int \hat{A}_x \, dt = -2\omega_2 = -2(1.72)_2 = -3.43 \text{ N} \cdot \text{s} \qquad \textit{Answer}$$

This agrees with the result given in Eq. (a). Note that it is legitimate to use B as the reference point, although it is neither a fixed point nor the mass center. As mentioned in Art. 18.7, there are no restrictions on the location of the reference point for impulsive motion (all points are, in effect, "fixed" during the infinitesimal period of impact).

Sample Problem *18.13*

The 20-kg uniform slender bar in Fig. (a) is moving to the right on frictionless rollers at A and B with the velocity v_1 when the roller at B strikes the small obstruction C without rebounding. Compute the minimum value of v_1 for which the bar will reach the vertical position after the impact.

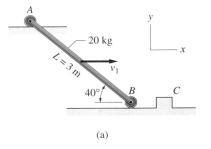

(a)

Solution

Figure (b) shows the FBD of the bar during the impact, and the momentum diagrams of the bar before and immediately after the impact, labeled 1 and 2,

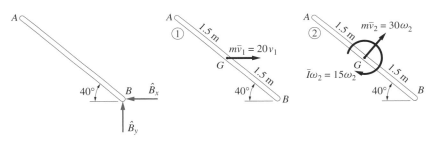

FBD during impact	Momentum diagram	Momentum diagram
(impulsive forces only)	before impact	after impact

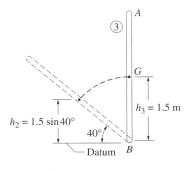

Final position
(at rest)

(b)

respectively. It also shows the final, vertical position 3, which is a rest position. We will neglect the time of impact, which means that the motion is assumed to be impulsive. Therefore, the bar occupies the same spatial position throughout the duration of the impact.

We must use the impulse-momentum method to analyze the impact that occurs between positions 1 and 2. To analyze the motion between positions 2 and 3, we employ the principle of conservation of mechanical energy.

The FBD in Fig. (b) shows the reactions $\hat{B}_x$ and $\hat{B}_y$, which are the only impulsive forces that act on the bar during impact. The weight of the bar is omitted from the FBD; it is a finite force whose impulse may be neglected because of the very small duration of impact. The roller reaction at A does not appear in the FBD for the same reason (the reaction is not impulsive because the roller is about to lift off the plane during the impact).

As shown in Fig. (b), the momentum diagram for position 1 consists only of the linear momentum vector, the magnitude of which is $mv_1 = 20\, v_1$ N $\cdot$ s, where v_1 is measured in m/s. The momentum diagram for position 2 contains both the linear and angular momenta. We have $\overline{I} = mL^2/12 = 20(3)^2/12 = 15$ N $\cdot$ m^2, so that the angular momentum is $\overline{I}\omega_2 = 15\omega_2$ N $\cdot$ m $\cdot$ s, where ω_2 is measured in rad/s. Because the bar is rotating about end B after the impact, $\overline{v}_2 = 1.5\omega_2$ m/s; consequently, the magnitude of the linear momentum vector is $m\overline{v}_2 = 20(1.5)\omega_2 = 30\omega_2$ N $\cdot$ s.

From the FBD in Fig. (b) we see that the angular impulse about B is zero, because both $\hat{B}_x$ and $\hat{B}_y$ pass through B. Because angular impulse equals the change in angular momentum about every point during the period of impact, we conclude that the angular momentum about B is conserved between positions 1 and 2. Referring to the momentum diagrams in Fig. (b), we thus obtain

$$(h_B)_1 = (h_B)_2$$

$$\overset{\curvearrowleft}{+}\quad 20v_1(1.5\sin 40°) = 15\omega_2 + 30\omega_2(1.5)$$

which gives the following relationship between v_1 and ω_2:

$$v_1 = 3.111\omega_2 \tag{a}$$

Because all the forces acting on the bar after the impact are conservative, the value of ω_2 can be computed by applying the conservation of mechanical energy principle between position 2 and position 3 (recall that the bar is at rest in the vertical position).

As shown in Fig. (b), we choose the datum for the gravitational potential energy to be the horizontal plane through B. Therefore, the potential energies in positions 2 and 3 are

$$V_2 = Wh_2 = 20(9.81)(1.5\sin 40°) = 189.2 \text{ J} \tag{b}$$

$$V_3 = Wh_3 = 20(9.81)(1.5) = 294.3 \text{ J} \tag{c}$$

The kinetic energies for the two positions are

$$T_2 = \frac{1}{2}\bar{I}\omega_2^2 + \frac{1}{2}m\bar{v}_2^2$$
$$= \frac{1}{2}(15)\omega_2^2 + \frac{1}{2}(20)(1.5\omega_2)^2$$
$$= 30.0\omega_2^2 \text{ J} \tag{d}$$

and

$$T_3 = 0 \tag{e}$$

Utilizing Eqs. (b)–(e), the conservation of mechanical energy yields

$$V_2 + T_2 = V_3 + T_3$$
$$189.2 + 30.0\omega_2^2 = 294.3 + 0$$

from which

$$\omega_2 = 1.872 \text{ rad/s}$$

Substituting this value into Eq. (a), we find that the smallest initial velocity for which the bar will reach the vertical position is

$$v_1 = 3.111\omega_2 = 3.111(1.872) = 5.82 \text{ m/s} \qquad \textit{Answer}$$

Problems

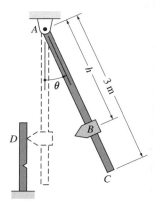

Fig. P18.83

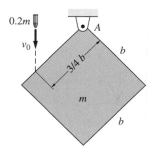

Fig. P18.84

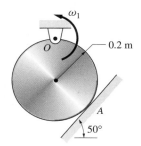

Fig. P18.85

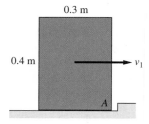

Fig. P18.88

18.83 The impact tester consists of the 20-kg striker B that is attached to the 16-kg uniform slender rod AC. The tester is released from an inclined position and breaks the test specimen D. Find the distance h for which the pin reaction at A will have no horizontal component during the impact with the specimen.

18.84 The uniform 15-kg disk rotates in the vertical plane about O. Immediately before it hits the inclined surface at A, its angular velocity is $\omega_1 = 4$ rad/s CCW. Just after the impact, the angular velocity of the disk is $\omega_2 = 2$ rad/s CW. Determine the magnitude of the impulse that acted during the impact on (a) the surface at A; and (b) the pin at O. Neglect friction.

18.85 The homogeneous square plate of mass m is suspended from a pin at A. The plate is at rest when it is struck by the small projectile of mass $0.2m$ traveling vertically with the velocity v_0. Assuming that the projectile becomes embedded in the plate, determine the angular velocity of the plate immediately after the impact.

18.86 The 10-kg uniform rod AB is stationary when it is hit by the 0.06 kg bullet D traveling horizontally at 540 m/s. Assuming that the bullet becomes embedded in the rod, calculate (a) the angular velocity of AB immediately after the impact; and (b) the maximum angular displacement of AB following the impact.

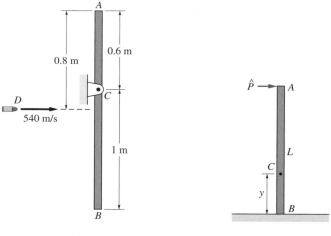

Fig. P18.86   **Fig. P18.87**

18.87 The uniform rod AB of mass m and length L is balanced on the smooth horizontal surface when it receives the horizontal blow at A. Determine the y-coordinate of the point C that has zero velocity immediately after the impact.

18.88 The homogeneous box of mass m is sliding on the horizontal surface at the speed v_1 when its corner A hits a small obstruction. Assuming that the box does not rebound, determine (a) the angular velocity of the box immediately after the impact in terms of v_1; and (b) the largest value of v_1 for which the box will not tip over after the impact.

18.89   The uniform slender bar of mass m is translating to the left at the speed $v_1 = 5\,\text{m/s}$ when the small roller at A hits the end of the slot. Assuming no rebound, calculate the angular velocity of the bar (a) immediately after the impact; and (b) when B is directly above A.

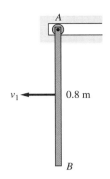

Fig. P18.89

18.90   The uniform cylinder of mass m is rolling without slipping on the horizontal surface when it strikes the incline. If the angular velocity of the cylinder before impact is 4 rad/s, what is its angular velocity immediately after impact? Assume that the cylinder does not rebound or slip on the incline.

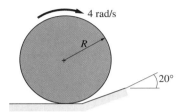

Fig. P18.90

18.91   The uniform rod AB of mass m and length L is released from rest with end A positioned in a corner. The rod strikes the obstruction at C with the clockwise angular velocity ω_1 and does not rebound. Determine the largest distance d for which the angular velocity of the rod will continue to be clockwise after the impact. Neglect friction.

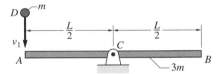

Fig. P18.91

18.92   The homogeneous rod AB of mass $3m$ is free to rotate about the pin at C. The rod is stationary when the ball D of mass m hits end A of the rod with the vertical velocity v_1. Knowing that the velocity of D immediately after the impact is zero, determine (a) the velocity of A immediately after the impact; and (b) the percentage of kinetic energy lost during the impact.

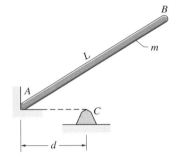

Fig. P18.92

18.93   The 4 kg uniform rod AB rotates freely about the pin at A. The rod is released from rest in the position shown and collides with the obstruction at C. Calculate the resulting impulses acting on the rod at A and C, assuming that (a) the rod does not rebound; and (b) no energy is lost during the impact.

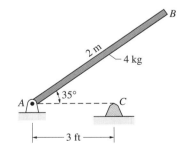

Fig. P18.93

18.94   The 1.2 kg uniform disk rolls without slipping with the angular velocity ω_1 prior to hitting the 1-in. curb. Assuming no rebound and no slipping between the disk and the curb, find the minimum value of ω_1 for which the disk will mount the curb.

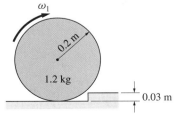

Fig. P13.94

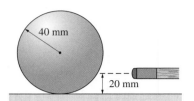

Fig. P18.95

18.95 The billiard ball of mass m is at rest on a table when it is struck horizontally by the cue. Immediately after the impact, the central velocity of the ball is $v = 1$ m/s. Find the angular velocity of the ball (a) immediately after the impact; and (b) after the ball stops slipping on the table.

18.96 The target on a rifle range consists of two identical thin disks, each with a mass of 2.5 kg, that are connected by a rod of negligible mass. The assembly is free to rotate about the z-axis. The target is at rest when a 9.7-g bullet enters the center of one of the disks with the speed $v_1 = 850$ m/s and exits at the speed v_2. If the resulting angular velocity of the target is 2.61 rad/s, determine v_2.

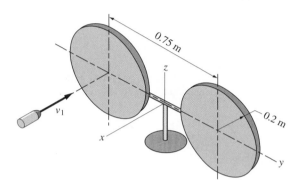

Fig. P18.96

18.97 The pendulum, consisting of the 1-kg slender rod AB and the 3-kg sphere C, hits the stationary sphere D of mass m with the angular velocity ω_1. All of the bodies are homogeneous. If the pendulum comes to a complete stop after the impact, determine m. Neglect friction and assume that no energy is lost during the impact.

18.98 The uniform 36-kg bar AB is initially at rest in the horizontal position, supported by the pin at B and the spring of stiffness $k = 4400$ N/m at A. The small (relative to the length of the bar) 18 kg sandbag C is tossed onto the bar with the initial velocity shown. Assuming that the sandbag does not rebound, calculate (a) the angular velocity of AB immediately after the impact; and (b) the maximum angular displacement of AB following the impact, assuming that it is a small angle.

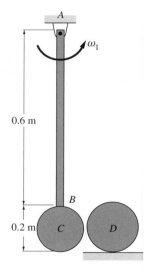

Fig. P18.97

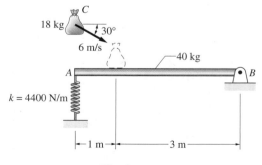

Fig. P18.98

Review of Equations

Work of a couple (plane motion)   The work of the couple C is

$$U_{1-2} = \int_{\theta_1}^{\theta_2} C\, d\theta$$

$$U_{1-2} = C(\theta_2 - \theta_1) \qquad \text{(constant } C\text{)}$$

Power of a couple (plane motion)   The power of couple C acting on a rigid body rotating with the angular velocity ω is

$$P = C\omega$$

Kinetic energy of a rigid body in plane motion

$$T = \frac{1}{2}m\bar{v}^2 + \frac{1}{2}\bar{I}\omega^2$$

$$T = \frac{1}{2}I_A\omega^2 \,(A \text{ is the instant center for velocities})$$

Principle of work and kinetic energy

$$(U_{1-2})_{\text{ext}} = T_2 - T_1 \quad \text{(single body)}$$

$$(U_{1-2})_{\text{ext}} + (U_{1-2})_{\text{int}} = T_2 - T_1 \quad \text{(connected bodies)}$$

Conservation of mechanical energy

$$V_1 + T_1 = V_2 + T_2$$

Impulse-momentum equations

$$\mathbf{L}_{1-2} = \mathbf{p}_2 - \mathbf{p}_1$$

$$(\mathbf{A}_A)_{1-2} = (\mathbf{h}_A)_2 - (\mathbf{h}_A)_1 \quad (A \text{ is a fixed point or mass center})$$

$\mathbf{L}$ = linear impulse of external forces
$\mathbf{p} = m\bar{\mathbf{v}}$ = linear momentum of the body
$\mathbf{A}_A$ = angular impulse of external forces about A

Review Problems

18.99 The uniform disk of mass m and radius R is attached to a spring of stiffness k. If the disk is released from rest with the spring undeformed, determine the expression for the maximum angular velocity of the disk. Assume that the disk rolls without slipping on the inclined surface.

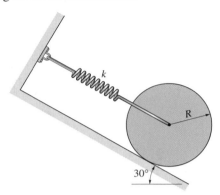

Fig. P18.99

18.100 The uniform bar of mass m and length L is attached to a vertical shaft of negligible mass. The angle θ between the bar and the shaft can be changed by an internal mechanism (not shown). If the assembly is rotating freely with the angular velocity ω when $\theta = 90°$, determine the angular velocity of the assembly when θ is changed to $45°$.

18.101 The 10-kg uniform bar AB is suspended from a roller at B, which travels on a horizontal rail. The bar is at rest in the vertical position when it is hit at end A by the 0.025 kg bullet moving horizontally at 900 m/s. Assuming that the bullet becomes embedded in the bar, determine (a) the angular velocity of the bar immediately after the impact; and (b) the maximum angular displacement of the bar after the impact. Neglect the mass of the roller.

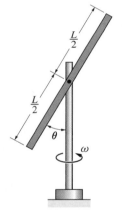

Fig. P18.100

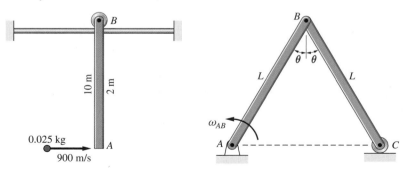

Fig. P18.101   **Fig. P18.102**

18.102 The mechanism consists of two uniform 30-kg bars, each of length $L = 2$ m. In the position $\theta = 30°$ the angular velocity of bar AB is $\omega_{AB} = 1.2$ rad/s.

Determine the kinetic energy of the mechanism in this position. Neglect the mass of the roller at C.

18.103   The mechanism is released from rest in the position shown. Determine the velocity of slider C when bar BC has become horizontal. Neglect friction.

18.104   The 15-kg uniform bar AB is in the position shown and moving with velocity $\bar{v} = 16$ m/s and angular velocity $\omega = 120$ rad/s when end A strikes a rigid obstruction. Determine the velocity of end B just after the impact, assuming that the bar does not rebound.

18.105   The central radius of gyration of the spool is $\bar{k}$, and its inner radius is R. If the spool is released from rest, determine the velocity of the spool as a function of its displacement x. Neglect the friction between the spool and the vertical wall.

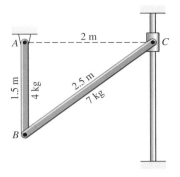

Fig. P18.103

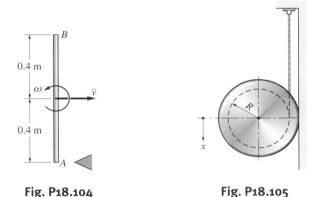

Fig. P18.104   **Fig. P18.105**

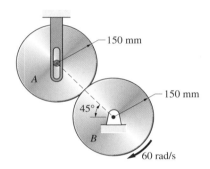

Fig. P18.106

18.106   Disk B is rotating at 60 rad/s when the identical, stationary disk A is lowered into contact with it. If the coefficient of kinetic friction between the disks is 0.25, determine the time when the slipping between the disks stops and the final angular velocities of the disks.

18.107   The 10-kg uniform disk is launched along the horizontal surface with the velocity 5 m/s and the backspin 45 rad/s. The coefficient of kinetic friction between the disk and the surface is 0.45. Determine (a) the final speed of the disk; and (b) the time required to reach the final speed.

18.108   The uniform bar AB of weight W is connected by a cable to the block C, also of weight W. If the system is released from rest in the position shown, determine the velocity of C when AB has rotated 90°.

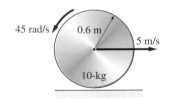

Fig. P18.107

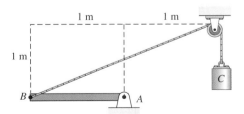

Fig. P18.108

18.109 The 15-kg homogeneous plank supported by cables at *A* and *B* is at rest when the 5-kg sandbag lands on it without rebounding. If the impact velocity of the sandbag is 6 m/s, determine the angular velocity of the plank immediately after the impact.

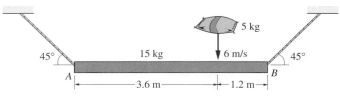

Fig. P18.109

18.110 The mechanism consisting of two uniform rods moves in the horizontal plane. The spring connected between the slider *A* and the pin *C* has a stiffness of 200 N/m and its free length is 150 mm. If the mechanism is released from rest in the position shown, determine the angular speed of rod *BC* when *A* is closest to *C*. Neglect friction and the mass of the slider.

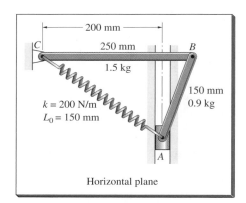

Fig. P18.110

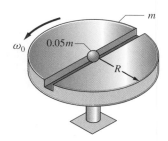

Fig. P18.111

18.111 The slotted, homogeneous disk of mass *m* rotates freely about the vertical axis with the angular velocity ω_0. A small marble of mass $0.05m$ is placed near the center of the disk. Because of centrifugal force, the marble travels outward along the slot. Determine the angular velocity of the disk after the marble has left the slot. Neglect the effect of the slot on the moment of inertia of the disk.

18.112 The 10-kg mass *B* is attached to a cable that passes over the small pulley *C* and is wrapped around the periphery of the unbalanced 40-kg disk *A*. The moment of inertia of the disk about its mass center *G* is 12 kg · m². When the disk is in the position shown, its angular velocity is $\omega_1 = 2.5$ rad/s clockwise.

Assuming that the disk rolls without slipping, determine its angular velocity after it has rotated 180°.

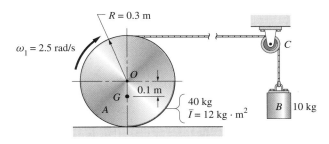

Fig. P18.112

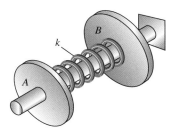

Fig. P18.113

18.113   The two disks can rotate freely on the horizontal shaft. The torsional stiffness of the spring connecting the disks is $k = 30$ N · m/rad. Initially, disk A is held stationary while disk B is rotated through two revolutions. Both disks are then released simultaneously. Determine the angular velocity of each disk at the instant when the spring has returned to its undeformed state. The moments of inertia of the disks about the shaft are $\bar{I}_A = 3.12$ kg · m^2 and $\bar{I}_B = 9.36$ kg · m^2.

18.114   The uniform 20-kg bar is suspended from the pin at O. The bar is at rest when the 0.2 kg bullet B is fired at end A with the velocity of 600 m/s. The bullet passes through the bar, emerging with the velocity of 450 m/s. Find the maximum angular displacement of the bar.

18.115   The uniform bars A and B are free to rotate about pins at their midpoints. Initially, the angular velocity of bar A is 5 rad/s CW and bar B is at rest. Determine the angular velocity of each bar immediately after the end of bar A strikes the end of bar B. Assume that the impact is plastic.

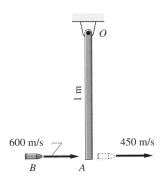

Fig. P18.114

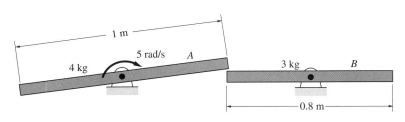

Fig. P18.115

18.116   The 12-kg homogeneous sphere is at rest in the corner when the constant couple $C = 5$ N · m is applied at $t = 0$. The coefficient of kinetic friction between the sphere and each of the contact surfaces is 0.4. Calculate the angular velocity of the sphere when $t = 3$ s.

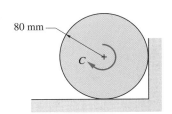

Fig. P18.116

18.117   As the homogenous cylinder of mass m enters the cylindrical surface, the velocity of its center is 1.5 m/s. Determine the angle θ at which the cylinder will come to a momentary stop. Assume that the cylinder rolls without slipping.

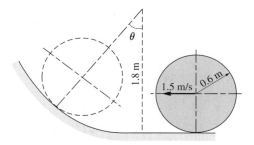

Fig. P18.117

18.118   The arm A, which is free to rotate about the vertical axis at O, supports a disk and an electric motor. The disk and the armature of the motor, labeled B in the figure, have a combined mass of 10 kg and a central radius of gyration of $\bar{k}_B = 0.3$ m. The motor housing C, which is attached to the arm A, mass 6 kg, and its central radius of gyration is $\bar{k}_C = 0.1$ m. The mass of arm A can be neglected. The system is at rest when the motor is turned on, which causes B to spin at 25 rad/s relative to the arm A. Determine the resulting angular speed of the arm A.

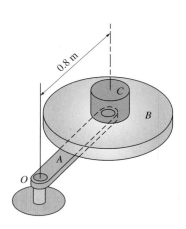

Fig. P18.118

19

Rigid-Body Dynamics in Three Dimensions

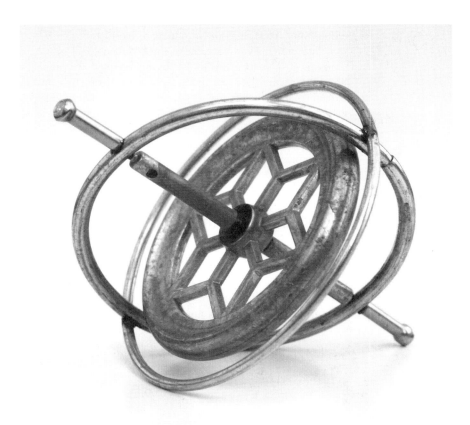

*19.1   Introduction

The kinematics of plane motion is based on the assumption that all points in the body move in planes that are parallel to each other. Consequently, our study of kinematics was reduced to a two-dimensional problem, in which only the plane of motion (usually the plane containing the mass center) had to be considered.

An additional constraint on plane motion was introduced in kinetics: The body had to be symmetric about the plane of motion.

If the conditions of plane motion are not applicable, two complications arise. First, the kinematics becomes more complicated due to the three-dimensional nature of the problem. This forces us to abandon scalar arithmetic and fully utilize the power of vector analysis.

The second complication is associated with kinetics. In plane motion, the kinetics involved only two inertial properties of the body: the mass m and the central moment of inertia $\bar{I}$. In three-dimensional problems, the description of the central moment of inertia requires *six* components (the moments of inertia about three axes plus three products of inertia).[*]

Comprehensive treatment of three-dimensional motion of rigid bodies is the domain of advanced textbooks. In this chapter, we present only an introduction that is sufficient for the analysis of several practical problems, such as unbalanced rotating machinery, flight of spinning bodies, and gyroscopic effects. This chapter begins with kinematics of three-dimensional motion. Subsequent articles extend the previously introduced methods of kinetic analysis (work-energy, impulse-momentum, and force-mass-acceleration methods) to spatial dynamics. The chapter concludes with a discussion of an important special case—the motion of an axisymmetric body.

*19.2   *Kinematics*

a.   *Relative motion of two points in a rigid body*

Figure 19.1 shows a rigid body undergoing three-dimensional motion. Points A and B are embedded in the body, and $\boldsymbol{\omega}$ is the angular velocity of the body at the instant shown. Let us now investigate the velocity of point B relative to A, denoted by $\mathbf{v}_{B/A}$. As was the case in plane motion, this relative velocity is due only to the rotation of the body. To a nonrotating observer attached to A, point B

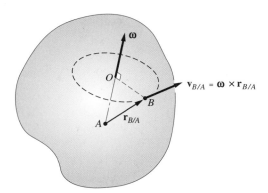

Fig. 19.1

---

[*]This complication arises even in problems that are kinematically two-dimensional if the body is not symmetric about the plane of motion.

appears to be traveling on a circular path centered at O. This is the same situation we encountered in plane motion, which led to the result

$$\mathbf{v}_{B/A} = \boldsymbol{\omega} \times \mathbf{r}_{B/A} \tag{19.1a}$$

where $\mathbf{r}_{B/A}$ is the position vector of B relative to A. A rigorous derivation of this equation appears in Appendix D.

The acceleration of B relative to A is obtained by differentiating Eq. (19.1a) with respect to time: $\mathbf{a}_{B/A} = \boldsymbol{\omega} \times \dot{\mathbf{r}}_{B/A} + \dot{\boldsymbol{\omega}} \times \mathbf{r}_{B/A}$. Substituting $\dot{\mathbf{r}}_{B/A} = \mathbf{v}_{B/A} = \boldsymbol{\omega} \times \mathbf{r}_{B/A}$ and introducing the notation $\dot{\boldsymbol{\omega}} = \boldsymbol{\alpha}$, where $\boldsymbol{\alpha}$ is the angular acceleration of the body, we get

$$\mathbf{a}_{B/A} = \boldsymbol{\omega} \times (\boldsymbol{\omega} \times \mathbf{r}_{B/A}) + \boldsymbol{\alpha} \times \mathbf{r}_{B/A} \tag{19.1b}$$

The two relative acceleration components are shown in Fig. 19.2. Note that the first component is perpendicular to $\boldsymbol{\omega}$ and directed toward O, as illustrated in Fig. 19.2(a). The second component, shown in Fig. 19.2(b), is perpendicular to $\boldsymbol{\alpha}$ and $\mathbf{r}_{B/A}$.

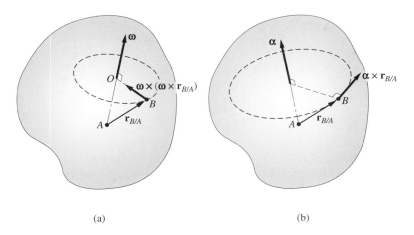

(a)   (b)

Fig. 19.2

Although Eqs. (19.1) have the same form as their counterparts for plane motion, there is a significant difference in their content. In plane motion, the direction of $\boldsymbol{\omega}$ does not change. Hence $\boldsymbol{\alpha}$ has the same direction as $\boldsymbol{\omega}$, and its magnitude is due only to the change in the magnitude of $\boldsymbol{\omega}$; that is, $\alpha = \dot{\omega}$. In three-dimensional motion, the direction of $\boldsymbol{\omega}$ changes continuously, so that $\boldsymbol{\alpha}$ reflects the changes in both the magnitude and direction of $\boldsymbol{\omega}$. Consequently, the directions of $\boldsymbol{\omega}$ and $\boldsymbol{\alpha}$ are generally different, and $\alpha = \dot{\omega}$ is no longer valid. For this reason, the two components of $\mathbf{a}_{A/B}$ in Fig. 19.2 are not mutually perpendicular; hence we can no longer call them "normal" and "tangential" components of acceleration.

The absolute velocity and acceleration of B can be obtained from the definition of relative motion:

$$\mathbf{v}_B = \mathbf{v}_A + \mathbf{v}_{B/A} = \mathbf{v}_A + \boldsymbol{\omega} \times \mathbf{r}_{B/A} \tag{19.2a}$$

$$\mathbf{a}_B = \mathbf{a}_A + \mathbf{a}_{B/A} = \mathbf{a}_A + \boldsymbol{\omega} \times (\boldsymbol{\omega} \times \mathbf{r}_{B/A}) + \boldsymbol{\alpha} \times \mathbf{r}_{B/A} \tag{19.2b}$$

b. Vector differentiation in a rotating reference frame

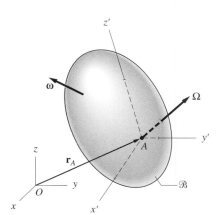

Fig. 19.3

As discussed in Art 16.8, it is sometimes convenient to describe the relative motion terms in Eqs. (19.1) within a reference frame that translates and rotates, rather than employing a frame that is fixed in space. Figure 19.3 shows two reference frames: the *xyz*-system, which is fixed in space, and the $x'y'z'$-system, the origin of which is attached to point A. If the $x'y'z'$-axes are embedded in body $\mathcal{B}$ (that is, the origin of these axes moves with A and the coordinate system rotates with the body), they are referred to as a *body frame,* whereas the fixed *xyz*-axes are called a *space frame.* Note that the angular velocity of the body frame is equal to the angular velocity $\boldsymbol{\omega}$ of the body. If the components of a vector $\mathbf{V}$ are described relative to the body frame, then the absolute derivative of $\mathbf{V}$ can be computed using the following identity derived in Art. 16.8.

$$\frac{d\mathbf{V}}{dt} = \left(\frac{d\mathbf{V}}{dt}\right)_{/\mathcal{B}} + \boldsymbol{\omega} \times \mathbf{V} \tag{19.3}$$

The notation "$/\mathcal{B}$" is used to indicate that the derivative is to be evaluated relative to the body frame—that is, as seen by an observer who translates and rotates with body $\mathcal{B}$.

If the $x'y'z'$-coordinate system rotates with the angular velocity $\boldsymbol{\Omega}$ that is not necessarily equal to the angular velocity $\boldsymbol{\omega}$ of the body, then Eq. (19.3) must be modified as follows:

$$\frac{d\mathbf{V}}{dt} = \left(\frac{d\mathbf{V}}{dt}\right)_{/x'y'z'} + \boldsymbol{\Omega} \times \mathbf{V} \tag{19.4}$$

By using the notation "$/x'y'z'$" we draw attention to the fact that the relative derivative is now referred to a coordinate system that is not necessarily a body frame.

c. Instant axis of rotation

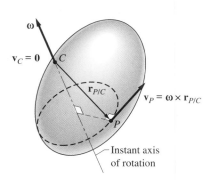

Fig. 19.4

Let the velocity of a point C in a rigid body (or body extended) be zero at a particular instant (in plane motion, this point is called the instant center for velocities). As shown in Fig. 19.4, the line passing through C that is parallel to the angular velocity $\boldsymbol{\omega}$ of the body is called the *instant axis of rotation,* or simply the *instant axis.*[*] With $\mathbf{v}_C = \mathbf{0}$, the velocity of any point P in the body is $\mathbf{v}_P = \mathbf{v}_{P/C} = \boldsymbol{\omega} \times \mathbf{r}_{P/C}$, as shown in Fig. 19.4. It follows that the velocities of all points behave as if the body were rotating about the instant axis, with the velocities of points on this axis being zero. A point that has zero velocity always exists in plane motion and for a

---

[*]In two-dimensional motion, $\boldsymbol{\omega}$ is perpendicular to the plane of the motion. Therefore, the instant axis of rotation is also perpendicular to this plane, and it passes through the instant center for velocities.

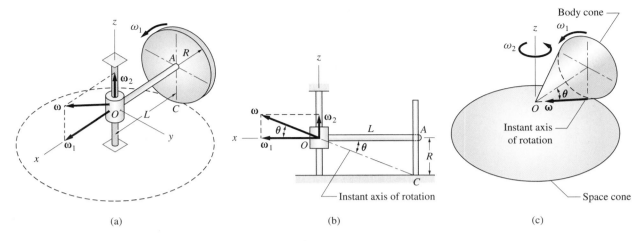

(a)   (b)   (c)

Fig. 19.5

body rotating about a fixed point. However, it can be shown that such a point need not exist for all motions.*

When analyzing velocities, it is advantageous to use the instant axis of rotation when its location can be determined by inspection. For example, Fig. 19.5(a) shows a wheel of radius R that is rolling without slipping along a circular path on a horizontal plane. The horizontal axle OA of length L is attached to a collar that rotates on the fixed vertical shaft. As shown in Fig. 19.5(a), the angular velocity of the wheel is given by $\boldsymbol{\omega} = \boldsymbol{\omega}_1 + \boldsymbol{\omega}_2$, where $\boldsymbol{\omega}_1$ is the *spin velocity* of the wheel (its angular velocity relative to the axle) and $\boldsymbol{\omega}_2$ is the angular velocity of the axle.

Figure 19.5(b) shows another view of the assembly in which the instant axis of the wheel has been identified. This axis is known to pass through points C and O, because both are points on the wheel and have zero velocity (the wheel must be "extended" to include O). Referring to Fig. 19.5(b), we see that the angle θ can be determined if L and R are known, which means that $\boldsymbol{\omega}$ and $\boldsymbol{\omega}_2$ can be found if the spin velocity $\boldsymbol{\omega}_1$ is known. (An alternate method for computing the angular velocities is to write the relative velocity equation using points O and C—see Sample Problem 19.3.)

As the wheel moves, the instant axis of rotation traces out a three-dimensional surface in space. Because the wheel undergoes rotation about the fixed point O, this surface is a cone, called the *space cone,* with its apex at O. The trace of the instant axes in the "extended" wheel is also a conical surface, known as the *body cone.* Both cones are shown in Fig. 19.5(c). Inspecting this figure, we see that the kinematics of the wheel and the body cone are identical if the latter is made to roll without slipping on the space cone with the spin velocity $\boldsymbol{\omega}_1$. The body-cone–space-cone analogy is often a convenient tool for visualizing the motion of a body.

*See, for example, *Advanced Engineering Dynamics,* J. H. Ginsberg, Harper & Row, 1988, p. 115.

Sample Problem 19.1

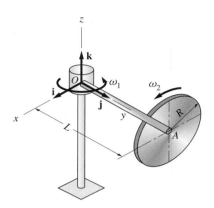

The arm *OA* of the system shown is rotating about the vertical shaft with the angular velocity ω_1. At the same time, the thin disk at *A* is spinning with the angular velocity ω_2 relative to *OA*. Determine the angular velocity $\boldsymbol{\omega}$ and the angular acceleration $\boldsymbol{\alpha}$ of the disk in the position shown. Assume ω_1 and ω_2 are not constants.

Solution

We assume that the origin of the *xyz*-reference frame shown in the figure is attached to point *O* and that the coordinate axes are embedded in arm *OA*. The angular motion of this frame is thus identical to that of *OA*. The base vectors for the rotating frame are denoted by $\mathbf{i}$, $\mathbf{j}$, and $\mathbf{k}$.

In vector representation, the two angular velocities shown in Fig. (a) are

$$\boldsymbol{\omega}_1 = \omega_1 \mathbf{k} \qquad \text{(angular velocity of arm } OA)$$

$$\boldsymbol{\omega}_2 = \omega_2 \mathbf{j} \qquad \text{(angular velocity of the disk relative to } OA)$$

The angular velocity of the disk is

$$\boldsymbol{\omega} = \boldsymbol{\omega}_1 + \boldsymbol{\omega}_2 \qquad\qquad\qquad \text{(a)}$$

or

$$\boldsymbol{\omega} = \omega_1 \mathbf{k} + \omega_2 \mathbf{j} \qquad \text{(valid for all time)} \qquad\qquad \textbf{\textit{Answer}} \quad \text{(b)}$$

Equation (b) is true for all time, because the *xyz*-reference frame rotates with arm *OA*, so that the *y*- and *z*-axes are always directed along *OA* and the vertical shaft, respectively. Indicating which expressions are valid for all time will help us avoid the common error of attempting to differentiate an expression that is true only at a particular instant.

The angular acceleration of the disk can be found by differentiating the expression for $\boldsymbol{\omega}$ given in Eq. (b). Using Eq. (19.4), we get

$$\boldsymbol{\alpha} = \dot{\boldsymbol{\omega}} = (\dot{\boldsymbol{\omega}})_{/xyz} + \boldsymbol{\Omega} \times \boldsymbol{\omega} \qquad\qquad\qquad \text{(d)}$$

where $(\dot{\boldsymbol{\omega}})_{/xyz}$ is the angular acceleration of the disk relative to the *xyz*-frame and $\boldsymbol{\Omega}$ is the angular velocity of the *xyz*-frame. From Eq. (b), we observe that $(\dot{\boldsymbol{\omega}})_{/xyz} = \dot{\omega}_1 \mathbf{k} + \dot{\omega}_2 \mathbf{j}$ (because the base vectors $\mathbf{k}$ and $\mathbf{j}$ are fixed in the *xyz*-frame, their derivatives relative to that frame vanish). Furthermore, because $\boldsymbol{\Omega} = \boldsymbol{\omega}_1$, we obtain $\boldsymbol{\Omega} \times \boldsymbol{\omega} = \omega_1 \mathbf{k} \times (\omega_1 \mathbf{k} + \omega_2 \mathbf{j}) = -\omega_1 \omega_2 \mathbf{i}$. Substituting this result into Eq. (d), we find that

$$\boldsymbol{\alpha} = -\omega_1 \omega_2 \mathbf{i} + \dot{\omega}_2 \mathbf{j} + \dot{\omega}_1 \mathbf{k} \quad \text{(valid for all time)} \qquad\qquad \textbf{\textit{Answer}}$$

Sample Problem 19.2

The figure repeats the system that was described in Sample Problem 19.1. Calculate the velocity and acceleration of point P on the disk in the position shown.

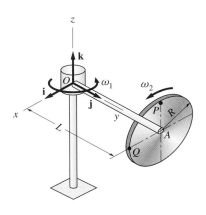

Solution

We assume that the xyz-reference frame shown in the figure is attached to point O and embedded in the arm OA. From the solution of Sample Problem 19.1, we know that the angular velocity and acceleration of the disk are

$$\boldsymbol{\omega} = \omega_1 \mathbf{k} + \omega_2 \mathbf{j} \qquad\qquad \text{(a)}$$

and

$$\boldsymbol{\alpha} = \dot{\boldsymbol{\omega}} = \dot{\omega}_1 \mathbf{k} + \dot{\omega}_2 \mathbf{j} - \omega_1 \omega_2 \mathbf{i} \qquad\qquad \text{(b)}$$

Because both expressions are valid for all time, we may use them for the position shown in the figure.

The motion of point P can be analyzed by relating it to the motion of point A. Referring to the figure, the position vector from the fixed point O to point A is $\mathbf{r}_{A/O} = L\mathbf{j}$. Because the angular velocity and acceleration of arm OA are ω_1 and $\dot{\omega}_1$, the velocity and acceleration of point A are

$$\mathbf{v}_A = \boldsymbol{\omega}_1 \times \mathbf{r}_{A/O} = \omega_1 \mathbf{k} \times L\mathbf{j} = -L\omega_1 \mathbf{i} \qquad\qquad \text{(c)}$$

and

$$\begin{aligned} \mathbf{a}_A &= \dot{\boldsymbol{\omega}}_1 \times \mathbf{r}_{A/O} + \boldsymbol{\omega}_1 \times (\boldsymbol{\omega}_1 \times \mathbf{r}_{A/O}) \\ &= \dot{\omega}_1 \mathbf{k} \times L\mathbf{j} + \omega_1 \mathbf{k} \times (-L\omega_1 \mathbf{i}) \\ &= -L\dot{\omega}_1 \mathbf{i} - L\omega_1^2 \mathbf{j} \end{aligned} \qquad\qquad \text{(d)}$$

The position vector of P relative to A is seen from the figure to be $\mathbf{r}_{P/A} = R\mathbf{k}$. Because both P and A belong to the disk, their relative velocity and acceleration vectors become

$$\mathbf{v}_{P/A} = \boldsymbol{\omega} \times \mathbf{r}_{P/A} = (\omega_1 \mathbf{k} + \omega_2 \mathbf{j}) \times R\mathbf{k} = R\omega_2 \mathbf{i} \qquad\qquad \text{(e)}$$

and

$$\begin{aligned} \mathbf{a}_{P/A} &= \boldsymbol{\alpha} \times \mathbf{r}_{P/A} + \boldsymbol{\omega} \times (\boldsymbol{\omega} \times \mathbf{r}_{P/A}) \\ &= (\dot{\omega}_1 \mathbf{k} + \dot{\omega}_2 \mathbf{j} - \omega_1 \omega_2 \mathbf{i}) \times R\mathbf{k} + (\omega_1 \mathbf{k} + \omega_2 \mathbf{j}) \times R\omega_2 \mathbf{i} \\ &= (R\dot{\omega}_2 \mathbf{i} + R\omega_1 \omega_2 \mathbf{j}) + (R\omega_1 \omega_2 \mathbf{j} - R\omega_2^2 \mathbf{k}) \end{aligned} \qquad\qquad \text{(f)}$$

The velocity of P is found by substituting Eqs. (c) and (e) into the equation $\mathbf{v}_P = \mathbf{v}_A + \mathbf{v}_{P/A}$:

$$\mathbf{v}_P = (-L\omega_1 + R\omega_2)\,\mathbf{i} \qquad\qquad \textit{Answer} \quad \text{(g)}$$

Similarly, substituting Eqs. (d) and (f) into $\mathbf{a}_P = \mathbf{a}_A + \mathbf{a}_{P/A}$, the acceleration vector of P becomes (after some rearrangement of terms)

$$\mathbf{a}_P = (-L\dot{\omega}_1 + R\dot{\omega}_2)\mathbf{i} + \left(-L\omega_1^2 + 2R\omega_1\omega_2\right)\mathbf{j} - R\omega_2^2\mathbf{k} \qquad \textit{Answer} \quad \text{(h)}$$

The velocity and acceleration of point P could also be calculated by recognizing that the disk is rotating about the fixed point O (note that the length of a line connecting point O to any point on the disk remains constant). In this case, it would be convenient to consider the disk to be extended to include point O. Utilizing this concept, the velocity and acceleration of point P could be computed using $\mathbf{v}_P = \boldsymbol{\omega} \times \mathbf{r}_{P/O}$ and $\mathbf{a}_P = \boldsymbol{\alpha} \times \mathbf{r}_{P/O} + \boldsymbol{\omega} \times (\boldsymbol{\omega} \times \mathbf{r}_{P/O})$. In this case $\boldsymbol{\omega}$ and $\boldsymbol{\alpha}$ are as given in Eqs. (a) and (b), respectively, and the position vector of P relative to O is seen from the figure to be $\mathbf{r}_{P/O} = L\mathbf{j} + R\mathbf{k}$. It may be verified that the results are identical to those given in Eqs. (g) and (h).

Sample Problem **19.3**

The gear A shown in Fig. (a) is rolling around the fixed gear B as it spins about arm C. If the arm is rotating about the vertical at the constant rate $\omega_C = 15$ rad/s, calculate (1) the angular velocity of gear A; and (2) the angular acceleration of gear A.

Solution

Let the reference frame shown in Fig. (a) be attached to arm C. The geometry of the system is shown in Fig. (b), where the distances and angles were obtained directly from Fig. (a), or were computed using trigonometry.

Part 1

From Fig. (a) we see that gear A is rotating about the fixed point O (note that a line from O to any point on gear A does not change length). Because the point of contact E has no velocity, we conclude that the instant axis of rotation of gear A is the line OE, as indicated in Fig. (b).

The angular velocity of gear A is

$$\boldsymbol{\omega}_A = \boldsymbol{\omega}_C + \boldsymbol{\omega}_{A/C} \qquad \text{(a)}$$

where $\boldsymbol{\omega}_C$ is the angular velocity of arm C, and $\boldsymbol{\omega}_{A/C}$ is the spin velocity of gear A—that is, the angular velocity of gear A relative to the arm C.

Referring to Fig. (a), we see that the angular velocity of arm C is

$$\boldsymbol{\omega}_C = \omega_C\mathbf{k} = 15\mathbf{k} \text{ rad/s} \quad \text{(valid for all time)} \qquad \text{(b)}$$

Because the direction of the spin velocity of gear A coincides with the unit vector $\boldsymbol{\lambda}_{FO}$ shown in Fig. (b), it may be written as

$$\boldsymbol{\omega}_{A/C} = \omega_{A/C}\boldsymbol{\lambda}_{FO} \quad \text{(valid for all time)} \qquad \text{(c)}$$

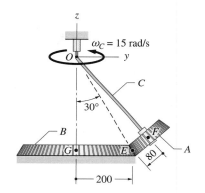

$\omega_C = 15$ rad/s

$30°$

Dimensions in mm

(a)

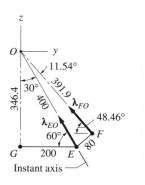

$11.54°$

391.9

346.4

$30°$

400

$\boldsymbol{\lambda}_{FO}$

$48.46°$

$\boldsymbol{\lambda}_{EO}$

$60°$

F

80

G 200 E

Instant axis

(b)

Substituting Eqs. (b) and (c) into Eq. (a), the angular velocity of gear A becomes

$$\boldsymbol{\omega}_A = 15\mathbf{k} + \omega_{A/C}\boldsymbol{\lambda}_{FO} \quad \text{(valid for all time)} \qquad \text{(d)}$$

The vector diagram representing Eq. (d) is shown in Fig. (c). Observe that $\boldsymbol{\omega}_A$ lies along the instant axis of rotation, and $\boldsymbol{\omega}_{A/C}$ is directed along the line FO, the axis of arm C. The angles α, β, and γ in Fig. (c) were computed from the angles shown in Fig. (b).

We see that Fig. (c) contains only two unknown variables, ω_A and $\omega_{A/C}$, which can be computed by trigonometry. Applying the law of sines yields

$$\frac{15}{\sin 11.54°} = \frac{\omega_{A/C}}{\sin 30°} = \frac{\omega_A}{\sin 138.46°}$$

from which we obtain

$$\omega_{A/C} = 37.49 \text{ rad/s} \quad \text{and} \quad \omega_A = 49.72 \text{ rad/s} \qquad \text{(e)}$$

Using the fact that the angular velocity vector of gear A coincides with the unit vector $\boldsymbol{\lambda}_{EO}$ in Fig. (b), it may be written as

$$\boldsymbol{\omega}_A = 49.72\boldsymbol{\lambda}_{EO} = 49.72(-\sin 30°\mathbf{j} + \cos 30°\mathbf{k})$$

or

$$\boldsymbol{\omega}_A = -24.9\mathbf{j} + 43.1\mathbf{k} \text{ rad/s} \quad \text{(valid for all time)} \qquad \textbf{\textit{Answer}} \quad \text{(f)}$$

An alternate method of calculating the angular and spin velocities of A is to solve the vector equation $\mathbf{v}_E = \boldsymbol{\omega}_A \times \mathbf{r}_{E/O} = \mathbf{0}$, where $\boldsymbol{\omega}_A$ is given in Eq. (d) and $\mathbf{r}_{E/O} = 200\mathbf{j} - 346.4\mathbf{k}$ mm.

Part 2

The angular acceleration of gear A can be computed by differentiating its angular velocity:

$$\boldsymbol{\alpha}_A = \dot{\boldsymbol{\omega}}_A = (\dot{\boldsymbol{\omega}}_A)_{/xyz} + \boldsymbol{\Omega} \times \boldsymbol{\omega}_A \qquad \text{(g)}$$

where $(\dot{\boldsymbol{\omega}}_A)_{/xyz}$ is the angular acceleration of A relative to the xyz-reference frame and $\boldsymbol{\Omega}$ is the angular velocity of the frame.

Using Eq. (f), $\boldsymbol{\omega}_A = -24.9\mathbf{j} + 43.1\mathbf{k}$ rad/s (which is valid for all time and can thus be differentiated), we obtain $(\dot{\boldsymbol{\omega}}_A)_{/xyz} = \mathbf{0}$. Furthermore, we have $\boldsymbol{\Omega} = \boldsymbol{\omega}_C$, and Eq. (g) becomes

$$\boldsymbol{\alpha}_A = \dot{\boldsymbol{\omega}}_A = \boldsymbol{\omega}_C \times \boldsymbol{\omega}_A = 15\mathbf{k} \times (-24.9\mathbf{j} + 43.1\mathbf{k})$$

which yields

$$\boldsymbol{\alpha}_A = 374\mathbf{i} \text{ rad/s}^2 \qquad \textbf{\textit{Answer}}$$

As shown in Fig. (d), the motion of gear A can be modeled as the body cone (representing gear A) rolling without slipping on the outside of the space cone (representing the fixed gear B). The figure also shows the angular velocity $\boldsymbol{\omega}_A$, its components $\boldsymbol{\omega}_C$ and $\boldsymbol{\omega}_{A/C}$, and the angular acceleration $\boldsymbol{\alpha}_A$. Note that $\boldsymbol{\alpha}_A$ is always perpendicular to $\boldsymbol{\omega}_A$.

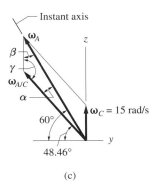

$\alpha = 60 - 48.46 = 11.54°$
$\beta = 30°$
$\gamma = 90 + 48.46 = 138.46°$

(c)

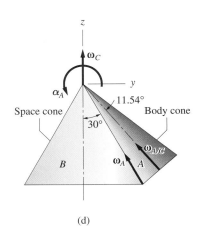

(d)

Sample Problem 19.4

The mechanism shown in Fig. (a) consists of the crank PQ, which rotates about axis OP, and the control rod B connected to the crank and the sliding collar C. (1) If the connections at Q and C are ball-and-socket joints, compute the velocity of collar C and the angular velocity of rod B, given that $\theta = 30°$ and $\dot{\theta} = 3.6$ rad/s. (2) Re-solve Part 1 assuming that the connection at C is a clevis, shown in Fig. (b).

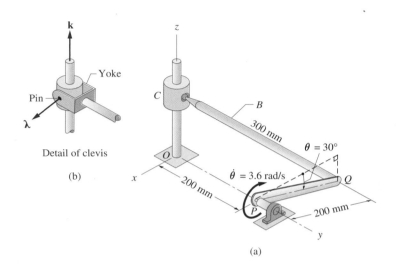

(a)

Solution

Part 1

The relative velocity equation between points C and Q (note that both are points on rod B) is

$$\mathbf{v}_C = \mathbf{v}_Q + \mathbf{v}_{C/Q} = \boldsymbol{\omega}_{PQ} \times \mathbf{r}_{Q/P} + \boldsymbol{\omega}_B \times \mathbf{r}_{C/Q} \qquad \text{(a)}$$

where $\boldsymbol{\omega}_{PQ}$ and $\boldsymbol{\omega}_B$ are the angular velocities of the crank and control rod, respectively, and the position vectors $\mathbf{r}_{Q/P}$ and $\mathbf{r}_{C/Q}$ are defined in Fig. (c).

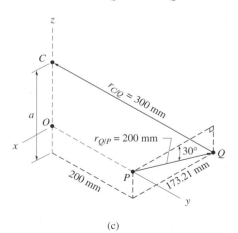

(c)

Assuming that the velocity of the collar is upward, we have

$$\mathbf{v}_C = v_C \mathbf{k} \tag{b}$$

With $\mathbf{r}_{Q/P} = -173.21\mathbf{i} - 100.0\mathbf{k}$ mm and $\boldsymbol{\omega}_{PQ} = -3.6\mathbf{j}$ rad/s, the velocity of Q is

$$\mathbf{v}_Q = \boldsymbol{\omega}_{PQ} \times \mathbf{r}_{Q/P} = -3.6\mathbf{j} \times (-173.21\mathbf{i} - 100.0\mathbf{k})$$
$$= 360.0\mathbf{i} - 623.6\mathbf{k} \text{ mm/s} \tag{c}$$

Before we can find $\mathbf{r}_{C/Q}$, it is necessary to compute the distance a shown in Fig. (c). From the geometry of that figure, we see that $a = [(300)^2 - (173.21)^2 - (200)^2]^{1/2} = 141.42$ mm, which gives

$$\mathbf{r}_{C/Q} = 173.21\mathbf{i} - 200.0\mathbf{j} + 141.42\mathbf{k} \text{ mm} \tag{d}$$

Substituting Eqs. (b)–(d) into Eq. (a), and using the determinant form for the second cross product, we obtain

$$v_C\mathbf{k} = (360.0\mathbf{i} - 623.6\mathbf{k}) + \begin{vmatrix} \mathbf{i} & \mathbf{j} & \mathbf{k} \\ \omega_x & \omega_y & \omega_z \\ 173.21 & -200.0 & 141.42 \end{vmatrix}$$

where $\boldsymbol{\omega}_B = \omega_x\mathbf{i} + \omega_y\mathbf{j} + \omega_z\mathbf{k}$ is the angular velocity of rod B. Expanding the determinant, and equating like components, yields the following three scalar equations:

$$
\begin{aligned}
0 &= \quad 360.0 & &+141.42\omega_y & &+200.0\omega_z \\
0 &= & -141.42\omega_x & & &+173.21\omega_z \quad \text{(e)} \\
v_C &= -623.6 & -200.0\omega_x & -173.21\omega_y &
\end{aligned}
$$

Since these equations contain four unknowns (v_C, ω_x, ω_y, and ω_z), a complete solution cannot be obtained without additional information. The physical reason for the indeterminacy lies in the ball-and-socket joints which allow the rod B to spin freely about its axis without affecting v_C. Therefore, the relative velocity equation, Eq. (a), is not capable of determining the spin velocity of rod B (the component of $\boldsymbol{\omega}_B$ along the axis of the rod). One way of overcoming this difficulty is to *specify* the spin velocity of B, thereby providing us with an additional equation. The chosen value of the spin velocity will affect $\boldsymbol{\omega}_B$, of course, but it does not affect v_C.

We will complete our solution by assuming that the spin velocity of rod B about its axis is zero; that is, $\boldsymbol{\omega}_B \cdot \mathbf{r}_{C/Q} = 0$, or

$$(\omega_x\mathbf{i} + \omega_y\mathbf{j} + \omega_z\mathbf{k}) \cdot (173.21\mathbf{i} - 200.0\mathbf{j} + 141.42\mathbf{k}) = 0$$
$$173.21\omega_x - 200.0\omega_y + 141.42\omega_z = 0 \tag{f}$$

Equations (e) and (f) represent four scalar equations in four unknowns, the solution of which gives $v_C = -182.7$ mm/s, $\omega_x = -0.980$ rad/s, $\omega_y = -1.414$ rad/s, and $\omega_z = -0.800$ rad/s. Expressed in vector form, the results are

$$\mathbf{v}_C = -182.7\mathbf{k} \text{ mm/s}$$
$$\boldsymbol{\omega}_B = -0.980\mathbf{i} - 1.414\mathbf{j} - 0.800\mathbf{k} \text{ rad/s}$$

Answer

Part 2

It is evident from Fig.(b) that the rotation of rod B relative to slider C can occur only about the pin of the clevis. In addition, the slider can rotate about the z-axis. Thus, the angular velocity $\boldsymbol{\omega}_B$ of the rod can have components only in the directions of the vectors $\boldsymbol{\lambda}$ and $\mathbf{k}$, where $\boldsymbol{\lambda}$ is a vector (not necessarily a unit vector) in the direction of the pin, as indicated in Fig. (b). The component of $\boldsymbol{\omega}_B$ that is perpendicular to both $\boldsymbol{\lambda}$ and $\mathbf{k}$ is zero; that is,

$$\boldsymbol{\omega}_B \cdot (\boldsymbol{\lambda} \times \mathbf{k}) = 0 \tag{g}$$

Equation (g) replaces the condition of zero spin $\boldsymbol{\omega}_B \cdot \mathbf{r}_{C/Q}$ used in Part 1 of the solution. Equations (e) are unchanged.

Because the pin of the clevis is perpendicular to the z-axis, we have $\boldsymbol{\lambda} = \lambda_x \mathbf{i} + \lambda_y \mathbf{j}$. The pin is also perpendicular to rod B, so that $\boldsymbol{\lambda} \cdot \mathbf{r}_{C/Q} = 0$. Substituting for $\mathbf{r}_{C/Q}$ from Eq. (d), we get $173.21\lambda_x - 200.0\lambda_y = 0$. Choosing $\lambda_x = 1$, we obtain $\lambda_y = 173.21/200.0 = 0.8661$ and Eq. (g) becomes

$$\begin{vmatrix} \omega_x & \omega_y & \omega_z \\ 1 & 0.8661 & 0 \\ 0 & 0 & 1 \end{vmatrix} = 0$$

which upon expansion yields

$$0.8661\omega_x - 1.0\omega_y = 0 \tag{h}$$

Solving Eqs. (e) and (h) simultaneously, we get $v = -182.7$ mm/s, $\omega_x = -1.260$ rad/s, $\omega_y = -1.091$ rad/s, and $\omega_z = -1.029$ rad/s. In vector form, the solution is

$$\mathbf{v}_C = 182.7\mathbf{k} \text{ mm/s} \qquad\qquad \textit{Answer}$$

$$\boldsymbol{\omega}_B = -1.260\mathbf{i} - 1.091\mathbf{j} - 1.029\mathbf{k} \text{ rad/s} \qquad\qquad \textit{Answer}$$

We see that the velocity of slider C is the same as in Part 1 of the solution, but the angular velocity of rod B is different, as expected.

Problems

19.1 Referring to Sample Problem 19.2, determine the velocity and acceleration of point Q on the disk in the position shown.

19.2 Compute the acceleration of the sliding collar C in Part 1 of Sample Problem 19.4.

19.3 Bar $OABC$ rotates about the ball-and-socket joint at O. In the position shown, the angular velocity and angular acceleration of the bar are $\omega = 2\mathbf{i} + 4\mathbf{j} - 3\mathbf{k}$ rad/s and $\alpha = 20\mathbf{i} - 30\mathbf{j}$ rad/s^2, respectively. Determine the velocity and acceleration vectors of point C.

19.4 Bar $OABC$ rotates about the ball-and-socket joint at O. In the position shown, the angular velocity vector ω of the bar is parallel to OC. Find the magnitude of ω if the speed of point B is 160 mm/s in this position.

19.5 Bar $OABC$ rotates about a ball-and-socket joint at O. In the position shown, the angular velocity vector of the rod is perpendicular to the line OC, and the velocity of point C is $\mathbf{v}_C = 3\mathbf{i} - 2\mathbf{j} + v_z\mathbf{k}$ m/s. For this position, determine v_z and the angular velocity vector.

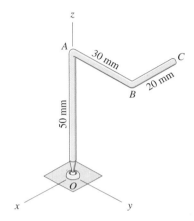

Fig. P19.3–P19.5

19.6 The cone rolls on the xy-plane without slipping. The spin velocity $\omega_1 = 2.4$ rad/s is constant. For the position shown, determine (a) the angular velocity of the cone; (b) the angular acceleration of the cone; and (c) the velocity and acceleration of point P.

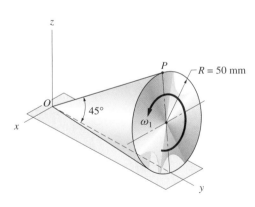

Fig. P19.6

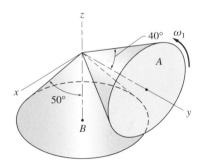

Fig. P19.7

19.7 Cone A rolls without slipping on the outside of the stationary cone B with the spin velocity ω_1. In the position shown, $\omega_1 = 2$ rad/s and $\dot\omega_1 = -7$ rad/s^2. For this position, determine (a) the angular velocity of cone A; and (b) the angular acceleration of cone A.

19.8 Cone A rolls without slipping inside the stationary cone B with the constant spin velocity $\omega_1 = 3.6$ rad/s. Calculate (a) the angular velocity of cone A; and (b) the angular acceleration of cone A.

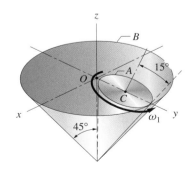

Fig. P19.8

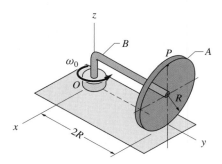

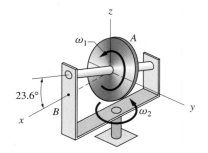

Fig. P19.9

Fig. P19.10

19.9 Disk A is free to spin about the bent axle B, which is rotating with the constant angular velocity ω_0 about the z-axis. Assuming that the disk rolls without slipping on the horizontal surface, determine the following for the position shown: (a) the angular velocity of the disk; (b) the angular acceleration of the disk; and (c) the velocity and acceleration of point P on the disk.

19.10 Disk A of the gyroscope spins about its axis, which is inclined at 23.6° relative to the xy-plane, at the constant angular speed $\omega_1 = 40$ rad/s. At the same time, frame B rotates about the z-axis with the variable angular velocity ω_2. At the instant when $\omega_2 = 16$ rad/s and $\dot{\omega}_2 = 400$ rad/s^2, calculate (a) the angular velocity of disk A; and (b) the angular acceleration of disk A.

19.11 Two bevel gears attached to the arm C roll on the fixed gear A. The arm rotates about the z-axis at the constant angular velocity of 25 rad/s. Compute (a) the angular velocity of gear B; and (b) the angular acceleration of gear B.

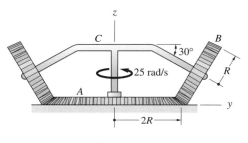

Fig. P19.11

19.12 Gears A and B spin freely on the bent shaft D, whereas gear C is fixed. The shaft D rotates about the y-axis with the constant angular velocity ω_0. For the position shown, calculate the angular velocity of (a) gear A; and (b) gear B.

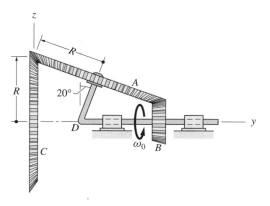

Fig. P19.12

19.13 The rod AB is connected by ball-and-socket joints to the sliding collars A and B. In the position shown, collar A is moving up with the velocity $v_A = 4$ m/s. Determine the speed of collar B in this position.

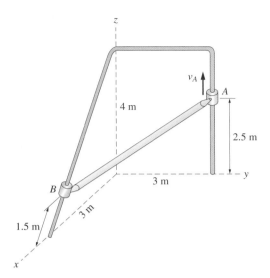

Fig. P19.13

19.14 Disk A spins with respect to arm B with the angular speed ω_1 as the arm rotates about the y-axis with the angular speed ω_2, neither speed being constant. When the assembly is in the position shown, $\omega_1 = 3$ rad/s, $\omega_2 = 4$ rad/s, $\dot{\omega}_1 = -16$ rad/s^2, and $\dot{\omega}_2 = 25$ rad/s^2. For this position, determine (a) the velocity of point Q on the disk; and (b) the acceleration of point Q.

19.15 Bar PQ spins about the axis OA with the constant angular velocity $\omega_1 = 20$ rad/s. At the same time, OA rotates about the z-axis with the constant angular velocity $\omega_2 = 12$ rad/s. For the position shown, determine (a) the velocity of end P; and (b) the acceleration of end P.

19.16 Cranks AB and CD rotate about axes that are parallel to the y-axis. Bar BD is attached to the cranks by ball-and-socket joints. If the angular speed of AB is constant at $\omega_0 = 12$ rad/s, determine for the position shown: (a) the angular velocities of CD and BD; and (b) the angular acceleration of CD. Assume that bar BD is not spinning about its axes; that is, assume the angular velocity vector of BD is perpendicular to BD.

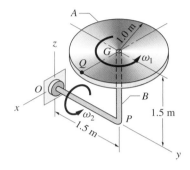

Fig. P19.14

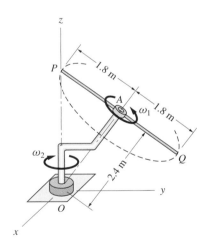

Fig. P19.15

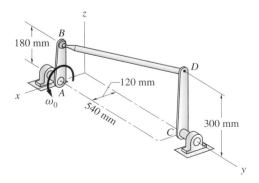

Fig. P19.16

19.17   The rod AB is attached to the rim of disk C and the sliding collar B with ball-and-socket joints. The disk rotates with the constant angular velocity 6 rad/s about the vertical axis at O. For the position shown, determine the velocity of collar B.

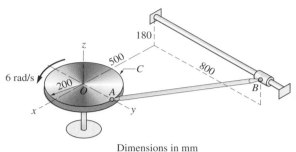

Dimensions in mm

Fig. P19.17

19.18   The ends of bar AB are connected to collars with ball-and-socket joints. The collars slide on the arms of the rigid frame. The frame rotates about the z-axis with the constant angular velocity $\omega_1 = 1.5$ rad/s. If the velocity of the collar at B relative to the frame is constant at $v_B = 0.2$ m/s, determine the angular acceleration of bar AB in the position $\theta = 30°$.

19.19   Rod C is connected to collar A by a clevis and to collar B by a ball-and-socket joint. In the position shown, collar A is moving to the right with the speed $v_A = 450$ mm/s. Find the speed of collar B for this position. (Note: It is not necessary to compute the angular velocity of rod C.)

19.20   For the mechanism described in Prob. 19.19, determine the angular velocities of collar A and rod C in the position shown.

19.21   Assume that the connections for the rod described in Prob. 19.19 are interchanged (that is, the ball-and-socket is at A and the clevis is at B). For the position shown, determine the angular velocity of rod C and the velocity of collar B.

19.22   Arm AB rotates around the z-axis with the constant angular velocity $\omega_1 = 3$ rad/s. The disk spins about the arm with the constant angular

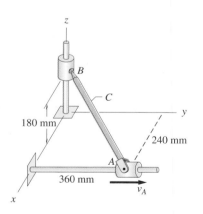

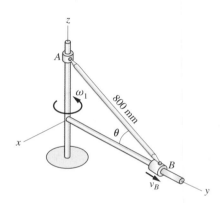

Fig. P19.18

Fig. P19.19–P19.21

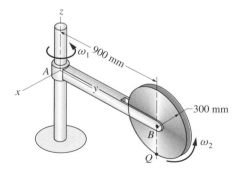

Fig. P19.22

velocity $\omega_2 = 10$ rad/s. For the position shown, compute (a) the angular acceleration of the disk; and (b) the acceleration of point Q on the disk.

19.23 The disk spins with the constant angular velocity $\omega_2 = 6$ rad/s relative to the arm OA. In the position shown, arm OA is rotating about the z-axis with the angular velocity $\omega_1 = 4$ rad/s and the angular acceleration $\dot\omega_1 = -15$ rad/s^2. For this position, determine (a) the angular acceleration of the disk; and (b) the velocity of P, the highest point on the disk.

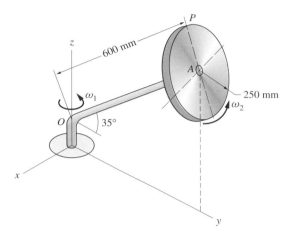

Fig. P19.23

19.24 A clevis is used to attach the rod AB to the vertical shaft. As the shaft rotates about the z-axis with the constant angular velocity $\omega_1 = 12$ rad/s, the rod rotates with the constant angular velocity $\omega_2 = 9$ rad/s relative to the clevis. For the position shown, calculate (a) the angular acceleration of the rod; and (b) the acceleration of end B of the rod.

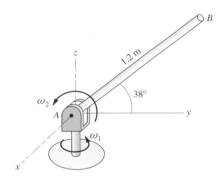

Fig. P19.24

Impulse-Momentum Method

a. *Angular momentum*

We introduced the angular momentum of a rigid body about a point in Chapter 17. It was pointed out that the angular momentum $\mathbf{h}_A$ about a point A is obtained by adding the angular momenta of the differential elements that make up the body:

$$\mathbf{h}_A = \int_{\mathcal{V}} \mathbf{r} \times (\mathbf{v}\, dm) \qquad (19.5)$$

As shown in Fig. 19.6, $\mathbf{v}\, dm$ is the linear momentum of a typical differential element of mass dm and velocity $\mathbf{v}$, and $\mathbf{r}$ is the position vector of the element relative to A. The integrand of Eq. (19.5) is thus angular momentum (that is, the moment of the linear momentum) of the element about A. The integral is taken over the region $\mathcal{V}$ occupied by the body.

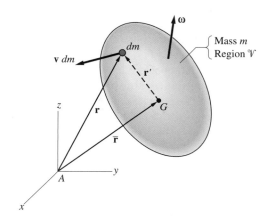

Fig. 19.6

We showed in Art. 17.3 that if we choose for A the mass center G of the body, the angular momentum takes the form

$$\mathbf{h}_G = \int_{\mathcal{V}} \mathbf{r}' \times (\boldsymbol{\omega} \times \mathbf{r}')\, dm \qquad (G\text{: mass center}) \qquad (19.6)$$

where $\boldsymbol{\omega}$ is the angular velocity of the body and $\mathbf{r}'$ is the position vector of dm relative to G, as illustrated in Fig. 19.6.

We also demonstrated that the angular momentum of a body of mass m about an arbitrary point A can be expressed in the form

$$\mathbf{h}_A = \mathbf{h}_G + \bar{\mathbf{r}} \times (m\bar{\mathbf{v}}) \qquad (A\text{: arbitrary point}) \qquad (19.7)$$

In Eq. (19.7), $\bar{\mathbf{r}}$ is the position vector of G relative to A, and $m\bar{\mathbf{v}}$ represents the linear momentum of the body. A convenient means of calculating $\mathbf{h}_A$ is the momentum diagram of the body in Fig. 19.7. From the diagram we see that $\mathbf{h}_A$ is the vector sum of $\mathbf{h}_G$ and the moment of $m\bar{\mathbf{v}}$ (acting at G) about A.

An important special case is the angular momentum of a body that rotates about a fixed point A. Then the velocity of the differential element in Fig. 19.6 is $\mathbf{v} = \boldsymbol{\omega} \times \mathbf{r}$, which upon substitution in Eq. (19.5) yields

$$\mathbf{h}_A = \int_{\mathcal{V}} \mathbf{r} \times (\boldsymbol{\omega} \times \mathbf{r})\, dm \qquad (A\text{: fixed in the body and in space}) \qquad (19.8)$$

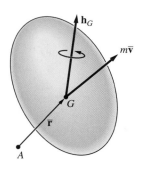

Fig. 19.7

b. *Inertial properties*

We introduce here the inertial properties of a rigid body described in three dimensions (this topic is discussed more fully in Appendix F). As we will see shortly, these properties arise in the computation of the angular momentum.

1. Moments and Products of Inertia and the Inertia Tensor The *mass moment of inertia* of a body about the a-axis was defined in Eq. (17.1):

$I_a = \int_V r^2 \, dm$, where V is the region occupied by the body and r is the perpendicular distance from the a-axis to the differential mass dm. It can be seen in Fig. 19.8 that the perpendicular distances from the x, y, z-axes to dm are

$$(y^2 + z^2)^{1/2} \qquad (z^2 + x^2)^{1/2} \quad \text{and} \quad (x^2 + y^2)^{1/2}$$

respectively. Therefore, the mass moments of inertia of the body about the three coordinate axes are

$$
\begin{aligned}
I_x &= \int_V (y^2 + z^2) \, dm \\[2mm]
I_y &= \int_V (z^2 + x^2) \, dm \\[2mm]
I_z &= \int_V (x^2 + y^2) \, dm
\end{aligned}
\tag{19.9}
$$

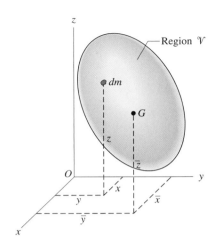

Fig. 19.8

The *products of inertia* of the body with respect to the rectangular axes shown in Fig. 19.8 are defined as

$$
\begin{aligned}
I_{xy} &= I_{yx} = \int_V xy \, dm \\[2mm]
I_{yz} &= I_{zy} = \int_V yz \, dm \\[2mm]
I_{zx} &= I_{xz} = \int_V zx \, dm
\end{aligned}
\tag{19.10}
$$

The dimensions of the inertial properties defined in Eqs. (19.9) and (19.10) are $[ML^2]$; hence the units are $\text{kg} \cdot \text{m}^2$ or $\text{slug} \cdot \text{ft}^2$ ($\text{lb} \cdot \text{ft} \cdot \text{s}^2$). Although moments of inertia are always positive, products of inertia may be positive, negative, or zero.

The following matrix of inertial properties is called the *inertia tensor* of the body *at point O* (the origin of the coordinate axes).

$$
\mathbf{I} = \begin{bmatrix} I_x & -I_{xy} & -I_{xz} \\ -I_{yx} & I_y & -I_{yz} \\ -I_{zx} & -I_{zy} & I_z \end{bmatrix}
\tag{19.11}
$$

Note that the off-diagonal terms are the negatives of the products of inertia. (The minus signs must be included for $\mathbf{I}$ to satisfy certain conventions of tensor algebra, a topic that is beyond the scope of this text.)

2. Parallel-Axis Theorem for Moments of Inertia The *parallel-axis theorem* for the moment of inertia was derived in Chapter 17: $I_a = \bar{I}_a + md^2$, where I_a is the moment of inertia about the a-axis, $\bar{I}_a$ is the moment of inertia about the central a-axis (the axis that passes through the mass center of the body and is parallel to the a-axis), m is the mass of the body, and d is the distance between the two axes. Referring to Fig. 19.9, we see that the distances between the x-, y-, and z-axes and the corresponding central axes are $(\bar{y}^2 + \bar{z}^2)^{1/2}$, $(\bar{z}^2 + \bar{x}^2)^{1/2}$, and

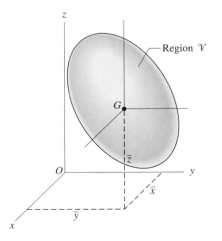

Fig. 19.9

$(\bar{x}^2 + \bar{y}^2)^{1/2}$, respectively. Letting $\bar{I}_x$, $\bar{I}_y$, and $\bar{I}_z$ denote the moments of inertia about the central axes, the parallel-axis theorem for moments of inertia becomes

$$
\begin{aligned}
I_x &= \bar{I}_x + m(\bar{y}^2 + \bar{z}^2) \\
I_y &= \bar{I}_y + m(\bar{z}^2 + \bar{x}^2) \\
I_z &= \bar{I}_z + m(\bar{x}^2 + \bar{y}^2)
\end{aligned}
\tag{19.12}
$$

3. Parallel-Plane Theorem for Products of Inertia Referring again to the body in Fig. 19.9, it can be shown that the products of inertia with respect to the two sets of axes are related by

$$
\begin{aligned}
I_{xy} &= \bar{I}_{xy} + m\bar{x}\bar{y} \\
I_{yz} &= \bar{I}_{yz} + m\bar{y}\bar{z} \\
I_{zx} &= \bar{I}_{zx} + m\bar{z}\bar{x}
\end{aligned}
\tag{19.13}
$$

where $\bar{I}_{xy}$, $\bar{I}_{yz}$, and $\bar{I}_{zx}$ denote the products of inertia about the central axes. This theorem, which is known as the *parallel-plane theorem,* is proved in Appendix F.

4. Principal Moments of Inertia It can be shown that it is always possible to find three perpendicular axes at any given point O such that the products of inertia with respect to these axes vanish. These axes are called the *principal axes at point O,* and the corresponding moments of inertia, denoted by I_1, I_2, and I_3, are known as the *principal moments of inertia* of the body *at point O.* Referred to the principal axes, the inertia tensor thus assumes the form of a diagonal matrix:

$$
\mathbf{I} = \begin{bmatrix} I_1 & 0 & 0 \\ 0 & I_2 & 0 \\ 0 & 0 & I_3 \end{bmatrix}
\tag{19.14}
$$

The determination of principal axes and the computation of principal moments of inertia are explained in Appendix F.

5. Body with a Plane of Symmetry Consider a homogeneous body that is symmetric about a plane, say the xy-plane. Let the mass center of the body, which lies in the plane of symmetry, be the origin of the xyz-frame. Symmetry implies that for every differential mass dm with coordinates (x, y, z), there exists another dm with coordinates $(x, y, -z)$. It follows that $\bar{I}_{yz} = \int_{\mathcal{V}} yz\, dm = 0$ and $\bar{I}_{zx} = \int_{\mathcal{V}} zx\, dm = 0$, because the integral over the region of $\mathcal{V}$ where $z < 0$ cancels the integral over the region where $z > 0$. Because $\bar{I}_{yz} = \bar{I}_{xz} = 0$ regardless of the orientation of the xy-axes, the z-axis must be a principal axis of inertia. In this special case, the parallel-plane theorem for products of inertia, Eqs. (19.13), becomes

$$
I_{xy} = \bar{I}_{xy} + m\bar{x}\bar{y} \qquad I_{yz} = m\bar{y}\bar{z} \qquad I_{zx} = m\bar{z}\bar{x}
\tag{19.15}
$$

In general, an axis that is perpendicular to a plane of symmetry and passes through the mass center G of the body is a principal axis of the body at point G.

c. Rectangular components of angular momentum

The angular momentum of a body can be expressed in terms of the angular velocity and the inertia tensor of the body. Because Eqs. (19.6) and (19.8) have the same form,

$$\mathbf{h} = \int_{\mathcal{V}} \mathbf{r} \times (\boldsymbol{\omega} \times \mathbf{r}) \, dm \qquad (19.16)$$

we can treat both cases simultaneously. Note that $\mathbf{r}$ represents the position vector of dm relative to the reference point (either a point in the body that is fixed or the mass center).

The rectangular representations of the vectors appearing in Eq. (19.16) are

$$\mathbf{h} = h_x \mathbf{i} + h_y \mathbf{j} + h_z \mathbf{k}$$
$$\boldsymbol{\omega} = \omega_x \mathbf{i} + \omega_y \mathbf{j} + \omega_z \mathbf{k}$$
$$\mathbf{r} = x \mathbf{i} + y \mathbf{j} + z \mathbf{k}$$

The cross product of $\boldsymbol{\omega}$ and $\mathbf{r}$ now becomes

$$\boldsymbol{\omega} \times \mathbf{r} = \mathbf{i}(\omega_y z - \omega_z y) + \mathbf{j}(\omega_z x - \omega_x z) + \mathbf{k}(\omega_x y - \omega_y x)$$

Consequently, the integrand in Eq. (19.16) is

$$\mathbf{r} \times (\boldsymbol{\omega} \times \mathbf{r}) = \begin{vmatrix} \mathbf{i} & \mathbf{j} & \mathbf{k} \\ x & y & z \\ \omega_y z - \omega_z y & \omega_z x - \omega_x z & \omega_x y - \omega_y x \end{vmatrix} \qquad (a)$$

Expansion of this determinant gives

$$\mathbf{r} \times (\boldsymbol{\omega} \times \mathbf{r}) = \mathbf{i}\left[(y^2 + z^2)\,\omega_x - xy\,\omega_y - xz\,\omega_z\right]$$
$$+ \mathbf{j}\left[-xy\,\omega_x + (z^2 + x^2)\,\omega_y - yz\,\omega_z\right]$$
$$+ \mathbf{k}\left[-zx\,\omega_x - zy\,\omega_y + (x^2 + y^2)\,\omega_z\right] \qquad (b)$$

Substituting Eq. (b) into Eq. (19.16), we obtain

$$h_x = \omega_x \int_{\mathcal{V}} (y^2 + z^2)\,dm - \omega_y \int_{\mathcal{V}} xy\,dm - \omega_z \int_{\mathcal{V}} xz\,dm$$

$$h_y = -\omega_x \int_{\mathcal{V}} xy\,dm + \omega_y \int_{\mathcal{V}} (z^2 + x^2)\,dm - \omega_z \int_{\mathcal{V}} yz\,dm \qquad (c)$$

$$h_z = -\omega_x \int_{\mathcal{V}} xz\,dm - \omega_y \int_{\mathcal{V}} yz\,dm + \omega_z \int_{\mathcal{V}} (x^2 + y^2)\,dm$$

Recognizing that the integrals in Eqs. (c) are the components of the inertia tensor *at the reference point*—see Eqs. (19.9) and (19.10)—we obtain

$$h_x = I_x \omega_x - I_{xy}\omega_y - I_{xz}\omega_z$$
$$h_y = -I_{yx}\omega_x + I_y\omega_y - I_{yz}\omega_z \qquad (19.17a)$$
$$h_z = -I_{zx}\omega_x - I_{zy}\omega_y + I_z\omega_z$$

Equations (19.17a) can be written concisely in matrix notation as

$$\mathbf{h}_A = \mathbf{I}_A \boldsymbol{\omega} \quad (A: \text{ fixed in the body and in space, or the mass center}) \quad \textbf{(19.17b)}$$

where $\mathbf{I}_A$ is the matrix in Eq. (19.11), where the inertia components are calculated about the axis passing through A. If the *xyz*-axes are the principal axes of inertia at the reference point, the products of inertia vanish, and we are left with

$$
\begin{aligned}
h_x &= I_x \omega_x \\
h_y &= I_y \omega_y \\
h_z &= I_z \omega_z
\end{aligned}
\qquad \textbf{(19.18)}
$$

In general, the angular momentum vector $\mathbf{h}$ is not in the same direction as the angular velocity vector $\boldsymbol{\omega}$. For example, if the *xyz*-axes are principal axes of inertia, we have

$$
\begin{aligned}
\mathbf{h} &= I_x \omega_x \mathbf{i} + I_y \omega_y \mathbf{j} + I_z \omega_z \mathbf{k} \\
\boldsymbol{\omega} &= \omega_x \mathbf{i} + \omega_y \mathbf{j} + \omega_z \mathbf{k}
\end{aligned}
\qquad \textbf{(19.19)}
$$

From Eqs. (19.19), we see that the directions of $\mathbf{h}$ and $\boldsymbol{\omega}$ coincide only for the following special cases.

1. The principal moments of inertia are equal (e.g., a homogeneous sphere with mass center being the origin of the coordinate system).
   Observe that with $I_x = I_y = I_z = I$, the first of Eqs. (19.19) gives $\mathbf{h} = I\boldsymbol{\omega}$.
2. The direction of $\boldsymbol{\omega}$ is parallel to one of the principal axes of inertia.
   If $\boldsymbol{\omega}$ is parallel to a principal axis, say the *z*-axis, then Eqs. (19.19) yield $\mathbf{h} = I_z \omega_z \mathbf{k}$ and $\boldsymbol{\omega} = \omega_z \mathbf{k}$. This case applies to plane motion of a rigid body. It also demonstrates that plane motion can exist only if the coordinate axis perpendicular to the plane of motion is a principal axis of inertia.

The rectangular components of the angular momentum about a fixed point, or the mass center, can be computed from Eqs. (19.17). If the angular momentum about some other point A is required, it can be obtained by first calculating $\mathbf{h}_G$ from Eqs. (19.17) and then applying Eq. (19.7): $\mathbf{h}_A = \mathbf{h}_G + \bar{\mathbf{r}} \times (m\bar{\mathbf{v}})$. Note again that Eqs. (19.17) are not directly applicable for an arbitrary reference point.

d.   *Impulse-momentum principles*

The impulse-momentum principles for plane motion, discussed in Art. 18.6, are also applicable to three-dimensional motion. Here we restate these principles without repeating the derivations given in Art. 18.6.

The linear impulse-momentum equation for a rigid body is

$$\mathbf{L}_{1-2} = \mathbf{p}_2 - \mathbf{p}_1 = \Delta \mathbf{p} \qquad \textbf{(19.20)}$$

where $\mathbf{L}_{1-2}$ is the linear impulse of the external forces acting on the body during the time interval t_1 to t_2, and $\Delta \mathbf{p}$ is the change in the linear momentum of the body during this time interval. Recall that the linear momentum of a body of mass m is $\mathbf{p} = m\bar{\mathbf{v}}$, where $\bar{\mathbf{v}}$ is the velocity of its mass center.

The angular impulse-momentum equation for a rigid body is

$$(\mathbf{A}_A)_{1-2} = (\mathbf{h}_A)_2 - (\mathbf{h}_A)_1 = \Delta\mathbf{h}_A \qquad \begin{array}{l} (A: \text{ fixed point}^*\\ \text{or mass center}) \end{array} \qquad (19.21)$$

where $(\mathbf{A}_A)_{1-2}$ is the angular impulse of the external forces about point A during the time interval t_1 to t_2, and $\Delta\mathbf{h}_A$ is the change in angular momentum about A during this time period. The angular momentum can be computed from Eqs. (19.17) or (19.18).

As mentioned in Art. 18.6, the impulse-momentum equations can also be applied to systems of rigid bodies provided that (1) the impulses refer only to forces that are external to the system; (2) the momenta are interpreted as the momenta of the system (obtained by summing the momenta of the bodies that constitute the system); and (3) the mass center referred to in Eq. (19.21) is the mass center of the system.

According to Eq. (19.20), linear momentum is conserved if the linear impulse is zero. Similarly, we conclude from Eq. (19.21) that the angular momentum about a fixed point or the mass center is conserved if the angular impulse about that point is zero.

If the motion is impulsive (infinite forces acting over infinitesimal time intervals), which is the approximation used in impact problems, Eq. (19.21) is valid for any reference point (a proof of this statement is given in Art. 18.7). Because the duration of the impact is assumed to be infinitesimal, only the impulsive forces need be taken into account, the impulses of finite forces being negligible.

*19.4   *Work-Energy Method*

a.   *Kinetic energy*

1.   General Case   In Chapter 18, we showed that the kinetic energy of a rigid body is

$$T = \frac{1}{2}m\bar{v}^2 + \frac{1}{2}\boldsymbol{\omega} \cdot \mathbf{h}_G \qquad \text{(18.6a, repeated)}$$

where m is the mass of the body, $\bar{v}$ is the speed of the mass center G, $\boldsymbol{\omega}$ represents the angular velocity of the body, and $\mathbf{h}_G$ is the angular momentum of the body about G. The first term on the right side of Eq. (18.6a) represents the kinetic energy of translation, and the second term is the kinetic energy of rotation.

Using rectangular representation for $\boldsymbol{\omega}$ and $\mathbf{h}_G$, where the components of $\mathbf{h}_G$ are obtained from Eq. (19.17), the kinetic energy can be written as[†]

$$T = \frac{1}{2}m\bar{v}^2 + \frac{1}{2}(\omega_x\mathbf{i} + \omega_y\mathbf{j} + \omega_z\mathbf{k}) \cdot \left[(\bar{I}_x\omega_x - \bar{I}_{xy}\omega_y - \bar{I}_{xz}\omega_z)\mathbf{i}\right.$$
$$\left. + (-\bar{I}_{yx}\omega_x + \bar{I}_y\omega_y - \bar{I}_{yz}\omega_z)\mathbf{j} + (-\bar{I}_{zx}\omega_x - \bar{I}_{zy}\omega_y + \bar{I}_z\omega_z)\mathbf{k}\right]$$

---

[*]Point A is fixed in space; it is not necessarily a point in the body.
[†]Overbars on the I's remind us that the inertial properties are to be computed at the mass center of the body.

After evaluating the dot products and noting that $\bar{I}_{yx} = \bar{I}_{xy}$ and so on, we get

$$T = \frac{1}{2}m\bar{v}^2 + \frac{1}{2}(\bar{I}_x\omega_x^2 + \bar{I}_y\omega_y^2 + \bar{I}_z\omega_z^2$$
$$- 2\bar{I}_{xy}\omega_x\omega_y - 2\bar{I}_{yz}\omega_y\omega_z - 2\bar{I}_{zx}\omega_z\omega_x)$$

(19.22)

Note that the components of the inertia tensor must be evaluated about the central axes.

2.   *Angular Velocity Parallel to an Axis*   If ω is parallel to one of the coordinate axes, say the z-axis, then $\omega_x = \omega_y = 0$, $\omega_z = \omega$, and Eq. (19.22) reduces to

$$T = \frac{1}{2}m\bar{v}^2 + \frac{1}{2}\bar{I}_z\omega^2 \qquad (\omega = \omega\mathbf{k})$$

(19.23)

This equation is analogous to Eq. (18.7a) used in plane motion. However, in plane motion, ω always remained parallel to the z-axis, whereas Eq. (19.23) requires ω to be in the z-direction only at the instant of concern.

3.   *Rotation about a Point of Zero Velocity*   We showed in Art. 18.3 that if A is a point in the body that has zero velocity, then the kinetic energy of the body can be written in the form

$$T = \frac{1}{2}\omega \cdot \mathbf{h}_A$$

(18.6b, repeated)

where $\mathbf{h}_A$ is the angular momentum of the body about A. Expressing ω and $\mathbf{h}_A$ in terms of their rectangular components and repeating the steps used in the derivation of Eq. (19.22), we obtain

$$T = \frac{1}{2}(I_x\omega_x^2 + I_y\omega_y^2 + I_z\omega_z^2 - 2I_{xy}\omega_x\omega_y$$
$$- 2I_{yz}\omega_y\omega_z - 2I_{zx}\omega_z\omega_x)$$
$$\text{(axes pass through point of zero velocity)}$$

(19.24)

Note that the inertial properties in Eq. (19.24) must be computed about axes that pass through the reference point A, a point with zero velocity.

4.   *Rotation about an Axis*   If the coordinate system is chosen so that the z-axis is the instant axis of rotation, then $\omega_x = \omega_y = 0$, $\omega_z = \omega$. Therefore, Eq. (19.24) becomes

$$T = \frac{1}{2}I_z\omega^2 \qquad (z\text{-axis is instant axis}) \qquad \text{(19.25)}$$

This equation is equivalent to Eq. (18.7b) used in plane motion.

b.   *Work-energy principle and the conservation of mechanical energy*

The principles of work-energy and conservation of mechanical energy for plane motion, presented in Art. 18.4, are also applicable to three-dimensional motion. Therefore, the following discussion is a review of the fundamental equations covered in Art. 18.4.

Letting the subscripts 1 and 2 refer to the initial and final positions of a rigid body, the work-energy principle is

$$(U_{1-2})_{\text{ext}} = \Delta T \qquad \text{(19.26)}$$

where $(U_{1-2})_{\text{ext}}$ is the work done by external forces and ΔT is the change in kinetic energy. The work can be computed using the methods described in Arts. 14.2 and 18.2, and Eqs. (19.22)–(19.25) can be used to calculate the kinetic energy.

For a system of connected rigid bodies, the work-energy principle is

$$(U_{1-2})_{\text{ext}} + (U_{1-2})_{\text{int}} = \Delta T \qquad \text{(19.27)}$$

where $(U_{1-2})_{\text{ext}}$ and $(U_{1-2})_{\text{int}}$ represent the work done on the system by external and internal forces, respectively.

If all the forces that act on a rigid body or a system of connected rigid bodies are conservative, then the mechanical energy is conserved. The principle of conservation of mechanical energy

$$V_1 + T_1 = V_2 + T_2 \qquad \text{(19.28)}$$

can then be used in place of the work-energy principle. In Eq. (19.28), V_1 and V_2 are the initial and final potential energies, and T_1 and T_2 are the initial and final kinetic energies.

Sample Problem 19.5

The uniform slender rods labeled as 1, 2, and 3 are welded together to form the rigid body shown in Fig. (a). The body, which is supported by bearings at A and B, is being driven at the constant angular velocity $\omega = 30$ rad/s. When the body is in the position shown, calculate the following: (1) the angular momentum about point C; and (2) the kinetic energy. The mass per unit length of each rod is $\rho = 600$ g/m.

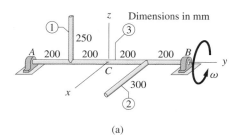

(a)

Solution

Preliminaries

Because the only nonzero component of the angular velocity vector is $\omega_y = \omega$, we see from Eq. (19.17a) that the components of the angular momentum vector about point C reduce to

$$h_x = -I_{xy}\omega \qquad h_y = I_y\omega \qquad h_z = -I_{zy}\omega \qquad \text{(a)}$$

Because the body rotates about the fixed y-axis, its kinetic energy, from Eq. (19.25), is

$$T = \frac{1}{2}I_y\omega^2 \qquad \text{(b)}$$

The computations of the moments and products of inertia required in Eqs. (a) and (b) are shown in the following table. Note that the results for the body are obtained by summing the properties of rods 1 and 2 only, because the relevant inertial properties of rod 3 are zero.

Part 1

Substituting the moments and products of inertia for point C and $\omega = 30$ rad/s into Eq. (a), the components of the angular momentum vector about C become

$$h_x = -I_{xy}\omega = -(54.00 \times 10^{-4})(30) = -16.20 \times 10^{-2}\, \text{N} \cdot \text{m} \cdot \text{s}$$

$$h_y = I_y\omega = (85.25 \times 10^{-4})(30) = 25.58 \times 10^{-2}\, \text{N} \cdot \text{m} \cdot \text{s}$$

$$h_z = -I_{zy}\omega = -(-37.5 \times 10^{-4})(30) = 11.25 \times 10^{-2}\, \text{N} \cdot \text{m} \cdot \text{s}$$

Therefore, the magnitude of the angular momentum about C is

$$h_C = (10^{-2})\sqrt{(-16.20)^2 + (25.58)^2 + (11.25)^2}$$

$$= 32.3 \times 10^{-2}\, \text{N} \cdot \text{m} \cdot \text{s} \qquad \textit{Answer}$$

	Rod 1	Rod 2	Totals
L	0.25 m	0.3 m	
$m = \rho L$	0.6(0.25) = 0.15 kg	0.6(0.3) = 0.18 kg	
$\bar{x}$	0	0.15 m	
$\bar{y}$	−0.2 m	0.2 m	
$\bar{z}$	0.125 m	0	
$I_y = \frac{1}{3}mL^2$	$\frac{1}{3}(0.15)(0.25)^2$ = 0.003 125 kg · m²	$\frac{1}{3}(0.18)(0.3)^2$ = 0.005 40 kg · m²	0.008 525 kg · m²
$I_{xy} = m\bar{x}\bar{y}$	0.15(0)(−0.2) = 0	0.18(0.15)(0.2) = 0.005 40 kg · m²	0.005 40 kg · m²
$I_{zy} = m\bar{z}\bar{y}$	0.15(0.125)(−0.2) = −0.003 75 kg · m²	0.18(0)(0.2) = 0	−0.003 75 kg · m²

Computation of Inertial Properties at Point C.

The vector $\mathbf{h}_C$ and its components are shown in Fig. (b).

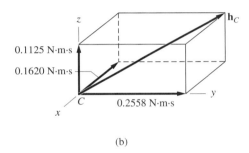

(b)

Part 2

The kinetic energy of the body is found by substituting I_y and $\omega = 30$ rad/s into Eq. (b), with the result being

$$T = \frac{1}{2}I_y\omega^2 = \frac{1}{2}(85.25 \times 10^{-4})(30)^2 = 3.84 \text{ J} \qquad \textit{Answer}$$

Sample Problem 19.6

The uniform slender rod AB shown in Fig. (a) has the mass $m = 8$ kg and the length $L = 1.2$ m. The weights of the sliding collars A and B, to which the rod is connected with ball-and-socket joints, may be neglected. When collar A is in the position $\theta = 0$, the system is slightly displaced and released from rest. Neglecting friction, determine the speed of A when it has moved to the position $\theta = 60°$.

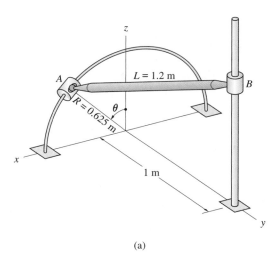

(a)

Solution

This problem is suited to the work-energy method of analysis, because it is concerned with the change in speed between two positions. Because the weight of the rod, which is a conservative force, is the only force that does work on the system, mechanical energy is conserved. Therefore, we can apply the principle of conservation of mechanical energy: $T_1 + V_1 = T_2 + V_2$, where T_1 and T_2 are the initial and final kinetic energies, and V_1 and V_2 are the initial and final gravitational potential energies. Figure (b) shows the reference plane that has been chosen for the potential energy V.

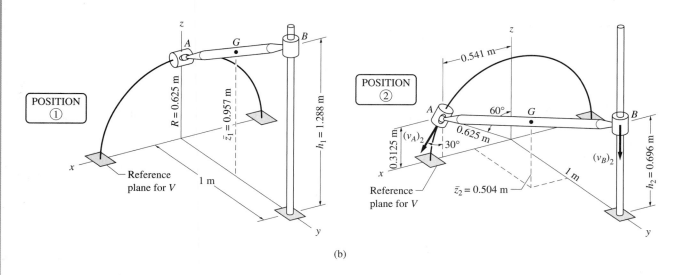

(b)

Position 1

Potential Energy Referring to position 1 in Fig. (b), the height h_1 of collar B above the reference plane must satisfy the geometric relation $(h_1 - 0.625)^2 + (1)^2 = (1.2)^2$, which gives $h_1 = 1.288$ m. Therefore, the height of the mass center

G is $\bar{z}_1 = (0.625 + 1.288)/2 = 0.957$ m, and the initial potential energy becomes

$$V_1 = W\bar{z}_1 = 8(9.8)(0.957) = 75.03 \text{ N} \cdot \text{m} \qquad \text{(a)}$$

Kinetic Energy   Because the system is released from rest, we have $T_1 = 0$.

Position 2

Potential Energy   When the system is in position 2 shown in Fig. (b), the coordinates of collar A are $x = 0.625 \sin 60° = 0.541$ m, $y = 0$, and $z = 0.625 \cos 60° = 0.3125$. Using these values, the height h_2 of B above the reference plane is obtained from the geometric relationship $(0.541)^2 + (1)^2 + (h_2 - 0.3125)^2 = (1.2)^2$, which gives $h_2 = 0.696$ m. The height of the mass center G above the reference plane is, therefore, $\bar{z}_2 = (0.3125 + 0.696)/2 = 0.504$ m. It follows that the gravitational potential energy of the rod is

$$V_2 = W\bar{z}_2 = 8(9.8)(0.504) = 39.51 \text{ N} \cdot \text{m} \qquad \text{(b)}$$

Kinematic Analysis   The purpose of the kinematic analysis is to relate $\boldsymbol{\omega}_2$ (the angular velocity of the rod) and $\bar{\mathbf{v}}_2$ (the velocity of its mass center) to $(\mathbf{v}_A)_2$ (the velocity of collar A). We utilize the relative velocity equation between A and B:

$$(\mathbf{v}_B)_2 = (\mathbf{v}_A)_2 + \boldsymbol{\omega}_2 \times (\mathbf{r}_{B/A})_2 \qquad \text{(c)}$$

where $(\mathbf{v}_B)_2$ is the velocity of B. The position vector of B relative to A in position 2 is, according to Fig. (b), $(\mathbf{r}_{B/A})_2 = -0.541\mathbf{i} + 1\mathbf{j} + (0.696 - 0.3125)\mathbf{k} = -0.541\mathbf{i} + 1\mathbf{j} + 0.3835\mathbf{k}$ ft. Again, referring to Fig. (b), we see that $(\mathbf{v}_A)_2 = (v_A)_2 \sin 30°\mathbf{i} - (v_A)_2 \cos 30°\mathbf{k}$. Expressing the angular velocity vector as $\boldsymbol{\omega}_2 = (\omega_x)_2\mathbf{i} + (\omega_y)_2\mathbf{j} + (\omega_z)_2\mathbf{k}$, Eq. (c) becomes

$$-(v_B)_2\mathbf{k} = (v_A)_2 \sin 30°\mathbf{i} - (v_A)_2 \cos 30°\mathbf{k}$$

$$+ \begin{vmatrix} \mathbf{i} & \mathbf{j} & \mathbf{k} \\ (\omega_x)_2 & (\omega_y)_2 & (\omega_z)_2 \\ -0.541 & 1 & 0.3835 \end{vmatrix} \qquad \text{(d)}$$

Expanding the determinant in Eq. (d), and equating like components, yields the following equations:

$$0 = (v_A)_2 \sin 30° + 0.3835(\omega_y)_2 - 1(\omega_z)_2$$
$$0 = -0.3835(\omega_x)_2 - 0.541(\omega_z)_2 \qquad \text{(e)}$$
$$-(v_B)_2 = -(v_A)_2 \cos 30° + 1(\omega_x)_2 + 0.541(\omega_y)_2$$

A fourth equation is obtained by setting the spin velocity of the rod to zero. This gives $\boldsymbol{\omega}_2 \cdot (\mathbf{r}_{B/A})_2 = 0$, or $[(\omega_x)_2\mathbf{i} + (\omega_y)_2\mathbf{j} + (\omega_z)_2\mathbf{k}] \cdot (-0.541\mathbf{i} + 1\mathbf{j} + 0.3835\mathbf{k}) = 0$, which becomes

$$-0.541(\omega_x)_2 + 1(\omega_y)_2 + 0.3835(\omega_z)_2 = 0 \qquad \text{(f)}$$

Equations (e) and (f) can be solved in terms of $(v_A)_2$, with the results being

$$\left.\begin{array}{l} (\omega_x)_2 = -0.1225(v_A)_2 \\ (\omega_y)_2 = -0.0996(v_A)_2 \\ (\omega_z)_2 = 0.0868(v_A)_2 \end{array}\right\} \ \omega_2 = 0.1802(v_A)_2 \qquad \text{(g)}$$

and

$$(v_B)_2 = 0.786(v_A)_2 \qquad \text{(h)}$$

The relationship between $\bar{\mathbf{v}}_2$ and $(\mathbf{v}_A)_2$ could be found from the relative velocity equation $\bar{\mathbf{v}} = (\mathbf{v}_A)_2 + \boldsymbol{\omega}_2 \times (\mathbf{r}_{G/A})_2$ and Eq. (g). However, it is simpler to use the fact that G is the midpoint of rod AB, so that its velocity is given by

$$\bar{\mathbf{v}}_2 = \frac{1}{2}[(\mathbf{v}_A)_2 + (\mathbf{v}_B)_2]$$

$$= \frac{1}{2}[(v_A)_2(\sin 30° \mathbf{i} - \cos 30° \mathbf{k}) - 0.786(v_A)_2 \mathbf{k}]$$

$$= (v_A)_2(0.250\mathbf{i} - 0.826\mathbf{j})$$

which yields

$$\bar{v}_2^2 = 0.7448(v_A)_2^2 \qquad \text{(i)}$$

Kinetic Energy Because $\boldsymbol{\omega}_2$ is perpendicular to AB, the kinetic energy of the rod can be obtained from Eq. (19.23):

$$T = \frac{1}{2}m\bar{v}^2 + \frac{1}{2}\bar{I}\omega_2^2$$

where $\bar{I} = mL^2/12$ is the moment of inertia of the rod about an axis perpendicular to the rod (parallel to $\boldsymbol{\omega}_2$) at G. Substituting the numerical values, we get

$$T_2 = \frac{1}{2}(8)(0.7448)(v_A)_2^2 + \frac{1}{2}\left(8 \frac{(1.2)^2}{12}\right)[0.1802(v_A)_2]^2$$

$$= (2.9792 + 0.0155)(v_A)_2^2 = 2.9947(v_A)_2^2 \qquad \text{(j)}$$

Conservation of Mechanical Energy

Substituting Eqs. (a), (b), and (j) into the conservation principle,

$$T_1 + V_1 = T_2 + V_2$$

we obtain

$$0 + 75.03 = 2.9947(v_A)_2^2 + 39.51$$

from which the speed of A in position 2 is found to be

$$(v_A)_2 = 3.44 \text{ m/s} \qquad\qquad \textit{Answer}$$

Sample Problem   19.7

Figure (a) shows a 2.0-kg homogeneous square plate that hangs from a ball-and-socket joint at O. The plate is at rest when it is struck by a hammer at A. The force $\hat{\mathbf{P}}$ applied by the hammer is impulsive, its impulse being $\int \hat{\mathbf{P}}\, dt = 1.20\mathbf{i}\, \text{N} \cdot \text{s}$. Determine the angular velocity and the kinetic energy of the plate immediately after the impact. Observe that the x-axis is perpendicular to the plate, and the y- and z-axes lie along the edges of the plate.

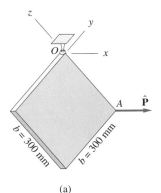

(a)

Solution

This problem must be analyzed by the impulse-momentum method, because the applied force is impulsive.

Figure (b) shows the free-body diagram (FBD) of the plate during the impact. Because the duration of the impact is negligible, the plate occupies essentially the same spatial position before, during, and after the impact. The FBD includes the applied force $\hat{\mathbf{P}}$ and the components of the impulsive reaction at O. The weight of the plate was omitted, because it is not an impulsive force.

We recall that for impulsive motion the angular impulse about any point equals the change in angular momentum of the body about the same point. In our case, point O is a convenient choice for the reference point, because the impulsive reaction passes through that point. Letting the subscripts 1 and 2 refer to the instants immediately before and immediately after the impact, respectively, the angular impulse-momentum principle becomes

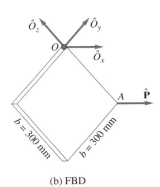

(b) FBD

$$(\mathbf{A}_O)_{1-2} = (\mathbf{h}_O)_2 - (\mathbf{h}_O)_1 = (\mathbf{h}_O)_2 \qquad \text{(a)}$$

Note that $(\mathbf{h}_O)_1 = \mathbf{0}$ because the plate is at rest before the impact.

From the FBD, we see that the angular impulse (that is, the moment of the linear impulse) about O is

$$(\mathbf{A}_O)_{1-2} = \mathbf{r}_{A/O} \times \int \hat{\mathbf{P}}\, dt = -0.3\mathbf{k} \times 1.2\mathbf{i} = -0.36\mathbf{j}\ \text{N} \cdot \text{m} \cdot \text{s} \qquad \text{(b)}$$

To determine the angular momentum of the plate about point O, we must first compute the inertial properties of the plate at that point. The central moments of inertia can be obtained from Table 17.1 (we must set $a = 0$ because the plate is thin): $\bar{I}_x = (1/12)m(2b^2) = mb^2/6$ and $\bar{I}_y = \bar{I}_z = mb^2/12$. Also, due to symmetry, we have $\bar{I}_{xy} = \bar{I}_{yz} = \bar{I}_{zx} = 0$. The moments and products of inertia about O can now be calculated from the parallel-axis theorem and Fig. (c):

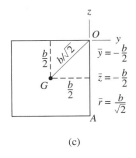

(c)

$$I_x = \bar{I}_x + m\bar{r}^2 = \frac{mb^2}{6} + m\left(\frac{b}{\sqrt{2}}\right)^2 = \frac{2mb^2}{3}$$

$$I_y = \bar{I}_y + m\bar{z}^2 = \frac{mb^2}{12} + m\left(-\frac{b}{2}\right)^2 = \frac{mb^2}{3}$$

$$I_z = \bar{I}_z + m\bar{y}^2 = \frac{mb^2}{12} + m\left(-\frac{b}{2}\right)^2 = \frac{mb^2}{3}$$

$$I_{xy} = \bar{I}_{xy} + m\bar{x}\bar{y} = 0 + m(0)\left(-\frac{b}{2}\right) = 0$$

$$I_{yz} = \bar{I}_{yz} + m\bar{y}\bar{z} = 0 + m\left(-\frac{b}{2}\right)\left(-\frac{b}{2}\right) = \frac{mb^2}{4}$$

$$I_{zx} = \bar{I}_{zx} + m\bar{z}\bar{x} = 0 + m\left(-\frac{b}{2}\right)(0) = 0$$

Substituting $m = 2.0$ kg and $b = 0.300$ m yields

$$I_x = \frac{2}{3}(2.0)(0.300)^2 = 0.1200 \text{ kg} \cdot \text{m}^2$$

$$I_y = I_z = \frac{2.0(0.300)^2}{3} = 0.0600 \text{ kg} \cdot \text{m}^2$$

$$I_{xy} = I_{zx} = 0$$

$$I_{yz} = \frac{2.0(0.300)^2}{4} = 0.0450 \text{ kg} \cdot \text{m}^2$$

The components of the angular momentum about O can now be obtained from Eqs. (19.17a), the result being

$$(\mathbf{h}_O)_2 = 0.12\omega_x \mathbf{i} + (0.06\omega_y - 0.045\omega_z)\mathbf{j} + (-0.045\omega_y + 0.06\omega_z)\mathbf{k} \quad \text{(c)}$$

Substituting Eqs. (b) and (c) into Eq. (a) and equating like components, we get

$$0 = 0.12\omega_x$$
$$-0.36 = 0.06\omega_y - 0.045\omega_z \quad \text{(d)}$$
$$0 = -0.045\omega_y + 0.06\omega_z$$

The solution of Eqs. (d) yields

$$\boldsymbol{\omega} = -13.71\mathbf{j} - 10.29\mathbf{k} \text{ rad/s} \qquad \textit{Answer}$$

Because O is a fixed point, the kinetic energy T_2 after the impact may be calculated using Eq. (18.6b).

$$T_2 = \frac{1}{2}\boldsymbol{\omega} \cdot \mathbf{h}_O = \frac{1}{2}(-13.71\mathbf{j} - 10.29\mathbf{k}) \cdot (-0.36\mathbf{j}) = 2.47 \text{ J} \qquad \textit{Answer}$$

If desired, the components of the impulsive reaction at O could now be calculated using the linear impulse-momentum equations.

Problems

19.25   The homogeneous rod of mass m and length L maintains the constant angle β with the horizontal as it rotates with the angular velocity ω about the vertical axis at its mid-point O. For the position shown, derive (a) the angular momentum of the rod about O; and (b) the kinetic energy of the rod.

19.26   Assume that the impulse acting on the plate in Sample Problem 19.7 is replaced by the impulse $\int \hat{\mathbf{P}}\,dt = 2.2\mathbf{i} + 1.4\mathbf{j} - 1.8\mathbf{k}$ N·s acting at the center of the plate. Calculate the angular velocity and the kinetic energy of the plate immediately after the impact.

19.27   The 12-kg crank, which was formed from a uniform slender rod, rotates about the y-axis with the angular velocity $\omega = 20$ rad/s in the direction shown. Determine (a) the angular momentum of the crank about point O in the position shown; and (b) the kinetic energy of the crank.

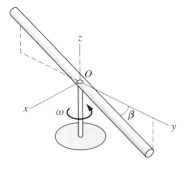

Fig. P19.25

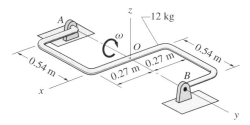

Fig. P19.27

19.28   The homogeneous 12-kg disk rotates about the y-axis with the angular velocity $\omega_1 = 60$ rad/s. At the same time, the assembly rotates about the vertical axis at A with the angular velocity $\omega_2 = 20$ rad/s. Determine (a) the angular momentum of the disk about the point O; and (b) the kinetic energy of the disk.

19.29   The two identical, thin circular plates each of mass m are welded to a shaft of negligible mass. The assembly rotates about the y-axis with the angular velocity ω_0. Calculate the angular momentum of the assembly about (a) point O; and (b) point A.

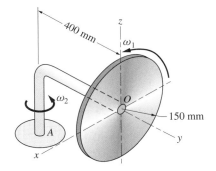

Fig. P19.28

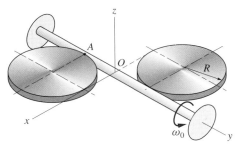

Fig. P19.29

19.30   The 2-kg slender rod is connected to the 6-kg uniform disk at A and to the sliding collar at B by ball-and-socket joints. The mass of the collar is negligible, and the spring attached to the collar has a stiffness of 2 kN/m. The assembly is released from rest in the position shown, where the spring is stretched 100 mm. Calculate the angular velocity of the disk after it has rotated 180°.

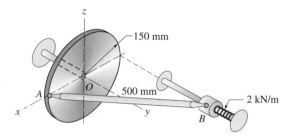

Fig. P19.30

19.31   Two uniform thin plates of mass m each are welded to a shaft of negligible mass that rotates at the angular velocity ω_0. Calculate the angular momentum of the assembly about (a) point O; and (b) point B.

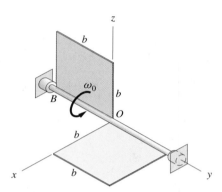

Fig. P19.31

19.32   The thin uniform disk of mass m and radius R spins about the bent shaft OG with the angular speed ω_2. At the same time, the shaft rotates about the z-axis with the angular speed ω_1. If the angle between the bent portion of the shaft and the z-axis is $\beta = 35°$, find the ratio ω_2/ω_1 for which the angular momentum of the disk about its mass center G is parallel to the z-axis.

19.33   Determine the kinetic energy of the disk described in Prob. 19.32 if $\omega_1 = \omega_2$ and $\beta = 35°$.

19.34   The uniform 6-kg thin disk spins about the axle OG as it rolls on the horizontal plane without slipping. End O of the axle is welded to a sliding collar that rotates about the fixed vertical rod with the constant angular velocity of 6 rad/s. Calculate the angular momentum of the disk about point O.

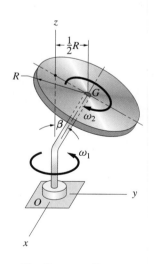

Fig. P19.32, P19.33

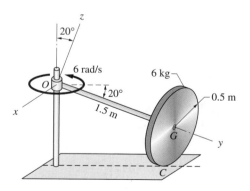

Fig. P19.34

19.35   The 12-kg uniform slender rod AB is connected to sliding collars at A and B by ball-and-socket joints. In the position shown, the velocity of collar A is $v_A = 1.4$ m/s. For this position, calculate (a) the angular momentum of the rod about its mass center; and (b) the kinetic energy of the rod.

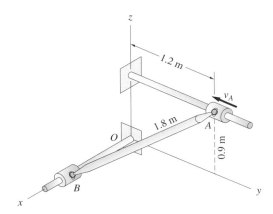

Fig. P19.35

19.36   The slender bar AB of length L and mass m is suspended from two strings, each of length L. If the bar is released from rest when $\theta = 90°$, find its maximum angular velocity.

19.37   The uniform thin disk of radius $R = 0.2$ m and mass 7 kg is attached to the bent axle OAB of negligible weight. The axle is attached to a vertical shaft at O with a clevis. The system is at rest when a constant couple $\mathbf{C}_0 = 0.45\,\text{N} \cdot \text{m}$ is applied to the axle. Assuming that the disk rolls without slipping, determine the angular velocity ω_1 of the axle after it has turned through two revolutions. Neglect friction in the bearings at O and B.

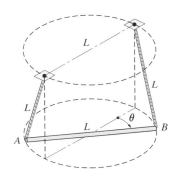

Fig. P19.36

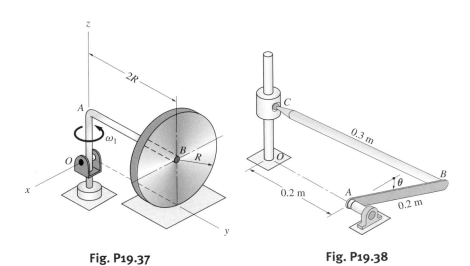

Fig. P19.37   **Fig. P19.38**

19.38   The mechanism consists of two homogeneous slender bars AB and BC of masses 1.8 kg and 1.2 kg, respectively, and the 2-kg slider C. The connections at B and C are ball-and-socket joints. If the mechanism is released from rest at $\theta = 0$, determine the angular velocity of bar AB when $\theta = 90°$. Neglect friction.

19.39   The thin rim of the flywheel *C* mass 8 kg, with the mass of the spokes and hub being negligible. The flywheel is joined to the 3-kg slider *B* with the 2-kg connecting rod *AB*. The joints at *A* and *B* are ball-and-sockets. A constant force $P = 270$ N acts on the slider *B* as shown. If the flywheel has an angular velocity $\omega = 20$ rad/s in the position shown, calculate its angular velocity when joint *A* reaches the position *A'*. Neglect friction.

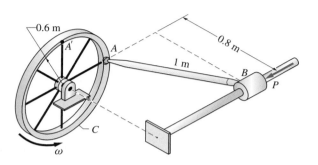

Fig. P19.39

19.40   The uniform bent rod of mass *m* is hanging from a cable at *O* when it receives the impulse $\mathbf{k} \int \hat{P} \, dt$ at *A*. Derive the expression for the angular velocity of the rod immediately after the impact.

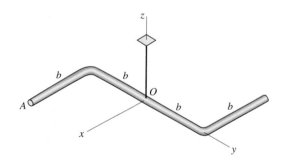

Fig. P19.40

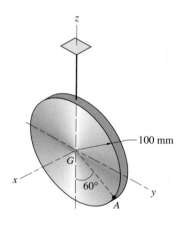

Fig. P19.41

19.41   The 1.2-kg thin circular plate of uniform thickness is suspended from a cable. The plate is at rest when it receives a sudden impulse of $-0.15\mathbf{i}$ N · s at point *A* on the rim of the disk. Determine the following at the instant after the impact: (a) the velocity of the mass center *G*; and (b) the kinetic energy of the plate.

19.42   A 0.4 m rod of negligible mass is welded to the 3.2-kg thin uniform plate and suspended from a ball-and-socket joint at *O*. The assembly is rotating about the *z*-axis with the angular speed of 6 rad/s when corner *A* of the plate hits a rigid obstruction. Assuming that *A* does not rebound, determine the angular velocity of the assembly immediately after the impact.

19.43   The uniform box of mass *m* is falling with a speed of 8 m/s and no angular velocity when corner *O* hits a rigid obstruction. Assuming plastic impact (that is,

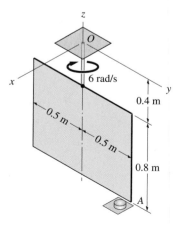

Fig. P19.42

no rebound), determine (a) the angular velocity of the box immediately after the impact; and (b) the percentage of kinetic energy lost during the impact.

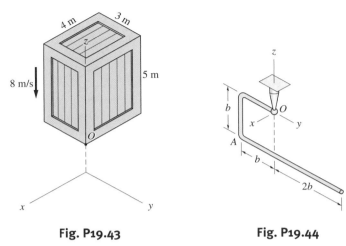

Fig. P19.43

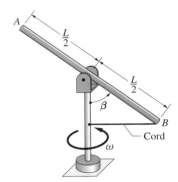

Fig. P19.44

Fig. P19.45

19.44   The uniform bent wire of total mass m is suspended from a ball-and-socket joint at O. The wire is stationary when a short impulse is applied at corner A in the x-direction. Find the unit vector in the direction of the instant axis of rotation immediately after the impulse is received.

19.45   The slender rod AB of mass m and length L is attached to a vertical shaft with a clevis. The shaft is rotating freely at the angular speed ω when the cord, which maintained the angle $\beta = \beta_1$ between the rod and the shaft, breaks. Determine the expressions for ω and $d\beta/dt$ when the rod reaches the position $\beta = 90°$.

<table>
<tr><td>***19.5**</td><td>*Force-Mass-Acceleration Method*</td></tr>
</table>

a.   *Equations of motion*

Three-dimensional motion of a rigid body is governed by the same basic equations that we used for particle systems in Chapter 15. Thus the force equation

$$\Sigma \mathbf{F} = m\bar{\mathbf{a}}$$ (15.19, repeated)

determines the motion of the mass center of the body, where $\Sigma\mathbf{F}$ is the sum of the external forces acting on the body, m is the mass of the body (assumed to be constant), and $\bar{\mathbf{a}}$ is the acceleration of its mass center. The rotational motion of the body is determined by the moment equation

$$\Sigma \mathbf{M}_A = \dot{\mathbf{h}}_A \qquad (A\text{: fixed point or mass center})$$ (15.35, repeated)

where $\Sigma\mathbf{M}_A$ is the resultant moment of the external forces acting on the body and $\mathbf{h}_A$ is the angular momentum of the body about point A. The latter must be a point in the body that is either *fixed in space* or the *mass center* of the body. (This restriction on the choice of *A* must be kept in mind; for the sake of brevity, we will not mention it continually.)

When using Eq. (15.35), any convenient reference frame can be used to describe the angular momentum vector $\mathbf{h}_A$. For example, if $\boldsymbol{\Omega}$ is the angular velocity of such a reference frame, the moment equation of motion can be written as

$$\Sigma \mathbf{M}_A = \frac{d\mathbf{h}_A}{dt} = \left(\frac{d\mathbf{h}_A}{dt}\right)_{/xyz} + \boldsymbol{\Omega} \times \mathbf{h}_A \qquad (19.29)$$

where the notation "/*xyz*" indicates that the derivative of $\mathbf{h}_A$ is to be evaluated relative to the *xyz*-reference frame.

The choice of reference frame should be such that $(d\mathbf{h}_A/dt)_{/xyz}$ can be easily evaluated. In particular, we wish to avoid having to evaluate the time derivatives of moments and products of inertia. These complications can be eliminated by letting the *xyz*-axes be a *body frame*—that is, a reference frame that is embedded in the body (has the same angular velocity as the body). If $\boldsymbol{\omega}$ is the angular velocity of the body, then for a body frame we have $\boldsymbol{\Omega} = \boldsymbol{\omega}$. Consequently, the moment equation, Eq. (19.29), becomes

$$\Sigma \mathbf{M}_A = \frac{d\mathbf{h}_A}{dt} = \left(\frac{d\mathbf{h}_A}{dt}\right)_{/xyz} + \boldsymbol{\omega} \times \mathbf{h}_A \qquad (19.30)$$

According to Eqs. (19.17a), the components of the angular momentum about a fixed point, or about the mass center, are

$$h_x = I_x \omega_x - I_{xy} \omega_y - I_{xz} \omega_z$$
$$h_y = -I_{yx} \omega_x + I_y \omega_y - I_{yz} \omega_z \qquad (19.31)$$
$$h_z = -I_{zx} \omega_x - I_{zy} \omega_y + I_z \omega_z$$

If the *xyz*-axes constitute a body frame, then the moments and products of inertia do not vary with time, and Eq. (19.30) becomes

$$\Sigma \mathbf{M}_A = (I_x \dot{\omega}_x - I_{xy} \dot{\omega}_y - I_{xz} \dot{\omega}_z)\mathbf{i}$$
$$+ (-I_{yx} \dot{\omega}_x + I_y \dot{\omega}_y - I_{yz} \dot{\omega}_z)\mathbf{j}$$
$$+ (-I_{zx} \dot{\omega}_x - I_{zy} \dot{\omega}_y + I_z \dot{\omega}_z)\mathbf{k}$$
$$+ \begin{vmatrix} \mathbf{i} & \mathbf{j} & \mathbf{k} \\ \omega_x & \omega_y & \omega_z \\ h_x & h_y & h_z \end{vmatrix}$$

Substituting for the components of the angular momentum from Eqs. (19.31) and expanding the determinant (also recalling that $I_{xy} = I_{yx}$, $I_{yz} = I_{zy}$, and $I_{xz} = I_{zx}$), the scalar components of the moment equation are

$$\Sigma M_x = I_x\dot{\omega}_x + \omega_y\omega_z(I_z - I_y) + I_{xy}(\omega_z\omega_x - \dot{\omega}_y)$$
$$- I_{xz}(\dot{\omega}_z + \omega_x\omega_y) - I_{yz}(\omega_y^2 - \omega_z^2)$$
$$\Sigma M_y = I_y\dot{\omega}_y + \omega_z\omega_x(I_x - I_z) + I_{yz}(\omega_x\omega_y - \dot{\omega}_z)$$
$$- I_{xy}(\dot{\omega}_x + \omega_z\omega_y) - I_{xz}(\omega_z^2 - \omega_x^2)$$
$$\Sigma M_z = I_z\dot{\omega}_z + \omega_x\omega_y(I_y - I_x) + I_{xz}(\omega_y\omega_z - \dot{\omega}_x)$$
$$- I_{yz}(\dot{\omega}_y + \omega_x\omega_z) - I_{xy}(\omega_x^2 - \omega_y^2)$$

(19.32)

Equations (19.32) are first-order, nonlinear differential equations, which are very difficult to solve analytically, except for a few special cases.

b.   *Euler's equations*

The moment equations simplify somewhat if we choose the *xyz*-axes to be *principal axes of inertia* at the reference point. Then the products of inertia vanish, and Eqs. (19.32) reduce to

$$\Sigma M_x = I_x\dot{\omega}_x + \omega_y\omega_z(I_z - I_y)$$
$$\Sigma M_y = I_y\dot{\omega}_y + \omega_z\omega_x(I_x - I_z)$$
$$\Sigma M_z = I_z\dot{\omega}_z + \omega_x\omega_y(I_y - I_x)$$

(19.33)

These equations, known as *Euler's equations,* are among the more useful equations in rigid-body dynamics. When using Eqs. (19.33), it must be remembered that the *xyz*-axes are a body frame and coincide with the principal axes of inertia at the reference point (fixed point or mass center).

If the angular velocity of the body and its time derivative are known, the resultant external moment applied to the body can be calculated in a straightforward manner from Eqs. (19.33). However, if the resultant moment is known, the equations must be integrated in order to find the angular velocities, a task that must be performed numerically in most problems. An exception is the special case where the angular velocity **ω** is constant, when Eqs. (19.33) take the form of algebraic rather than differential equations.

c.   *Modified Euler's equations*

An important problem in dynamics is the motion of an *axisymmetric* rigid body, such as a spinning top or gyroscope. Consider a body that possesses an axis of rotational symmetry, and let one of the coordinate axes be embedded in the body so that it will always coincide with the axis of symmetry. We allow the coordinate axes to rotate with an angular velocity that is different from that of the body. Because only one of the coordinate axes is embedded in the body, the axes are not a body frame. However, due to the symmetry of the body, each of the coordinate axes will always be a principal axis of inertia.

As an illustration, consider the axisymmetric top in Fig. 19.10, where the *z*-axis was chosen to be the axis of symmetry. It is seen that, no matter how the

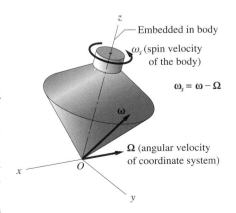

Fig. 19.10

coordinate system rotates, the *xyz*-axes will always be principal axes of inertia at point *O* if the *z*-axis remains embedded in the body.

As before, we let $\mathbf{\Omega} = \Omega_x\mathbf{i} + \Omega_y\mathbf{j} + \Omega_z\mathbf{k}$ be the angular velocity of the *xyz*-reference frame, and we let $\mathbf{\omega} = \omega_x\mathbf{i} + \omega_y\mathbf{j} + \omega_z\mathbf{k}$ be the angular velocity of the body. These angular velocities will differ by the angular velocity of the body relative to the *xyz*-frame, called the *spin velocity*. The spin velocity vector will be parallel to the axis that is embedded in the body—that is, the axis of symmetry. For example, if *z* is the axis of symmetry of the body, as in Fig. 19.10, the spin velocity is $\mathbf{\omega}_s = \omega_s\mathbf{k}$. Consequently, $\omega_x = \Omega_x$, $\omega_y = \Omega_y$, and $\omega_z = \Omega_z + \omega_s$.

The moment equation of motion given in Eq. (19.29) is

$$\Sigma\mathbf{M}_A = \left(\frac{d\mathbf{h}_A}{dt}\right)_{/xyz} + \mathbf{\Omega} \times \mathbf{h}_A \qquad \text{(19.29, repeated)}$$

Because the *xyz*-axes are principal axes, the components of the angular momentum of the body about the reference point *A* (fixed point or mass center) are

$$h_x = I_x\omega_x \qquad h_y = I_y\omega_y \qquad h_z = I_z\omega_z \qquad \text{(a)}$$

Then the cross product in Eq. (19.29) becomes

$$\mathbf{\Omega} \times \mathbf{h}_A = \begin{vmatrix} \mathbf{i} & \mathbf{j} & \mathbf{k} \\ \Omega_x & \Omega_y & \Omega_z \\ I_x\omega_x & I_y\omega_y & I_z\omega_z \end{vmatrix} \qquad \text{(b)}$$

Using Eqs. (a) and (b), and the fact that the moments of inertia are independent of time, the scalar components of Eq. (19.29) are

$$\Sigma M_x = I_x\dot{\omega}_x + I_z\Omega_y\omega_z - I_y\Omega_z\omega_y$$
$$\Sigma M_y = I_y\dot{\omega}_y - I_z\Omega_x\omega_z + I_x\Omega_z\omega_x \qquad \text{(19.34)}$$
$$\Sigma M_z = I_z\dot{\omega}_z + I_y\Omega_x\omega_y - I_x\Omega_y\omega_x$$

These equations are called the *modified Euler's equations*. Observe that the moments ΣM_x, ΣM_y, and ΣM_z are referred to *xyz*-axes, which do not represent a body frame.

d.   *A note on angular acceleration*

The terms $\dot{\omega}_x$, $\dot{\omega}_y$, and $\dot{\omega}_z$ that appear in the various moment equations are not necessarily the components of the angular acceleration vector of the body. In order to clarify this statement, let us consider first the case in which the *xyz*-axes are a body frame, and then examine the situation in which the body spins relative to the *xyz*-frame.

1. The *xyz*-axes form a body frame: $\boldsymbol{\Omega} = \boldsymbol{\omega}$ (Euler's equations).

   Recall that the absolute derivative of a vector $\mathbf{V}$ is $(d\mathbf{V}/dt) = (d\mathbf{V}/dt)_{/xyz} + \boldsymbol{\omega} \times \mathbf{V}$, where $\boldsymbol{\omega}$ is the angular velocity of the *xyz*-frame. Therefore, the absolute derivative of the angular velocity vector $\boldsymbol{\omega}$ (that is, the angular acceleration) of the body becomes

$$\frac{d\boldsymbol{\omega}}{dt} = \left(\frac{d\boldsymbol{\omega}}{dt}\right)_{/xyz} + \boldsymbol{\omega} \times \boldsymbol{\omega} = \left(\frac{d\boldsymbol{\omega}}{dt}\right)_{/xyz}$$

This result means that the absolute derivative of the angular velocity vector is identical to its derivative relative to the body frame. It follows that if $\boldsymbol{\omega} = \omega_x \mathbf{i} + \omega_y \mathbf{j} + \omega_z \mathbf{k}$, then

$$\dot{\boldsymbol{\omega}} = \dot{\omega}_x \mathbf{i} + \dot{\omega}_y \mathbf{j} + \dot{\omega}_z \mathbf{k}$$

which gives

$$(\dot{\boldsymbol{\omega}})_x = \dot{\omega}_x \qquad (\dot{\boldsymbol{\omega}})_y = \dot{\omega}_y \qquad (\dot{\boldsymbol{\omega}})_z = \dot{\omega}_z \qquad (19.35)$$

Therefore, $\dot{\omega}_x$, $\dot{\omega}_y$, and $\dot{\omega}_z$ are the components of the angular acceleration $\dot{\boldsymbol{\omega}}$ of the body, a conclusion that can be very useful when applying Euler's equations.

2. The body spins relative to the *xyz*-axes: $\boldsymbol{\Omega} \neq \boldsymbol{\omega}$ (modified Euler's equations).

   In this case, the angular acceleration vector of the body becomes $d\boldsymbol{\omega}/dt = (d\boldsymbol{\omega}/dt)_{/xyz} + \boldsymbol{\Omega} \times \boldsymbol{\omega}$, which may be written in the form

$$\dot{\boldsymbol{\omega}} = \dot{\omega}_x \mathbf{i} + \dot{\omega}_y \mathbf{j} + \dot{\omega}_z \mathbf{k} + \begin{vmatrix} \mathbf{i} & \mathbf{j} & \mathbf{k} \\ \Omega_x & \Omega_y & \Omega_z \\ \omega_x & \omega_y & \omega_z \end{vmatrix}$$

Expanding the above determinant, and equating like components, yields

$$(\dot{\boldsymbol{\omega}})_x = \dot{\omega}_x + (\Omega_y \omega_z - \Omega_z \omega_y)$$

$$(\dot{\boldsymbol{\omega}})_y = \dot{\omega}_y - (\Omega_x \omega_z - \Omega_z \omega_x) \qquad (19.36)$$

$$(\dot{\boldsymbol{\omega}})_z = \dot{\omega}_z + (\Omega_x \omega_y - \Omega_y \omega_x)$$

It can be seen here that the components of the angular acceleration vector $\dot{\boldsymbol{\omega}}$ are not equal to the time derivatives of ω_x, ω_y, and ω_z.

e.   *Plane motion*

Let the *xyz*-axes be a body frame with the *z*-axis remaining perpendicular to a fixed plane. In this case, $\boldsymbol{\omega} = \omega_z \mathbf{k}$ and $\omega_x = \omega_y = 0$, and the moment equations, Eqs. (19.32), become

$$\sum M_x = -I_{xz}\dot{\omega}_z + I_{yz}\omega_z^2$$
$$\sum M_y = -I_{yz}\dot{\omega}_z - I_{xz}\omega_z^2 \qquad\qquad (19.37)$$
$$\sum M_z = I_z\dot{\omega}_z$$

If, in addition, the *z*-axis is a principal axis of inertia of the body, these equations further simplify to

$$\sum M_x = 0 \qquad \sum M_y = 0 \qquad \sum M_z = I_z\dot{\omega}_z \qquad (19.38)$$

These equations are identical to the moment equations for plane motion discussed in Chapter 17.

f.   *Rotation about a fixed axis*

Equations (19.37) and (19.38) are, of course, also valid for the special case in which a body is rotating about an axis that is fixed in space. It is instructive to consider the case in which a body is mounted on a shaft that is supported by a bearing at each end. Let the shaft coincide with the *z*-axis, assumed to be fixed in space, and the body rotate with *constant* angular speed ω_z. If the *xyz*-body axes are not principal axes of inertia, Eqs. (19.37) yield (with $\dot{\omega}_z = 0$)

$$\sum M_x = I_{yz}\omega_z^2 \qquad \sum M_y = -I_{xz}\omega_z^2 \qquad \sum M_z = 0 \qquad (19.39)$$

In this case, we see that the reactions at the two bearings must provide the moments $\sum M_x$ and $\sum M_y$. These bearing reactions, which rotate with the *xyz*-reference frame (that is, with the body), are called *dynamic bearing reactions,* and the body is said to be *dynamically unbalanced.* Because the dynamic bearing reactions are proportional to the square of the angular speed, they can reach large magnitudes and cause severe vibrations of the body. To dynamically balance the body, the products of inertia in Eq. (19.39) must be made to vanish. This can be accomplished by redistributing the mass of the body or by adding additional mass to the body at appropriate locations. A common example of the latter method is the addition of small weights to the rim of an automobile wheel when it is being "spin balanced."

Sample Problem 19.8

The uniform slender bar AB in Fig. (a) mass 10 kg. It is connected to the vertical shaft OC by a ball-and-socket joint at A. The assembly rotates with the constant angular velocity ω, about OC, which causes AB to be inclined at 40° with the vertical. Determine ω and the magnitude of the reaction at A.

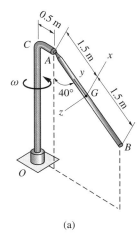

(a)

Solution

Our solution will employ the xyz-coordinate system shown in Fig. (a). This reference frame is embedded in bar AB with the origin located at the mass center G of the bar. The x- and z-axes are perpendicular to the bar (the x-axis lies in the plane CAB), whereas the y-axis is directed along the bar. Hence the coordinate axes are principal axes of the bar at G.

The free-body diagram of the bar, displaying only the forces acting in the xy-plane, is shown in Fig. (b). The only forces acting on the bar are its 10-kg weight and the ball-and-socket reactions at A (the component A_z is perpendicular to the paper).

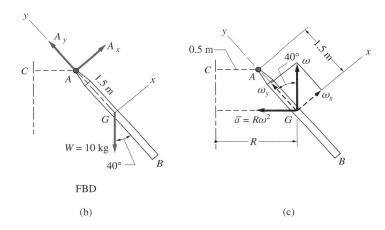

FBD

(b)

(c)

The kinematics of the bar are illustrated in Fig. (c). We see that the components of the angular velocity vector of the bar are

$$\omega_x = \omega \sin 40° = 0.6428\omega \qquad \omega_y = \omega \cos 40° = 0.7660\omega \qquad \omega_z = 0$$

We also observe that the path of G is a horizontal circle of radius $R = 0.5 + 1.5 \sin 40° = 1.4642$ m. Therefore, the acceleration of G is $\bar{a} = R\omega^2 = 1.4642\omega^2$, directed as shown in Fig. (c).

Referring to Figs. (b) and (c), the force equations of motion are

$$\Sigma F_x = m\bar{a}_x \quad +\nearrow \quad A_x - W \sin 40° = -mR\omega^2 \cos 40°$$

$$A_x - 10(9.8) \sin 40° = -10(1.4642)\omega^2 \cos 40°$$

$$A_x = 62.99 - 11.22\omega^2 \text{ N} \qquad\qquad \text{(a)}$$

$$\Sigma F_y = m\bar{a}_y \quad \nwarrow \quad A_y - W\cos 40° = mR\omega^2 \sin 40°$$

$$A_y - 10(9.8)\cos 40° = 10(1.4642)\omega^2 \sin 40°$$

$$A_y = 75.07 + 9.4117\omega^2 \text{ N} \qquad \text{(b)}$$

$$\Sigma F_z = m\bar{a}_z \qquad A_z = 0 \qquad \text{(c)}$$

For the moment equations of motion, we use the Euler's equations in Eqs. (19.33). The principal moments of inertia at G are

$$I_x = I_z = \frac{mL^2}{12} = \frac{10(3)^2}{12} = 7.5\,\text{kg}\cdot\text{m}^2 \qquad I_y \approx 0$$

Because $\dot{\omega}_x = \dot{\omega}_y = \dot{\omega}_z = \omega_z = 0$, and because there are no moments acting about either the x- or y-axis, the first two Euler's equations,

$$\Sigma M_x = I_x\dot{\omega}_x + \omega_y\omega_z(I_z - I_y)$$
$$\Sigma M_y = I_y\dot{\omega}_y + \omega_z\omega_x(I_x - I_z)$$

are trivially satisfied (each yields $0 = 0$). The third Euler's equation is

$$\Sigma M_z = I_z\dot{\omega}_z + \omega_x\omega_y(I_y - I_x)$$
$$-1.5A_x = 0 + (0.6428\omega)(0.7660\omega)(0 - 7.5)$$
$$A_x = 2.4619\omega^2 \qquad \text{(d)}$$

Solution of Eqs. (a) and (d) is $A_x = 11.38$ N and

$$\omega = 2.15 \text{ rad/s} \qquad \qquad \textit{Answer}$$

Substituting this result in Eq. (b) and solving for A_y yields $A_y = 118.58$ N. Therefore, the magnitude of the reaction at A is

$$A = \sqrt{A_x^2 + A_y^2 + A_z^2} = \sqrt{(11.38)^2 + (118.58)^2 + 0} = 119.12 \text{ N} \quad \textit{Answer}$$

Sample Problem *19.9*

Figure (a) shows a body that is formed by welding together three uniform rods labeled 1, 2, and 3, each with the mass shown. (This body also appeared in Sample Problem 19.5.) Support for the body is provided by smooth bearings at A and B, with only the bearing at A being capable of providing an axial thrust. A small motor at A (not shown) is driving the body at the constant angular velocity ω_0 rad/s. For the position shown in Fig. (a), determine (1) the output torque T; (2) the bearing reactions at A and B; and (3) the dynamic bearing reactions (the reactions caused by the rotation of the body).

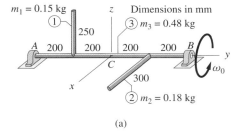

(a)

Solution

Part 1

The free-body diagram of the body, shown in Fig. (b), displays the bearing reactions at A and B; the weight of each rod: $W_1 = 0.15(9.8) = 1.47$ N, $W_2 = 0.18(9.8) = 1.764$ N, $W_3 = 0.48(9.8) = 4.704$ N; and the output torque T of the motor required to drive the body at constant angular velocity. The origin of the xyz-reference frame, which is assumed to rotate with the body, is attached to C, the midpoint of rod 3.

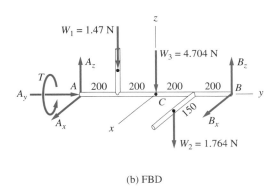

(b) FBD

Because the rotation occurs about a fixed axis, we can use Eqs. (19.39). However, we must first convert Eqs. (19.39) from rotation about the z-axis, which was assumed in their derivation, to rotation about the y-axis. This conversion can be accomplished with the aid of the following diagram.

From this diagram, we see that Eqs. (19.39) must be changed by replacing x by z, y by x, and z by y. The results are

$$\Sigma M_x = -I_{yz}\omega_y^2 \qquad \Sigma M_y = 0 \qquad \Sigma M_z = I_{xy}\omega_y^2$$

The products of inertia about point C were computed in Sample Problem 19.5: $I_{xy} = 0.005\,40$ kg $\cdot$ m^2 and $I_{yz} = -0.003\,75$ kg $\cdot$ m^2. Consequently, referring to the FBD in Fig. (b) and using $\omega_y = \omega_0$, the moment equations of motion become

$$\Sigma M_x = -I_{yz}\omega_0^2 \qquad -A_z(0.400) + B_z(0.400) + 1.47(0.200)$$
$$-1.764(0.200) = -\left(-0.003\,75\omega_0^2\right) \qquad \text{(a)}$$

$$\Sigma M_y = 0 \qquad T + 1.764(0.150) = 0 \qquad \text{(b)}$$

$$\Sigma M_z = I_{xy}\omega_0^2 \qquad A_x(0.400) - B_x(0.400) = 0.005\,40\omega_0^2 \qquad \text{(c)}$$

From Eq. (b), the output torque of the motor for the position shown is

$$T = -0.2646 \text{ N} \cdot \text{m} \qquad\qquad \textit{Answer}$$

Note that the magnitude of this torque will vary as the position of the body changes.

Part 2

We next apply the force equation of motion $\Sigma \mathbf{F} = m\bar{\mathbf{a}}$ to the body, where m is the mass of the body and $\bar{\mathbf{a}}$ is the acceleration of its mass center. The inertia vector for the body is the sum of the inertia vectors of the individual rods. Because the mass center of rod 1 moves with constant speed on a circle centered on the y-axis, the acceleration of its mass center consists of the normal acceleration $\bar{a}_n = r\omega_0^2$; that is, $\bar{\mathbf{a}}_1 = -0.125\omega_0^2\mathbf{k}$. Similarly, for rod 2 we get $\bar{\mathbf{a}}_2 = -0.150\omega_0^2\mathbf{i}$. The mass center of rod 3 is stationary. Therefore, the inertia vector for the body becomes

$$m\bar{\mathbf{a}} = m_1\bar{\mathbf{a}}_1 + m_2\bar{\mathbf{a}}_2$$
$$= 0.15\left(-0.125\omega_0^2\right)\mathbf{k} + 0.18\left(-0.150\omega_0^2\right)\mathbf{i}$$
$$= \left(-0.018\,75\omega_0^2\right)\mathbf{k} - \left(0.027\,00\omega_0^2\right)\mathbf{i} \qquad \text{(d)}$$

Using these results and the FBD in Fig. (b), the force equations of motion become

$$\Sigma F_x = m\bar{a}_x \qquad A_x + B_x = -0.027\,00\omega_0^2 \qquad \text{(e)}$$

$$\Sigma F_y = m\bar{a}_y \qquad A_y = 0 \qquad\qquad \textit{Answer} \quad \text{(f)}$$

$$\Sigma F_z = m\bar{a}_z \qquad A_z + B_z - 1.47 - 1.764 - 4.704$$
$$= -0.018\,75\omega_0^2 \qquad \text{(g)}$$

Solving Eqs. (a) and (g), and Eqs. (c) and (e), the x- and z-components of the bearing reactions in Fig. (b) are found to be

$$A_x = -0.006\,75\omega_0^2 \text{ N}$$
$$B_x = -0.020\,25\omega_0^2 \text{ N}$$
$$A_z = -0.014\,06\omega_0^2 + 3.900 \text{ N}$$
$$B_z = -0.004\,688\omega_0^2 + 4.047 \text{ N}$$

Answer

Part 3

The components of the bearing reactions apply only when the body is in the given position. However, it can be seen that the terms with ω_0^2 are valid for all positions of the body. These terms, called the dynamic bearing reactions, are caused by the rotation of the body. Figure (c) shows the dynamic bearing reactions together with the inertia vectors for bars 1 and 2. Each vector shown in Fig. (c) rotates with the xyz-axes—that is, with the body.

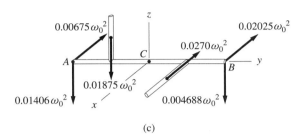

$0.00675\,\omega_0^2$

$0.02025\,\omega_0^2$

$0.0270\,\omega_0^2$

$0.01875\,\omega_0^2$

$0.01406\,\omega_0^2$

$0.004688\,\omega_0^2$

(c)

The so-called *static bearing reactions* (the terms that are independent of ω_0) support the weight of the body and do not rotate with the coordinate system. However, their magnitudes vary as the position of the body changes. The total reaction at a bearing is found by adding the static and dynamic bearing reactions.

Sample Problem 19.10

The platform A is rotating about the fixed vertical axis with an angular velocity ω_2 and angular acceleration $\dot{\omega}_2$. At the same time, an internal motor (not shown) spins the uniform disk B about its axle, which is rigidly mounted to the platform, at the angular velocity ω_1 and angular acceleration $\dot{\omega}_1$. Determine the components of the couple that is applied to the disk B by its axle.

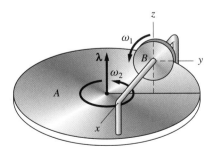

Solution

We solve this problem using two different reference frames. First, the xyz-axes are embedded in disk B (Euler's equations); then the xyz-axes are embedded in the axle—that is, in the platform extended (modified Euler's equations).

Method I: *xyz* Rotating with the Disk

The *xyz*-axes shown in the figure are principal axes at the mass center of disk B. If these axes are assumed to rotate with the disk, they constitute a body frame, which means that Euler's equations can be used.

It is convenient to write the angular velocity $\boldsymbol{\omega}$ of disk B as

$$\boldsymbol{\omega} = \omega_1 \mathbf{i} + \omega_2 \boldsymbol{\lambda} \qquad \text{(valid for all time)} \qquad \text{(a)}$$

where $\boldsymbol{\lambda}$ is a unit vector directed perpendicular to the platform. Because Eq. (a) is valid for all time, it can be differentiated, resulting in the angular acceleration of disk B

$$\dot{\boldsymbol{\omega}} = \dot{\omega}_1 \mathbf{i} + \omega_1 \dot{\mathbf{i}} + \dot{\omega}_2 \boldsymbol{\lambda} + \omega_2 \dot{\boldsymbol{\lambda}} \qquad \text{(b)}$$

We note that $\dot{\boldsymbol{\lambda}} = \mathbf{0}$. Furthermore, at the instant shown, we have $\boldsymbol{\lambda} = \mathbf{k}$, and $\dot{\mathbf{i}} = \boldsymbol{\omega} \times \mathbf{i} = (\omega_1 \mathbf{i} + \omega_2 \mathbf{k}) \times \mathbf{i} = \omega_2 \mathbf{j}$. Therefore, Eq. (b) becomes

$$\dot{\boldsymbol{\omega}} = \dot{\omega}_1 \mathbf{i} + \dot{\omega}_2 \mathbf{k} + \omega_1 \omega_2 \mathbf{j} \qquad \text{(at this instant)} \qquad \text{(c)}$$

Because the *xyz*-axes constitute a body frame, we have, according to Eqs. (19.35), $\dot{\omega}_x = (\dot{\boldsymbol{\omega}})_x$, $\dot{\omega}_y = (\dot{\boldsymbol{\omega}})_y$, and $\dot{\omega}_z = (\dot{\boldsymbol{\omega}})_z$. It follows from Eqs. (a) and (c) that the components of $\boldsymbol{\omega}$ and $\dot{\boldsymbol{\omega}}$ that are to be substituted into Euler's equations are

$$\omega_x = \omega_1 \qquad \dot{\omega}_x = \dot{\omega}_1$$

$$\omega_y = 0 \qquad \dot{\omega}_y = \omega_1 \omega_2 \qquad \text{(d)}$$

$$\omega_z = \omega_2 \qquad \dot{\omega}_z = \dot{\omega}_2$$

Substituting Eqs. (d) into Eqs. (19.33) and recognizing that $I_y = I_z$ for disk B, we obtain the moments acting on the disk about its mass center (the origin of the coordinate system).

$$\Sigma M_x = I_x \dot{\omega}_x + \omega_y \omega_z (I_z - I_y) = I_x \dot{\omega}_1$$

$$\Sigma M_y = I_y \dot{\omega}_y + \omega_z \omega_x (I_x - I_z) = I_y \omega_1 \omega_2 + \omega_1 \omega_2 (I_x - I_z) = I_x \omega_1 \omega_2 \quad \textit{Answer}$$

$$\Sigma M_z = I_z \dot{\omega}_z + \omega_x \omega_y (I_y - I_x) = I_z \dot{\omega}_2$$

Because the only support for disk B is provided by its axle, these moments represent the components of the couple that is exerted on the disk by the axle. The moment about the *x*-axis must be applied by the motor, whereas the moments about the *y*- and *z*-axes are provided by bearings that support the disk. Because the coordinate axes rotate with the disk, the above expressions for the moments are valid only in the position shown in the figure.

Method II: *xyz* Rotating with the Platform

Because disk B is spinning about its axis of symmetry (the x-axis), modified Euler's equations are applicable. Assuming that the xyz-axes in the figure are embedded in the platform A, the angular velocity of the reference frame is

$$\boldsymbol{\Omega} = \omega_2 \mathbf{k} \qquad \text{(valid for all time)} \qquad \text{(e)}$$

and the angular velocity of disk B becomes

$$\boldsymbol{\omega} = \omega_1 \mathbf{i} + \omega_2 \mathbf{k} \qquad \text{(valid for all time)} \qquad \text{(f)}$$

The components of $\boldsymbol{\omega}$, namely $\omega_x = \omega_1$, $\omega_y = 0$, and $\omega_z = \omega_2$, are differentiable because Eq. (f) is valid for all time. Therefore, the components of $\boldsymbol{\omega}$, $\dot{\boldsymbol{\omega}}$, and $\boldsymbol{\Omega}$ that are to be substituted into the modified Euler's equations—Eqs. (19.34)—are

$$
\begin{array}{lll}
\omega_x = \omega_1 & \dot{\omega}_x = \dot{\omega}_1 & \Omega_x = 0 \\[2mm]
\omega_y = 0 & \dot{\omega}_y = 0 & \Omega_y = 0 \qquad \text{(g)} \\[2mm]
\omega_z = \omega_2 & \dot{\omega}_z = \dot{\omega}_2 & \Omega_z = \omega_2
\end{array}
$$

Completing the substitutions, Eqs. (19.34) yield

$$\Sigma M_x = I_x \dot{\omega}_x + I_z \Omega_y \omega_z - I_y \Omega_z \omega_y = I_x \dot{\omega}_1$$

$$\Sigma M_y = I_y \dot{\omega}_y - I_z \Omega_x \omega_z + I_x \Omega_z \omega_x = I_x \omega_1 \omega_2 \qquad \textit{Answer}$$

$$\Sigma M_z = I_z \dot{\omega}_z + I_y \Omega_x \omega_y - I_x \Omega_y \omega_x = I_z \dot{\omega}_2$$

These moments are, of course, identical to those obtained previously by Method I. The difference is that the foregoing expressions are valid for all time, not just for the position shown in the figure.

Problems

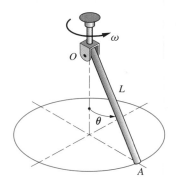

Fig. P19.46

19.46   The uniform slender bar *OA* of mass *m* and length *L* is attached to the vertical shaft with a clevis. The assembly is rotating about the shaft with the constant angular velocity ω. Determine the angle θ between *OA* and the vertical.

19.47   The uniform slender rod *AB*, mass 16 kg, is welded at a 60° angle to the midpoint of the thin shaft *CD*. Determine the magnitudes of the dynamic bearing reactions at *C* and *D* when the assembly is rotating at the constant angular velocity $\omega = 6$ rad/s.

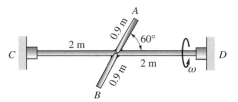

Fig. P19.47

19.48   The thin uniform disk of mass *m* and radius *R* is mounted at *O* on the end of a vertical shaft. The plane of the disk is inclined at the angle β to the horizontal. Determine the dynamic reactions acting on the disk at *O* when it is rotating about the *z*-axis with the constant angular velocity ω_0.

19.49   Repeat Prob. 19.48 assuming that a small mass $m_A = m/16$ is attached to the rim of the disk at *A*.

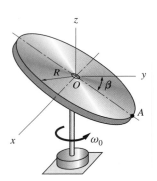

Fig. P19.48, P19.49

19.50   The uniform rod *AB* of mass *m* is rigidly attached to the arm *OA*. The assembly is rotating with the angular velocity ω about the vertical axis at *O*. Determine the magnitude of the couple that is exerted by arm *OA* on the rod *AB* at *A*.

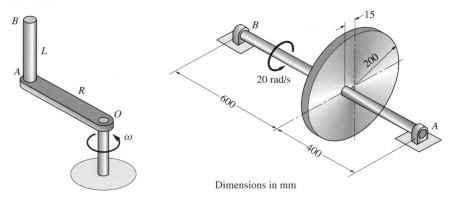

Fig. P19.50

Fig. P19.51

19.51   The 24-kg homogeneous disk is mounted on the shaft *AB* with an eccentricity of 15 mm. If the disk is rotating at the angular speed of 20 rad/s, determine the magnitudes of the dynamic bearing reactions at *A* and *B*.

19.52   The crank is made of a uniform slender rod of total mass 8 kg. If the crank rotates about the *y*-axis with the constant angular speed $\omega = 40$ rad/s, find the dynamic bearing reactions acting on the crank at *A* and *B*.

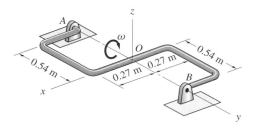

Fig. P19.52

19.53   The homogeneous 1.8-kg bent thin plate can rotate freely about the axis *AB*. A couple $\mathbf{C} = -0.8\mathbf{k}$ N · m is applied to the plate when it is at rest in the position shown. Determine the angular acceleration of the plate and the bearing reactions at *A* and *B* immediately after the couple is applied.

19.54   The plate described in Prob. 19.53 is rotating about the axis *AB* with the constant angular velocity $\boldsymbol{\omega} = 10\mathbf{k}$ rad/s. Calculate the dynamic bearing reactions at *A* and *B*.

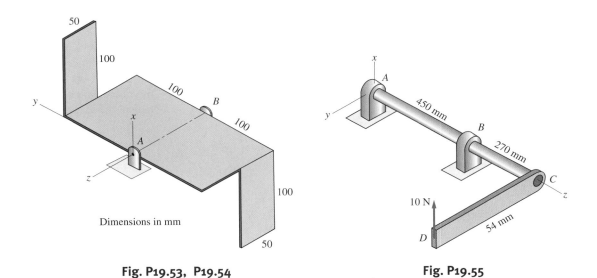

Fig. P19.53, P19.54   **Fig. P19.55**

19.55   The slender 2-kg uniform bar *CD* is rigidly attached to the shaft *ABC*. The assembly is at rest in the position shown when the 10 N vertical force is applied at *D*. Determine the bearing reactions at *A* and *B*, and the angular acceleration of the shaft for this position. Neglect the weight of the shaft.

19.56   The 0.15-kg mass of the wheel is concentrated primarily in its thin, uniform rim of mean radius $R = 80$ mm. The wheel spins about the axle at *O* at the constant angular speed $\omega_1 = 180$ rad/s. At the same time, the mounting fork is

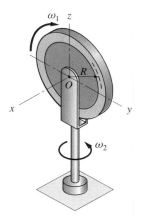

Fig. P19.56

rotating about the z-axis at the constant angular speed $\omega_2 = 40$ rad/s. Calculate the couple exerted on the wheel by the axle.

19.57 The 9-kg uniform disk spins about the axle AG with the constant angular velocity of 20 rad/s. The axle is supported by a ball-and-socket joint at A, and it rotates about the vertical axis with the constant angular velocity ω_1. Find the value of ω_1 for which the axle will remain horizontal during the motion. Neglect the weight of the axle.

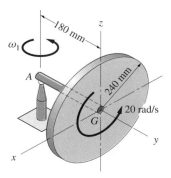

Fig. P19.57

19.58 The uniform slender rod OA of mass m is attached to the vertical shaft with a clevis. Determine the constant angular velocity ω_0 of the shaft for which the rod will maintain a constant angle $\beta = 25°$ with the vertical.

19.59 The position of the 1.2-kg uniform slender rod OA is controlled by two small electric motors. A motor at B rotates the vertical shaft at the constant angular speed of 1.8 rad/s, and a motor at O increases the angle β between OA and the vertical at the constant rate $d\beta/dt = 1.5$ rad/s. Compute the output torque of each motor when the rod is in the position $\beta = 30°$. Neglect the masses of the motors.

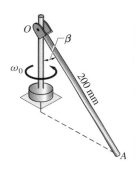

Fig. P19.58

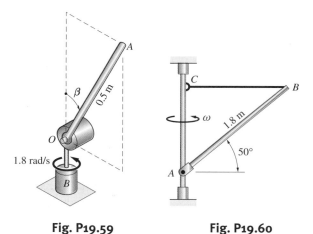

Fig. P19.59   **Fig. P19.60**

19.60 The homogeneous slender rod AB of mass 20 kg is connected to the vertical shaft with a clevis at A and the horizontal cable BC. Find the tension in the cable when the assembly is rotating at the constant angular velocity $\omega = 6$ rad/s.

19.61   A clevis at A connects the uniform bar AB of mass m to the L-shaped arm. As the arm rotates about the vertical axis, bar AB maintains a 45° angle with the vertical. Determine the angular velocity ω of the arm. Use $L = 1$ m.

19.62   The homogeneous disk of mass m rotates about the L-shaped axle with the angular speed $\omega_1 = 2\omega_0$ relative to the axle. The axle in turn rotates about the vertical axis at A with the angular speed $\omega_2 = \omega_0$. Determine the force-couple system that represents the dynamic bearing reaction at A. Neglect the mass of the axle.

19.63   The uniform 16-kg slender rod PQ spins about the axis BD at the constant angular speed $\dot\theta = 12$ rad/s. At the same time, the bracket AB is rotating about the vertical axis AD at the constant rate of 4 rad/s. The angular speeds are maintained by small electric motors (not shown) at A and D. Calculate the torque that each motor must develop as a function of the angle θ. Neglect the masses of the bracket and the motors.

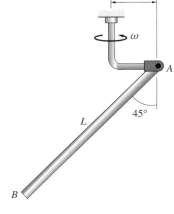

Fig. P19.61

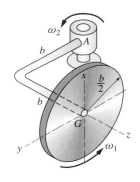

Fig. P19.62

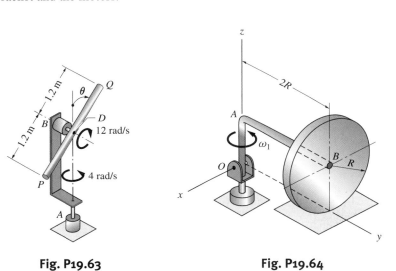

Fig. P19.63   Fig. P19.64

19.64   The uniform disk of radius $R = 0.2$ m and mass $m = 6$ kg is attached to the bent axle OAB of negligible weight. The axle is joined to a vertical shaft with a clevis at O. Assuming that the axle rotates freely about the z-axis with the constant angular velocity $\omega_1 = 6$ rad/s, and that the disk rolls without slipping, determine the vertical force exerted on the wheel by the horizontal surface.

*19.6   Motion of an Axisymmetric Body

In this article we discuss the motion of axisymmetric bodies, which includes several important applications, such as gyroscopes, satellites, and projectiles. The equations of motion employed here are the modified Euler equations, but these equations will be reformulated by introducing new kinematic variables known as Euler's angles.

a.   *Euler's angles and angular velocity*

We recall that in the modified Euler's equations, the rotation of a body has two components: (1) the spin of the body relative to an *xyz*-reference frame, and (2) the rotation of the *xyz*-axes relative to a fixed *XYZ*-coordinate system. If the body is axisymmetric, it is convenient to orient the *xyz*-axes in a special way, as described below and illustrated in Fig. 19.11(a).

- The *z*-axis is chosen as the axis of symmetry of the body. The *Z*-axis (its direction can be chosen arbitrarily) is known as the *invariable line*. The angle θ between the *Z*- and the *z*-axes is called the *nutation angle*.
- The *x*-axis, called the *nodal line*, is oriented so that it always lies in the *XY*-plane. The angle ϕ between the *X*- and the *x*-axes is known as the *precession angle*.

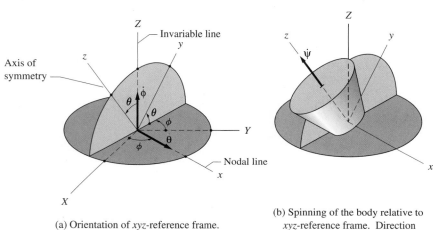

(a) Orientation of *xyz*-reference frame.
Directions of $\dot{\phi}$ and $\dot{\theta}$ shown are positive.

(b) Spinning of the body relative to *xyz*-reference frame. Direction of $\dot{\psi}$ shown is positive.

Fig. 19.11

The angular velocity vectors $\dot{\theta}$ and $\dot{\phi}$ shown in Fig. 19.11(a) are called the rates of nutation and precession, respectively.

Because the *z*-axis is embedded in the body, the rotation of the body relative to the *xyz*-frame is confined to a rotation, or spin, about the *z*-axis. The rate of this rotation, called the rate of spin and denoted by $\dot{\psi}$, is shown in Fig. 19.11(b). The *spin angle* ψ (not shown) can be measured from any convenient reference. The angles ϕ, θ, and ψ, which are called *Euler's angles,* are useful kinematic variables for describing the motion of an axisymmetric body.

Before proceeding, we must reiterate that the *xyz*-axes shown in Fig. 19.11 are not a body frame, because the body is allowed to spin about the *z*-axis. From Fig. 19.11(a), we see that the angular velocity of the *xyz*-reference frame is

$$\mathbf{\Omega} = \dot{\mathbf{\phi}} + \dot{\mathbf{\theta}} \tag{19.40}$$

Because the body spins at the rate $\dot{\psi}$ relative to the *xyz*-frame, its angular velocity is $\boldsymbol{\omega} = \mathbf{\Omega} + \dot{\boldsymbol{\psi}}$, or

$$\boldsymbol{\omega} = \dot{\boldsymbol{\phi}} + \dot{\boldsymbol{\theta}} + \dot{\boldsymbol{\psi}} \tag{19.41}$$

Utilizing Fig. 19.11, we deduce that the components of $\boldsymbol{\Omega}$ and $\boldsymbol{\omega}$ relative to the *xyz*-axes are

$$
\begin{aligned}
\Omega_x &= \dot{\theta} & \omega_x &= \dot{\theta} \\
\Omega_y &= \dot{\phi}\sin\theta & \omega_y &= \dot{\phi}\sin\theta \\
\Omega_z &= \dot{\phi}\cos\theta & \omega_z &= \dot{\phi}\cos\theta + \dot{\psi}
\end{aligned}
\tag{19.42}
$$

b. Moment equations of motion

Because the *z*-axis is an axis of symmetry for the body, we have $I_{xy} = I_{yz} = I_{zx} = 0$ and $I_x = I_y$. From now on, we will use the following notation:

$$
I_x = I_y = I \tag{19.43}
$$

Substituting Eqs. (19.42) and (19.43) into the modified Euler equations, Eqs. (19.34), we obtain

$$
\begin{aligned}
\Sigma M_x &= I\ddot{\theta} + (I_z - I)\dot{\phi}^2\sin\theta\cos\theta + I_z\dot{\phi}\dot{\psi}\sin\theta \\
\Sigma M_y &= I\ddot{\phi}\sin\theta + 2I\dot{\theta}\dot{\phi}\cos\theta - I_z\dot{\theta}(\dot{\psi} + \dot{\theta}\cos\theta) \\
\Sigma M_z &= I_z(\ddot{\psi} + \ddot{\phi}\cos\theta - \dot{\phi}\dot{\theta}\sin\theta) \\
&= I_z\frac{d}{dt}(\dot{\psi} + \dot{\phi}\cos\theta)
\end{aligned}
\tag{19.44}
$$

Sometimes it is convenient to use the following equations, obtained by substituting $\omega_z = \dot{\phi}\cos\theta + \dot{\psi}$ from Eqs. (19.42) into Eqs. (19.44).

$$
\begin{aligned}
\Sigma M_x &= I\ddot{\theta} + I_z\omega_z\dot{\phi}\sin\theta - I\dot{\phi}^2\sin\theta\cos\theta \\
\Sigma M_y &= I\ddot{\phi}\sin\theta + 2I\dot{\theta}\dot{\phi}\cos\theta - I_z\dot{\theta}\omega_z \\
\Sigma M_z &= I_z\dot{\omega}_z
\end{aligned}
\tag{19.45}
$$

When using Eqs. (19.44) or (19.45), recall from Art. 19.5 that the modified Euler's equations are valid only if the origin of the *xyz*-axes is located at the *mass center* of the body, or at a *fixed point* (fixed in the body and in space).

c. Steady precession

The special motion that arises when $\dot{\psi}$, $\dot{\phi}$, and θ are constants is known as *steady precession*. In this case, Eqs. (19.44) simplify considerably:

$$
\Sigma M_x = (I_z - I)\dot{\phi}^2\sin\theta\cos\theta + I_z\dot{\phi}\dot{\psi}\sin\theta
$$
$$
\Sigma M_y = 0 \qquad \Sigma M_z = 0 \tag{19.46}
$$

whereas the equivalent equations of Eqs. (19.45) reduce to

$$
\Sigma M_x = I_z\omega_z\dot{\phi}\sin\theta - I\dot{\phi}^2\sin\theta\cos\theta
$$
$$
\Sigma M_y = 0 \qquad \Sigma M_z = 0 \tag{19.47}
$$

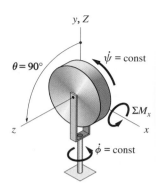

Fig. 19.12

From either Eqs. (19.46) or (19.47) we see that for a body to undergo steady precession, it must be acted upon by forces that provide a constant moment about the x-axis (the nodal line) with no moments acting about the other two axes. Therefore, the direction of the moment vector must be perpendicular to both the precession axis (Z) and the spin axis (z).

An interesting special case of steady precession occurs when the precession axis (Z) is perpendicular to the spin axis (z), as depicted in Fig. 19.12. Setting $\theta = 90°$ in the first of Eqs. (19.46), we find that the moment required to maintain the steady precession is

$$\Sigma M_x = I_z \dot{\phi} \dot{\psi} \tag{19.48}$$

d.   Torque-free motion

If the resultant moment of the external forces about the mass center is zero, the body is said to undergo torque-free motion. *Torque-free motion* is thus characterized by $\Sigma \mathbf{M}_G = d\mathbf{h}_G/dt = \mathbf{0}$, from which we conclude that the angular momentum of the body about its mass center G remains constant in both magnitude and direction. Examples of torque-free motion are projectiles (with air resistance neglected) and space vehicles in unpowered flight.

Figure 19.13 shows an axisymmetric projectile in free flight. For mathematical convenience, it is customary to choose the fixed Z-axis to be in the direction of $\mathbf{h}_G$. The xyz-coordinate system is attached to G, with z being the axis of symmetry of the body. In accordance with Fig. 19.11, the x-axis is perpendicular to the plane formed by the Z- and z-axes. Although the z-axis is embedded in the body, it is important to recall that the xyz-axes do not constitute a body frame, because the body can rotate (spin) about the z-axis relative to the xyz-frame.

As shown in Fig. 19.13, we let β be the angle between the angular velocity vector $\boldsymbol{\omega}$ of the body and the z-axis. The angle θ between the Z- and z-axes is the Euler angle that was defined in Fig. 19.11. The components of $\mathbf{h}_G$ relative to the xyz-axes thus are

$$h_x = 0 \qquad h_y = h_G \sin \theta \qquad h_z = h_G \cos \theta \tag{19.49}$$

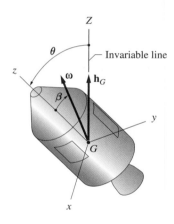

Fig. 19.13

Due to symmetry, the xyz-axes are principal axes of inertia of the projectile at G. Using the notation $\bar{I}_x = \bar{I}_y = \bar{I}$, the components of $\mathbf{h}_G$ take the form [see Eq. (19.18)]

$$h_x = \bar{I} \omega_x \qquad h_y = \bar{I} \omega_y \qquad h_z = \bar{I}_z \omega_z \tag{19.50}$$

Comparing Eqs. (19.49) and (19.50), we conclude that

$$\omega_x = 0 \qquad \omega_y = \frac{h_G \sin \theta}{\bar{I}} \qquad \omega_z = \frac{h_G \cos \theta}{\bar{I}_z} \tag{19.51}$$

Because $\omega_x = \dot{\theta}$ according to Eq. (19.42), we conclude that θ remains constant, which indicates that the motion is a steady precession about the Z-axis.

A useful relationship between the angles θ and β can be derived by noting that (see Fig. 19.13) $\tan \beta = \omega_y / \omega_z$. Using Eqs. (19.51), we obtain

$$\tan \beta = \frac{\omega_y}{\omega_z} = \frac{h_G \sin \theta / \bar{I}}{h_G \cos \theta / \bar{I}_z} = \frac{\bar{I}_z}{\bar{I}} \tan \theta$$

or

$$\tan \theta = \frac{1}{\lambda} \tan \beta \qquad \text{where } \lambda = \frac{\bar{I}_z}{\bar{I}} \tag{19.52}$$

Using Eq. (19.52), it is possible to obtain the following relationships between the angular velocity and the rates of spin and precession (see Sample Problem 19.11 for the derivation):

$$\dot{\psi} = \omega(1 - \lambda) \cos \beta \tag{19.53a}$$

$$\dot{\phi} = \omega \cos \beta \sqrt{\lambda^2 + \tan^2 \beta} \tag{19.53b}$$

$$\lambda \dot{\psi} = (1 - \lambda)\dot{\phi} \cos \theta \tag{19.53c}$$

It is customary to distinguish between cases of steady precession, depending on whether λ in Eq. (19.52) is greater or less than one.

Case 1: Direct (regular) precession: $\lambda < 1$.

For $\lambda < 1$, it follows from Eq. (19.52) that $\bar{I} > \bar{I}_z$, which is the case for an elongated body such as the rocket shown in Fig. 19.14. From Eq. (19.52), we also see that $\theta > \beta$, indicating that the angular velocity vector $\boldsymbol{\omega}$ lies within the angle formed by the positive Z- and z-axes, as shown in Fig. 19.14(a). Note that the projection of $\dot{\boldsymbol{\psi}}$ onto $\dot{\boldsymbol{\phi}}$ is in the same direction as $\dot{\boldsymbol{\phi}}$, which is a characteristic of direct precession. The system of vectors shown in Fig. 19.14(a) precesses at a constant rate $\dot{\phi}$ about the Z-axis, while the angular velocity $\boldsymbol{\omega}$ sweeps out a cone in space, the axis of the cone being Z.

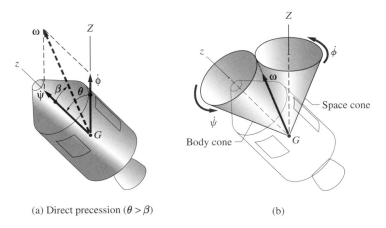

(a) Direct precession $(\theta > \beta)$ (b)

Fig. 19.14

This motion may be represented by the geometric model shown in Fig. 19.14(b), which consists of two right circular cones. The body cone is fixed in the body, with its axis being the axis of symmetry of the body (z-axis). The space cone is stationary, with Z being its axis. The angular velocity $\boldsymbol{\omega}$ lies along the line of contact between the two cones. Because the instant axis of rotation is the locus of points that have zero velocity, we see that the body cone rolls without slipping on the outside of the space cone. Therefore, the z-axis and angular velocity $\boldsymbol{\omega}$ precess about the Z-axis at the rate $\dot{\phi}$, and the body cone spins about the z-axis at the rate $\dot{\psi}$. By direct comparison of parts (a) and (b) of Fig. 19.14, we conclude that the motions of the body cone and the physical body (that is, the rocket) are identical.

Case 2: Retrograde precession: $\lambda > 1$.

If $\lambda > 1$, we see from Eq. (19.52) that $\bar{I} < \bar{I}_z$, which would be the case for a flattened body such as the orbiting space vehicle in Fig. 19.15. Because $\lambda > 1$ implies that $\theta < \beta$, the angular velocity vector $\boldsymbol{\omega}$ lies outside the angle formed by the positive Z- and z-axes, as shown in Fig. 19.15(a). We see also that the direction of the projection of $\dot{\psi}$ onto $\dot{\phi}$ is opposite to the direction of $\dot{\phi}$. The cone model for retrograde precession in Fig. 19.15(b) shows that the space cone is on the inside of the body cone.

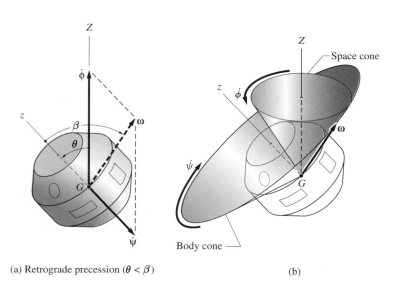

(a) Retrograde precession ($\theta < \beta$)   (b)

Fig. 19.15

For a body for which all three principal moments of inertia are equal to $\bar{I}$—that is, for $\lambda = 1$ (as is the case for a uniform sphere)—it can be seen from Eqs. (19.18) that $\mathbf{h}_G = \bar{I}\boldsymbol{\omega}$. Therefore, once launched with an initial rotation about a given axis, the body simply continues to rotate about that axis with constant angular velocity.

e. *Gyroscopes*

A gyroscope consists of an axisymmetric rotor, or disk, that is mounted in such a way that the rotor is free to spin about its axis of symmetry. Figure 19.16 shows a gyroscope that is mounted in a so-called *Cardan's suspension*, a widely used

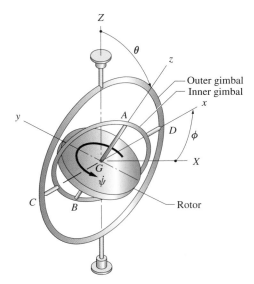

Fig. 19.16

design for inertial guidance systems, gyrostabilizers, and so on. The elements of this suspension are as follows: the rotor of mass m, which spins about the axis AB, which is fixed in the inner gimbal, or ring; the inner gimbal, which is free to rotate about the axis CD relative to the outer gimbal; and the outer gimbal, which can rotate about the Z-axis. Figure 19.16 also shows the manner in which the orientations of the gimbals correspond to the Eulerian angles ϕ and θ, and to the rate of spin $\dot{\psi}$ of the rotor. Assuming that the Z-axis has a fixed orientation in space, the rotor has three degrees of rotational freedom, and it can, therefore, assume all possible angular positions. Of particular interest is the fact that the three axes of rotation intersect at the mass center G of the rotor. Consequently, the motion of the rotor is torque-free, assuming that all of the bearings have negligible friction and that no external forces are applied to the gimbals.

If the rotor in an otherwise stationary gyroscope is set spinning about the z-axis, its initial angular momentum $\mathbf{h}_G$ will also be directed along the z-axis. Because the motion is torque-free, the direction of that axis will remain fixed ($\Sigma\mathbf{M}_G = d\mathbf{h}_G/dt = \mathbf{0}$). The capability of a gyroscope rotor to maintain a fixed direction serves as the principle of operation in many navigational instruments.

A force applied to one of the gimbals can result in a moment about the mass center of the rotor. In this case, the rotor will undergo steady precession provided that suitable initial conditions were present. The direction of the precession can be deduced from $\Sigma\mathbf{M}_G = d\mathbf{h}_G/dt$, where $\Sigma\mathbf{M}_G$ is the moment of the applied force about the mass center G.

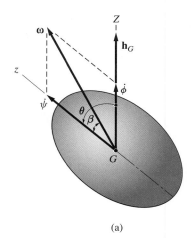

(a)

Sample Problem 19.11

Figure (a) shows an axisymmetric body (z-axis is the axis of symmetry) that is undergoing torque-free motion. The fixed Z-axis is chosen to coincide with $\mathbf{h}_G$, the angular momentum about the mass center G. The angular velocity vector $\boldsymbol{\omega}$ is inclined at the angle β from the z-axis.

(1) Letting $\lambda = \bar{I}_z/\bar{I}$, derive Eqs. (19.53a–c):

$$\dot{\psi} = \omega(1 - \lambda)\cos\beta \tag{a}$$

$$\dot{\phi} = \omega\cos\beta\sqrt{\lambda^2 + \tan^2\beta} \tag{b}$$

$$\lambda\dot{\psi} = (1 - \lambda)\dot{\phi}\cos\theta \tag{c}$$

where $\dot{\psi}$ and $\dot{\phi}$ are the spin and precession rates, respectively, and θ is the Euler angle shown in Fig. (a).

(2) Sketch the body and space cones, and calculate the rates of spin and precession for a satellite given that $\lambda = 1.8$, $\omega = 1.2$ rad/s, and $\beta = 25°$. Repeat the procedure for a rocket for which $\lambda = 0.2$, $\omega = 0.8$ rad/s, and $\beta = 15°$.

Solution

Part 1

Applying the law of sines to the velocity diagram shown in Fig. (b), we obtain

$$\frac{\dot{\psi}}{\sin(\theta - \beta)} = \frac{\omega}{\sin(\pi - \theta)}$$

which yields

$$\dot{\psi} = \omega\frac{\sin\theta\cos\beta - \cos\theta\sin\beta}{\sin\theta} = \omega\left(\cos\beta - \frac{\sin\beta}{\tan\theta}\right)$$

Substituting $\tan\theta = (1/\lambda)\tan\beta$ [see Eq. (19.52)], we obtain

$$\dot{\psi} = \omega\left(\cos\beta - \frac{\sin\beta}{(1/\lambda)\tan\beta}\right) = \omega(1 - \lambda)\cos\beta$$

which agrees with Eq. (a).

Application of the law of sines to Fig. (b) also yields

$$\frac{\dot{\phi}}{\sin\beta} = \frac{\omega}{\sin(\pi - \theta)}$$

or

$$\dot{\phi} = \omega\frac{\sin\beta}{\sin\theta} \tag{d}$$

Utilizing, as before, the relationship $\tan\theta = (1/\lambda)\tan\beta$, $\sin\theta$ may be written as

$$\sin\theta = \frac{\tan\theta}{\sqrt{1 + \tan^2\theta}} = \frac{(1/\lambda)\tan\beta}{\sqrt{1 + [(1/\lambda)\tan\beta]^2}} = \frac{\tan\beta}{\sqrt{\lambda^2 + \tan^2\beta}} \tag{e}$$

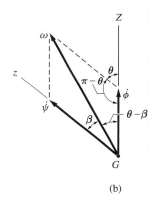

(b)

Substituting Eq. (e) into Eq. (d) gives

$$\dot{\phi} = \omega \sin \beta \frac{\sqrt{\lambda^2 + \tan^2 \beta}}{\tan \beta} = \omega \cos \beta \sqrt{\lambda^2 + \tan^2 \beta}$$

which agrees with Eq. (b).

Dividing Eq. (a) by Eq. (b), we obtain

$$\frac{\dot{\psi}}{\dot{\phi}} = \frac{\omega(1 - \lambda)\cos \beta}{\omega \cos \beta \sqrt{\lambda^2 + \tan^2 \beta}} = \frac{1 - \lambda}{\sqrt{\lambda^2 + \lambda^2 \tan^2 \theta}}$$

$$= \frac{1 - \lambda}{\lambda \sec \theta} = \frac{(1 - \lambda)\cos \theta}{\lambda}$$

or

$$\lambda \dot{\psi} = (1 - \lambda)\dot{\phi} \cos \theta$$

which is identical to Eq. (c).

Part 2

The Euler angle θ between the fixed Z-axis and the axis of symmetry is found by using Eqs. (19.52) and (19.53). Equations (a) and (b) can be used to calculate the spin rate $\dot{\psi}$ and the precession rate $\dot{\phi}$. When the given parameters for the satellite and the rocket are substituted into these equations, the results are as follows.

For the satellite:

$$\theta = \tan^{-1}[(1/\lambda)\tan \beta] = \tan^{-1}[(1/1.8)\tan 25°] = 14.52°$$

$$\dot{\psi} = \omega(1 - \lambda)\cos \beta = 1.2(1 - 1.8)\cos 25° = -0.870 \text{ rad/s}$$

$$\dot{\phi} = \omega \cos \beta \sqrt{\lambda^2 + \tan^2 \beta}$$

$$= 1.2 \cos 25° \sqrt{(1.8)^2 + \tan^2 25°} = 2.022 \text{ rad/s}$$

For the rocket:

$$\theta = \tan^{-1}[(1/\lambda)\tan \beta] = \tan^{-1}[(1/0.2)\tan 15°] = 53.26°$$

$$\dot{\psi} = \omega(1 - \lambda)\cos \beta = 0.8(1 - 0.2)\cos 15° = 0.618 \text{ rad/s}$$

$$\dot{\phi} = \omega \cos \beta \sqrt{\lambda^2 + \tan^2 \beta}$$

$$= 0.8 \cos 15° \sqrt{(0.2)^2 + \tan^2 15°} = 0.258 \text{ rad/s}$$

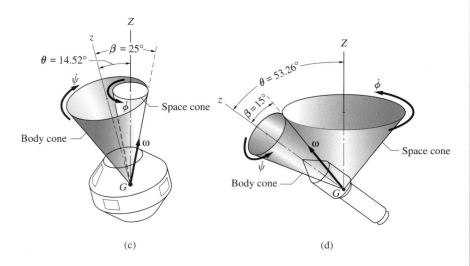

(c) (d)

The space and body cones for the satellite and rocket are shown in Figs. (c) and (d), respectively. Observe that the satellite undergoes retrograde precession ($\dot{\psi}$ is negative), whereas the precession of the rocket is direct ($\dot{\psi}$ is positive).

Sample Problem *19.12*

Figure (a) shows a uniform sphere of radius R and mass m that is welded to the rod AB of length L (the mass of AB may be neglected). The clevis at B connects the rod to the vertical shaft BC. The assembly is initially rotating about the vertical at the angular velocity ω, with the sphere resting against the shaft. Assuming that ω is gradually increased, determine the critical angular velocity ω_{cr} at which contact between the sphere and rod is lost. Neglect friction.

Solution

The free-body diagram of the sphere and the rod, drawn at the instant when the assembly is rotating at $\omega = \omega_{cr}$, is shown in Fig. (b). The xyz-axes are assumed to be attached to rod AB with the origin at B (the x-axis is out of the paper). In addition to the weight mg of the sphere, the FBD also contains the reactions provided by the clevis at B: the pin force $\mathbf{B}$ and the two moment components M_y and M_z ($M_x = 0$ because the pin of the clevis is frictionless). If the angular velocity were less than the critical angular velocity, the FBD would also contain the normal force N that is exerted on the sphere by the vertical shaft. However, when $\omega = \omega_{cr}$, then $N = 0$.

It is convenient to choose the Z-axis to coincide with the vertical shaft, as shown in Fig. (b). The Euler angle θ, which was defined as the angle between the Z- and z-axes, is also shown in the figure. When the sphere is about to lose contact with the vertical shaft, its motion consists of a rotation about the Z-axis at the rate ω_{cr}. The spin rate is zero, because the sphere cannot rotate relative to the rod AB. Therefore, the motion of the sphere can be described as a steady

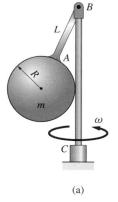

(a)

precession with no spin; in other words, $\dot{\phi} = \omega_{cr}$, $\dot{\psi} = \dot{\theta} = 0$. As a result, the steady precession equation, Eq. (19.46), becomes

$$\Sigma M_x = (I_z - I)\omega_{cr}^2 \sin\theta \cos\theta \qquad \text{(a)}$$

The inertial properties of the sphere about point B are

$$I_z = \frac{2}{5}mR^2 \quad \text{and} \quad I = I_y = \frac{2}{5}mR^2 + m(L + R)^2$$

from which we obtain

$$I_z - I = -m(L + R)^2 \qquad \text{(b)}$$

Referring to Fig. (b), we find that the moment of the external forces (the weight) about the x-axis is

$$\Sigma M_x = mgR \qquad \text{(c)}$$

From the same figure we also deduce that

$$\sin\theta = \sin(\pi - \theta) = \frac{R}{L + R} \qquad \text{(d)}$$

and

$$\cos\theta = -\cos(\pi - \theta) = -\frac{\sqrt{(L + R)^2 - R^2}}{L + R} \qquad \text{(e)}$$

Substituting Eqs. (b)–(e) into Eq. (a) and solving for the critical angular velocity, we obtain

$$\omega_{cr} = \frac{\sqrt{g}}{\sqrt[4]{(L + R)^2 - R^2}} \qquad \textit{Answer}$$

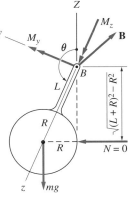

FBD (x-axis out of paper)

(b)

Sample Problem 19.13

Figure (a) shows the same assembly that was described in Sample Problem 19.12. The assembly is initially stationary with the sphere resting against the shaft BC. A motor at C is then activated that drives the shaft with the constant angular acceleration α (the resulting angular velocity of the shaft is $\omega = \alpha t$). Letting t_0 be the time when the sphere loses contact with the shaft, (1) derive the equation of motion for the sphere in terms of the angle β for the period $t \geq t_0$, and state the initial conditions; and (2) solve the equations numerically for the time interval $t = t_0$ to $t = t_0 + 2$ s and plot β versus t. Use $m = 7$ kg, $L = R = 60$ mm, and $\alpha = 5.5$ rad/s^2.

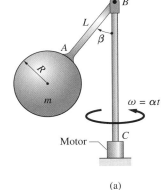

(a)

Solution

Part 1

The free-body diagram (FBD) of the rigid unit containing the sphere and the rod AB is shown in Fig. (b). The xyz-axes are assumed to be attached to rod AB with the origin at B (the x-axis is out of the paper). This FBD displays the weight of the sphere and the reactions at B: the pin force $\mathbf{B}$ and the moment components

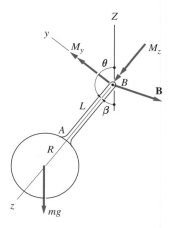

(b) FBD (*x*-axis out of paper)

M_y and M_z (the pin of the clevis does not provide a moment component about the *x*-axis). The fixed *Z*-axis is assumed to be directed along the vertical shaft as shown in Fig. (b), with θ being the Euler angle between the *Z*- and *z*-axes. Because the sphere and rod *AB* rotate as a rigid unit, the spin rate $\dot{\psi}$ is zero. Furthermore, comparing Figs. (a) and (b) with Fig. 19.11, the precession rate is found to be $\dot{\phi} = \omega = \alpha t$.

The modified Euler equation that governs β is the first of Eqs. (19.44) (the remaining two equations could be used to find M_y and M_z).

$$\Sigma M_x = I\ddot{\theta} + (I_z - I)\dot{\phi}^2 \sin\theta \cos\theta + I_z \dot{\phi}\dot{\psi} \sin\theta \qquad \text{(a)}$$

Using $L = R$ and $\theta = \pi - \beta$, the various terms in Eq. (a) become

$$\Sigma M_x = 2mgR\sin\beta \qquad \text{(from the FBD)}$$

$$I_z = \frac{2}{5}mR^2$$

$$I = I_y = \frac{2}{5}mR^2 + m(R+R)^2 = \frac{22}{5}mR^2$$

$$\ddot{\theta} = -\ddot{\beta} \qquad \dot{\phi} = \alpha t \qquad \dot{\psi} = 0$$

Substituting these expressions into Eq. (a), we get

$$2mgR\sin\beta = \frac{22}{5}mR^2(-\ddot{\beta}) + (-4mR^2)(\alpha t)^2 \sin\beta(-\cos\beta) + 0$$

which, on canceling the mass m and rearranging terms, reduces to

$$\ddot{\beta} = \frac{10}{11}\alpha^2 t^2 \sin\beta \cos\beta - \frac{10g}{22R}\sin\beta \qquad \text{(b)}$$

When the numerical values $\alpha = 5.5$ rad/s², $g = 9.81$ m/s², and $R = 0.060$ m are substituted into Eq. (b), the equation of motion becomes

$$\ddot{\beta} = 27.50t^2 \sin\beta \cos\beta - 74.32 \sin\beta \qquad \text{(for } t \geq t_0\text{)} \qquad \textbf{\textit{Answer}} \qquad \text{(c)}$$

From the solution to Sample Problem 19.12, we know that ω_{cr}, the critical angular velocity at which contact between the sphere and vertical shaft is lost, is

$$\omega_{cr} = \frac{\sqrt{g}}{\sqrt[4]{(L+R)^2 - R^2}} = \frac{\sqrt{9.81}}{\sqrt[4]{(0.120)^2 - (0.060)^2}}$$

$$= 9.716 \text{ rad/s}$$

Consequently, the time at which contact is lost is $t_0 = \omega_{cr}/\alpha = 9.716/5.5 = 1.7665$ s. The initial value of β (the value when the sphere touches the vertical shaft) is $\beta_0 = \tan^{-1}[R/(L+R)] = \tan^{-1}(R/2R) = \tan^{-1}(1/2) = 30° = 0.5236$ rad. Therefore, the initial conditions are

$$t_0 = 1.7665 \text{ s} \qquad \beta_0 = 0.5236 \text{ rad} \qquad \dot{\beta}_0 = 0 \qquad \textbf{\textit{Answer}} \qquad \text{(d)}$$

Part 2

The equivalent first-order equations and the initial conditions are (with $x_1 = \beta$ and $x_2 = \dot{\beta}$)

$$\dot{x}_1 = x_2 \qquad\qquad\qquad x_1(1.7665) = 0.5236$$

$$\dot{x}_2 = 27.50t^2 \sin\beta\cos\beta - 74.32\sin\beta \qquad x_2(1.7665) = 0$$

The MATLAB program that produced the plot in Fig. (c) is:

```
function example19_13
[t,x] = ode45(@f,[1.7665:0.01:3.7665],[0.5236 0]);
axes('fontsize',14)
plot(t,x(:,1)*180/pi,'linewidth',1.5)
grid on
xlabel('time (s)'); ylabel('beta (deg)')
      function dxdt = f(t,x)
      s = sin(x(1)); c = cos(x(1));
      dxdt =[x(2); 27.5*t^2*s*c - 74.32*s];
      end
end
```

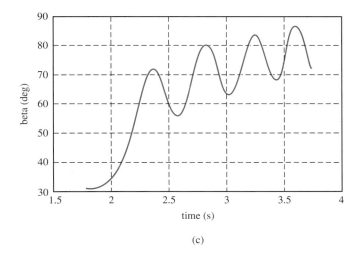

(c)

Referring to Fig. (c), we see that β is made up of two parts—an average value that increases with time, and oscillations about the average value. As the angular speed ω of the shaft increases, we expect the average value of β to approach 90° and the amplitude of the oscillations to decrease (due to the increasing centrifugal force on the sphere). Both of these trends can be observed in Fig. (c).

Problems

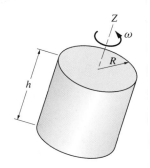

Fig. P19.65

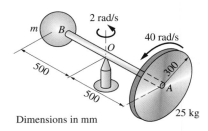

Dimensions in mm

Fig. P19.67

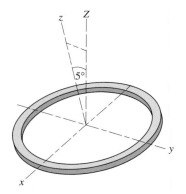

Fig. P19.68

19.65 The homogeneous cylinder is launched into space with the spin velocity ω. Find the ratio h/R for which no precession will occur.

19.66 The uniform 2.5 kg disk rotates about the shaft OA which is attached to a vertical axle with a clevis. If the disk rolls without slipping on the horizontal surface with the angular speed of 20 rad/s, determine the vertical reaction between the disk and the horizontal surface. Note that the path of the disk is a circle of radius 4 m that is centered at O. Neglect the weight of shaft OA.

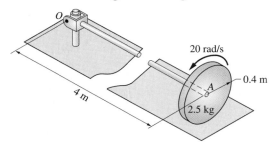

Fig. P19.66

19.67 The 25-kg homogeneous disk is spinning about the shaft AB at 40 rad/s. At the same time, the shaft is rotating about the vertical axis through O at 2 rad/s. Determine the mass m of the counterweight at B that will keep the shaft horizontal. Neglect the mass of the shaft.

19.68 The axis of the thin ring is observed to precess at the rate $\dot{\phi} = 10$ rad/s, the precession angle being 5° from the Z-axis. Determine the spin velocity vector and the angular velocity vector of the ring.

19.69 The jet airplane is flying at 972 km/h on the circular path of 3 km radius. The rotor of the jet engine weighs 250 kg and has a radius of gyration of 0.4 m about its axis. The rotor is turning at 15×10^3 rev/min with its angular velocity vector being parallel to the velocity of the airplane. Calculate the gyroscopic moment acting on the rotor.

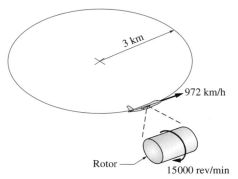

Fig. P19.69

19.70 The two thin uniform disks, each of radius R and mass m, are rigidly connected by a shaft of length L and negligible mass. When the assembly spins freely in the sleeve at O with the angular velocity of $\dot{\psi} = 3$ rad/s, it is observed to precess about the vertical axes at the rate $\dot{\phi} = 2$ rad/s with $\theta = 32°$. Determine the ratio L/R.

19.71 The homogeneous cylinder of mass m, radius R, and length R spins with angular velocity ω_1 relative to its axle, which is inclined at the angle θ from the vertical. The axle rotates in the bearing at O at an angular speed ω_2. Determine the ratio ω_1/ω_2 for which no moment is exerted on the cylinder by the axle. Note that the mass center G of the cylinder is directly above O, and assume that $\theta \neq 0$ and $\omega_2 \neq 0$.

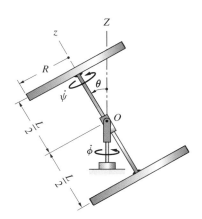

Fig. P19.70

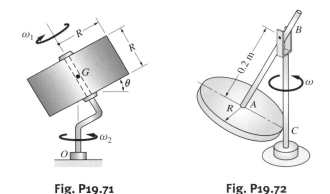

Fig. P19.71   **Fig. P19.72**

19.72 The weight of the thin uniform disk is 4 kg, and its radius is $R = 0.1$ m. The light rod AB is rigidly attached to the disk at A and connected by a clevis to the vertical shaft BC. The entire assembly rotates about the vertical with the constant angular velocity $\omega = 4$ rad/s. Calculate the normal force that acts between the disk and the shaft at C.

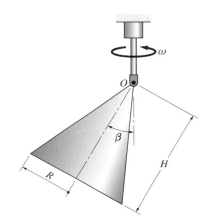

Fig. P19.73

19.73 The 2-kg homogeneous cone of radius $R = 62.5$ mm and height $H = 125$ mm is attached to the vertical shaft by a clevis. The system is at rest, except for a very small oscillation of the cone about the clevis, when the shaft starts rotating with the constant angular acceleration $\alpha = 10$ rad/s² (the corresponding angular velocity is $\omega = \alpha t$). (a) Show that the equation of motion for the angle β is

$$\ddot{\beta} = 88.24t^2 \sin \beta \cos \beta - 92.33 \sin \beta \text{ rad/s}^2$$

and state the initial conditions (do not neglect the small initial oscillation of the cone). (b) Integrate the equation of motion from $t = 0$ to $t = 3$ s. Use the results to determine the time when the cone reaches the position $\beta = \pi/2$. (c) Plot β versus t for the period of integration. (d) What would happen to the numerical solution if the small initial motion of the cone were neglected?

19.74 The slender rod AB of mass m and length L is pinned at O to the fork that is attached to the vertical shaft. The masses of the fork and shaft may be neglected. The cord keeps rod AB at the angle $\theta = 30°$ with the vertical. The shaft is rotating

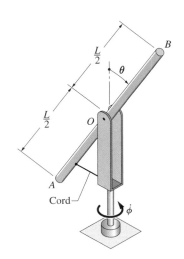

Fig. P19.74

freely with the angular velocity $\dot{\phi} = 200$ rad/s when the cord suddenly breaks. (a) Show that the equations that govern the motion of the rod after the break are

$$\ddot{\theta} = \dot{\phi}^2 \sin\theta \cos\theta \quad \text{and} \quad \ddot{\phi} = -2\dot{\theta}\dot{\phi} \cot\theta$$

and state the initial conditions. (b) Integrate the equations of motion numerically for a time period during which the rod completes at least two full oscillations about the pin. From the numerical solution, find the period of oscillation of the rod, and the range of values of $\dot{\theta}$ and $\dot{\phi}$. (c) Plot θ and $\dot{\phi}$ versus time over the period of integration.

19.75 The homogeneous 200-kg cylinder spins about its axis AB at the constant angular speed of 200 rad/s. At the same time, the mounting table is rotated about the vertical Z-axis at the constant rate of 1.0 rad/s. Determine the bearing reactions at A and B.

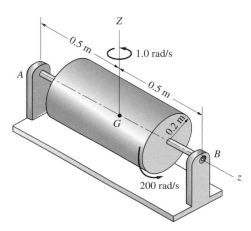

Fig. P19.75

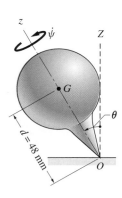

Fig. P19.76–P19.78

19.76 The moments of inertia of the top about axes passing through O are I_z and I. The weight of the top is W, and d is the distance between the mass center G and O. Show that the top can have a steady precession at a given angle $\theta \neq 0$ only when $\omega_z \geq (\omega_z)_{\text{cr}}$, where $(\omega_z)_{\text{cr}} = (2/I_z)\sqrt{IWd\cos\theta}$. [Note: The top is said to be *spin-stabilized* if $\omega_z \geq (\omega_z)_{\text{cr}}$.]

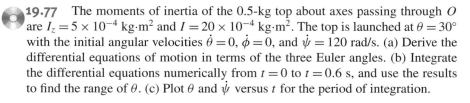

19.77 The moments of inertia of the 0.5-kg top about axes passing through O are $I_z = 5 \times 10^{-4}$ kg·m² and $I = 20 \times 10^{-4}$ kg·m². The top is launched at $\theta = 30°$ with the initial angular velocities $\dot{\theta} = 0$, $\dot{\phi} = 0$, and $\dot{\psi} = 120$ rad/s. (a) Derive the differential equations of motion in terms of the three Euler angles. (b) Integrate the differential equations numerically from $t = 0$ to $t = 0.6$ s, and use the results to find the range of θ. (c) Plot θ and $\dot{\psi}$ versus t for the period of integration.

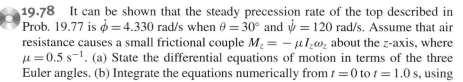

19.78 It can be shown that the steady precession rate of the top described in Prob. 19.77 is $\dot{\phi} = 4.330$ rad/s when $\theta = 30°$ and $\dot{\psi} = 120$ rad/s. Assume that air resistance causes a small frictional couple $M_z = -\mu I_z \omega_z$ about the z-axis, where $\mu = 0.5$ s⁻¹. (a) State the differential equations of motion in terms of the three Euler angles. (b) Integrate the equations numerically from $t = 0$ to $t = 1.0$ s, using the steady precession values as the initial conditions. From the results, determine the initial and final values of ω_z. (c) Show that the analytical solution for ω_z is $\omega_z = (\omega_z)_0 e^{-\mu t}$, where $(\omega_z)_0$ is its initial value. Verify that the values of ω_z found in part (b) agree with this result. (d) Plot θ and $\dot\phi$ versus t.

19.79   Because the axis of the projectile is not aligned with its velocity vector, the resultant aerodynamic force P passes through the point C that is located a distance d from the mass center G. Determine the expression for the smallest value of ω_z for which the projectile will be spin-stabilized. (Hint: Refer to the result stated in Prob. 19.76.)

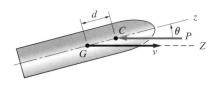

P19.79

19.80   The gyroscope consists of a thin disk of radius R and mass m that is supported by two gimbals of negligible mass. A constant vertical force F is applied to the inner gimbal. Assuming steady precession with $\theta = 90°$, derive an expression for $\dot\phi$ in terms of $\dot\psi$.

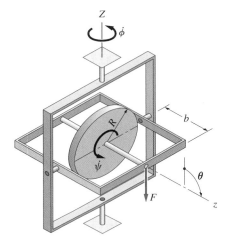

Fig. P19.80, P19.81

19.81   The thin homogeneous disk of the gyroscope weighs 26 N, and its radius is $R = 12$ cm. The weights of the two gimbals may be neglected. Before the constant vertical force $F = 80$ N was applied to the inner gimbal with $b = 10$ cm, the disk was spinning at $\dot\psi = 500$ rev/min with its axis horizontal ($\theta = 90°$) and no precession ($\dot\phi = 0$). (a) Derive the differential equations of motion in terms of Euler angles, and state the initial conditions. (b) Integrate the equations of motion from $t = 0$ (the time when F was applied) to $t = 0.12$ s. From the result determine the range of values of θ, $\dot\phi$, and $\dot\psi$. (c) Plot $\dot\phi$ versus θ and $\dot\psi$ versus t.

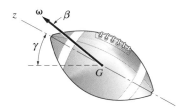

Fig. P19.82

19.82   The football is thrown with the angular velocity $\omega = 12$ rad/s, directed at the angle $\beta = 5.2°$ with the z-axis. The z-axis makes an initial angle of $\gamma = 15°$ with the horizontal. (a) Assuming that $\bar I_z = \bar I/4$ and neglecting air resistance, find the angle θ that locates the precession axis, and calculate the precession and spin rates. (b) Employing a sketch of the body and space cones, find the range of γ during the flight.

19.83   The axisymmetric satellite is rotating at the angular velocity $\omega = 0.6$ rad/s, directed at $\beta = 30°$ from the z-axis. Given that $\bar I_z = 2\bar I$, (a) determine the angle θ that locates the axis of precession, and find the rates of spin and precession; and (b) sketch the space and body cones.

19.84   The axisymmetric satellite is rotating with the angular velocity vector ω, where the angle β between ω and the z-axis is very small. Show that the precession and spin rates of the satellite are $\dot\phi = \omega\bar I_z/\bar I$ and $\dot\psi = \omega(1 - \bar I_z/\bar I)$, respectively.

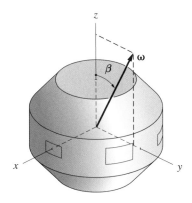

Fig. P19.83, P19.84

19.85 The thin homogeneous disk of radius R is thrown with the spin velocity of 60 rad/s in the direction shown. During the flight, the axis of the disk wobbles at 30° about the vertical Z-axis. (a) Determine the rate of precession of the disk. (b) Sketch the space and body cones.

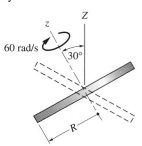

Fig. P19.85

19.86 During its free flight, the rocket is observed to precess steadily about the horizontal Z-axis at the rate of 1 cycle every 3 minutes. Knowing that the moments of inertia about axes passing through the mass center G are related by $\bar{I} = 8\bar{I}_z$, calculate the magnitude and direction of the angular velocity vector of the rocket.

Fig. P19.86

19.87 Because the earth is slightly flattened at the poles, its moment of inertia $\bar{I}_z$ about the z-axis (the axis of symmetry) is slightly larger than its moment of inertia $\bar{I}$ about an equatorial diameter. In addition, the polar axis of the earth, about which the earth rotates at the rate of one revolution per day, forms a small angle β with the z-axis. Knowing that the poles precess about the z-axis at the approximate rate of one complete cycle every 430 days, estimate the ratio $\bar{I}_z/\bar{I}$. (Hint: When using Eqs. (19.53), be certain to correctly identify the variable that represents the rate of precession of the poles about the z-axis.)

Fig. P19.87

Review of Equations

Differentiation of a vector function in a rotating reference frame

$$\frac{d\mathbf{V}}{dt} = \left(\frac{d\mathbf{V}}{dt}\right)_{/\mathcal{B}} + \boldsymbol{\omega} \times \mathbf{V} \quad \text{(frame fixed in body } \mathcal{B})$$

$$\frac{d\mathbf{V}}{dt} = \left(\frac{d\mathbf{V}}{dt}\right)_{/x'y'z'} + \boldsymbol{\Omega} \times \mathbf{V} \quad (x'y'z'\text{rotating frame})$$

Rectangular components of angular momentum

$$h_x = I_x\omega_x - I_{xy}\omega_y - I_{xz}\omega_z$$

$$h_y = -I_{yx}\omega_x + I_y\omega_y - I_{yz}\omega_z \quad \text{(About fixed point or mass center)}$$

$$h_z = -I_{zx}\omega_x - I_{zy}\omega_y + I_z\omega_z$$

Kinetic energy of a rigid body

$$T = \frac{1}{2}m\bar{v}^2 + \frac{1}{2}(\bar{I}_x\omega_x^2 + \bar{I}_y\omega_y^2 + \bar{I}_z\omega_z^2 - 2\bar{I}_{xy}\omega_x\omega_y - 2\bar{I}_{yz}\omega_y\omega_z - 2\bar{I}_{zx}\omega_z\omega_x)$$

Euler's moment equations of motion

$$\Sigma M_x = I_x\dot{\omega}_x + \omega_y\omega_z(I_z - I_y)$$

$$\Sigma M_y = I_y\dot{\omega}_y + \omega_z\omega_x(I_x - I_z) \quad (xyz\text{-axes are principal axes})$$

$$\Sigma M_z = I_z\dot{\omega}_z + \omega_x\omega_y(I_y - I_x)$$

Modified Euler's equations

$$\Sigma M_x = I_x\dot{\omega}_x + I_z\Omega_y\omega_z - I_y\Omega_z\omega_y$$

$$\Sigma M_y = I_y\dot{\omega}_y - I_z\Omega_x\omega_z + I_x\Omega_z\omega_x \quad (z\text{-axis is axis of symmetry})$$

$$\Sigma M_z = I_z\dot{\omega}_z + I_y\Omega_x\omega_y - I_x\Omega_y\omega_x$$

Steady precession of an axisymmetric body

$$\Sigma M_x = (I_z - I)\dot{\phi}^2 \sin\theta \cos\theta + I_z \dot{\phi}\dot{\psi} \sin\theta \qquad \Sigma M_y = \Sigma M_z = 0$$
$$\Sigma M_x = I_z \omega_z \dot{\phi} \sin\theta - I\dot{\phi}^2 \sin\theta \cos\theta \qquad \Sigma M_y = \Sigma M_z = 0$$

$\theta = $ nutation angle
$\dot{\phi} = $ rate of precession
$\dot{\psi} = $ spin velocity

20
Vibrations

An automobile suspension is an example of a damped mass-spring system, where the shock absorber provides the damping. Damped vibrations is one of the topics discussed in this chapter. (© Leslie Garland Picture Library/Alamy)

20.1 Introduction

Vibration refers to the oscillation of a body or a mechanical system about its equilibrium position. Some vibrations are desirable, such as the oscillation of the pendulum that controls the movement of a clock, or the vibration of a string on a musical instrument. The majority of vibrations, however, are deemed to be objectionable or harmful, ranging from the annoying (vibration-induced noise) to the catastrophic (structural failure of aircraft). Excessive vibration of machines or structures can cause loosening of joints and connections, premature wear, and metal fatigue (breakage due to cyclic loading).

The study of vibrations is so extensive that entire textbooks are devoted to the subject. It is our intention here to introduce the fundamentals of vibrations that should be understood by all engineers and that will serve as the basis for further study. We consider only the simplest case: vibration of one-degree-of-freedom systems—that is, problems in which the motion can be described in terms of a single position coordinate.

The two basic components of all vibratory systems are the mass and the restoring force. The restoring force, often provided by an elastic mechanism, such as a spring, tends to return the mass to its equilibrium position. When the mass is displaced from its equilibrium position and released, it overshoots the equilibrium position, comes to a momentary stop, and then reverses direction. This oscillation between two stationary positions is a simple example of vibratory motion. If the restoring force is linear, the resulting equation of motion will also be linear, and its solution can be found by analytical means. Nonlinear restoring forces result in nonlinear differential equations that can usually be solved only by numerical methods.

In general terms, vibrations are categorized as forced or free, and damped or undamped. When the vibration of a system is maintained by an external force, the vibration is said to be *forced*. If no external forces are driving the system, the motion is referred to as *free* vibration. *Damped* vibrations refer to a system in which energy is being removed by friction or a viscous damper (resistance caused by the viscous drag of a fluid). If damping is absent, the motion is called *undamped*.

In free vibrations that are undamped, no energy is supplied to or dissipated from the system; consequently, the motion will continue forever, at least in theory. In reality, there is always some damping present, however small, that will eventually stop the vibration. In forced vibration, the oscillation can continue even if there is damping present, because the applied force provides energy to the system that can compensate for any energy that is removed by the damping.

This chapter begins with a discussion of free vibrations of particles, both damped and undamped, followed by forced vibrations of particles. The chapter concludes with rigid-body vibrations and the application of energy methods.

20.2   *Free Vibrations of Particles*

a.   *Undamped free vibrations*

Consider the vertical motion of the mass-spring system shown in Fig. 20.1(a). The position coordinate x of the mass is measured downward from the static equilibrium position. Note that at $x = 0$ the elongation of the spring is

$$\Delta = mg/k \qquad \text{(a)}$$

Figure 20.1(b) displays the free-body diagram (FBD) and the mass-acceleration diagram (MAD) of the mass in an arbitrary position. The forces in the FBD are the weight mg of the mass and the spring force $k(x + \Delta)$. The MAD contains the the inertia vector $m\ddot{x}$ of the mass. The equation of motion of the mass is

$$\Sigma F_x = ma_x \quad +\!\!\downarrow \quad mg - k(x + \Delta) = m\ddot{x}$$

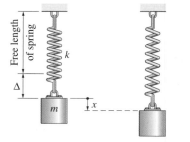

Equilibrium position   Arbitrary position

(a)

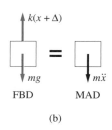

(b)

Fig. 20.1

Substituting for Δ from Eq. (a), the equation of motion reduces to

$$m\ddot{x} + kx = 0 \tag{20.1a}$$

The force kx is called the *restoring force* because it tends to return the mass to its equilibrium position. Dividing Eq. (20.1a) by m, we obtain

$$\ddot{x} + p^2 x = 0 \tag{20.1b}$$

where

$$p = \sqrt{\frac{k}{m}} \tag{20.2}$$

Equation (20.1b) is a linear, second-order differential equation. The solution of this equation is

$$x = A \cos pt + B \sin pt \tag{20.3}$$

where A and B are constants of integration, to be determined by the initial conditions. Successive differentiations of Eq. (20.3) yield

$$\dot{x} = p(-A \sin pt + B \cos pt) \tag{b}$$

$$\ddot{x} = -p^2(A \cos pt + B \sin pt) = -p^2 x \tag{c}$$

Substituting Eq. (c) into Eq. (20.1b), we get $-p^2 x + p^2 x = 0$, thereby verifying the solution.

Another convenient form of the solution is obtained by letting

$$A = E \sin \alpha \quad B = E \cos \alpha \tag{d}$$

Equation (20.3) then becomes

$$x = E(\cos pt \sin \alpha + \sin pt \cos \alpha)$$

which can be written as

$$x = E \sin(pt + \alpha) \tag{20.4}$$

Consider now the case where the initial conditions are given at time $t = 0$. Substituting $t = 0$, $x(0) = x_0$, and $\dot{x}(0) = v_0$ into Eqs. (20.3) and (b), we obtain

$$A = x_0 \quad B = v_0/p \tag{20.5}$$

Substituting Eqs. (20.5) into Eqs. (d) and solving for α and E yields

$$\tan \alpha = \frac{x_0 p}{v_0} \quad E = \sqrt{x_0^2 + (v_0/p)^2} \tag{20.6}$$

A graphical representation of Eq. (20.4) is shown in Fig. 20.2. Consider the motion of point c along the circle of radius E, the radial line ac having the constant

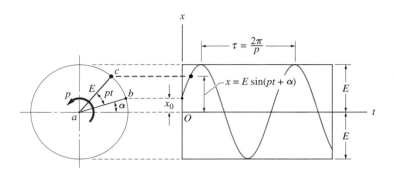

Fig. 20.2

angular velocity p. If the point starts at b at time $t = 0$, its vertical position at time t is $x = E \sin(pt + \alpha)$, which is identical to Eq. (20.4). Note that $E \sin \alpha = x_0$ represents the position at $t = 0$.

The plot of x versus t in Fig. 20.2 shows that the mass oscillates, or vibrates, about its equilibrium position $x = 0$. Because the motion repeats itself over equal intervals of time, it is called *periodic motion*. Furthermore, motion that is described in terms of the circular functions, sine and cosine, is known as *harmonic motion*. (All harmonic motion is periodic, but not all periodic motion is harmonic.) The parameter p is referred to as the (natural) *circular frequency, E* is called the *amplitude,* and α is known as the *phase angle.* As shown in Fig. 20.2, τ denotes the *period* of the motion—that is, the time taken by one complete cycle of the motion. Therefore, $p\tau = 2\pi$, which gives

$$\tau = \frac{2\pi}{p} \tag{20.7}$$

The *frequency* of the motion is the number of cycles completed per unit time:

$$f = \frac{1}{\tau} = \frac{p}{2\pi} \tag{20.8}$$

The differential equation that describes the motion of the spring-mass system, Eq. (20.1a), is linear because the restoring force kx is a linear function of the displacement x. Vibrations that are described by linear differential equations are called *linear vibrations*.

Example of Nonlinear Free Vibration

As an example consider the simple pendulum shown in Fig. 20.3(a), which consists of a bob of mass m attached to the end of a string of length L and negligible mass. The angular displacement of the pendulum from the vertical is measured by the angle θ. The free-body diagram in Fig. 20.3(b) shows that the forces acting on the bob are the tension T and its weight mg. The normal and tangential (n-t) components of the inertia vector are shown in the mass-acceleration diagram in Fig. 20.3(b). Note that the restoring force—that is, the force tending to return the pendulum to its equilibrium position—is $mg \sin \theta$, which is a nonlinear function of

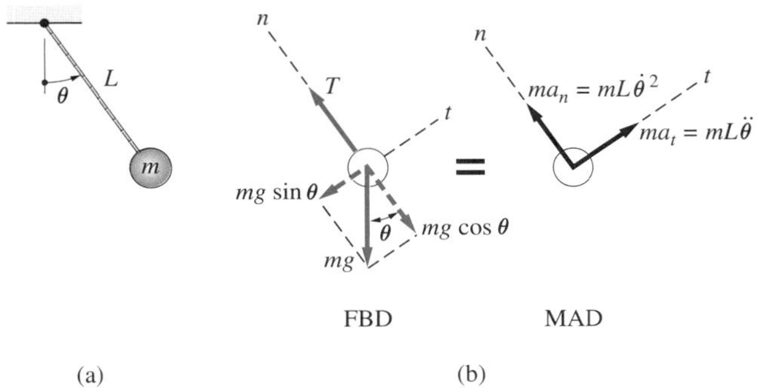

Fig. 20.3

the angular displacement θ. Summing forces in the tangential direction, we obtain

$$\Sigma F_t = ma_t \quad +\nearrow \quad -mg\sin\theta = ma_t = mL\ddot{\theta}$$

which becomes

$$\ddot{\theta} + \frac{g}{L}\sin\theta = 0 \qquad (20.9)$$

The solution of this nonlinear differential equation must be obtained numerically. Although the motion of the pendulum is periodic, it is not harmonic; that is, the solution of Eq. (20.9) cannot be expressed in terms of sine and cosine functions. The motion of the pendulum can be approximated by a harmonic solution only if the amplitude of the vibration is assumed to be small. Using $\sin\theta \approx \theta$, an approximation that is sufficiently accurate for $\theta < 6°$ for most applications, Eq. (20.9) becomes

$$\ddot{\theta} + \frac{g}{L}\theta = 0 \qquad (20.10)$$

which has the same form as Eq. (20.1a). Therefore, the motion of the simple pendulum is harmonic for small oscillations, the circular frequency being $p = \sqrt{g/L}$.

Many vibration problems are nonlinear if the amplitude is large, but simplify to a linear form if the amplitude is assumed to be sufficiently small. But be forewarned: Not all vibration problems can be linearized in this fashion—it must be demonstrated that a linear equation of motion is a valid approximation for small amplitudes.

b.   Damped free vibrations

When energy is dissipated from a vibrating system, the motion is said to be *damped.* The common forms of damping are viscous, Coulomb, and solid damping. Viscous damping describes resistance to motion that is proportional to the first power of the velocity. (Incidental damping, such as air resistance, is often assumed to be viscous. However, a damping force proportional to the square of the velocity would be a more accurate description.) Coulomb damping arises from dry friction between sliding surfaces. Solid damping is caused by internal friction within the body itself. In this text, we consider only viscous damping.

In viscous damping, the damping force F_d is proportional to the velocity; that is, $F_d = -cv$, where v is the velocity and c is a constant of proportionality, called the *coefficient of viscous damping*. The negative sign indicates that the damping force always opposes the velocity. The dimension of c is $[FT/L]$ with the units being $N \cdot s/m$.

A common example of a viscous damper (also known as a *dashpot*) is the automobile shock absorber. When an automobile hits a bump, the shock absorbers carry most of the impact loading, thus preventing the springs from "bottoming out." The shock absorbers are also responsible for damping out the ensuing up-and-down oscillations of the vehicle. An automobile shock absorber consists of a piston that is encased in oil and mounted between the wheel and the frame of the automobile. As the piston moves, oil is forced to flow through a hole from one side of the piston to the other. The amount of damping (due to the viscous resistance of the oil) depends largely on the size of the orifice being used.

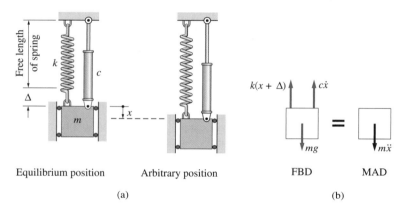

Equilibrium position	Arbitrary position	FBD	MAD
(a)		(b)	

Fig. 20.4

Figure 20.4(a) shows a spring-mass system. We have added a dashpot with damping coefficient c. Choosing x to be the downward displacement of the mass, measured from its equilibrium position, results in the FBD shown in Fig. 20.4(b), where Δ is the static deflection of the spring. The mass-acceleration diagram in Fig. 20.4(b) consists of the inertia vector $m\ddot{x}$. The differential equation of motion is

$$\Sigma F_x = ma_x \quad +\!\!\downarrow \quad mg - k(x + \Delta) - c\dot{x} = m\ddot{x}$$

Utilizing the equilibrium equation $mg - k\Delta = 0$, we obtain

$$m\ddot{x} + c\dot{x} + kx = 0 \tag{20.11}$$

A linear differential equation with constant coefficients, such as Eq. (20.11), admits a solution of the form

$$x = Ae^{\lambda t}$$

where A and λ are constants. Substituting this into Eq. (20.11) and dividing each term by $Ae^{\lambda t}$ results in the *characteristic equation*

$$m\lambda^2 + c\lambda + k = 0 \tag{20.12}$$

the roots of which are

$$\left.\begin{array}{c}\lambda_1\\\lambda_2\end{array}\right\} = -\frac{c}{2m} \pm \sqrt{\left(\frac{c}{2m}\right)^2 - \frac{k}{m}} \qquad (20.13)$$

The *critical damping coefficient* c_{cr} is defined as the value of c for which the radical in Eq. (20.13) vanishes. Therefore, we find that

$$c_{cr} = 2mp \qquad (20.14)$$

where $p = \sqrt{k/m}$, the undamped circular frequency of the system. It is convenient to introduce the *damping factor* ζ, defined as the ratio of the actual damping to the critical damping; that is,

$$\zeta = \frac{c}{c_{cr}} = \frac{c}{2mp} = \frac{c}{2\sqrt{km}} \qquad (20.15)$$

Now Eq. (20.13) can be written as

$$\left.\begin{array}{c}\lambda_1\\\lambda_2\end{array}\right\} = p\left(-\zeta \pm \sqrt{\zeta^2 - 1}\right) \qquad (20.16)$$

The general solution of Eq. (20.11) is any linear combination of the two solutions corresponding to λ_1 and λ_2:

$$x = A_1 e^{\lambda_1 t} + A_2 e^{\lambda_2 t}$$

where A_1 and A_2 are arbitrary constants. After substitution for the λ's from Eq. (20.16), the solution becomes

$$x = A_1 e^{\left(-\zeta + \sqrt{\zeta^2 - 1}\right)pt} + A_2 e^{\left(-\zeta - \sqrt{\zeta^2 - 1}\right)pt} \qquad (20.17)$$

There are three categories of damping, determined by the value of the damping factor ζ.[*]

1   *Overdamping:* $\zeta > 1$ The roots λ_1 and λ_2 in Eq. (20.16) are real and distinct. Consequently, the motion is nonoscillatory and decaying with time, as shown in Fig. 20.5. Motion of this type is called *aperiodic,* or *dead-beat,* motion.

2   *Critical Damping:* $\zeta = 1$ The roots λ_1 and λ_2 in Eq. (20.16) are both equal to $-p$, and the solution in Eq. (20.17) can be shown to take the form

$$x = (A_1 + A_2 t)e^{-pt} \qquad (20.18)$$

where A_1 and A_2 are arbitrary constants. The motion is again aperiodic, as shown in Fig. 20.5.

---

[*]The three forms of solutions are stated here without proof. For a complete discussion, see a textbook on differential equations.

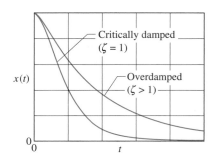

Both curves drawn for the initial condition $\dot{x}_0 = 0$

Fig. 20.5

3   *Underdamping:* $\zeta < 1$ The roots λ_1 and λ_2 in Eq. (20.16) are complex conjugates. It can be shown that Eq. (20.17) can be written in the form

$$x = Ee^{-\zeta pt} \sin(\omega_d t + \alpha) \tag{20.19}$$

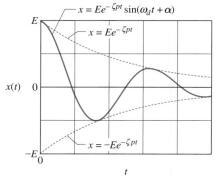

Fig. 20.6

where E and α are arbitrary constants, and

$$\omega_d = p\sqrt{1 - \zeta^2} \tag{20.20}$$

The motion represented by Eq. (20.19) is oscillatory with decreasing amplitude, as shown in Fig. 20.6 (the plot is drawn for $\alpha = 0$). Observe that the plot is tangent to the curves $x = \pm Ee^{-\zeta pt}$. Although the motion does not repeat itself, ω_d is called the *damped circular frequency,* and the corresponding *damped period* is given by

$$\tau_d = \frac{2\pi}{\omega_d} \tag{20.21}$$

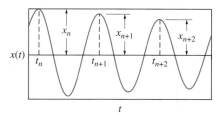

Fig. 20.7

According to Eq. (20.20), the damped circular frequency ω_d is smaller than the circular frequency p. Consequently, the damped period τ_d is larger than the period of the undamped free vibration.

Let x_n, x_{n+1}, x_{n+2}, ... be the peak displacements (measured from the equilibrium position) of an underdamped system, as shown in Fig. 20.7. It can be proven that the ratio x_{n+1}/x_n of two successive peaks is constant. The natural logarithm of this ratio, called the *logarithmic decrement,* is (see Prob. 20.16)

$$\ln\left(\frac{x_{n+1}}{x_n}\right) = -\frac{2\pi\zeta}{\sqrt{1 - \zeta^2}} \tag{20.22}$$

Sample Problem 20.1

Three identical springs, each of stiffness k, support a block of mass m, as shown in Fig. (a). Deformation of bar AB may be neglected. (1) Find the equivalent spring stiffness k_0—that is, the stiffness of a single spring as shown in Fig. (c)—that can replace the original springs without changing the displacement characteristics of the block. (2) If the block mass $0.2\,\text{kg}$ and $k = 875\,\text{N/m}$, find the circular frequency, the frequency, and the period of free vibration.

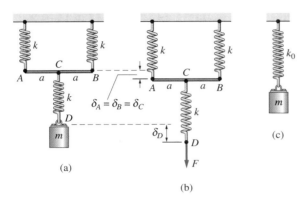

(a)   (b)   (c)

Solution

Part 1

Figure (b) shows a method for calculating the equivalent spring stiffness. First, a static vertical force F is applied at point D, and the vertical movement δ_D is computed. The equivalent spring stiffness k_0 is then calculated using $F = k_0\delta_D$. Following this procedure, the mass in Fig. (c) will have the same displacement characteristics as the mass in Fig. (a).

From equilibrium analysis of Fig. (b), we see that the spring forces are equal to F in the lower spring and $F/2$ in each of the upper springs (C is the midpoint of bar AB). The elongations of the upper springs are, therefore, identical, and the vertical displacements δ_A, δ_B, and δ_C are each equal to $(F/2)/k$ (note that bar AB remains horizontal). Next we observe that the vertical movement of D equals the vertical movement of C plus the elongation of the lower spring; that is, $\delta_D = \delta_C + (F/k)$. The equivalent spring stiffness thus becomes

$$k_0 = \frac{F}{\delta_D} = \frac{F}{\delta_C + (F/k)} = \frac{F}{(F/2k) + (F/k)}$$

$$= \frac{1}{(1/2k) + (1/k)} = \frac{2k}{3} \qquad \textit{Answer}$$

Part 2

Using the given data and the equivalent system shown in Fig. (c), the circular frequency, frequency, and period of free vibration are

$$p = \sqrt{\frac{k_0}{m}} = \sqrt{\frac{2k/3}{m}} = \sqrt{\frac{2(875)/3}{0.2}} = 54\,\text{rad/s} \qquad \textit{Answer}$$

$$f = \frac{p}{2\pi} = \frac{54}{2\pi} = 8.6\,\text{Hz} \qquad\qquad \textit{Answer}$$

$$\tau = \frac{1}{f} = \frac{1}{8.6} = 0.116\,\text{s} \qquad\qquad \textit{Answer}$$

Sample Problem   20.2

The simple pendulum consists of a small bob of mass m that is attached to the end of a string. The pendulum is released from rest when $\theta = 30°$. Using numerical integration of the differential equation of motion, calculate (1) the period of oscillation; and (2) the maximum angular velocity. Compare the maximum angular velocity to the exact value obtained by the work-energy method.

Solution

Part 1

Because the pendulum is a conservative system, it oscillates between $\theta = \pm 30°$. The motion is periodic, but not harmonic, because θ is not a small angle—that is, not less than $6°$. We calculate the period of the pendulum by noting that the time taken by the pendulum to travel from the initial position ($\theta = 30°$) to the vertical position ($\theta = 0$) equals one-fourth of the period.

From Eq. (20.9), the differential equation of motion is $\ddot{\theta} = -(g/L)\sin\theta = -(9.81/0.750)\sin\theta = -13.080\sin\theta$. The equivalent first-order equations are

$$\dot{\theta} = \omega \qquad \dot{\omega} = -13.080\sin\theta$$

with the initial conditions $\theta = 30° = \pi/6\,\text{rad}$ and $\omega = 0$ at $t = 0$ (the time of release). The integration interval extends from $t = 0$ until the time when $\theta = 0$ for the first time. An estimate of the period can be obtained from the linearized differential equation, Eq. (20.10): $\ddot{\theta} + (g/L)\theta = 0$. We get from Eq. (20.7) $\tau = 2\pi/p = 2\pi/\sqrt{g/L} = 2\pi/\sqrt{9.81/0.75} = 1.737\,\text{s}$. To be on the safe side, the integration period for the nonlinear problem should be somewhat larger than $1.737/4$. We used $0.45\,\text{s}$ in the MATLAB program listed below.

```
function example20_3
[t,x] = ode45(@f,[0:0.01:0.45],[pi/6 0]);
printSol(t,x)
    function dxdt = f(t,x)
    dxdt = [x(2); -13.080*sin(x(1))];
    end
end
```

The last two lines of the ouput (x1 corresponds to θ) were

```
    t              x1              x2
4.4000e-001   3.5331e-003    -1.8721e+000
4.5000e-001   -1.5186e-002   -1.8713e+000
```

The value of t when $\theta = 0$ now can be obtained by linear interpolation:

$$\frac{-0.015\ 186 - 0.003\ 533\ 1}{0.45 - 0.44} = \frac{0 - 0.003\ 533\ 1}{t - 0.44}$$

which yields $t = 0.4419$ s for the quarter period. Therefore, the period is

$$\tau = 4(0.4419) = 1.768\,\text{s} \qquad \textit{Answer}$$

Part 2

The maximum angular velocity (x2 in the printout) occurs when $\theta = 0$. By inspection, we see that

$$\left|\dot{\theta}\right|_{\max} = 1.872\,\text{rad/s} \qquad \textit{Answer}$$

To calculate the exact value for $\dot{\theta}_{\max}$ using the work-energy method, we let the subscripts 1 and 2 denote the positions $\theta = \pi/6$ rad and $\theta = 0$, respectively. The corresponding kinetic energies are $T_1 = 0$ (pendulum is stationary in position 1) and $T_2 = (1/2)m(L\dot{\theta}_{\max})^2$, where $L\dot{\theta}_{\max}$ is the velocity of the pendulum in position 2. The work done by the weight of the bob as it moves from position 1 to position 2 is $U_{1-2} = mg(L - L\cos\theta) = mgL[1 - \cos(\pi/6)]$, because the vertical distance between the two positions equals $L - L\cos\pi/6$. Applying the work-energy principle, we obtain

$$U_{1-2} = T_2 - T_1$$

$$mgL\left(1 - \cos\frac{\pi}{6}\right) = \frac{1}{2}m(L\dot{\theta}_{\max})^2 - 0$$

from which the value of $\dot{\theta}_{\max}$ is found to be

$$\dot{\theta}_{\max} = \sqrt{\frac{2g}{L}\left(1 - \cos\frac{\pi}{6}\right)} = \sqrt{\frac{2(9.81)}{0.750}\left(1 - \cos\frac{\pi}{6}\right)}$$

$$= 1.872\,\text{rad/s} \quad \text{(clockwise)}$$

We see that the value for $\dot{\theta}_{\max}$ obtained by numerical integration agrees with the above result within the four significant digits used, thereby verifying the accuracy of the numerical integration.

It is instructive to note that although the angular velocity at any position can be obtained analytically, the time of travel between two positions must be computed numerically.

Sample Problem 20.3

The block of mass m shown in Fig. (a) is at rest in the equilibrium position at $x = 0$ when it receives an impulse that results in the initial velocity $\dot{x}(0) = v_0$. (1) Derive the differential equation of motion for the block. Assuming that the system is critically damped, determine (2) the damping coefficient c; and (3) the maximum displacement of the block.

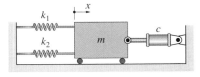

(a)

Solution

Part 1

The free-body (FBD) and mass-acceleration (MAD) diagrams of the block in an arbitrary position are shown in Fig. (b). The forces acting on the mass are its weight mg, the resultant normal reaction N, the two spring forces, and the force exerted by the damper. The differential equation of motion in the x-direction is

$$\Sigma F_x = ma_x \quad \xrightarrow{+} \quad -k_1 x - k_2 x - c\dot{x} = m\ddot{x}$$

or

$$m\ddot{x} + c\dot{x} + (k_1 + k_2)x = 0 \qquad \textit{Answer} \quad \text{(a)}$$

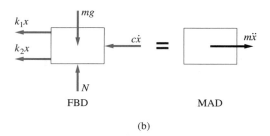

(b)

Part 2

By comparing Eq. (a) with Eq. (20.11), we deduce that the undamped circular frequency is

$$p = \sqrt{\frac{k_1 + k_2}{m}} \qquad \text{(b)}$$

Therefore, the critical damping coefficent is—see Eq. (20.14)—

$$c = c_{\text{cr}} = 2mp = 2m\sqrt{\frac{k_1 + k_2}{m}} = 2\sqrt{m(k_1 + k_2)} \qquad \textit{Answer}$$

Part 3

The displacement of the block is given by Eq. (20.18)

$$x = (A_1 + A_2 t)e^{-pt} \qquad \text{(c)}$$

Hence, the velocity is

$$\dot{x} = A_2 e^{-pt} - p(A_1 + A_2 t)e^{-pt} \qquad \text{(d)}$$

Substituting the initial conditions $x = 0$ and $\dot{x} = v_0$ at $t = 0$ into Eqs. (c) and (d) yields $A_1 = 0$ and $A_2 = v_0$. Consequently,

$$x = v_0 t e^{-pt} \quad \dot{x} = v_0(1 - pt)e^{-pt}$$

The maximum displacement occurs when $\dot{x} = 0$, which happens at time $t = 1/p$. Therefore, the maximum displacement is

$$x_{max} = \frac{v_0}{p}e^{-1} = 0.368v_0\sqrt{\frac{m}{k_1 + k_2}} \qquad \textit{Answer}$$

Sample Problem 20.4

The system shown in Fig. (a) of Sample Problem 20.3 is underdamped with the damping factor being $\zeta = 0.25$. If the initial conditions on the motion of the block are $x = 0$ and $\dot{x} = 4$ m/s, determine the displacement of the block at time $t = 0.1$ s. Use the data $m = 0.2$ kg, $k_1 = 20$ N/m, and $k_2 = 30$ N/m.

Solution

The motion of an underdamped system is described by Eq. (20.19):

$$x = Ee^{-\zeta pt}\sin(\omega_d t + \alpha)$$

where E and α are to be determined from the initial conditions. From Eq. (b) of Sample Problem 20.3, we have

$$p = \sqrt{\frac{k_1 + k_2}{m}} = \sqrt{\frac{20 + 30}{0.2}} = 15.811 \text{ rad/s}$$

Hence $\zeta p = 0.25(15.811) = 3.953$ rad/s. The damped circular frequency is given by Eq. (20.20):

$$\omega_d = p\sqrt{1 - \zeta^2} = 15.811\sqrt{1 - 0.25^2} = 15.309 \text{ rad/s}$$

Therefore, the displacement and the velocity of the block are

$$x = Ee^{-3.953t}\sin(15.309t + \alpha)$$
$$\dot{x} = Ee^{-3.953t}[15.309\cos(15.309t + \alpha) - 3.953\sin(15.309t + \alpha)]$$

Applying the initial conditions yields

$$x(0) = E\sin\alpha = 0$$
$$\dot{x}(0) = E(15.309\cos\alpha - 3.953\sin\alpha) = 4 \text{ m/s}$$

The solution is $\alpha = 0$ and $E = 4/15.309 = 0.2613$ m. Consequently, the displacement is

$$x = 0.2613e^{-3.953t}\sin 15.309t \text{ m}$$

Substituting $t = 0.1$ s, we obtain

$$x(0.1) = 0.2613e^{-0.3953}\sin 1.5309 = 0.1758 \text{ m} \qquad \textit{Answer}$$

Problems

Fig. P20.1, P20.2

20.1 The mass $m = 10\,\text{kg}$ is suspended from an ideal spring of stiffness $k = 250\,\text{N/m}$. If the mass is set into motion at $t = 0$ with the initial conditions $x_0 = 40\,\text{mm}$ and $v_0 = -80\,\text{mm/s}$, calculate (a) the amplitude of the motion; and (b) the time when the mass stops for the first time. Assume that x is measured from the equilibrium position of the mass.

20.2 The mass m is suspended from an ideal spring of stiffness k and is set into motion with the initial conditions $x_0 = 7\,\text{mm}$ and $v_0 = 1.5\,\text{m/s}$. Assume that x is measured from the equilibrium position of the mass. If the amplitude of the vibration is $12\,\text{mm}$, determine (a) the frequency; and (b) x as a function of time.

20.3 The mass m is suspended from two springs of stiffnesses k_1 and k_2. Determine the expression for the circular frequency of the mass if the springs are arranged as shown in (a); and in (b).

20.4 The block of mass m is suspended from three springs with stiffnesses k_1 and k_2 as shown. The frequency of vibration of this system is $5\,\text{Hz}$. After the middle spring is removed, the frequency drops to $3.6\,\text{Hz}$. Determine the ratio k_2/k_1.

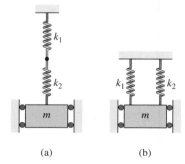

(a)   (b)

Fig. P20.3

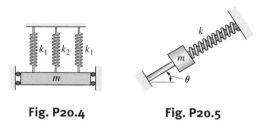

Fig. P20.4   **Fig. P20.5**

20.5 The mass m of the spring-mass system slides on the inclined rod with negligible friction. By deriving the differential equation of motion for the mass, show that the frequency of vibration is independent of the angle θ.

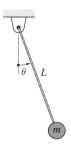

Fig. P20.6, P20.7

20.6 When only mass B is attached to the ideal spring, the frequency of the system is $3.90\,\text{Hz}$. When mass C is added, the frequency decreases to $2.55\,\text{Hz}$. Determine the ratio m_B/m_C of the two masses.

20.7 The ideal spring of stiffness $k = 120\,\text{N/m}$ is connected to mass B through a hole in mass C. The mass of B and C are $0.4\,\text{kg}$ and $0.8\,\text{kg}$, respectively. Find the smallest vibrational amplitude for which C will lose contact with B. Would the result change if the mass of B and C were interchanged?

Fig. P20.8

20.8 The simple pendulum is released from rest at $\theta = \theta_0$. Determine the expressions for the maximum values of $\dot{\theta}$ and $\ddot{\theta}$ (a) assuming that θ is small (simple harmonic motion); and (b) without making any simplifying assumptions. (c) Compare the expressions found in (a) and (b) for $\theta_0 = 5°$, $10°$, and $15°$.

20.9   The block of mass m is supported by a small pulley. One end of the cable running around the pulley is attached to a fixed support, while the other end is connected to a spring. Determine the circular frequency p of the system. Neglect the mass of the pulley.

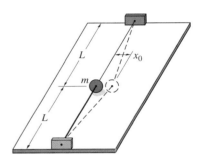

Fig. P20.9

20.10   The figure shows two elastic cords, each of length L, connected to a small ball of mass m that slides on a smooth horizontal surface. The cords are stretched to an initial tension T between rigid supports. The ball is given a small displacement $x = x_0$ perpendicular to the cords and then released from rest at time $t = 0$. Derive the equation of motion for the ball, and show that the ball undergoes simple harmonic motion.

20.11   The metronome consists of a 0.2-kg small mass attached to the arm OA. The metronome's period of oscillation can be adjusted by changing the distance L. Determine the value of L for which the period will be 1.0 s for small amplitudes. Neglect the mass of arm OA.

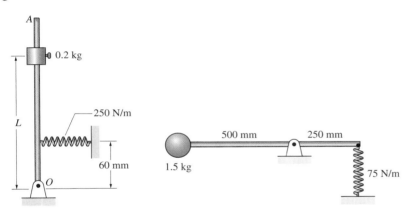

Fig. P20.10

Fig. P20.11   **Fig. P20.12**

20.12   The system is in equilibrium in the position shown. Find the period of vibration for small amplitudes. Neglect the mass of the rod and size of the 1.5 kg mass.

20.13   Two identical springs of free length L_0 and stiffness k are attached to the collar of weight W. Friction between the collar and the horizontal rod may be neglected. Assuming small amplitudes, (a) derive the differential equation of motion; and (b) find the frequency given that $W = 1$ N, $L_0 = 0.1$ m, $b = 0.15$ m, and $k = 80$ N/m.

20.14   For the collar described in Prob. 20.13, (a) show that the differential equation of motion for large amplitudes is

$$\ddot{x} = -\frac{2gk}{W}\left(1 - \frac{L_0}{\sqrt{b^2 + x^2}}\right)x$$

(b) Using the data given in Prob. 20.13, determine the period if the amplitude is 6 in.

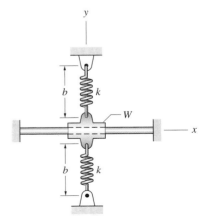

Fig. P20.13, P20.14

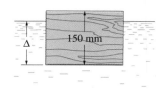

Fig. P20.15

*20.15 A block of wood floats in water in the stable equilibrium position shown. The block is displaced slightly in the vertical direction and released. Derive the differential equation of motion, and show that the motion of the block is simple harmonic. Find the period, knowing that the densities of wood and water are 600 kg/m³ and 1000 kg/m³, respectively.

20.16 Derive Eq. (20.22): $\ln(x_{n+1}/x_n) = -2\pi\zeta / \sqrt{1 - \zeta^2}$.

20.17 (a) Using the expression for the logarithmic decrement, show that

$$\ln(x_{n+k}/x_n) = -\frac{2k\pi\zeta}{\sqrt{1 - \zeta^2}}$$

where k is a positive integer greater than 1. (b) Using the result of part (a), estimate the damping factor ζ for a system that has the displacement-time curve shown in the figure.

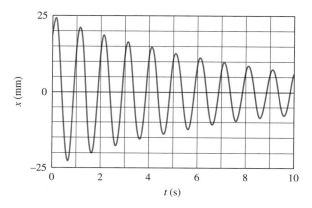

Fig. P20.17

20.18 An oscillator consists of a 7 kg mass that is connected to a spring of stiffness $k = 3100\,\text{N/m}$ and a viscous damper with the damping constant $c = 117\,\text{N} \cdot \text{s/m}$. (a) Show that the oscillator is underdamped. (b) Determine the ratio x_{n+1}/x_n of two successive peak displacements.

20.19 (a) Use the logarithmic decrement to show that the relationship between ΔE, the percentage of energy lost per cycle, and the damping factor ζ of an underdamped oscillator is

$$\Delta E = \left(1 - e^{-4\pi\zeta/\sqrt{1-\zeta^2}}\right) \times 100\%$$

(b) Using the results of part (a), compute the damping factor that would cause 10% energy loss per cycle.

20.20 Calculate the damping coefficient c if the system shown is to be critically damped.

20.21 The system shown is underdamped with a damped period of τ_d. When the dashpot is removed, the period changes to $0.5\tau_d$. What is the damping coefficient c?

20.22 The 3-kg mass has an initial displacement of $x(0) = 0.01$ m and an initial velocity of $\dot{x}(0) = -0.2$ m/s. Knowing that the system is critically damped, determine the displacement of the mass when $t = 0.1$ s.

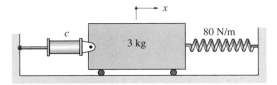

Fig. P20.20–P20.22

20.23 The system is released from rest at $x = 25$ mm, where x is measured from the position where the springs are unstretched. (a) Determine if the system is underdamped, critically damped, or overdamped. (b) Derive the expression for $x(t)$.

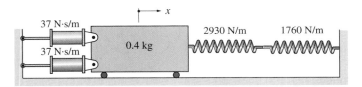

Fig. P20.23

20.24 The mass m of the underdamped system is displaced slightly and then released. Determine the damped period τ_d of the resulting oscillation. Neglect the mass of the bar and use the following data: $b = 0.3$ m, $m = 4$ kg, $k = 2$ kN/m, and $c = 36$ N · s/m.

20.25 The system is released from rest at time $t = 0$ with the initial displacement $x_0 = 50$ mm. Both springs are unstretched when $x = 0$. Determine the expression for $x(t)$.

20.26 Each of the two bumpers mounted on the end of a 81 500-kg railroad car has a spring stiffness of $k = 1.76 \times 10^5$ N/m and a coefficient of viscous damping equal to 6.6×10^5 N · s/m. Recognizing that a bumper will "bottom out" when its deformation exceeds 304 mm, find the largest velocity v_0 with which the car can safely hit a rigid wall.

20.27 A critically damped oscillator is released from rest with the initial displacement x_0 (measured from the equilibrium position). (a) Derive the expressions for the displacement and the velocity of the oscillator in terms of x_0, p, and t. (b) Determine the expression for the maximum speed of the oscillator in terms of x_0 and p.

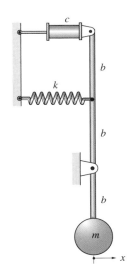

Fig. P20.24

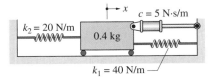

Fig. P20.25

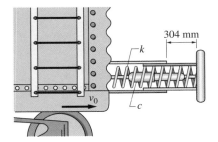

Fig. P20.26

20.28 The mass of the pendulum is $m = 0.5\,\text{kg}$, and its length is $L = 1.5\,\text{m}$. At a certain time, the amplitude of the pendulum was measured to be 5°; one hour later, it was 3°. Assuming that the damping responsible for the decay of amplitude is viscous, calculate the damping coefficient.

Fig. P20.28

20.29 Viscous damping is not an accurate representation of the resistance experienced by a body that is moving through a low-viscosity fluid, such as air or water. Experiments indicate that the damping force is actually proportional to the square of the velocity: $F_d = cv^2$. For a one-inch diameter sphere moving in air, the approximate value of the damping constant is $c = 1.24 \times 10^{-4}\,\text{N} \cdot \text{s}^2/\text{m}^2$. (a) Use this information to derive the differential equation of motion for the pendulum shown (the bob is made of steel). Neglect the damping effect of the string. (b) Integrate the equation of motion numerically over two periods of vibration, assuming that the pendulum is released from rest when $\theta = 30°$. (c) Use the results of the integration to calculate the percent loss of amplitude over the first two periods. (Note: To verify that the loss of amplitude is real and not due to numerical errors in the integration procedure, it is recommended that you repeat parts (b) and (c) with zero damping—there should be no loss of amplitude.)

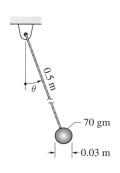

Fig. P20.29, P20.30

20.30 Repeat Prob. 20.29 if the bob of the pendulum is replaced with a 0.03 m diameter styrofoam ball that mass $2 \times 10^{-4}\,\text{kg}$.

20.31 The system shown is released from rest with the initial displacement $x_0 = 0.1$ m, measured from the equilibrium position. (a) Assuming that the damping force exerted by the dashpot is $F_d = c_2 \dot{x}^2$, where $c_2 = 50\,\text{N} \cdot \text{s}^2/\text{m}^2$, determine the ratio x_1/x_0. (b) If the damping force were $F_d = c_1 \dot{x}$, determine the value of c_1 that would give the same ratio x_1/x_0 as in part (a). (c) Plot x versus t over the period $t = 0$ to $1.0\,\text{s}$ for the two cases of damping. What is the main difference in the amplitude decay between the two types of damping?

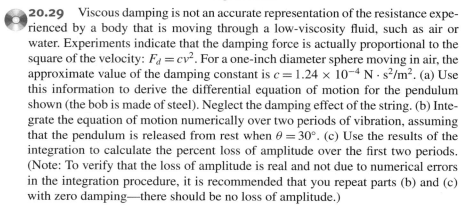

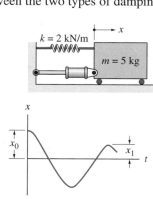

Fig. P20.31

20.3 *Forced Vibrations of Particles*

In free vibrations, the oscillations are initiated by a disturbance that gives rise to an initial displacement, initial velocity, or both. No external forces are required to maintain the motion. In forced vibration, a sustained external source is responsible for maintaining the vibration. A common example of forced vibration is the automobile "rattle," caused either by the reciprocation of the engine or irregularities of the road surface. Here we consider forced vibrations that arise from either a harmonic (that is, sinusoidally varying) forcing function, or a harmonic support displacement. Apart from being important by itself, harmonic input is also useful in the analysis of more general forced motion, because any forcing function can be decomposed into harmonic components by expressing it as a Fourier series.

a. *Harmonic forcing function*

Figure 20.8(a) shows a damped spring-mass system that is subjected to a time-dependent force $P = P_0 \sin \omega t$, where P_0 is the magnitude of the force and ω is its circular frequency. The force P is referred to as the *harmonic forcing function*, and ω is called the *forcing frequency*. The static deflection of the spring is Δ, and x is the displacement of the mass from its equilibrium position. From the free-body and mass-acceleration diagrams shown in Fig. 20.8(b), we obtain for the differential equation of motion

$$\Sigma F_x = ma_x \quad +\downarrow \quad mg + P_0 \sin \omega t - k(x + \Delta) - c\dot{x} = m\ddot{x}$$

Using the equilibrium equation $mg = k\Delta$, the equation of motion simplifies to

$$m\ddot{x} + c\dot{x} + kx = P_0 \sin \omega t \tag{20.23}$$

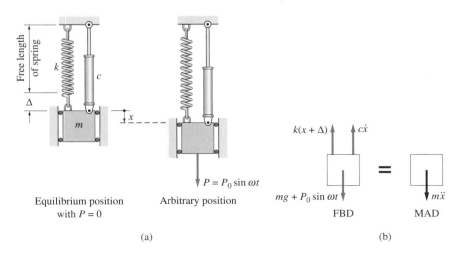

Equilibrium position with $P = 0$ — Arbitrary position

$k(x + \Delta)$ — $c\dot{x}$

$mg + P_0 \sin \omega t$ — FBD

$m\ddot{x}$ — MAD

$P = P_0 \sin \omega t$

(a) (b)

Fig. 20.8

Equation (20.23) is a nonhomogeneous (right-hand side is not zero), second-order linear differential equation. The general solution of this equation can be represented as the sum of the *complementary solution* x_c and a *particular solution* x_p; that is,

$$x = x_c + x_p \qquad (20.24)$$

The complementary solution of Eq. (20.23) is a solution of the homogeneous equation (obtained by setting the right-hand side equal to zero), and the particular solution is any solution of the complete equation.

We note that the homogeneous equation is identical to Eq. (20.11). Therefore, the complementary solution of Eq. (20.23) is given by Eqs. (20.17), (20.18), or (20.19), depending on the damping factor ζ.

The complementary solution, also called the *transient vibration,* is generally not interesting from a practical viewpoint, because it decays with time. From here on, we restrict our attention to the particular solution, which represents the steady-state vibration.

By direct substitution, it can be shown that a particular solution of Eq. (20.23) is

$$x_p = X \sin(\omega t - \phi) \qquad (20.25)$$

where the amplitude X is given by

$$X = \frac{P_0/k}{\sqrt{[1 - (\omega/p)^2]^2 + (2\zeta\omega/p)^2}} \qquad (20.26)$$

and the phase angle ϕ (the angle by which x_p lags P) is

$$\phi = \tan^{-1}\left[\frac{2\zeta\omega/p}{1 - (\omega/p)^2}\right] \qquad (20.27)$$

where $p = \sqrt{k/m}$ is the undamped circular frequency and $\zeta = c/(2mp)$ represents the damping factor. The numerator P_0/k in Eq. (20.26) is called the *zero-frequency deflection*—the deflection caused by the constant force P_0 (not to be confused with the static deflection $\Delta = mg/k$). The term ω/p (ratio of the forcing frequency to the undamped frequency) is known as the *frequency ratio.*

The *magnification factor* is defined as the ratio of the amplitude of the steady-state vibration divided by the zero-frequency deflection:

$$\text{Magnification factor} = \frac{X}{P_0/k} = \frac{1}{\sqrt{[1 - (\omega/p)^2]^2 + (2\zeta\omega/p)^2}} \qquad (20.28)$$

The magnification factor is plotted versus the frequency ratio in Fig. 20.9. As expected, the magnification factor is largest for frequencies near *resonance* ($\omega = p$) if the damping coefficient is sufficiently small (the magnification factor becomes infinite at resonance in the absence of damping).

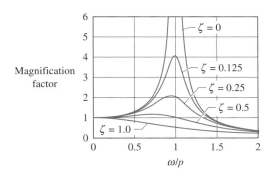

Fig. 20.9

Additional information about the plots in Fig. 20.9 can be obtained by differentiating Eq. (20.28) with respect to ω/p and setting the result equal to zero. This procedure yields the following information: (1) all curves in Fig. 20.9 are tangent to the horizontal at $\omega/p = 0$ and as $\omega/p \to \infty$; (2) if $\zeta > 0.707$, the maximum magnification factor is 1.0, occurring at $\omega/p = 0$; (3) if $\zeta < 0.707$, the magnification factor attains its maximum value at $\omega/p = \sqrt{1 - 2\zeta^2}$, not at resonance.

b. *Harmonic support displacement*

Figure 20.10(a) shows a viscously damped spring-mass system with the support undergoing the prescribed harmonic displacement $y = Y \sin \omega t$. The free-body and mass-acceleration diagrams are shown in Fig. 20.10(b). Note that x is the

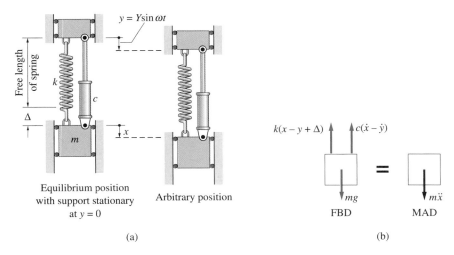

Fig. 20.10

displacement of the mass from its equilibrium position (with support stationary at $y = 0$), and Δ is the static deflection of the spring. We see that the relative displacement $x - y$ and the relative velocity $\dot{x} - \dot{y}$ determine the spring and damping forces, respectively. Applying Newton's second law to the FBD and MAD, we obtain

$$\Sigma F_x = ma_x \quad +\!\downarrow \quad mg - c(\dot{x} - \dot{y}) - k(x - y + \Delta) = m\ddot{x} \tag{a}$$

Noting that $mg = k\Delta$ (the equilibrium equation) and introducing the relative position coordinate $z = x - y$, Eq. (a) simplifies to

$$m\ddot{z} + c\dot{z} + kz = -m\ddot{y} \tag{b}$$

Substituting $y = Y \sin \omega t$ into Eq. (b), the differential equation of motion becomes

$$m\ddot{z} + c\dot{z} + kz = mY\omega^2 \sin \omega t \tag{20.29}$$

Comparing Eqs. (20.23) and (20.29), we see that our previous analysis with a harmonic force is also applicable to harmonic support displacement, provided that P_0 is replaced by $mY\omega^2$ and x by z. Making these substitutions, Eqs. (20.25)–(20.27) describing the steady-state vibration become

$$z = Z \sin(\omega t - \phi) \tag{20.30}$$

where the amplitude Z and phase angle ϕ (the angle by which z lags y) are given by

$$Z = \frac{mY\omega^2/k}{\sqrt{[1 - (\omega/p)^2]^2 + (2\zeta\omega/p)^2}}$$

$$= Y\frac{(\omega/p)^2}{\sqrt{[1 - (\omega/p)^2]^2 + (2\zeta\omega/p)^2}} \tag{20.31}$$

and, as before,

$$\phi = \tan^{-1}\left[\frac{2\zeta\omega/p}{1 - (\omega/p)^2}\right] \tag{20.32}$$

where $p = \sqrt{k/m}$ and $\zeta = c/(2mp)$.

Sample Problem 20.5

The electric motor and its frame shown in Fig. (a) have a combined mass M. The unbalance of the rotor is equivalent to a mass m (included in M), located at the distance e from the center O of the shaft. The frame is supported by vertical guides, a spring of stiffness k, and a pin at A. (The pin at A was inserted when the motor was at rest in the static equilibrium position.) The motor is running at the constant angular speed ω when the pin is withdrawn at the instant when the mass m is in the position shown. (1) Derive the differential equation of motion of the assembly. (2) Plot the transient, steady-state, and total displacement versus time, using the data $M = 40$ kg, $m = 1.2$ kg, $e = 150$ mm, $k = 900$ N/m, and $\omega = 8$ rad/s.

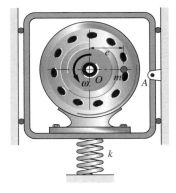

(a)

Solution

The rotating unbalance causes the position of the mass center of the assembly to vary with time, thereby giving rise to vibrations in the vertical direction (the vertical guides prevent horizontal motion of the frame).

Part 1

The free-body diagram (FBD) of the assembly is shown in Fig. (b). Note that x was chosen to be the displacement of O from its static equilibrium position. The FBD contains the total weight Mg of the assembly, acting at its mass center G, the spring force $k(x - \Delta)$, where Δ is the static deflection of the spring, and the resultant horizontal force N exerted by the vertical guides. The mass-acceleration diagram in Fig. (b) displays the inertia vectors $(M - m)\ddot{x}$ and $m\ddot{x}$ associated with the vertical motion of the assembly, and the radial inertia vector $me\omega^2$ caused by the rotation of the mass m about O. Equating the resultant vertical forces on the FBD and the MAD, we get

$$+\uparrow \quad -Mg - k(x - \Delta) = (M - m)\ddot{x} + m\ddot{x} - me\omega^2 \sin \omega t \qquad \text{(a)}$$

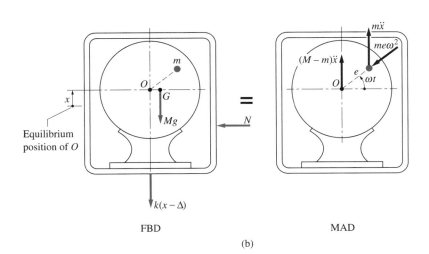

(b)

569

After eliminating Δ by utilizing the equilibrium equation $Mg - k\Delta = 0$, the differential equation of motion becomes

$$M\ddot{x} + kx = me\omega^2 \sin \omega t \qquad \qquad \textit{Answer} \quad \text{(b)}$$

Comparing Eq. (b) with Eq. (20.23), we see that the equations are identical if we let $c = 0$ and replace the magnitude P_0 of the forcing function by $me\omega^2$ (the term $me\omega^2$ is sometimes referred to as the centrifugal force due to the unbalanced mass).

Part 2

With $P_0 = me\omega^2$ and $\zeta = 0$ (no damping), Eqs. (20.26) and (20.27) yield

$$X = \frac{me\omega^2/k}{1 - (\omega/p)^2} \quad \text{and} \quad \phi = 0$$

so that the particular solution of Eq. (b) is—see Eq. (20.25)—

$$x_p = \frac{me\omega^2/k}{1 - (\omega/p)^2} \sin \omega t$$

Since the system is undamped, the complementary solution is given by Eq. (20.4): $x_c = E \sin(\omega t + \alpha)$. Therefore, the complete solution is

$$x = x_c + x_p = E \sin(\omega t + \alpha) + \frac{me\omega^2/k}{1 - (\omega/p)^2} \sin \omega t \qquad \text{(c)}$$

Using the given numerical values, we obtain

$$p = \sqrt{k/M} = \sqrt{900/40} = 4.743 \text{ rad/s}$$

$$\frac{me\omega^2/k}{1 - (\omega/p)^2} = \frac{1.2(0.15)(8)^2/900}{1 - (8/4.743)^2} = -0.006\,938 \text{ m} = -6.938 \text{ mm}$$

Substituting these values into Eq. (c) yields

$$x(t) = E \sin(4.743t + \alpha) - 6.938 \sin 8t \text{ mm} \qquad \text{(d)}$$

Taking the time derivative, we obtain for the velocity

$$\dot{x}(t) = 4.743 E \cos(4.743t + \alpha) - 55.50 \cos 8t \text{ mm/s} \qquad \text{(e)}$$

Substituting the initial conditions ($x = 0$ and $\dot{x} = 0$ when $t = 0$) into Eqs. (d) and (e) gives $E \sin \alpha = 0$ and $4.743E \cos \alpha - 55.50 = 0$, which yields $\alpha = 0$ and $E = 11.701$ mm. Therefore the description of the motion is

$$x(t) = 11.70 \sin 4.74t - 6.94 \sin 8t \text{ mm} \qquad \qquad \textit{Answer} \quad \text{(f)}$$

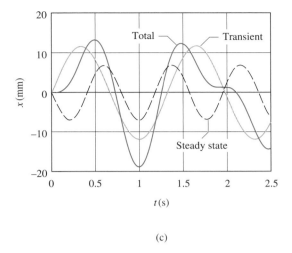

(c)

Figure (c) shows plots of the transient vibration (with the period $\tau_t = 2\pi/p = 2\pi/4.743 = 1.325$ s), the steady-state vibration (period $\tau_s = 2\pi/\omega = 2\pi/8 = 0.785$ s), and the superposition of the two. Because the coefficient of $\sin 8t$ in Eq. (f) is negative, the steady-state vibration is 180° out of phase with the rotation of the unbalanced mass.

Sample Problem **20.6**

The block of weight W is connected in a rigid frame between a linear spring and a viscous damper. The frame is subjected to the time-dependent vertical displacement $y(t) = Y \sin \omega t$. The displacement x of the block is measured from its static equilibrium position (with support stationary at $y = 0$). Determine the steady-state solution for (1) the relative displacement $z = x - y$; and (2) the absolute displacement x. Use $Y = 40$ mm, $\omega = 400$ rad/s, $M = 3$ kg, $k = 2.63 \times 10^5$ N/m, and $c = 585$ N $\cdot$ s/m.

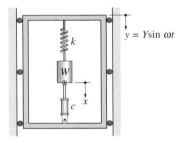

Solution

Part 1

Because the given system is equivalent to that shown in Fig. 20.10, Eqs. (20.30)–(20.32) can be used to determine the relative displacement z. The parameters appearing in these equations are

$$p = \sqrt{\frac{k}{m}} = \sqrt{\frac{2.63 \times 10^5}{3}} = 296 \text{ rad/s}$$

$$\zeta = \frac{c}{2mp} = \frac{585}{2(3)(296)} = 0.329 \quad \text{(underdamped)}$$

$$\frac{\omega}{p} = \frac{400}{296} = 1.351$$

The relative amplitude of the steady-state solution is now obtained from Eq. (20.31)

$$Z = Y \frac{(\omega/p)^2}{\sqrt{[1 - (\omega/p)^2]^2 + (2\zeta\omega/p)^2}}$$

$$= 0.04 \frac{(1.287)^2}{\sqrt{[1 - (1.351)^2]^2 + [2(0.329)(1.351)]^2}}$$

$$= 0.07 \text{ m}$$

and the phase angle can be computed from Eq. (20.32)

$$\phi = \tan^{-1}\left[\frac{2\zeta\omega/p}{1 - (\omega/p)^2}\right] = \tan^{-1}\left[\frac{2(0.329)(1.351)}{1 - (1.351)^2}\right]$$

$$= -0.8226$$

The relative displacement thus becomes

$$z(t) = 2.249 \sin(400t + 0.8226) \text{ m} \qquad \textit{Answer}$$

Note that $z(t)$ leads $y(t)$ by 0.8226 rad (47.1°).

Part 2

Using the trigonometric identity $\sin(a + b) = \sin a \cos b + \cos a \sin b$, the expression for $z(t)$ in Part 1 can be written as

$$z(t) = 0.07(\cos 0.8226 \sin 400t + \sin 0.8226 \cos 400t)$$

$$= 0.048 \sin 400t + 0.0512 \cos 400t \text{ m}$$

Therefore, the expression for the absolute displacement of the mass becomes

$$x = z + y$$

$$= (0.048 \sin 400t + 0.0512 \cos 400t) + 0.04 \sin 400t$$

$$= 0.088 \sin 400t + 0.0512 \cos 400t \text{ m} \qquad \textit{Answer}$$

Problems

20.32   The spring-mass system is at rest in the equilibrium position $x = 0$ when the harmonic force $P(t) = P_0 \sin \omega t$ is applied at $t = 0$, where $P_0 = 100$ N and $\omega = 25$ rad/s. Determine (a) the expression for the displacement $x(t)$ (include both the transient and steady-state solutions); (b) the magnification factor; and (c) the maximum value of $x(t)$.

20.33   The spring-mounted mass is driven by the force $P(t) = P_0 \sin \omega t$, where $P_0 = 100$ N. Calculate the two values of ω for which the amplitude of the steady-state vibration is 50 mm.

20.34   The 0.2 kg mass is suspended from a rigid frame. Pin A at the end of arm OA engages a slot in the frame, causing the frame to oscillate in the vertical direction as the arm turns. If the angular velocity of OA is $\omega = 35$ rad/s, determine the amplitude of the steady-state vibration of the weight relative to the frame.

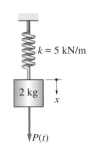

Fig. P20.32, P20.33

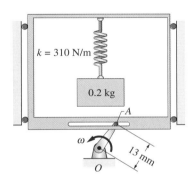

Fig. P20.34

20.35   The block of mass $m = 4$ kg is attached to a spring of stiffness k. When the harmonic force $P_0 \sin \omega t$ is applied to the block where $P_0 = 0.25$ kg and $\omega = 6$ rad/s, the steady-state amplitude is 7.5 mm. Determine the two possible values of k.

20.36   The spring-mass system is being driven by the harmonic force $P_0 \sin \omega t$. When $\omega = 5$ rad/s, the steady-state amplitude of the block is 20 mm. When the frequency is increased to $\omega = 10$ rad/s, with P_0 unchanged, the amplitude becomes 4 mm. Determine (a) the natural frequency p of the system; and (b) the zero-frequency deflection P_0/k.

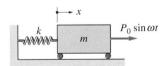

Fig. P20.35–P20.37

20.37   The system shown consists of the mass $m = 6$ kg and a nonlinear spring. The force-extension relationship of the spring is $F = kx[1 + (x/b)^2]$, where x is measured in meters, $k = 1200$ N/m, and $b = 0.25$ m. The amplitude and frequency of the driving force are $P_0 = 60$ N and $\omega = 14.142$ rad/s, respectively, the latter being equal to the natural circular frequency of the system for small amplitudes. (Note that if $x \ll b$, then $F \approx kx$, and $p = \sqrt{k/m} = \sqrt{1200/6} = 14.142$ rad/s.)

The system starts from rest in the equilibrium position $x = 0$ when $t = 0$. (a) Show that the differential equation of motion is

$$\ddot{x} = -200x(1 + 16x^2) + 10\sin 14.142t \text{ m/s}^2$$

(b) Integrate the equation of motion numerically from $t = 0$ to $t = 2$ s, and plot x versus t.

20.38 The block of mass $m = 1$ kg is connected to the shaker table by a spring of stiffness $k = 2100$ N/m. When the spring is unstretched, the distance between the block and a stop attached to the table is $b = 25$ mm. If the table is being driven at $y = 6\sin\omega t$ mm, determine the range of ω for which the block will not hit the stop. Consider only the steady-state vibration.

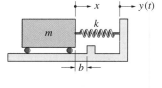

Fig. P20.38

20.39 The enclosure of the system shown undergoes the vertical displacement $y(t) = Y\sin\omega t$. Knowing that $Z = 5Y$, where Z is the steady-state amplitude of the mass m relative to the support, determine the forcing frequency ω.

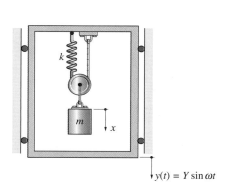

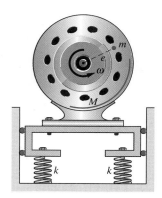

Fig. P20.39   **Fig. P20.40**

20.40 An electric motor and its base, with a combined mass of $M = 12$ kg, are supported by four identical springs, each of stiffness $k = 480$ kN/m. The unbalance of the rotor is equivalent to a mass $m = 0.005$ kg located at a distance $e = 90$ mm from its axis. (a) Calculate the angular speed of the motor that would cause resonance. (b) Compute the maximum steady-state displacement of the motor when its angular speed is 99 percent of the speed at resonance.

20.41 The pendulum of length L and mass m is suspended from a sliding collar. If the horizontal displacement $y(t) = Y\sin\omega t$ is imposed on the collar, show that for sufficiently small θ, the steady-state amplitude of the pendulum is

$$\theta_{\max} = \frac{\omega^2 L}{|g - \omega^2 L|}\frac{Y}{L}$$

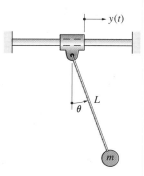

Fig. P20.41, P20.42

20.42 The pendulum of length $L = 1.09$ m is suspended from a sliding collar, the displacement of which is given by $y(t) = 0.545\sin 5t$ m, where t is in seconds. It is known that at time $t = 0$, $\theta = 30°$ and $\dot{\theta} = 0$. (a) Derive the differential equation of motion for the pendulum, choosing θ as the independent coordinate (do not restrict θ to small angles). (b) Integrate the equation of motion numerically from $t = 0$ to $t = 10$ s, and plot θ versus t for the period of integration. (c) Describe the motion of the pendulum during this period.

20.43 When the mass m is attached to the end of a light elastic rod, its static deflection is 20 mm. The sliding collar supporting the rod is then given a vertical harmonic displacement of amplitude 5 mm at a circular frequency of 18 rad/s. Determine the steady-state amplitude of the mass relative to the collar.

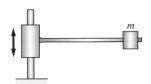

Fig. P20.43

20.44 A slightly unbalanced motor of mass m is attached to the middle of a light elastic beam. The unbalance is equivalent to a mass $m/400$ located at a distance $e = 0.2$ m from the axis of the motor. When the motor is rotating at $\omega = 1280$ rev/min, which is known to be less than the speed at resonance, its steady-state amplitude is $x_{max} = 0.001$ m. Find the angular speed of the motor at which resonance will occur.

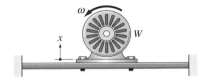

Fig. P20.44

20.45 For the damped system shown, prove that the maximum steady-state amplitude for a given damping factor ζ occurs at the frequency ratio $\omega/p = \sqrt{1 - 2\zeta^2}$ for $\zeta^2 \leq 1/2$. Also show that the corresponding maximum amplitude is

$$x_{max} = \frac{P_0/k}{2\zeta\sqrt{1 - \zeta^2}}$$

20.46 For the system shown, $m = 0.2$ kg and $k = 2880$ N/m. When the system is driven by the harmonic force of amplitude P_0, it is observed that the amplitude of the steady-state vibration is the same at $\omega = 96.0$ rad/s and $\omega = 126.4$ rad/s. Calculate the damping coefficient c.

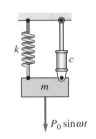

Fig. P20.45–P20.48

20.47 For the system shown, $mg = 14$ kg, $k = 880$ N/m, and $c = 117$ N $\cdot$ s/m. The frequency ω of the applied force is one-half of the resonant frequency. Determine the amplitude P_0 of the applied force if the amplitude of the steady-state vibration is 75 mm.

20.48 The mass $m = 7$-kg is suspended from a nonlinear spring and a viscous damper with the damping coefficient $c = 270$ N $\cdot$ s/m. The force-deformation relationship of the spring is $F = kx[1 + (x/b)^2]$ lb, where $k = 1750$ N/m, $b = 50$ mm, and x is measured in inches from the undeformed position of the spring. The amplitude of the applied force is $P_0 = 45$ N, and its circular frequency is $\omega = 12$ rad/s. (a) Show that the differential equation of motion of the mass is

$$\ddot{x} = -250x(1 + 400x^2) - 38.57\dot{x} + 9.8 + 6.43 \sin 12t \text{ m/s}^2$$

where x is the displacement in feet measured from the undeformed position of the spring (not the equilibrium position) and t is the time in seconds. (b) Use numerical integration to determine the maximum and minimum values of x for the steady-state motion. (Hint: Start with the system at rest in the equilibrium position and integrate until the transient term has been damped out.) Is the motion symmetric about the equilibrium position?

20.49 (a) Derive the differential equation of motion for the system shown. (b) Compute the amplitude of the steady-state vibration and the phase angle if $m = 3$ kg, $k = 7300$ N/m, $c = 265$ N · s/m, $Y = 38$ mm, and $\omega = 20$ rad/s.

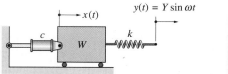

Fig. P20.49   **Fig. P20.50**

20.50 (a) Derive the differential equation of motion for the system shown. (b) Determine the amplitude of the steady-state vibration and the angle by which x lags y if $m = 6$ kg, $k = 8$ kN/m, $c = 40$ N · s/m, $Y = 80$ mm, and $\omega = 30$ rad/s.

20.51 Determine the expression for the steady-state displacement $x(t)$ of the block if $P_0 = 0.5$ N and $\omega = 60$ rad/s. Does $x(t)$ lead or lag the applied force?

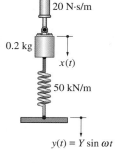

Fig. P20.51

20.52 Find the expression for the steady-state response $x(t)$ of the block if $Y = 10$ mm and $\omega = 600$ rad/s. Does $x(t)$ lead or lag the imposed displacement $y(t)$?

20.53 Block A is connected to the shaker table B with a spring and a viscous damper. When the horizontal displacement $y(t) = Y \sin \omega t$ is imposed on the table, the resulting steady-state displacement of the block relative to the table is $z(t) = Z \sin(\omega t - \phi)$. Show that if the damping factor $\zeta \geq 1/\sqrt{2}$, then Z never exceeds Y regardless of the value of ω.

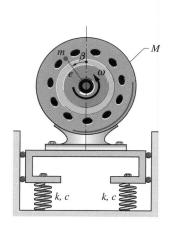

Fig. P20.52

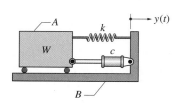

Fig. P20.53–P20.55

20.54 For the system described in Prob. 20.53, determine the largest possible value of Z and the corresponding frequency ratio ω/p if $\zeta = 1/2$.

20.55 For the system shown, $m = 5.5$ kg, $k = 2300$ N/m, and $c = 37$ N · s/m. The horizontal displacement imposed on the shaker table B is $y(t) = 2.5 \sin 18t$ mm, where the time t is measured in seconds. Determine the steady-state amplitude of the block.

20.56 An electric motor and its base have a combined mass of $M = 12$ kg. Each of the four springs attached to the base has a stiffness $k = 480$ kN/m and a viscous damping coefficient c. The unbalance of the motor is equivalent to a mass $m = 0.005$ kg located at the distance $e = 90$ mm from the center of the shaft. When the motor is running at $\omega = 400$ rad/s, its steady-state amplitude is 1.8 mm.

Fig. P20.56

Determine (a) the damping coefficient of each spring; and (b) the phase angle between the displacement of the motor and ωt.

20.57 When the system is in the position shown, the spring is undeformed. Determine the amplitude of the steady-state vibration of the weight C caused by the harmonic forcing function acting at A. Neglect all weights except that of C.

20.58 (a) Derive the differential equation of motion for the 10-kg block in terms of its absolute displacement $x(t)$. (b) Determine the expression for the steady-state response if the forcing frequency ω equals the natural frequency p of the system.

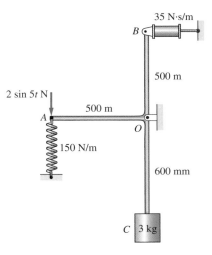

Fig. P20.57

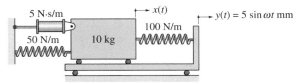

Fig. P20.58

20.59 The two masses are attached to the L-shaped bar that is free to rotate about the pin at O. (a) Neglecting the mass of the bar, derive the differential equation of motion for small angular displacements θ of the bar. (b) Compute the amplitude Θ and the phase angle ϕ of the steady-state vibration if $m = 0.6\,\text{kg}$, $k = 150\,\text{N/m}$, $c = 4\,\text{N} \cdot \text{s/m}$, $b = 0.5\,\text{m}$, $P_0 = 2\,\text{N}$, and $\omega = 8\,\text{rad/s}$.

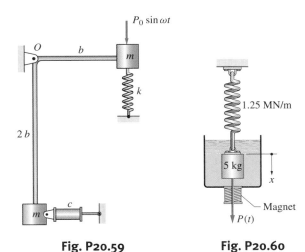

Fig. P20.59   **Fig. P20.60**

20.60 An iron cylinder is suspended from a spring and placed in a container filled with fluid. An electromagnet at the base of the container applies the force $P(t) = P_0 \sin \omega t$ to the cylinder, where $P_0 = 1.0\,\text{kN}$ and $\omega = 500\,\text{rad/s}$. The damping force acting on the cylinder due to the fluid is $F_d = c\dot{x}^2$, where $c = 250\,\text{N} \cdot \text{s}^2/\text{m}^2$. (a) Derive the differential equation of motion of the cylinder. (b) Estimate the steady-state amplitude of the cylinder by numerical integration of the equation of motion. [Hint: Start with the initial condition $x(0) = \dot{x}(0) = 0$, and integrate until the transient motion has been damped out.]

20.61   The 0.2-kg mass is suspended from a rigid frame as shown. Pin A at the end of the rotating arm OA engages a slot in the frame, causing the frame to oscillate in the vertical direction. The arm is accelerated uniformly from rest at $t = 0$ and $\theta = 0$ at the rate $\dot{\omega} = 100$ rad/s^2. (a) Show that the differential equation of motion of the weight in terms of the relative displacement $z = x - y$ is

$$\ddot{z} = -1750z - 10\dot{z} - 1.27(\cos 50t^2 - 100t^2 \sin 50t^2) \text{ m/s}^2$$

where z and t are measured in feet and seconds, respectively. (b) Use numerical integration to obtain the plot of z versus t from $t = 0$ to $t = 1.0$ s. (c) Use the numerical results of part (b) to determine the largest value of z.

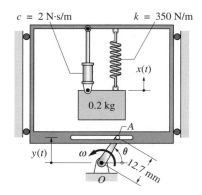

Fig. P20.61

20.4   *Rigid-Body Vibrations*

The analysis of rigid-body vibrations is fundamentally no different from the analysis of vibrating particles. We first derive the equation of motion utilizing the free-body and mass-acceleration diagrams of the body, and then seek a solution of this equation. If the system is linear, the equation of motion will be a second-order linear differential equation with constant coefficients. For a harmonically driven system with a single degree of freedom, this equation will have the general form

$$M\ddot{q} + C\dot{q} + Kq = F_0 \sin \omega t \tag{20.33}$$

where q is a variable that defines the position of the body (either a linear or an angular position coordinate). Because Eq. (20.33) has the same form as the equation of particle motion, $m\ddot{x} + c\dot{x} + kx = P_0 \sin \omega t$, we can write its solution by analogy. For example, the solution of forced, steady-state motion is

$$q = Q \sin(\omega t - \phi) \tag{20.34a}$$

where

$$Q = \frac{F_0/K}{\sqrt{[1 - (\omega/p)^2]^2 + (2\zeta\omega/p)^2}} \tag{20.34b}$$

and

$$\phi = \tan^{-1}\left[\frac{2\zeta\omega/p}{1-(\omega/p)^2}\right] \qquad (20.34c)$$

The parameters appearing in the above equations for Q and ϕ are

$$\left.\begin{array}{lll}
\text{Undamped circular frequency} & p = \sqrt{\dfrac{K}{M}} \\[2mm]
\text{Critical damping constant} & C_{\text{cr}} = 2Mp \\[2mm]
\text{Damping factor} & \zeta = \dfrac{C}{C_{\text{cr}}} = \dfrac{C}{2Mp}
\end{array}\right\} \qquad (20.34d)$$

The expression for the transient vibration can be written in the same manner (its form would depend on whether the motion is over-, under-, or critically damped).

As an example, consider the nonhomogeneous disk of radius R and mass m shown in Fig. 20.11(a). The mass center G of the disk is located at the distance e from the pin O, and its moment of inertia about O is I_O. A linear spring and a viscous damper are attached to the periphery of the disk at A and B, respectively. A clockwise couple $M_0 \sin \omega t$ also acts on the disk. We assume that the spring tension is adjusted so that the line OG is horizontal when the disk occupies its static equilibrium position. Therefore, the static deflection Δ of the spring at equilibrium satisfies the equation $mge = kR\Delta$.

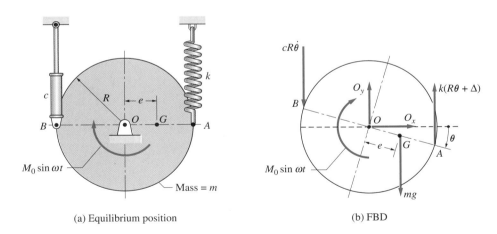

(a) Equilibrium position   (b) FBD

Fig. 20.11

To derive the equation for rotational motion of the disk, we draw the FBD shown in Fig. 20.11(b). The angular position of the disk is indicated by the angle θ, with clockwise rotation assumed to be positive. Throughout the analysis we assume θ to be small, so that we can use $\sin\theta \approx \theta$ and $\cos\theta \approx 1$. With these approximations, the displacement of point A is $R\theta$ (downward) and the velocity of point B is $R\dot{\theta}$ (upward). This gives the spring and damping forces shown in the

FBD. Without needing to draw the mass-acceleration diagram of the disk, we can sum moments about O to obtain the equation of motion

$$\Sigma M_O = I_O\ddot{\theta} \quad \text{(+)} \quad M_0 \sin\omega t - k(R\theta + \Delta)R - (cR\dot{\theta})R + mge = I_O\ddot{\theta}$$

Using the equilibrium equation $mge = kR\Delta$, and rearranging the terms, yields

$$I_O\ddot{\theta} + cR^2\dot{\theta} + kR^2\theta = M_0 \sin\omega t \qquad \text{(a)}$$

Comparing Eq. (a) with Eq. (20.33), we find that they are of the same form, the coefficients of Eq. (20.33) being

$$M = I_O \qquad C = cR^2 \qquad K = kR^2 \qquad F_0 = M_0 \qquad \text{(b)}$$

We conclude, therefore, that the steady-state vibration of the disk is harmonic for small oscillations, and the motion can be determined directly from Eqs. (20.34). For example, the undamped circular frequency is

$$p = \sqrt{\frac{K}{M}} = \sqrt{\frac{kR^2}{I_O}}$$

The critical damping coefficient can be obtained from $C_{cr} = 2Mp$, which on substitution from Eq. (b) becomes

$$(cR^2)_{cr} = 2I_O p$$

yielding

$$c_{cr} = \frac{2I_O p}{R^2}$$

Sample Problem   20.7

The homogeneous slender bar of mass m and length L in Fig. (a) is supported by a pin at O. The bar is also connected to an ideal spring and viscous damper at points A and B, respectively. The bar is initially in equilibrium in the position shown with the spring undeformed. (1) Derive the differential equation of motion for small angular displacements of the bar. (2) Determine whether the bar is overdamped or underdamped, given that $m = 12$ kg, $L = 800$ mm, $a = 400$ mm, $k = 80$ N/m, and $c = 20$ N $\cdot$ s/m.

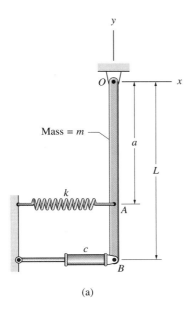

(a)

Solution

Part 1

Figure (b) shows the free-body diagram (FBD) of the bar when it is displaced a small counterclockwise angle θ from the vertical. The horizontal displacements x_G (G is the mass center), x_A, and x_B shown in the figure were obtained from the small angle approximation $\sin\theta \approx \theta$. The forces acting on the bar are its weight mg, the spring force $k(a\theta)$ (recall that the spring is undeformed when $x_A = 0$), and the damping force $c(L\dot{\theta})$ due to the dashpot.

We derive the differential equation of motion by summing moments about point O, thereby eliminating the pin reaction. Because O is a fixed point, a valid equation of motion is $\Sigma M_O = I_O\ddot{\theta}$. From the FBD in Fig. (b) and the approximation $\cos\theta \approx 1$, we obtain

$$\Sigma M_O = I_O\ddot{\theta} \quad \overset{\curvearrowleft}{+} \quad -mg\frac{L}{2}\theta - (ka\theta)a - (cL\dot{\theta})L = I_O\ddot{\theta}$$

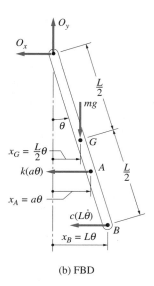

(b) FBD

which, on rearranging the terms, becomes

$$I_O\ddot{\theta} + cL^2\dot{\theta} + \left(ka^2 + \frac{mgL}{2}\right)\theta = 0$$ ***Answer***   (a)

Part 2

Comparing Eq. (a) with Eq. (20.33), we conclude that

$$M = I_O \qquad C = cL^2 \qquad K = ka^2 + \frac{mgL}{2}$$

Therefore, the undamped circular frequency is [see Eq. (20.34d)]

$$p = \sqrt{\frac{K}{M}} = \sqrt{\frac{ka^2 + (mgL/2)}{I_O}}$$

With $I_O = mL^2/3 = 12(0.8)^2/3 = 2.560$ kg $\cdot$ m^2, this becomes

$$p = \sqrt{\frac{80(0.4)^2 + [12(9.81)(0.8)/2]}{2.560}} = 4.837 \text{ rad/s}$$

From Eq. (20.34d), we also obtain $C_{\text{cr}} = 2Mp$, which in our case becomes $c_{\text{cr}}L^2 = 2I_O p$, yielding

$$c_{\text{cr}} = \frac{2I_O p}{L^2} = \frac{2(2.560)(4.837)}{(0.8)^2} = 38.70 \text{ N} \cdot \text{s/m}$$

Because the given damping constant $c = 20$ N $\cdot$ s/m is less than c_{cr}, we conclude that the bar is *underdamped*.

Problems

20.62   The uniform slender bar of mass m is in equilibrium in the horizontal position shown. Calculate the frequency f of small oscillations about the equilibrium position.

Fig. P20.62

20.63   Each of the two thin homogeneous plates is attached to an identical elastic rod that acts as a linear torsion spring (that is, the restoring torque is proportional to the angular displacement). If the frequency of torsional oscillation of the circular plate of mass m_1 is f_1, find the frequency f_2 of the square plate of mass m_2.

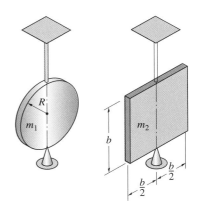

Fig. P20.63

20.64   The rigid body that is suspended from a pin at A is displaced slightly from its equilibrium position and released. (a) Show that the circular frequency of the resulting vibration is

$$p = \sqrt{\frac{gy}{\overline{k}^2 + y^2}}$$

where y is the distance from A to the mass center G and $\overline{k}$ is the radius of gyration of the body about G. (b) Determine the largest possible circular frequency and the corresponding value of y.

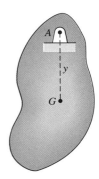

Fig. P20.64

20.65   The mass center of the pendulum shown is at G. When the pendulum is suspended from the pin at A, the period for small oscillations is 1.6 s. Determine the period if the pin is moved to B. (Hint: See Prob. 20.64.)

20.66   The uniform bar of mass m is in equilibrium in the horizontal position. (a) Derive the differential equation of motion for small oscillations of the bar. (b) Determine the damping factor given that $m = 20$ kg, $c_1 = 25$ N · s/m, $c_2 = 16$ N · s/m, and $k = 80$ N/m.

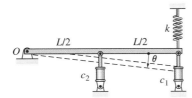

Fig. P20.66

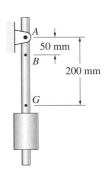

Fig. P20.65

20.67   The uniform slender bar of mass m is rigidly attached to the homogeneous disk of mass $3m$. The assembly is free to rotate about the pin at O.

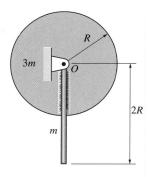

Fig. P20.67

Determine the frequency f of small oscillations about the equilibrium position shown.

20.68 The 8-kg uniform plate supported by the pin at O is in equilibrium in the position shown. Find the frequency f of small oscillations about this position.

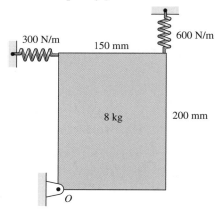

Fig. P20.68

20.69 The radius of gyration of the 14 kg disk about its center of gravity G is $\bar{k} = 400$ mm. One end of the spring is attached to the disk at A and the other end is connected to the block B. The disk is in static equilibrium in the position shown when the displacement $y(t) = 10 \sin 25t$ mm is imposed on B (t is the time in seconds). Assuming that the disk does not slip on the horizontal surface, (a) derive the differential equation of motion of the disk in terms of its angular displacement θ; and (b) determine the amplitude of the steady-state oscillation at point G.

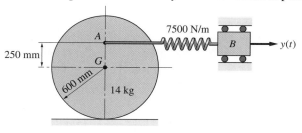

Fig. P20.69

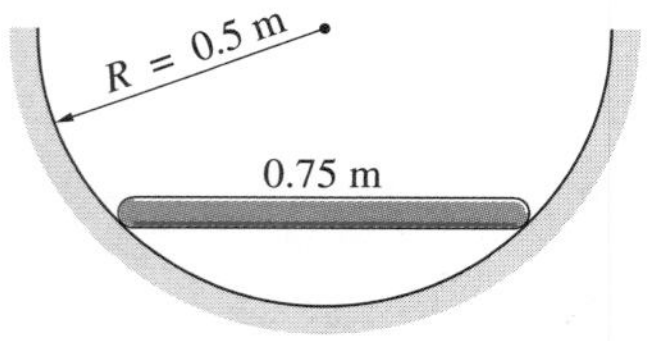

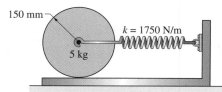

Fig. P20.70

20.70 The uniform bar of length 0.75 m rests in a frictionless circular trough of radius 0.5 m. If the bar is displaced slightly from the equilibrium position and released, calculate the natural frequency of the ensuing oscillation.

20.71 When the equilibrium of the 5 kg uniform disk is disturbed, it rolls back and forth on the rigid base without slipping. If the maximum displacement of the center of the disk from its equilibrium position is 25 mm, find (a) the circular frequency of the oscillation; and (b) the minimum coefficient of friction between the disk and the base.

Fig. P20.71

20.72 When the 0.8 kg uniform rod AB is in the vertical position, the two springs of stiffness $k = 1450$ N/m are unstretched. The weight of the horizontal bar attached to the rod at *A* is negligible. If the support *C* undergoes the harmonic angular displacement $\beta(t) = \beta_0 \sin \omega t$, where $\beta_0 = 2°$ and $\omega = 7.5\,\text{rad/s}$, find the steady-state angular amplitude of bar *AB*.

20.73   The uniform slender bar of mass *m* is initially in static equilibrium in the position shown. The bar is then rotated slightly and released. (a) Derive the differential equation of motion in terms of the angular displacement θ of the bar. (b) If $m = 6\,\text{kg}$ and $k = 3\,\text{kN/m}$, determine the damping coefficient *c* that will critically damp the vibration.

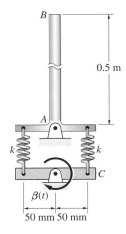

Fig. P20.72

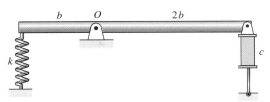

Fig. P20.73

20.74   The platform, supported by a pin at *B* and a spring at *C*, is in equilibrium in the position shown. When the viscous damper at *A* is disconnected, the frequency of the system for small amplitudes is 2.22 Hz. Determine the damping coefficient *c* that would critically damp the system.

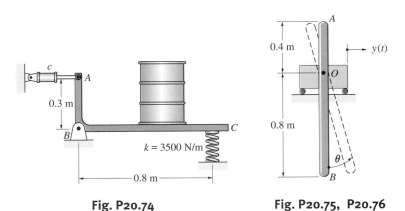

Fig. P20.74   **Fig. P20.75, P20.76**

20.75   The 12-kg uniform bar *AOB* is pinned to the carriage at *O*. The carriage is undergoing the harmonic displacement $y(t) = Y \sin \omega t$, where $Y = 10\,\text{mm}$ and $\omega = 2\,\text{rad/s}$. Calculate the steady-state angular amplitude of the bar for small amplitudes.

20.76   The 12-kg uniform bar *AOB* is pinned to the carriage at *O*. The bar is at rest in the vertical position at $t = 0$ when the harmonic displacement $y(t) = Y \sin pt$ is imposed on the carriage, where *p* is the resonant frequency for small amplitudes and $Y = 10\,\text{mm}$. (a) Show that the differential equation of motion for the bar is

$$\ddot{\theta} = -12.263 \sin \theta + 0.153\,31 \cos \theta \sin 3.502t \ \text{rad/s}^2$$

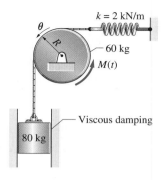

Fig. P20.77

(b) Integrate the equation of motion numerically from $t = 0$ to $t = 50$ s, and plot θ versus t. (c) By inspection of the plot, estimate the maximum value of θ.

20.77 The radius of the 60-kg uniform disk is $R = 500$ mm. The space between the 80-kg mass and the vertical slot is lubricated, providing viscous damping with a damping factor of $\zeta = 0.15$. If the couple $M(t) = M_0 \sin \omega t$ is acting on the disk, where $M_0 = 40$ N · m and $\omega = 5$ rad/s, determine the steady-state angular displacement $\theta(t)$ of the disk. Assume that the cable remains taut and does not slip on the disk.

20.78 The 3-kg uniform disk oscillates about the pin at A. The pin is lubricated with heavy grease, which gives rise to rotational resistance equivalent to a viscous damping couple $M_d = -c\dot\theta$, where c is a constant. If the period of oscillation is measured to be 1.468 s, calculate (a) the damping ratio; and (b) the constant c.

Fig. P20.78

****20.79** The differential equation of motion for an undamped system with constant external forces, from Eq. (20.45), is $M\ddot{q} + Kq = F$, where F is the constant generalized force. Prove that the plot of $\dot{q}/p$ versus q is the circle shown, where p is the natural circular frequency of the system. (Note: The circle is called the *phase plane* plot of the system. Plots of this type can be very helpful in describing the behavior of complex, single-degree-of-freedom systems. See, for example, Prob. 20.80.)

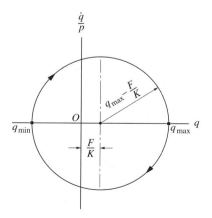

Fig. P20.79

*20.80   The differential equation of motion for a system that is subjected to Coulomb damping (dry friction) of constant magnitude is

$$M\ddot{q} + Kq = \begin{cases} -F & \text{if } \dot{q} > 0 \\ F & \text{if } \dot{q} < 0 \end{cases}$$

where F is the generalized constant friction force. Assume that the system is released from rest at $q = q_0$, and let $F = 0.15Kq_0$. (a) Draw the phase plane plot for the system. (Hint: See Prob. 20.79.) (b) From the phase plane plot, determine the reduction of amplitude in each cycle of oscillation, and the point where the system comes to rest.

*20.5   Methods Based on Conservation of Energy

Here we examine two convenient methods for analyzing undamped, free vibrations with a single degree of freedom: the energy method and Rayleigh's principle. Both methods are based on the principle of conservation of mechanical energy.

a.   Energy method

Up to now, our analysis of a vibrating system started with the derivation of the differential equation of motion using a free-body diagram and Newton's second law. When a system is conservative (damping is absent or negligible), we can bypass the equation of motion and obtain the natural frequency directly from the principle of conservation of mechanical energy.

If all of the forces that act on a mechanical system are conservative, the total mechanical energy of the system is conserved. That is,

$$T + V = \text{constant} \tag{20.35}$$

where T is the kinetic energy of the system and V represents the potential energy of the forces acting on the system. Let us assume that the system is linear (its response is proportional to the applied forces) and has a single degree of freedom with q being the generalized position coordinate. The kinetic energy of such a system has the form

$$T = \frac{1}{2}M\dot{q}^2 \tag{20.36}$$

where the constant M is the generalized mass of the system.

The potential energy of a linear system is always a quadratic form in q:

$$V = V_0 + F_0 q + \frac{1}{2} K q^2 \tag{20.37}$$

where V_0, F_0, and K are constants. To facilitate the derivations that follow, we assume that $q = 0$ corresponds to the static equilibrium position of the system. By inspection of Eq. (20.37), we see that V_0 is the the potential energy of the system in the equilibrium position. Moreover, because the condition for static equilibrium is $dV/dq|_{q=0} = 0$, we conclude that $F_0 = 0$, so that the potential energy becomes

$$V = V_0 + \frac{1}{2} K q^2 \tag{20.38}$$

Substituting Eqs. (20.36) and (20.38) into Eq. (20.35) and differentiating with respect to time, we get

$$\frac{d}{dt} \left(\frac{1}{2} M \dot{q}^2 + V_0 + \frac{1}{2} K q^2 \right) = 0$$

which yields $(M\ddot{q} + Kq)\dot{q} = 0$. Discarding the trivial solution $\dot{q} = 0$, we obtain the equation of motion

$$\ddot{q} + p^2 q = 0 \tag{20.39}$$

where

$$p = \sqrt{\frac{K}{M}} \tag{20.40}$$

Since Eq. (20.39) has the same form as Eq. (20.1b), we conclude that p is the circular frequency of the system.

In summary, the procedure for computing the circular frequency of an undamped system using conservation of energy is:

- Choose a position coordinate q so that $q = 0$ at static equilibrium.
- Derive the expression for the potential energy V in an arbitrary position. (When making small-displacement approximations, such as $\sin q \approx q$ and $\cos q \approx 1 - q^2/2$, be sure to keep all quadratic as well as linear terms in q.) Identify K by comparing your expression for V with Eq. (20.37).
- Derive the expression for the kinetic energy T in an arbitrary position and identify M by comparing the expression with Eq. (20.36).
- Compute the circular frequency from $p = \sqrt{K/M}$.

b.   *Rayleigh's principle*

In some cases, geometric complications make it difficult to compute the kinetic energy in an arbitrary position. These cases are best analyzed using Rayleigh's principle, which is a variant of the principle of conservation of energy. Rayleigh's principle requires the kinetic energy only at a specific position of the system, which simplifies the geometry. The method also can be used for an approximate analysis of systems whose motion is not purely harmonic.

As before, we assume the system to be linear and have a single degree of freedom. Let q be the generalized position coordinate, which is chosen so that $q = 0$ in the static equilibrium position. From the conservation of mechanical energy in Eq. (20.35) we draw the following conclusions:

- The position of maximum kinetic energy is also the position of minimum potential energy. According to Eq. (20.38), $V_{min} = V_0$ occurs at $q = 0$; hence T_{max} also occurs at $q = 0$.
- The position of maximum potential energy is also the position of minimum kinetic energy. Note that in this position $\dot{q} = 0$, so that $T_{min} = 0$.

Conservation of mechanical energy requires that $T_{max} + V_{min} = T_{min} + V_{max}$. Since we found that $V_{min} = V_0$ and $T_{min} = 0$, the energy balance becomes

$$T_{max} = V_{max} - V_0 \qquad (20.41)$$

which is known as *Rayleigh's principle*. If the datum for potential energy is chosen such that $V_0 = 0$, this principle assumes the more conventional form $T_{max} = V_{max}$. In Eq. (20.41) we have

$$V_{max} = V_0 + \frac{1}{2}Kq_{max}^2 \qquad T_{max} = \frac{1}{2}M\dot{q}_{max}^2 \qquad \text{(a)}$$

When applying Rayleigh's principle, we must *assume* that the motion is simple harmonic: $q = A\sin(pt + \alpha)$. Therefore, $q_{max} = A$ and $\dot{q}_{max} = Ap$, or

$$\dot{q}_{max} = pq_{max} \qquad (20.42)$$

Substituting Eqs. (a) together with Eq. (20.42) into Eq. (20.41), we obtain

$$\frac{1}{2}p^2 M q_{max}^2 = V_0 + \frac{1}{2}Kq_{max}^2 - V_0$$

from which $p = \sqrt{K/M}$ as before.

The procedure for applying Rayleigh's method is:

- Choose a generalized position coordinate q so that $q = 0$ at static equilibrium.
- Derive the expression for the potential energy V in the position $q = q_{max}$, thus obtaining V_{max} as a function of q_{max}. (When making small-displacement approximations, be sure to keep all quadratic as well as linear terms in q_{max}.)
- Derive the expression for the kinetic energy T in the static equilibrium position $q = 0$. The result is T_{max} as a function of $\dot{q}_{max}$.
- Substitute $\dot{q}_{max} = pq_{max}$ into the expression for T_{max}, obtaining T_{max} as a function of q_{max}.
- Substitute the expressions for T_{max} and V_{max} into Eq. (20.41) and solve for p (q_{max} will invariably cancel out).

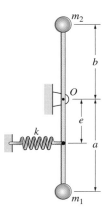

(a)

Sample Problem 20.8

The rigid body in Fig. (a) consists of the small masses m_1 and m_2, which are connected to the ends of a light bar that can rotate about a pin at O. The ideal spring is undeformed when the bar is vertical. Determine the circular frequency of small vibrations.

Solution

Because the kinetic energy of the system is computed easily, this problem is well suited for solution using conservation of energy. We choose the angle θ shown in Fig. (b) as the position coordinate. Note that $\theta = 0$ in the static equilibrium position, as required by the energy method. Taking the equilibrium position as the datum for potential energy and referring to Fig. (b), we obtain

$$V = m_1 g h_1 - m_2 g h_2 + \frac{1}{2}k(e\theta)^2$$

where $h_1 = a(1 - \cos\theta)$ and $h_2 = b(1 - \cos\theta)$. With the approximation $\cos\theta \approx 1 - \theta^2/2$, we get $h_1 \approx a\theta^2/2$ and $h_2 \approx b\theta^2/2$. Therefore, the potential energy becomes

$$V = \frac{1}{2}(m_1 g a - m_2 g b + ke^2)\theta^2$$

Comparison with Eq. (20.37) yields

$$K = m_1 g a - m_2 g b + ke^2$$

Because O is a fixed point, the kinetic energy of the system is

$$T = \frac{1}{2}I_O \dot{\theta}^2 = \frac{1}{2}(m_1 a^2 + m_2 b^2)\dot{\theta}^2$$

Comparing this with Eq. (20.36) we conclude that

$$M = m_1 a^2 + m_2 b^2$$

Hence, the circular frequency of the system is

$$p = \sqrt{\frac{K}{M}} = \sqrt{\frac{m_1 g a - m_2 g b + ke^2}{m_1 a^2 + m_2 b^2}} \qquad \textit{Answer}$$

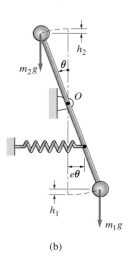

(b)

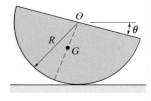

(a)

Sample Problem 20.9

Figure (a) shows a homogeneous semicylinder that rocks back and forth without slipping on the horizontal plane. The angular position of the semicylinder is defined by the angle θ. Determine the natural circular frequency of the oscillations for small amplitudes, assuming the motion to be simple harmonic. The mass center G is located at the distance $4R/3\pi = 0.4244R$ from point O.

Solution

This problem is not well suited for solution using conservation of energy, because the kinetic energy cannot be easily computed in an *arbitrary position*. The source of the difficulty is the displacement of the contact point between the semicylinder and the horizontal surface, which gives rise to geometric complications. We can overcome this problem by using Rayleigh's principle.

The position of maximum potential energy is shown in Fig. (b). Choosing the horizontal surface as the datum, the potential energy is

$$V_{\text{max}} = mg(R - 0.4244R\cos\theta_{\text{max}}) \approx mg\left(R - 0.4244R\left(1 - \frac{\theta_{\text{max}}^2}{2}\right)\right)$$

$$= 0.5756mgR + 0.2122mgR\theta_{\text{max}}^2$$

Note that $V_0 = 0.5756mgR$ is the potential energy in the equilibrium ($\theta = 0$) position.

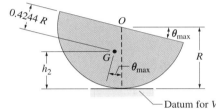

(b) Position of maximum displacement ($V = V_{\text{max}}$, $T = 0$)

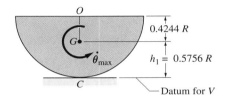

(c) Passing through equilibrium position ($V = V_0$, $T = T_{\text{max}}$)

The maximum kinetic energy of the semicylinder occurs in the position $\theta = 0$, which is shown in Fig. (c). Because the contact point C is the instant center for velocity, the kinetic energy is $T_{\text{max}} = (1/2)I_C\dot\theta_{\text{max}}^2$. By the parallel-axis theorem

$$\bar{I} = I_O - m(\overline{OG})^2 = 0.5mR^2 - m(0.4244R)^2 = 0.3199mR^2$$

$$I_C = \bar{I} + m(\overline{GC})^2 = 0.3199mR^2 + m(0.5756R)^2 = 0.6512mR^2$$

so that

$$T_{\text{max}} = \frac{1}{2}(0.6512mR^2)\dot\theta_{\text{max}}^2 = 0.3256mR^2\dot\theta_{\text{max}}^2$$

$$= 0.3256mR^2p^2\theta_{\text{max}}^2$$

In the last step, we substituted $\dot\theta_{\text{max}}^2 = p^2\theta_{\text{max}}^2$, thereby assuming the motion to be harmonic.

Rayleigh's principle $T_{\text{max}} = V_{\text{max}} - V_0$ now becomes

$$0.3256mR^2p^2\theta_{\text{max}}^2 = 0.2122mgR\theta_{\text{max}}^2$$

yielding for the circular frequency

$$p = 0.807\sqrt{\frac{g}{R}} \qquad\qquad \textit{Answer}$$

Problems

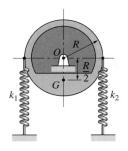

Fig. P20.81

20.81 The mass center of the unbalanced disk of weight W is located at G. In the equilibrium position shown, the two springs attached to the rim of the disk are undeformed. Determine the natural circular frequency of the disk for small oscillations by the energy method. Use the following data: $W = 180\,\text{N}$, $R = 0.6\,\text{m}$, $\bar{k} = 0.5\,\text{m}$ (radius of gyration about G), $k_1 = 450\,\text{N/m}$, and $k_2 = 600\,\text{N/m}$.

20.82 The uniform bars AB and BC, each of length L and mass m, are connected with a pin at B. End C of bar BC is attached to the roller support. The system is in equilibrium in the position shown. Calculate the period of small oscillations about this position.

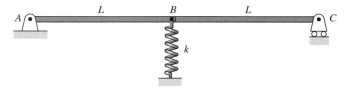

Fig. P20.82

20.83 One end of the L-shaped arm is connected to the block of mass m, while the other end is attached to the linear spring. The mass of the arm can be neglected. (a) Determine the natural circular frequency of small oscillations about the equilibrium position shown. (b) What is the range of values of the spring constant k for which the oscillations will be stable?

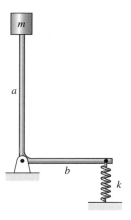

Fig. P20.83

20.84 The homogeneous semicircular hoop rocks back and forth on the horizontal surface without slipping. Determine the natural circular frequency of oscillations for small amplitudes. Assume simple harmonic motion.

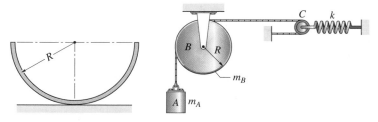

Fig. P20.84　　　　　　**Fig. P20.85**

20.85 Pulley B can be approximated as a uniform disk, and the mass of pulley C can be neglected. Determine the expression for the natural frequency of the system.

20.86 The hoop to which the pendulum is rigidly attached rolls without slipping on the horizontal surface. Neglecting the masses of the hoop and the rod, calculate the period for small oscillations about the equilibrium position shown. Assume simple harmonic motion.

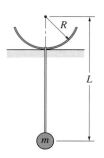

Fig. P20.86

20.87 The uniform rod of mass m and length R is attached to a circular base of radius R and negligible mass. The system is released from rest in the position shown, where θ is a small angle. Determine the circular frequency of the ensuing oscillations, assuming that the motion is harmonic and that the base does not slip on the horizontal surface.

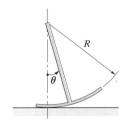

Fig. P20.87

20.88 The pendulum consists of a uniform disk attached to a rod of negligible weight. Determine the two values of the distance L for which the period of the pendulum is 1.4 s for small amplitudes.

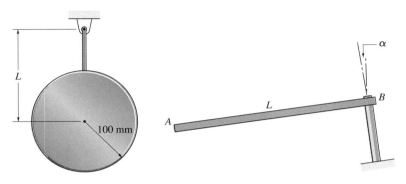

Fig. P20.88 **Fig. P20.89**

20.89 The uniform slender bar AB of mass m is free to rotate about the axis that is inclined at the angle α to the vertical. The bar is given a small angular displacement from the equilibrium position shown and then released. Determine the frequency of the resulting vibration.

20.90 If the uniform slender bar is in equilibrium in the position shown, calculate the period of vibration for small amplitudes.

20.91 The homogeneous disk of mass m and radius R rolls without slipping on the inclined surface. Determine the frequency of small oscillations of the disk about its equilibrium position.

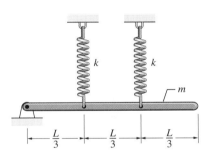

Fig. P20.90

20.92 The uniform block of wood is floating in water when it is displaced slightly in the vertical direction and then released. Find the period of the resulting oscillation. Use the following weight densities: $\rho_{\text{water}} = 1000 \text{ kg/m}^3$ and $\rho_{\text{wood}} = 600 \text{ kg/m}^3$.

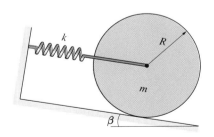

Fig. P20.91

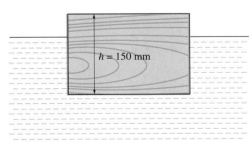

$h = 150 \text{ mm}$

Fig. P20.92

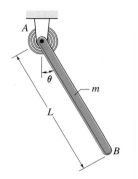

Fig. P20.93

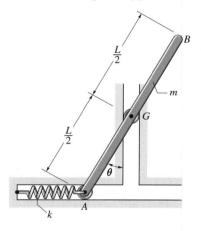

Fig. P20.94, P20.95

20.93 The torsional spring at end A of the uniform slender bar is adjusted so that the bar is in equilibrium in the position $\theta = \theta_0$, where θ_0 is not necessarily small. If the rotational stiffness of the spring is k (torque/rad), determine the expression for the natural frequency of the bar for small oscillations about the equilibrium position. Assume that the plane of the figure is (a) vertical; and (b) horizontal.

20.94 The rollers at A and G are pinned to the uniform bar AB and move freely in fixed slots. The spring at A is initially unstretched, and the bar is at rest when $\theta = 0$. (a) Derive the differential equation of motion for the bar, assuming that the angle θ remains small. (b) Compute the period of oscillation if $m = 15$ kg, $L = 1.2$ m, and $k = 490.5$ N/m. (c) Using the values of L and k given in part (b), find the largest value of m for which the angle θ will remain small.

20.95 (a) Show that the differential equation of motion for the bar described in Prob. 20.94 is

$$\ddot{\theta} = -\left[\left(\dot{\theta}^2 + \frac{k}{m}\right)\cos\theta - \frac{2g}{L}\right]\frac{3\sin\theta}{1 + 3\sin^2\theta}$$

(b) Use numerical integration with $m = 15$ kg, $L = 1.2$ m, and $k = 490.5$ N/m to calculate the period of oscillation if the amplitude is (i) small; (ii) 30°.

20.96 The homogeneous sphere of mass m and radius r rolls without slipping on a cylindrical surface of radius R. Calculate the frequency of small oscillations about the equilibrium position shown.

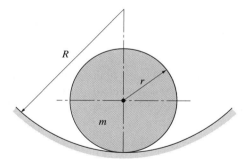

Fig. P20.96

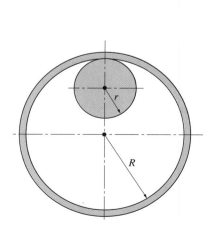

Fig. P20.97

20.97 A thin ring of radius R and mass m is suspended from a peg of radius r. Determine the circular frequency of small oscillations about the equilibrium position shown if the ring does not slip on the peg. Assume simple harmonic motion.

Review of Equations

Undamped free vibration

Equation of motion: $m\ddot{x} + kx = 0$

General solution: $x = A \cos pt + B \sin pt = E \sin(pt + \alpha)$

Circular frequency: $p = \sqrt{k/m}$

Damped free vibration

Equation of motion: $m\ddot{x} + c\dot{x} + kx = 0$

General solution: $x = A_1 e^{\left(-\zeta + \sqrt{\zeta^2 - 1}\right)pt} + A_2 e^{\left(-\zeta - \sqrt{\zeta^2 - 1}\right)pt}$

Damping factor: $\zeta = \dfrac{c}{c_{cr}} \qquad c_{cr} = 2mp$

Damped circular frequency: $\omega_d = p\sqrt{1 - \zeta^2}$

Forced vibration

Equation of motion: $m\ddot{x} + c\dot{x} + kx = P_0 \sin \omega t$

Particular solution: $X \sin(\omega t - \phi)$

Amplitude: $X = \dfrac{P_0/k}{\sqrt{[1 - (\omega/p)^2]^2 + (2\zeta\omega/p)^2}}$

Phase angle: $\phi = \tan^{-1}\left[\dfrac{2\zeta\omega/p}{1 - (\omega/p)^2}\right]$

Harmonic support displacement

$$y = Y \sin \omega t$$

Equation of motion: $m\ddot{z} + c\dot{z} + kz = mY\omega^2 \sin \omega t$

$z = x - y =$ position coordinate of mass relative to support

Energy method

$$V = V_0 + F_0 q + \frac{1}{2}Kq^2 \qquad T = \frac{1}{2}M\dot{q}^2 \qquad p = \sqrt{K/M}$$

Rayleigh's principle

$T_{max} = V_{max} - V_0$

Review Problems

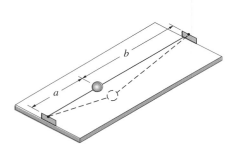

Fig. P20.99

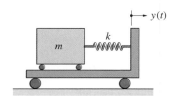

Fig. P20.100

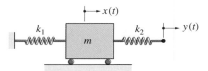

Fig. P20.101

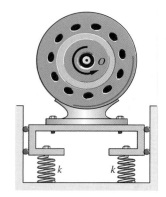

Fig. P20.102

20.98   The 10-kg mass is set into motion at time $t = 0$ with the initial conditions $x = 20$ mm and $\dot{x} = -110$ mm/s, where x is measured from the position where the springs are undeformed. (a) Derive the differential equation of motion for the mass. (b) Compute the frequency of vibration. (c) Determine the expression for $x(t)$.

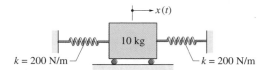

Fig. P20.98

20.99   The two elastic cords are connected to the ball of mass m and stretched to initial tension T. If the ball is given a small lateral displacement and released, determine the frequency of the ensuing vibration.

20.100   The block of mass $m = 1$ kg is connected to the shaker table by a spring of stiffness $k = 2100$ N/m. The system is at rest with the spring undeformed when the harmonic displacement $y(t) = 10 \sin 60t$ mm is imposed on the table. Determine the displacement of the weight relative to the table as a function of time.

20.101   Two springs of stiffnesses k_1 and k_2 are attached to the mass m. One of the springs is connected to a rigid support, whereas the free end of the other spring undergoes the harmonic displacement $y(t) = Y \sin \omega t$. (a) Derive the differential equation of motion for the mass. (b) Determine the amplitude of the steady-state vibration. (c) Find the magnification factor.

20.102   The electric motor of total mass of 30 kg is supported by four identical springs of stiffness $k = 8$ kN/m each. The mass center of the 8-kg armature is 250 mm from the axis O of the armature. (a) Determine the resonant speed of the motor. (b) Find the amplitude of the steady-state vibration if the motor is running at twice the resonant speed.

20.103   The figure shows a damped oscillator and the plot of its displacement during free vibration (x is measured from the equilibrium position). Knowing that

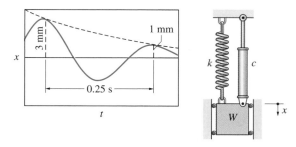

Fig. P20.103

the weight of the oscillator is $m = 0.2\,\text{kg}$, find the spring stiffness k and the damping coefficient c. (Hint: Use the logarithmic decrement.)

20.104   The 5 kg mass is released from rest at time $t = 0$ in the position $x = 50\,\text{mm}$, where x is measured from the position where the spring is unde-formed. If the damping factor is 0.5, determine the expression for $x(t)$.

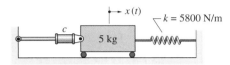

Fig. P20.104

20.105   The system is set into motion at time $t = 0$, with $x = 0$ and $\dot{x} = 2\,\text{m/s}$, where x is measured from the undeformed position of the spring. Determine the expression for $x(t)$.

20.106   Solve Prob. 20.105 if c_2 is changed to $3\,\text{N} \cdot \text{s/m}$.

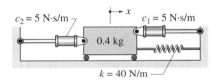

Fig. P20.105, P20.106

20.107   For the system shown, $m = 7$ kg, $k = 1750\,\text{N/m}$, and $c = 270\,\text{N} \cdot \text{s/m}$. (a) Determine the magnification factor if the circular frequency ω of the applied force equals the resonant frequency p of the system. (b) Find the maximum possible magnification factor and the corresponding value of ω.

20.108   Repeat Prob. 20.107 if the damping coefficient is changed to $c = 90\,\text{N} \cdot \text{s/m}$.

20.109   Determine the steady-state displacement $x(t)$ of the 6-kg mass if $Y = 28$ mm and $\omega = 20\,\text{rad/s}$. Does $x(t)$ lead or lag $y(t)$?

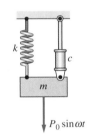

Fig. P20.107, P20.108

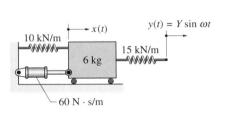

Fig. P20.109

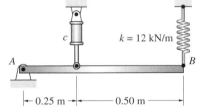

Fig. P20.110, P20.111

20.110   The rigid 24-kg bar AB is in equilibrium in the position shown. Deter-mine the damping coefficient c for which the bar would be critically damped for small oscillations.

20.111   The rigid bar AB has a mass of 24 kg, and the damping coefficient of the dashpot is $c = 3\,\text{kN} \cdot \text{s/m}$. The bar is driven by a vertical force (not shown) $P(t) = 0.5 \sin 25t$ kN acting at B. Determine the displacement amplitude of end B of the steady-state vibration.

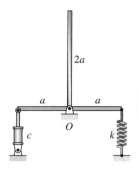

Fig. P20.112

20.112   The T-shaped body is made of uniform bar that has a mass per unit length ρ. In the position shown, the body is in equilibrium. Determine whether the system is overdamped or underdamped for small oscillations, given that $\rho = 0.5$ kg/m, $a = 500$ mm, $c = 20$ N · s/m, and $k = 800$ N/m.

20.113   The figure shows the inside of an instrument that is used to measure the amplitude Y of the vertical ground motion $y(t) = Y \sin \omega t$. The L-shaped arm is in equilibrium in the position shown. Derive the equation of motion for the mass m in terms of z, the displacement of the mass relative to the frame. Assume small displacements and neglect all masses except m.

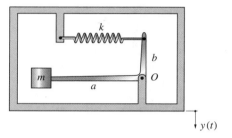

Fig. P20.113

20.114   For the damped spring-mass system shown, the damping factor is $\zeta = 0.25$ and the undamped natural frequency is $f = 3$ Hz. When the support undergoes the vertical movement $y(t) = Y \sin 18t$ (t is the time in seconds), the relative amplitude of the steady-state vibration of the mass m is $Z = 10$ mm. Determine the amplitude Y of the support displacement.

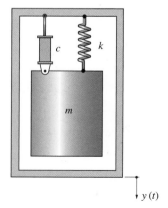

Fig. P20.114

D

Proof of the Relative Velocity Equation for Rigid-Body Motion

Here we prove Eq. (19.1a), $\mathbf{v}_{B/A} = \boldsymbol{\omega} \times \mathbf{r}_{B/A}$, for rigid-body motion—where $\boldsymbol{\omega}$ is called the *angular velocity* of the body. In a three-dimensional setting, this result is neither intuitively obvious, nor is its proof trivial.

Figure D.1 shows four points A, B, C, and D belonging to the same rigid body. These points may be chosen arbitrarily, except for the following restrictions: (1) the reference point A must not lie on the line BC or on the line CD, and (2) the four points must not lie in the same plane.

The rigidity of the body imposes the following constraints on the motion: (1) the lengths of the relative position vectors shown in Fig. D.1 remain constant, and (2) the angles β_1, β_2, and β_3 between the relative position vectors remain constant. It will be demonstrated that these conditions can be satisfied only if the relative velocity between any two points on the body has the form of Eq. (19.1a).

Consider first the requirement that the magnitude of vector $\mathbf{r}_{B/A}$ is constant: $\mathbf{r}_{B/A} \cdot \mathbf{r}_{B/A} = |\mathbf{r}_{B/A}|^2 = \text{constant}$. Taking the time derivative of both sides of this equation, we get $(\mathbf{r}_{B/A} \cdot \dot{\mathbf{r}}_{B/A}) + (\dot{\mathbf{r}}_{B/A} \cdot \mathbf{r}_{B/A}) = 0$, or $2\mathbf{r}_{B/A} \cdot \mathbf{v}_{B/A} = 0$. A similar argument can be applied to $\mathbf{r}_{C/A}$ and $\mathbf{r}_{D/A}$. Consequently, we obtain

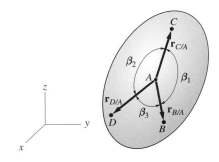

Fig. D.1

$$\mathbf{r}_{B/A} \cdot \mathbf{v}_{B/A} = 0$$
$$\mathbf{r}_{C/A} \cdot \mathbf{v}_{C/A} = 0 \tag{D.1}$$
$$\mathbf{r}_{D/A} \cdot \mathbf{v}_{D/A} = 0$$

The first of these equations is satisfied if either $\mathbf{v}_{B/A} = \mathbf{0}$ or if $\mathbf{r}_{B/A}$ is perpendicular to $\mathbf{v}_{B/A}$. In either case, we conclude that the relative velocity vector must be of the form $\mathbf{v}_{B/A} = \boldsymbol{\omega}_1 \times \mathbf{r}_{B/A}$, where $\boldsymbol{\omega}_1$ is a vector. Observe that if $\boldsymbol{\omega}_1 = \mathbf{0}$ or if $\boldsymbol{\omega}_1$ is parallel to $\mathbf{r}_{B/A}$, we obtain $\mathbf{v}_{B/A} = \mathbf{0}$. Otherwise $\mathbf{v}_{B/A}$ is perpendicular to $\mathbf{r}_{B/A}$ (this follows from the properties of the cross product). Applying similar arguments to the remaining two cases in Eqs. (D.1), we obtain

$$\mathbf{v}_{B/A} = \boldsymbol{\omega}_1 \times \mathbf{r}_{B/A}$$
$$\mathbf{v}_{C/A} = \boldsymbol{\omega}_2 \times \mathbf{r}_{C/A} \tag{D.2}$$
$$\mathbf{v}_{D/A} = \boldsymbol{\omega}_3 \times \mathbf{r}_{D/A}$$

So far, we have placed no restrictions on the vectors $\boldsymbol{\omega}_1$, $\boldsymbol{\omega}_2$, and $\boldsymbol{\omega}_3$. Equations (D.2) are necessary and sufficient for the magnitudes of relative position vectors to remain constant during the motion of the body.

Consider next the requirement that the angle β_1, between $\mathbf{r}_{B/A}$ and $\mathbf{r}_{C/A}$, remains constant. Utilizing the properties of the dot product, we have $\mathbf{r}_{B/A} \cdot \mathbf{r}_{C/A} = |\mathbf{r}_{B/A}||\mathbf{r}_{C/A}| \cos \beta_1$. Taking the time derivative of this equation and imposing the condition that $|\mathbf{r}_{B/A}|$, $|\mathbf{r}_{C/A}|$, and β_1 are constants, we obtain

$$(\mathbf{r}_{B/A} \cdot \dot{\mathbf{r}}_{C/A}) + (\dot{\mathbf{r}}_{B/A} \cdot \mathbf{r}_{C/A}) = 0$$

or

$$(\mathbf{r}_{B/A} \cdot \mathbf{v}_{C/A}) + (\mathbf{r}_{C/A} \cdot \mathbf{v}_{B/A}) = 0$$

Similar arguments can also be applied to β_2 and β_3, the results being

$$(\mathbf{r}_{B/A} \cdot \mathbf{v}_{C/A}) + (\mathbf{r}_{C/A} \cdot \mathbf{v}_{B/A}) = 0$$
$$(\mathbf{r}_{C/A} \cdot \mathbf{v}_{D/A}) + (\mathbf{r}_{D/A} \cdot \mathbf{v}_{C/A}) = 0 \qquad \text{(D.3)}$$
$$(\mathbf{r}_{D/A} \cdot \mathbf{v}_{B/A}) + (\mathbf{r}_{B/A} \cdot \mathbf{v}_{D/A}) = 0$$

Using Eqs. (D.2) to eliminate the relative velocities from Eqs. (D.3), we obtain

$$(\mathbf{r}_{B/A} \cdot \boldsymbol{\omega}_2 \times \mathbf{r}_{C/A}) + (\mathbf{r}_{C/A} \cdot \boldsymbol{\omega}_1 \times \mathbf{r}_{B/A}) = 0$$
$$\text{(etc.)}$$

Using the properties of the scalar triple product, these equations simplify to

$$(\boldsymbol{\omega}_1 - \boldsymbol{\omega}_2) \cdot (\mathbf{r}_{B/A} \times \mathbf{r}_{C/A}) = 0$$
$$(\boldsymbol{\omega}_2 - \boldsymbol{\omega}_3) \cdot (\mathbf{r}_{C/A} \times \mathbf{r}_{D/A}) = 0 \qquad \text{(D.4)}$$
$$(\boldsymbol{\omega}_3 - \boldsymbol{\omega}_1) \cdot (\mathbf{r}_{D/A} \times \mathbf{r}_{B/A}) = 0$$

These three scalar equations contain nine unknowns: the three components of $\boldsymbol{\omega}_1$, $\boldsymbol{\omega}_2$, and $\boldsymbol{\omega}_3$.

The general solution to Eqs. (D.4) is (this may be verified by substitution)

$$\boldsymbol{\omega}_1 = \boldsymbol{\omega} + k_1 \mathbf{r}_{B/A}$$
$$\boldsymbol{\omega}_2 = \boldsymbol{\omega} + k_2 \mathbf{r}_{C/A} \qquad \text{(D.5)}$$
$$\boldsymbol{\omega}_3 = \boldsymbol{\omega} + k_3 \mathbf{r}_{D/A}$$

where $\boldsymbol{\omega}$ is called the angular velocity of the body and the k's are undetermined constants.

When Eqs. (D.5) are substituted into Eqs. (D.2), the k's vanish (because $\mathbf{r}_{B/A} \times \mathbf{r}_{B/A} = \mathbf{0}$, etc.), and the final results are

$$\mathbf{v}_{B/A} = \boldsymbol{\omega} \times \mathbf{r}_{B/A}$$
$$\mathbf{v}_{C/A} = \boldsymbol{\omega} \times \mathbf{r}_{C/A} \qquad \text{(D.6)}$$
$$\mathbf{v}_{D/A} = \boldsymbol{\omega} \times \mathbf{r}_{D/A}$$

Equations (D.6) not only prove Eq. (19.1a), but also show that $\boldsymbol{\omega}$ (the angular velocity) is a property of the body that does not depend on the chosen points (recall that the choice of points A, B, C, and D was arbitrary).

E

Numerical Solution of Differential Equations

E.1 Introduction

In general, the acceleration of a particle depends upon its velocity, position, and time. For example, if a particle is moving in the xy plane, its acceleration components are

$$a_x = f_x(x, y, v_x, v_y, t) \quad a_y = f_y(x, y, v_x, v_y, t) \tag{E.1}$$

where f_x and f_y are known functions determined by kinetic analysis. When the equations are rewritten as

$$\ddot{x} = f_x(x, y, \dot{x}, \dot{y}, t) \quad \ddot{y} = f_y(x, y, \dot{x}, \dot{y}, t) \tag{E.2}$$

we see that they represent a coupled set of second-order differential equations. The term *coupled* means that the motions in the three coordinate directions depend on each other. Coupled differential equations are difficult, or impossible, to solve analytically. If the equations are linear, it may be possible to find a closed-form solution, but the labor involved is seldom worthwhile. Nonlinear differential equations do not, as a rule, have analytical solutions. Even the single differential equation encountered in rectilinear motion

$$\ddot{x} = f(x, \dot{x}, t)$$

may be impossible to solve analytically, unless it belongs to one of the special cases discussed in Art. 12.4.

E.2 Numerical Methods

Because analytical solutions are seldom available, numerical methods are the main tool for solving Eqs. (E.2). There is no shortage of software for the solution of ordinary differential equations. Most software packages come with graphics programs that can be used to plot the results. Nearly all existing programs work with sets of first-order differential equations of the form

$$\dot{x}_1 = f_1(x_1, x_2, \ldots, x_n, t)$$
$$\dot{x}_2 = f_2(x_1, x_2, \ldots, x_n, t)$$
$$\vdots$$
$$\dot{x}_n = f_n(x_1, x_2, \ldots, x_n, t)$$

(E.3)

rather than the second-order equations in Eqs. (E.2). This not a problem, because Eqs. (E.2) easily can be transformed into *equivalent first-order equations* of the form shown in Eqs. (E.3). Using the notation

$$x = x_1 \quad y = x_2 \quad \dot{x} = x_3 \quad \dot{y} = x_4$$

(E.4)

the equivalent first-order equations are

$$\dot{x}_1 = x_3$$
$$\dot{x}_2 = x_4$$
$$\dot{x}_3 = f_x(x_1, x_2, x_3, x_4, t)$$
$$\dot{x}_4 = f_y(x_1, x_2, x_3, x_4, t)$$

(E.5)

In vector notation, these equations are

$$\dot{\mathbf{x}} = \mathbf{f}(\mathbf{x}, t)$$

(E.6)

All numerical methods that solve differential equations work in a similar manner: starting with the initial conditions, they step forward in time increments Δt, computing $\mathbf{x}$ at each step. The output is thus a table of $\mathbf{x}$ versus t. The required input consists of the vector $\mathbf{f}(\mathbf{x}, t)$ that defines the first-order differential equations, the initial conditions, and the time span of the solution. There is seldom a need to specify the time increment Δt, because most programs automatically compute the most efficient value of Δt. The essential difference between various software packages lies in the syntax that is used.

E.3 Application of MATLAB

a. MATLAB function ode45

MATLAB® is one of many software packages that are capable of solving differential equations. It is not the easiest program to use, but since it is widely available at academic institutions, we adopt it for numerical solutions in this book. MATLAB has several functions for solving ordinary differential equations. The most popular of these is called ode45, which uses an adaptive 5th-order Runge-Kutta method ("adaptive" methods adjust Δt to keep errors within prescribed limits). The function call has the form

```
[t,x] = ode45(@f,[t0 t1 t2...tEnd],[x1 x2 x3...xn])
```

The input arguments are:

@f—the handle of the user-supplied function f that returns the vector $\mathbf{f}(\mathbf{x}, t)$ in Eq. (E.6). A *function handle* is a MATLAB data type that contains the information required to find and execute a function. The name of a function handle is simply the character @ followed by the function name.

[t0 t1 t2...tEnd]—times at which the solution will be obtained. If solution is to be obtained from t0 to tEnd in steps of dt, use [t0:dt:tEnd].

[x1 x2 x3...xn]—the initial values of $x_1, x_2, \ldots, x_n$

The two items in the output are:

t—a vector containing the times at which the solution was obtained.

x—the solution matrix whose columns contain the values of $x_1, x_2, \ldots, x_n$.

As an example, consider the numerical solution of the second-order differential equation

$$\ddot{x} = -0.5(x^2 - 1)\dot{x} - x$$

where x is in feet and t is in seconds. The initial conditions are $x(0) = 1$ ft and $\dot{x}(0) = 0$. The solution is to be obtained from $t = 0$ to $t = 10$ s in increments of 0.2 s. With the notation $x = x_1$ and $\dot{x} = x_2$, the equivalent first-order differential equations are

$$\dot{x}_1 = x_2$$

$$\dot{x}_2 = -0.5(x_1^2 - 1)x_2 - x_1$$

subject to the initial conditions $x_1(0) = 1$ ft and $x_2(0) = 0$. The MATLAB program that solves this problem is

```
function exampleE_1
[t,x] = ode45(@f,[0:0.2:10],[1 0]);
    function dxdt = f(t,x)
    dxdt = [x(2)
            -0.5*(x(1)^2 - 1)*x(2) - x(1)];
    end
end
```

Here the primary function is exampleE_1 (an arbitrarily chosen name). The function f(t,x), which defines the differential equations, is known as an embedded function (it can be called only by the primary function). MATLAB provides other ways of incorporating the function f(t,x) into the program, but the method used in the program listed previously is the most straightforward.

b.   *Plotting the solution*

The output of ode45 can be displayed graphically by calling the MATLAB function plot. The plot of x_1 (first column of matrix x) versus t in Fig. E.1 was obtained by including the following line in exampleE_1:

```
plot(t,x(:,1))
```

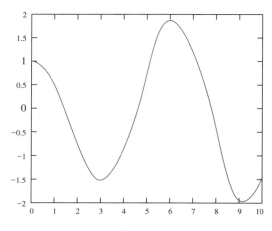

Fig. E.1

The plot can be improved by increasing the font size, using a thicker line, labelling the axis, and adding a grid, as shown in Fig. E.2. These can be achieved by replacing `plot(t,x(:,1))` with the following commands:

```
axes('fontsize',14)
plot(t,x(:,1),'linewidth',1.5)
xlabel('t (s)'); ylabel('x (ft)')
grid on
```

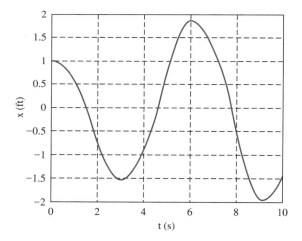

Fig. E.2

c.   *Printing the solution*

MATLAB has no functions that can print the output of `ode45` in a formatted table. Here is our own function that we use in Sample Problems throughout the text:

```
function printSol(t,x)
[m,n] = size(x);
head ='    t';
```

```
for i = 1:n
    head = strcat(head,'        x',num2str(i));
end
fprintf(head); fprintf('\n')
for i = 1:m
    fprintf('%13.4e',t(i),x(i,:)); fprintf('\n')
end
```

By calling this function in `exampleE_1`, we obtain the following printout (only the first four and last four lines are shown):

```
      t              x1              x2
0.0000e+000    1.0000e+000    0.0000e+000
2.0000e-001    9.8006e-001   -1.9886e-001
4.0000e-001    9.2084e-001   -3.9245e-001
6.0000e-001    8.2356e-001   -5.7915e-001

9.4000e+000   -1.9139e+000    4.3391e-001
9.6000e+000   -1.8026e+000    6.7054e-001
9.8000e+000   -1.6482e+000    8.6656e-001
1.0000e+001   -1.4572e+000    1.0428e+000
```

E.4　*Linear Interpolation*

The results obtained from numerical solutions seldom coincide exactly with the points of interest. For example, assume that we have obtained a printout of x versus t and wish to determine t at the instant when $x = 0$. Since we cannot control the values of x on the printout, it is very unlikely that we will encounter a line where $x = 0$. The best we can obtain from the printout is the time interval, say t_1 to t_2, where x changes its sign. To obtain a more accurate estimate of t, interpolation must be used.

Linear interpolation assumes that over a short time span the plot of x versus t can be approximated by a straight line. Since any segment of a straight line has the same slope, we have

$$\frac{x(t_2) - x(t_1)}{t_2 - t_1} = \frac{x - x(t_1)}{t - t_1} \qquad (E.7)$$

which is the interpolation formula between the points $[t_1,\ x(t_1)]$ and $[t_2,\ x(t_2)]$. By specifying the value of x, we can solve the equation for the corresponding t and *vice versa*.

As an illustration, let us compute t when $x_1 = -1.75$ ft using the printout of `exampleE_1`. By inspection, we see that $9.6 < t < 9.8$. Based on these time values, the interpolation formula becomes

$$\frac{-1.6482 - (-1.8026)}{9.8 - 9.6} = \frac{-1.75 - (-1.8026)}{t - 9.6}$$

The solution is $t = 9.6681$ s.

F

Mass Moments and Products of Inertia

F.1 Introduction

The concept of mass moment of inertia was first introduced in Art. 17.2. In that article, we considered only those inertial properties that were required for a discussion of the plane motion of a rigid body—specifically, the definitions of mass moment of inertia and the radius of gyration, the parallel-axis theorem, and the method of composite bodies. We used these concepts extensively throughout the analysis of plane motion in Chapters 17 and 18. The inertial properties of a rigid body described in three dimensions were introduced in Art. 19.3, including the definition of mass product of inertia, the parallel-axis theorem for products of inertia, and the principal moments of inertia. The techniques for determining moment of inertia by integration, which were not discussed, are included in this appendix. This appendix concludes with a discussion of the inertia tensor, including the principal moments of inertia and the principal directions.

F.2 Review of Mass Moment of Inertia

The mass moment of inertia of a rigid body of mass m about an axis, such as the axis a-a in Fig. F.1, was defined as

$$I_a = \int_V r^2 \, dm \qquad \text{(17.1, repeated)}$$

where r is the distance from the axis to the mass element dm, and the integral is taken over the region V occupied by the body. Methods for computing the integral in Eq. (17.1) are discussed in Art. F.4.

The radius of gyration of the body with respect to the axis a-a was defined as

$$k_a = \sqrt{I_a/m} \qquad \text{(17.2, repeated)}$$

The parallel-axis theorem (proved in Art. 17.2) states that

$$I_a = \bar{I}_a + md^2 \qquad \text{(17.3, repeated)}$$

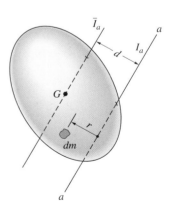

Fig. F.1

As indicated in Fig. F.1, $\bar{I}_a$ is the moment of inertia of the body about the axis that is parallel to the axis *a-a* and passes through the mass center *G* of the body (we refer to this axis as a *central axis*), and *d* is the distance between the two axes. In Art. 17.2, the use of this theorem was restricted to the computation of moments of inertia of composite bodies. As we see in Art. F.4, the parallel-axis theorem is also very useful when computing the moment of inertia of a body by integration.

F.3   *Moments of Inertia of Thin Plates*

Here we present a convenient method of calculating the mass moments of inertia of homogeneous thin plates from the second moments of their surface areas. Not only are thin plates important in their own right, but their inertial properties are also useful in the calculation of the inertial properties of solids by integration.

Figure F.2 shows a thin plate of thickness *t*. The surface of the plate is the plane region $\mathscr{A}$ of area *A*, and the solid region enclosing the plate is denoted by $\mathscr{V}$. Letting ρ be the mass density (mass per unit volume), the mass of the plate is

$$m = \rho t A \tag{F.1}$$

Using the definition in Eq. (17.1) and referring to Fig. F.2, the moments of inertia of the differential mass element *dm* about the coordinate axes are

$$dI_x = y^2\, dm \qquad dI_y = x^2\, dm \qquad dI_z = r^2\, dm \tag{a}$$

where *r* is measured from the origin *O* of the coordinate system. Substituting $dm = \rho t\, dA$ and integrating over the region $\mathscr{A}$ yield

$$I_x = \rho t \int_{\mathscr{A}} y^2\, dA \qquad I_y = \rho t \int_{\mathscr{A}} x^2\, dA \qquad I_z = \rho t \int_{\mathscr{A}} r^2\, dA \tag{b}$$

The integrals in Eqs. (b) represent the moments of inertia of the surface area of the plate:[*] $\int_{\mathscr{A}} y^2\, dA = (I_x)_{\text{area}}$, $\int_{\mathscr{A}} x^2\, dA = (I_y)_{\text{area}}$, and $\int_{\mathscr{A}} r^2\, dA = (J_O)_{\text{area}}$.

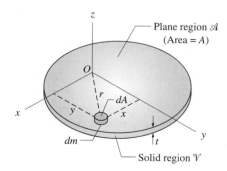

Fig. F.2

---

[*]In this appendix, we use the subscript "area" to distinguish the moment of inertia of the plate area from the mass moment of inertia of the plate.

Therefore, mass moments of inertia of the plate are related to the area moments of inertia by

$$I_x = \rho t (I_x)_{\text{area}} \qquad I_y = \rho t (I_y)_{\text{area}}$$
$$I_z = \rho t (J_O)_{\text{area}}$$

(F.2)

Substituting $\rho t = m/A$ from Eq. (F.1), we obtain the following alternative forms of Eqs. (F.2):

$$I_x = \frac{m}{A}(I_x)_{\text{area}} \qquad I_y = \frac{m}{A}(I_y)_{\text{area}}$$
$$I_z = \frac{m}{A}(J_O)_{\text{area}}$$

(F.3)

Because $r^2 = x^2 + y^2$, Eqs. (b) yield the identity

$$I_z = I_x + I_y$$

(F.4)

It is important to note that Eq. (F.4) is valid only for thin plates. In general, it is not true for bodies of arbitrary shape.

As an illustration, consider the homogeneous thin disk of mass m shown in Fig. F.3(a). The surface area of the disk is the circle of radius R in Fig. F.3(b). From the table on page 652, the moments of inertia of a circular area are $(I_x)_{\text{area}} = (I_y)_{\text{area}} = \pi R^4/4$. Therefore, the mass moments of inertia of the plate can be computed as follows.

$$I_x = I_y = \frac{m}{A}(I_x)_{\text{area}} = \frac{m}{\pi R^2} \frac{\pi R^4}{4} = \frac{mR^2}{4}$$

and

$$I_z = I_x + I_y = 2\left(\frac{mR^2}{4}\right) = \frac{mR^2}{2}$$

The above result for I_z could also have been found using $I_z = (m/A)(J_O)_{\text{area}}$, where $(J_O)_{\text{area}} = (I_x)_{\text{area}} + (I_y)_{\text{area}}$.

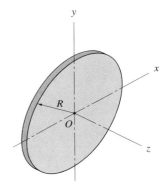

(a) Homogeneous thin
circular disk

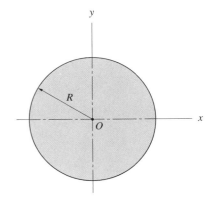

(b) Surface area
of disk

Fig. F.3

F.4   *Mass Moment of Inertia by Integration*

According to Eq. (17.1), the mass moment of inertia of a body that occupies a region $\mathcal{V}$ is obtained by evaluating an integral of the form $\int_{\mathcal{V}} r^2 \, dm$, which, in general, represents a triple integral. Using rectangular coordinates, for example, we have $dm = \rho \, dV = \rho \, dx \, dy \, dz$, where ρ is the mass density at the point whose coordinates are (x, y, z). Techniques of evaluating multiple integrals are presented in introductory calculus texts (you will find that most of those texts use moment of inertia as a practical application of integration performed over spatial regions).

Here we consider only bodies whose symmetry permits us to evaluate their moments of inertia with a single integration. As will be seen in Sample Problem F.3, the single integration technique is based on the inertial properties of thin plates that have been discussed in the preceding article.

Sample Problem   *F.1*

Figure (a) shows a homogeneous slender rod of mass m and length L. Determine the moments of inertia of the rod about the x-, y-, and z-axes that pass through its mass center G.

Solution

The term *slender* implies that the cross-sectional dimensions of the rod are negligible compared with its length. Therefore, the mass of the rod may be considered as being distributed along the x-axis, which means that its moment of inertia about that axis is negligible; that is,

$$I_x \approx 0 \qquad\qquad \textit{Answer}$$

The differential mass dm chosen for integration is shown in Fig. (b). Because x is the perpendicular distance from both the y-axis and the z-axis to dm, the moments of inertia about these two axes are identical. Letting ρ be the (constant) mass of the rod *per unit length,* we have $dm = \rho\, dx$. The definition of moment of inertia gives

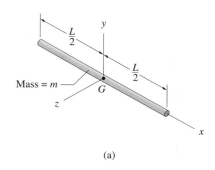

(a)

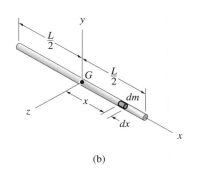

(b)

$$I_y = I_z = \int x^2\, dm = \rho \int_{-L/2}^{L/2} x^2\, \rho\, dx = \frac{\rho}{3}\left(\frac{L^3}{8} + \frac{L^3}{8}\right) = \frac{\rho L^3}{12}$$

Because the mass of the rod is $m = \rho L$, this result may be written as

$$I_y = I_z = \frac{mL^2}{12} \qquad\qquad \textit{Answer}$$

Sample Problem   *F.2*

As shown in the figure, an assembly is formed by joining a 0.3-kg rectangular plate to a 0.2-kg triangular plate. Assuming that both plates are thin and homogeneous, calculate the moment of inertia of the assembly about each of the three coordinate axes.

Solution

We use Eqs. (F.3) to compute the mass moment of inertia of each plate from the properties of areas listed in the table on page 652. Summing the results for the two plates then gives the moments of inertia for the assembly.

Rectangular Plate

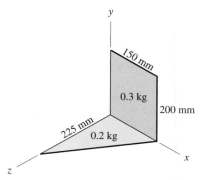

$$m = 0.3\,\text{kg}$$

$$A = 0.15 \times 0.20 = 0.03\,\text{m}^2$$

$$(I_x)_{\text{area}} = \frac{0.15(0.2)^3}{3} = 4 \times 10^{-4}\,\text{m}^4$$

$$(I_y)_{\text{area}} = \frac{0.2 \times (0.15)^3}{3} = 2.25 \times 10^{-4}\,\text{m}^4$$

Because the plate lies in the xy-plane,

$$(J_O)_{\text{area}} = (I_x)_{\text{area}} + (I_y)_{\text{area}}$$
$$= (4 + 2.25)(10^{-4}) = 6.25 \times 10^{-4} \text{ m}^4$$

Substituting the above results into Eqs. (F.3) gives

$$I_x = \frac{m}{A}(I_x)_{\text{area}} = \frac{0.3}{0.03}(4 \times 10^{-4}) = 4 \times 10^{-3} \text{ kg} \cdot \text{m}^2$$

$$I_y = \frac{m}{A}(I_y)_{\text{area}} = \frac{0.3}{0.03}(2.25 \times 10^{-4}) = 2.25 \times 10^{-3} \text{ kg} \cdot \text{m}^2$$

$$I_z = \frac{m}{A}(J_O)_{\text{area}} = \frac{0.3}{0.03}(6.25 \times 10^{-4}) = 6.25 \times 10^{-3} \text{ kg} \cdot \text{m}^2$$

Triangular Plate

$$m = 0.2 \text{ kg}$$

$$A = \frac{0.225 \times 0.15}{2} = 1.69 \times 10^{-2} \text{ m}^2$$

$$(I_x)_{\text{area}} = \frac{0.15 \times 0.225^3}{12} = 1.42 \times 10^{-4} \text{ m}^4$$

$$(I_z)_{\text{area}} = \frac{0.225 \times 0.15^3}{12} = 6.33 \times 10^{-5} \text{ m}^4$$

Noting that the plate lies in the xz-plane, we have

$$(J_O)_{\text{area}} = (I_x)_{\text{area}} + (I_z)_{\text{area}}$$
$$= (14.2 + 6.33)10^{-5} \text{ m}^4 = 20.53 \times 10^{-5} \text{ m}^4$$

Substituting the above results into Eqs. (F.3) (modified to take into account the fact that the area lies in the xz-plane) gives

$$I_x = \frac{m}{A}(I_x)_{\text{area}} = \frac{0.2}{1.69 \times 10^{-2}}(1.42 \times 10^{-4}) = 1.68 \times 10^{-3} \text{ kg} \cdot \text{m}^2$$

$$I_y = \frac{m}{A}(J_O)_{\text{area}} = \frac{0.2}{1.69 \times 10^{-2}}(20.53 \times 10^{-5}) = 2.43 \times 10^{-3} \text{ kg} \cdot \text{m}^2$$

$$I_z = \frac{m}{A}(I_z)_{\text{area}} = \frac{0.2}{1.69 \times 10^{-2}}(6.33 \times 10^{-5}) = 7.49 \times 10^{-4} \text{ kg} \cdot \text{m}^2$$

Assembly

Summing the results obtained for the two plates gives the following values for the inertial properties of the assembly.

$$I_x = (4 + 1.68)(10^{-3}) = 5.68 \times 10^{-3} \text{ kg} \cdot \text{m}^2$$

$$I_y = (2.25 + 2.43)(10^{-3}) = 4.68 \times 10^{-3} \text{ kg} \cdot \text{m}^2 \qquad \textit{Answer}$$

$$I_z = (6.25 + 0.749)(10^{-3}) = 6.999 \times 10^{-3} \text{ kg} \cdot \text{m}^2$$

Sample Problem   *F.3*

Figure (a) shows a homogeneous block of mass m. Using integration, calculate its mass moments of inertia about each of the coordinate axes shown. The origin O is located at the center of the bottom face.

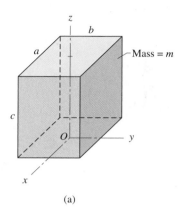

(a)

Solution

We select for the differential element the plate of mass dm that is shown in Fig. (b). Because the thickness dz of this element is infinitesimal, all parts of the plate are a distance z from the xy-plane. The x'- and y'-axes shown in Fig. (b) are central axes of the element.

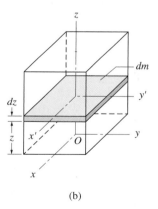

(b)

Applying the parallel-axis theorem to the element, we get

$$dI_x = dI_{x'} + z^2\, dm \qquad \text{(a)}$$

where dI_x and $dI_{x'}$ are the mass moments of inertia of dm about the x- and x'-axes, respectively. According to Eqs. (F.3), we have

$$dI_{x'} = \rho\, dz\, (I_{x'})_{\text{area}} = \rho\, dz\, \frac{ab^3}{12} \qquad \text{(b)}$$

where ρ is the (constant) mass density of the block; the expression for $(I_{x'})_{\text{area}}$ was taken from the table on page 652.

Substituting Eq. (b) and $dm = \rho ab\, dz$ into Eq. (a), we obtain

$$dI_x = \rho\, dz \frac{ab^3}{12} + z^2(\rho ab\, dz) = \rho ab \left(\frac{b^2}{12} + z^2 \right) dz \qquad \text{(c)}$$

Integrating Eq. (c) from $z = 0$ to $z = c$ yields

$$I_x = \rho ab \int_0^c \left(\frac{b^2}{12} + z^2 \right) dz = \rho abc \left(\frac{b^2}{12} + \frac{c^2}{3} \right)$$

Recognizing that the mass of the block is $m = \rho abc$, the mass moment of inertia about the x-axis may be written as

$$I_x = m \left(\frac{b^2}{12} + \frac{c^2}{3} \right) \qquad \textit{Answer} \quad \text{(d)}$$

The computation of I_y is identical to that of I_x, except that the dimensions a and b are interchanged. Therefore, we can deduce from Eq. (d) that

$$I_y = m \left(\frac{a^2}{12} + \frac{c^2}{3} \right) \qquad \textit{Answer} \quad \text{(e)}$$

From Eqs. (F.2), the mass moment of inertia of the element dm about the z-axis is

$$dI_z = \rho\, dz\, (J_O)_{\text{area}} \qquad \text{(f)}$$

where $(J_O)_{\text{area}}$ is the polar moment of inertia of the area of the element. Substituting

$$(J_O)_{\text{area}} = (I_{x'})_{\text{area}} + (I_{y'})_{\text{area}} = \frac{ab^3}{12} + \frac{a^3b}{12} \qquad \text{(g)}$$

into Eq. (f), we obtain

$$dI_z = \rho \frac{ab}{12} \left(a^2 + b^2 \right) dz$$

Integrating between $z = 0$ and $z = c$, we find that

$$I_z = \frac{\rho ab}{12} \left(a^2 + b^2 \right) \int_0^c dz = \frac{\rho abc}{12} \left(a^2 + b^2 \right)$$

which, on substituting $m = \rho abc$, may be written as

$$I_z = \frac{m}{12} \left(a^2 + b^2 \right) \qquad \textit{Answer}$$

Problems

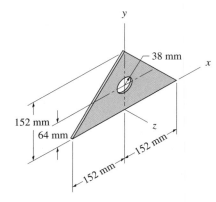

Fig. PF.1

F.1 The thin plate mass 113 g. Calculate its moment of inertia about each coordinate axis.

F.2 The thin plate of mass m has the shape of a circular segment. Determine its moment of inertia about the z-axis.

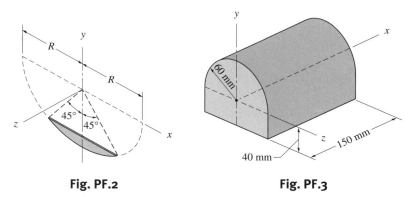

Fig. PF.2   **Fig. PF.3**

F.3 Calculate the moment of inertia of the iron casting about the x-axis. The mass density of cast iron is 7200 kg/m³.

F.4 The bracket of mass m has a uniform thickness. Calculate its moment of inertia about each coordinate axis.

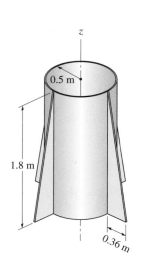

Fig. PF.6

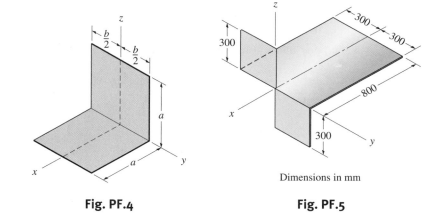

Dimensions in mm

Fig. PF.4   **Fig. PF.5**

F.5 The part shown is formed by slitting and bending a thin plate. If the total mass of the plate is 7.5 kg, determine its moment of inertia about the three coordinate axes.

F.6 The rocket casing consists of a 120-kg cylindrical shell and four triangular fins, each of mass 15 kg. Assuming all components to be thin and of uniform thickness, determine the moment of inertia of the casing about the z-axis.

F.7 Without integrating, find the moment of inertia of the wire about the y'-axis, which passes through the mass center G.

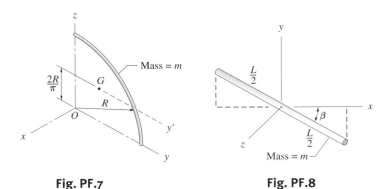

Fig. PF.7   **Fig. PF.8**

F.8 (a) Find the moment of inertia of the slender rod about the x-axis by integration. (b) Using the results of part (a), determine the moments of inertia about the other two coordinate axes.

F.9 The slender rod of mass m lies in the xy-plane. Using integration, determine its moment of inertia about the x-axis.

F.10 (a) Determine the moment of inertia for the homogeneous cylinder about the z-axis by integration. (b) Obtain the moment of inertia about the y-axis using Eq. (F.3).

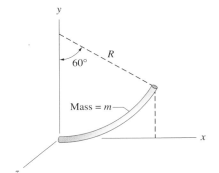

Fig. PF.9

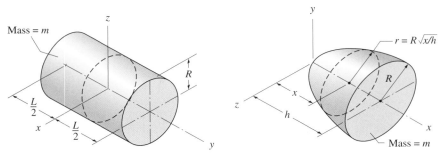

Fig. PF.10   **Fig. PF.11**   **Fig. PF.12**

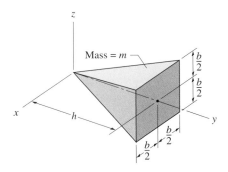

F.11 Using integration, find the moment of inertia of the paraboloid of revolution about the x-axis.

F.12 Use integration to determine the moment of inertia of the rectangular pyramid about (a) the z-axis; and (b) the y-axis.

F.13 (a) Determine the moment of inertia of the homogeneous hemisphere of mass m about the y-axis using integration. (b) Using the result found in part (a), find the moment of inertia of a sphere of mass M about a diameter.

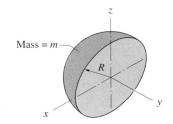

Fig. PF.13

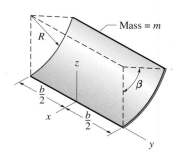

Fig. PF.14

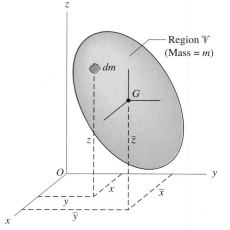

Fig. F.4

F.14 The mass of the truncated conical shell of constant wall thickness is 1.5 kg. Use integration to find the moment of inertia about the axis of the shell.

F.15 Determine the moment of inertia of the thin cylindrical panel about the z-axis using integration.

F.16 The cover of the football may be approximated by a thin homogeneous shell of mass m. Determine its moment of inertia about the z-axis.

F.17 A thin steel plate of mass 80 kg is cut from the pattern shown. Calculate the moment of inertia of the plate about the y-axis.

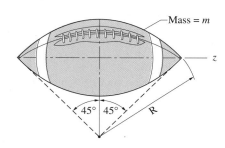

Fig. PF.15

Fig. PF.16

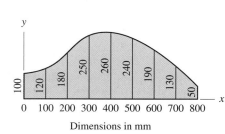

Fig. PF.17, PF.18

F.18 An axisymmetric cavity is formed in a sand mold by rotating the pattern shown about the x-axis. A casting is then made by filling the cavity with aluminum of density 2650 kg/m³. Calculate the moment of inertia of the casting about the x-axis.

F.5 **Mass Products of Inertia; Parallel-Axis Theorems**

Figure F.4 shows a body of mass m that occupies a region $\mathcal{V}$. The mass products of inertia of this body relative to the coordinate axes shown are defined as

$$I_{xy} = I_{yx} = \int_{\mathcal{V}} xy\, dm$$

$$I_{yz} = I_{zy} = \int_{\mathcal{V}} yz\, dm \qquad \text{(F.5)}$$

$$I_{zx} = I_{xz} = \int_{\mathcal{V}} zx\, dm$$

The dimensions for product of inertia are the same as those for moment of inertia, that is, $[ML^2]$ (slug · ft² or kg · m²). Whereas the moment of inertia of a body is always positive, its product of inertia can be positive, negative, or zero.

Products of inertia satisfy the following parallel-axis theorems, which are similar to the parallel-axis theorems for moment of inertia.

$$I_{xy} = \bar{I}_{xy} + m\bar{x}\bar{y}$$
$$I_{yz} = \bar{I}_{yz} + m\bar{y}\bar{z} \qquad \text{(F.6)}$$
$$I_{zx} = \bar{I}_{zx} + m\bar{z}\bar{x}$$

In the first of Eqs. (F.6), $\bar{I}_{xy}$ is the product of inertia with respect to central axes that are parallel to the x- and y-axes, respectively; and $\bar{x}$ and $\bar{y}$ are the coordinates of the mass center G—see Fig. F.4. The terms in the other two equations are defined in an analogous manner.

To prove the parallel-axis theorem, we consider Fig. F.5, which shows a body viewed along the positive z-axis. The $x'y'$-coordinate system has its origin at G, and its axes are parallel to the xy-axes. From Fig. F.5, we see that $x = \bar{x} + x'$ and $y = \bar{y} + y'$, which, when substituted into the definition $I_{xy} = \int_{\mathcal{V}} xy\, dm$, gives

$$I_{xy} = \int_{\mathcal{V}} (\bar{x} + x')(\bar{y} + y')\, dm$$

Carrying out the multiplication, we obtain

$$I_{xy} = \bar{x}\bar{y} \int_{\mathcal{V}} dm + \bar{y} \int_{\mathcal{V}} x'\, dm + \bar{x} \int_{\mathcal{V}} y'\, dm + \int_{\mathcal{V}} x'y'\, dm$$

Note that because the x'-and y'-axes pass through G, we have $\int_{\mathcal{V}} x'y'\, dm = \bar{I}_{xy}$, $\int_{\mathcal{V}} x'dm = 0$, and $\int_{\mathcal{V}} y'dm = 0$. Consequently, the above equation reduces to $I_{xy} = \bar{I}_{xy} + m\bar{x}\bar{y}$. This completes the proof of the parallel-axis theorem.

The method of composite bodies for mass products of inertia is equivalent to the same method for moments of inertia—the product of inertia of a composite body equals the sum of the products of inertia of its parts. The proof of this statement follows directly from the definition of product of inertia: The integral of a sum equals the sum of the integrals.

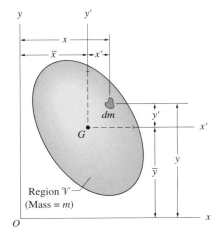

Fig. F.5

F.6 Products of Inertia by Integration; Thin Plates

The evaluation of the integrals that define the mass products of inertia of a body, such as $\int_{\mathcal{V}} xy\, dm$, generally requires triple integration. As in Art. F.4 for mass moments, we restrict our attention here to bodies whose symmetry permits their products of inertia to be evaluated with only a single integration utilizing the properties of thin plates.

The mass products of inertia of the homogeneous thin plate depicted in Fig. F.2 are

$$I_{xy} = \int_{\mathcal{V}} xy\, dm \qquad I_{yz} = \int_{\mathcal{V}} yz\, dm \qquad I_{zx} = \int_{\mathcal{V}} zx\, dm \qquad \text{(a)}$$

Using $dm = \rho t\, dA$, Eqs. (a) become

$$I_{xy} = \rho t \int_{\mathcal{A}} xy\, dA \qquad I_{yz} = \rho t \int_{\mathcal{A}} yz\, dA \qquad I_{zx} = \rho t \int_{\mathcal{A}} zx\, dA \qquad \text{(b)}$$

The integrals in Eqs. (b) are the products of inertia of the plane region $\mathcal{A}$ with respect to the coordinate axes. Again using the label "area" to refer to area properties, Eqs. (b) become

$$I_{xy} = \rho t (I_{xy})_{\text{area}} \qquad I_{yz} = \rho t (I_{yz})_{\text{area}}$$
$$I_{zx} = \rho t (I_{zx})_{\text{area}} \tag{F.7}$$

Substituting $\rho t = m/A$ yields the alternative forms of Eqs. (F.7):

$$I_{xy} = \frac{m}{A}(I_{xy})_{\text{area}} \qquad I_{yz} = \frac{m}{A}(I_{yz})_{\text{area}}$$
$$I_{zx} = \frac{m}{A}(I_{zx})_{\text{area}} \tag{F.8}$$

The products of inertia for commonly encountered plane areas are given in the table on page 652.

F.7 Inertia Tensor; Moment of Inertia about an Arbitrary Axis

As defined in Art. 19.3, the inertia tensor of a body at point O (the origin of the coordinate axes) is the matrix

$$\mathbf{I} = \begin{bmatrix} I_x & -I_{xy} & -I_{xz} \\ -I_{yx} & I_y & -I_{yz} \\ -I_{zx} & -I_{zy} & I_z \end{bmatrix} \tag{19.11, repeated}$$

In this article, we show that the inertia tensor at point O completely determines the moment of inertia about any axis that passes through O.

Figure F.6 shows a rigid body of mass m that occupies the region $\mathcal{V}$. Point O, a point on the body or body extended, is chosen as the origin of the xyz-coordinate system. The axis OM is an arbitrary axis that passes through O. The angle between the position vector $\mathbf{r}$ of the differential mass dm and the axis OM is denoted by θ. Furthermore, we let $\boldsymbol{\lambda}$ be a unit vector in the direction of OM. Note that the magnitude of the cross product of $\mathbf{r}$ and $\boldsymbol{\lambda}$ is $|\mathbf{r} \times \boldsymbol{\lambda}| = r\sin\theta = a$, the perpendicular distance between OM and dm, as shown in Fig. F.6.

The moment of inertia of the body with respect to the axis OM is

$$I_{OM} = \int_{\mathcal{V}} a^2\, dm = \int_{\mathcal{V}} (\mathbf{r} \times \boldsymbol{\lambda}) \cdot (\mathbf{r} \times \boldsymbol{\lambda})\, dm \qquad \text{(a)}$$

where we have used the fact that the dot product of a vector with itself equals the square of the magnitude of the vector; that is, $(\mathbf{r} \times \boldsymbol{\lambda}) \cdot (\mathbf{r} \times \boldsymbol{\lambda}) = |\mathbf{r} \times \boldsymbol{\lambda}|^2 = a^2$.

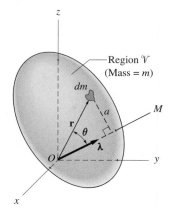

Fig. F.6

We let x, y, and z be the coordinates of the location of dm and λ_x, λ_y, and λ_z be the components of $\boldsymbol{\lambda}$. The cross product of $\mathbf{r} = x\mathbf{i} + y\mathbf{j} + z\mathbf{k}$ and $\boldsymbol{\lambda} = \lambda_x\mathbf{i} + \lambda_y\mathbf{j} + \lambda_z\mathbf{k}$ then becomes

$$\mathbf{r} \times \boldsymbol{\lambda} = (y\lambda_z - z\lambda_y)\mathbf{i} + (\lambda_x z - x\lambda_z)\mathbf{j} + (x\lambda_y - y\lambda_x)\mathbf{k}$$

and the expression for a^2 is

$$a^2 = (\mathbf{r} \times \boldsymbol{\lambda}) \cdot (\mathbf{r} \times \boldsymbol{\lambda}) = (y\lambda_z - z\lambda_y)^2 + (z\lambda_x - x\lambda_z)^2 + (x\lambda_y - y\lambda_x)^2$$

Expanding the squares and collecting terms, we get

$$a^2 = (y^2 + z^2)\lambda_x^2 + (x^2 + z^2)\lambda_y^2 + (x^2 + y^2)\lambda_z^2$$
$$- 2xy\lambda_x\lambda_y - 2xz\lambda_x\lambda_z - 2yz\lambda_y\lambda_z \qquad \text{(b)}$$

Substituting Eq. (b) into Eq. (a), and identifying the expressions for the moments and products of inertia, the moment of inertia about the axis OM becomes

$$I_{OM} = I_x\lambda_x^2 + I_y\lambda_y^2 + I_z\lambda_z^2 - 2I_{xy}\lambda_x\lambda_y - 2I_{yz}\lambda_y\lambda_z - 2I_{zx}\lambda_z\lambda_x \qquad \text{(F.9)}$$

We have now arrived at the following important conclusion: If the components of the inertia tensor (I_x, I_{xy}, I_{xz}, etc.) are known at a point, the moment of inertia about any axis through the point can be computed from Eq. (F.9).

F.8 Principal Moments and Principal Axes of Inertia

In general, the components of the inertia tensor (I_x, I_{xy}, etc.) vary with the location of the reference point O and with the orientation of the xyz-axes. Throughout this article, we assume that the reference point does not change, and we study the effect of changing the orientation of the coordinate axes.

It can be shown[*] that there exists at least one orientation of the xyz-axes for which the inertia tensor has the following *diagonal form:*

$$\mathbf{I} = \begin{bmatrix} I_1 & 0 & 0 \\ 0 & I_2 & 0 \\ 0 & 0 & I_3 \end{bmatrix} \qquad \text{(F.10)}$$

I_1, I_2, and I_3 are called the *principal moments of inertia* at point O, and the corresponding coordinate axes are called the *principal axes* (or *principal directions*) *of inertia* at point O. Note that the products of inertia are zero with respect to the principal axes.

---

[*]In this article, we state some important properties of inertia tensors. Proofs may be found in most advanced dynamics texts—see, for example, *Advanced Engineering Dynamics,* J. H. Ginsberg, Harper & Row, 1988.

The principal moments of inertia and the direction cosines of each principal axis (denoted λ_x, λ_y, and λ_z) can be found by solving the following four equations for λ_x, λ_y, and λ_z.

$$
\begin{aligned}
(I_x - I)\lambda_x \quad &-I_{xy}\lambda_y \quad &-I_{xz}\lambda_z \quad &= 0 \\
-I_{yx}\lambda_x \quad &(I_y - I)\lambda_y \quad &-I_{yz}\lambda_z \quad &= 0 \\
-I_{zx}\lambda_x \quad &-I_{zy}\lambda_y \quad &(I_z - I)\lambda_z \quad &= 0 \\
&\lambda_x^2 + \lambda_y^2 + \lambda_z^2 = 1
\end{aligned}
\tag{F.11}
$$

The first, second, and third equations can be shown to represent the conditions for zero products of inertia with respect to the principal axes, and the fourth equation must be satisfied for λ_x, λ_y, and λ_z to be direction cosines.

Note that the first three equations of Eqs. (F.11) are linear and homogeneous (right side equal to zero) in the unknowns λ_x, λ_y, and λ_z. Therefore, the three values of I (representing I_1, I_2, and I_3) can be obtained by solving the equation that results from setting the determinant of the coefficients of λ_x, λ_y, and λ_z equal to zero; that is,

$$
\begin{vmatrix}
I_x - I & -I_{xy} & -I_{xz} \\
-I_{yx} & I_y - I & -I_{yz} \\
-I_{zx} & -I_{zy} & I_z - I
\end{vmatrix} = 0
\tag{F.12}
$$

Once the principal moments of inertia have been found, the direction cosines of the principal axes can be obtained from Eqs. (F.11).

In linear algebra, the computation of the principal moments of inertia and the principal axes is an example of a *matrix eigenvalue problem,* in which the principal moments are the *eigenvalues* of the inertia tensor, and the unit vectors in the direction of the principal axes are its *eigenvectors.* Equation (F.12) is referred to as the *characteristic equation* of the eigenvalue problem. Because matrix eigenvalue problems occur in many branches of the physical sciences, their properties have been thoroughly studied, and several numerical methods have been developed for their solution (e.g., the Jacobi method).

Using the known characteristics of eigenvalue problems, it can be shown that the solutions of Eqs. (F.11) possess the following properties.

1. The eigenvalues—that is, I_1, I_2, and I_3—are real and positive.
2. Assuming that the eigenvalues are ordered so that $I_3 > I_2 > I_1$, then I_3 and I_1 are the maximum and minimum moments of inertia, respectively, at point O. In other words, I_1 and I_3 are the extrema (extreme values) of I_{OM} in Eq. (F.9) with respect to changes in the direction of OM.
3. If the eigenvalues are distinct—that is, if Eq. (F.12) has no double roots—then the eigenvectors (principal axes) are mutually perpendicular.

Sample Problem   F.4

Using integration, compute the products of inertia of the homogeneous body in Fig. (a) with respect to the axes shown. Express the results in terms of the mass m of the body.

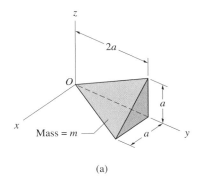

(a)

Solution

We choose the integration element shown in Fig. (b). This element can be considered to be a thin triangular plate of thickness dy, with the surface area A, as shown in Fig. (c). Recognizing the similar triangles in Fig. (b), we see that $(x/y) = (a/2a)$ and that $(z/y) = (a/2a)$, which gives

$$x = z = \frac{y}{2} \tag{a}$$

Next we use Eqs. (F.7) to relate the products of inertia of the mass of the element to the properties of its area. Therefore, we turn our attention to finding the products of inertia of the area shown in Fig. (c).

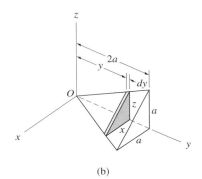

(b)

Products of Inertia of Area A

Because the plane of area A is parallel to the xz-plane, $(\bar{I}_{xy})_{\text{area}} = (\bar{I}_{yz})_{\text{area}} = 0$. And from the table on page 652, we find that $(\bar{I}_{zx})_{\text{area}} = x^2z^2/24$. Using the relationships in Eq. (a), the products of inertia of A become

$$(\bar{I}_{xy})_{\text{area}} = 0 \qquad (\bar{I}_{yz})_{\text{area}} = 0$$

$$(\bar{I}_{zx})_{\text{area}} = \frac{x^2z^2}{24} = \frac{1}{24}\left(\frac{y}{2}\right)^2\left(\frac{y}{2}\right)^2 = \frac{y^4}{384} \tag{b}$$

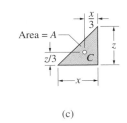

(c)

In terms of the coordinate y, the area A is

$$A = \frac{xz}{2} = \frac{1}{2}\left(\frac{y}{2}\right)\left(\frac{y}{2}\right) = \frac{y^2}{8} \tag{c}$$

and the coordinates of its centroid C are

$$\bar{x} = \frac{x}{3} = \frac{y}{6} \qquad \bar{y} = y \qquad \bar{z} = \frac{z}{3} = \frac{y}{6} \tag{d}$$

We now use the parallel-axis theorem and Eqs. (b) through (d) to compute the products of inertia of A with respect to the coordinate axes.

$$(I_{xy})_{\text{area}} = (\bar{I}_{xy})_{\text{area}} + A\bar{x}\bar{y} = 0 + \frac{y^2}{8}\left(\frac{y}{6}\right)(y) = \frac{y^4}{48} \tag{e}$$

$$(I_{yz})_{\text{area}} = (\bar{I}_{yz})_{\text{area}} + A\bar{y}\bar{z} = 0 + \frac{y^2}{8}(y)\left(\frac{y}{6}\right) = \frac{y^4}{48} \tag{f}$$

$$(I_{zx})_{\text{area}} = (\bar{I}_{zx})_{\text{area}} + A\bar{z}\bar{x} = \frac{y^4}{384} + \frac{y^2}{8}\left(\frac{y}{6}\right)\left(\frac{y}{6}\right) = \frac{7y^4}{1152} \tag{g}$$

Products of Inertia of Mass

We let ρ be the mass density of the body, and we reconsider the differential mass element—the thin plate of thickness dy shown in Fig. (b). Substituting Eqs. (e) through (g) into Eqs. (F.7), the inertial properties of the mass element become

$$dI_{xy} = \rho \, dy \, (I_{xy})_{\text{area}} = \rho \, dy \, \frac{y^4}{48} \tag{h}$$

$$dI_{yz} = \rho \, dy \, (I_{yz})_{\text{area}} = \rho \, dy \, \frac{y^4}{48} \tag{i}$$

$$dI_{zx} = \rho \, dy \, (I_{zx})_{\text{area}} = \rho \, dy \, \frac{7y^4}{1152} \tag{j}$$

Integrating with respect to y between the limits 0 and $2a$ yields

$$I_{xy} = I_{yx} = \frac{\rho}{48} \int_0^{2a} y^4 \, dy = \frac{\rho}{48} \frac{(2a)^5}{5} = \frac{2\rho a^5}{15} \tag{k}$$

$$I_{zx} = \frac{7\rho}{1152} \int_0^{2a} y^4 \, dy = \frac{7\rho}{1152} \frac{(2a)^5}{5} = \frac{7\rho a^5}{180} \tag{l}$$

We note that $dm = \rho \, dV = \rho A \, dy = \rho(y^2/8) \, dy$, which, when integrated between the limits $y = 0$ and $2a$, gives $m = \rho a^3/3$. Therefore, Eqs. (k) and (l) become

$$I_{xy} = I_{yz} = \frac{2\rho a^5}{15} \cdot \frac{m}{\rho a^3/3} = \frac{2}{5}ma^2 \qquad \textit{Answer}$$

$$I_{zx} = \frac{7\rho a^5}{180} \cdot \frac{m}{\rho a^3/3} = \frac{7}{60}ma^2 \qquad \textit{Answer}$$

Sample Problem **F.5**

The assembly shown is formed by joining two pieces of 1.5 mm steel plate. Determine its products of inertia with respect to the axes shown. The weight density of steel is $\gamma = 7900$ kg/m^3.

Solution

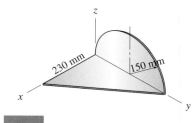

In this problem, we are justified in using the thin plate approximation of Eqs. (F.2) or (F.3), because the thickness of each component plate is much smaller than its in-plane dimensions.

Triangular Plate

Because the triangular area lies in the xy-plane, $(I_{yz})_{\text{area}} = (I_{zx})_{\text{area}} = 0$. From the table on page 652, we obtain

$$(I_{xy})_{\text{area}} = \frac{(0.3)^2(0.23)^2}{24} = 1.98 \times 10^{-4}\ \text{m}^4$$

Using Eqs. (F.2), the mass products of inertia of the triangular plate become

$$I_{xy} = \rho t\, (I_{xy})_{\text{area}} = 7900[0.0015]\left(1.98 \times 10^{-4}\right) = 2.346 \times 10^{-3}\ \text{kg} \cdot \text{m}^2$$
$$I_{yz} = I_{zx} = 0$$

(a)

Semicircular Plate

Noting that the semicircular area lies in the yz-plane, we have $(I_{xy})_{\text{area}} = (I_{zx})_{\text{area}} = 0$. We use the parallel-axis theorem to compute $(I_{yz})_{\text{area}}$. Referring to the table on page 652, the centroid of the semicircular area is located at $\bar{x} = 0$, $\bar{y} = 0.15$ m, $\bar{z} = 4(0.15)/3\pi$ m. Also observe that $(\bar{I}_{yz})_{\text{area}} = 0$, because the centroidal z-axis is an axis of symmetry. The parallel-axis theorem thus yields

$$(I_{yz})_{\text{area}} = (\bar{I}_{yz})_{\text{area}} + A\bar{y}\bar{z}$$
$$= 0 + \frac{\pi}{2}(0.15)^2(0.15)\frac{4}{3\pi}(0.15) = 3.375 \times 10^{-4}\ \text{m}^4$$

The mass products of inertia of the semicircular plate can now be computed from Eqs. (F.2)—remembering to convert inches to feet and weight density to mass density:

$$I_{xy} = I_{zx} = 0$$
$$I_{yz} = \rho t\, (I_{yz})_{\text{area}} = 7900[0.0015]\left(3.375 \times 10^{-4}\right) = 4 \times 10^{-3}\ \text{kg} \cdot \text{m}^2$$

(b)

Assembly

The mass products of inertia of the assembly are found by adding the results for the triangular and semicircular plates in Eqs. (a) and (b), which gives

$$I_{xy} = 23.46 \times 10^{-4}\ \text{kg} \cdot \text{m}^2$$
$$I_{yz} = 40 \times 10^{-4}\ \text{kg} \cdot \text{m}^2 \qquad\qquad \textit{Answer}$$
$$I_{zx} = 0$$

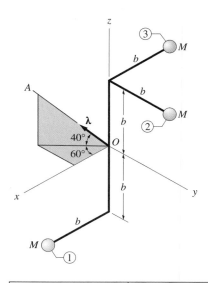

Sample Problem F.6

The assembly consists of three small balls, each of mass M, that are attached to slender rods of negligible mass. Calculate the moment of inertia of the assembly about the axis OA.

Solution

The moment of inertia about the axis OA can be found from Eq. (F.9). In order to use this equation, we must first calculate the inertia tensor at point O and the direction cosines of the axis OA.

The following table shows the computation of the inertia tensor at point O of the assembly, obtained by summing the inertia tensors of the individual balls. Because each ball is small, its moments and products of inertia with respect to axes passing through its mass center may be neglected. Therefore, the parallel-axis theorems simplify to $I_x = \bar{I}_x + M(\bar{y}^2 + \bar{z}^2) = M(y^2 + z^2)$, and so on, and to $I_{xy} = \bar{I}_{xy} + M\bar{x}\bar{y} = Mxy$, and so on.

	Ball 1	**Ball 2**	**Ball 3**	**Totals**
	$x = b,\ y = 0,\ z = -b$	$x = 0,\ y = b,\ z = b$	$x = -b,\ y = 0,\ z = b$	
$I_x = M(y^2 + z^2)$	$M[0 + (-b)^2] = Mb^2$	$M(b^2 + b^2) = 2Mb^2$	$M(0 + b^2) = Mb^2$	$4Mb^2$
$I_y = M(z^2 + x^2)$	$M[(-b)^2 + b^2] = 2Mb^2$	$M(b^2 + 0) = Mb^2$	$M[b^2 + (-b)^2] = 2Mb^2$	$5Mb^2$
$I_z = M(x^2 + y^2)$	$M(b^2 + 0) = Mb^2$	$M(0 + b^2) = Mb^2$	$M[(-b)^2 + 0] = Mb^2$	$3Mb^2$
$I_{xy} = Mxy$	$M(b)(0) = 0$	$M(0)(b) = 0$	$M(-b)(0) = 0$	0
$I_{yz} = Myz$	$M(0)(-b) = 0$	$M(b)(b) = Mb^2$	$M(0)(b) = 0$	Mb^2
$I_{zx} = Mzx$	$M(-b)(b) = -Mb^2$	$M(b)(0) = 0$	$M(b)(-b) = -Mb^2$	$-2Mb^2$

The unit vector $\boldsymbol{\lambda}$ that is directed along the axis OA is

$$\boldsymbol{\lambda} = (\cos 40° \cos 60°)\mathbf{i} - (\cos 40° \sin 60°)\mathbf{j} + \sin 40°\mathbf{k}$$

$$= 0.3830\mathbf{i} - 0.6634\mathbf{j} + 0.6428\mathbf{k}$$

Therefore, the direction cosines of OA are $\lambda_x = 0.3830$, $\lambda_y = -0.6634$, and $\lambda_z = 0.6428$.

Substituting the inertial properties of the assembly (computed in the table) and the direction cosines into Eq. (F.9), we get

$$I_{OA} = I_x\lambda_x^2 + I_y\lambda_y^2 + I_z\lambda_z^2 - 2I_{xy}\lambda_x\lambda_y - 2I_{yz}\lambda_y\lambda_z - 2I_{zx}\lambda_z\lambda_x$$

$$= Mb^2[4(0.3830)^2 + 5(-0.6634)^2 + 3(0.6428)^2$$

$$- 2(0)(0.3830)(-0.6634) - 2(1)(-0.6634)(0.6428)$$

$$- 2(-2)(0.6428)(0.3830)]$$

$$= 5.86Mb^2 \qquad\qquad Answer$$

Sample Problem F.7

The assembly in Fig. (a) consists of two identical, thin plates, each of mass M and thickness t. For point O, determine (1) the inertia tensor with respect to the axes shown; and (2) the principal moments of inertia and the principal axes.

(a)

Solution

Part 1

The inertia tensor at point O for the assembly is found by summing the moments and products of inertia of the two plates about the xyz-axes. We use the thin-plate approximations described in Arts. F.3 and F.6.

Plate 1 Referring to Fig. (a) and the table on page 652, the moments and products of inertia for the area of Plate 1 are

$$(I_x)_{\text{area}} = \frac{a(2a)^3}{12} = \frac{2a^4}{3}$$

$$(I_y)_{\text{area}} = \frac{2a(a)^3}{3} = \frac{2a^4}{3}$$

$$(J_O)_{\text{area}} = (I_x)_{\text{area}} + (I_y)_{\text{area}} = \frac{2a^4}{3} + \frac{2a^4}{3} = \frac{4a^4}{3}$$

$$(I_{xy})_{\text{area}} = 0 \qquad (x\text{-axis is an axis of symmetry})$$

$$(I_{yz})_{\text{area}} = (I_{zx})_{\text{area}} = 0 \qquad (\text{area lies in } xy\text{-plane})$$

Using the thin plate equations [Eqs. (F.2) and (F.7)] and $M = \rho A t = \rho 2a^2 t$, where $A = 2a^2$ is the plate area and ρ the mass density, the mass properties become

$$I_x = \rho t (I_x)_{\text{area}} = \rho t \frac{2a^4}{3} = \frac{1}{3} M a^2$$

$$I_y = \rho t (I_y)_{\text{area}} = \rho t \frac{2a^4}{3} = \frac{1}{3} M a^2$$

$$I_z = \rho t (J_O)_{\text{area}} = \rho t \frac{4a^4}{3} = \frac{2}{3} M a^2$$

$$I_{xy} = I_{yz} = I_{zx} = 0$$

Plate 2 The area moments and products of inertia for Plate 2 are

$$(I_y)_{\text{area}} = \frac{a(2a)^3}{3} = \frac{8a^4}{3}$$

$$(I_z)_{\text{area}} = \frac{2a(a)^3}{3} = \frac{2a^4}{3}$$

625

$$(J_O)_{\text{area}} = (I_y)_{\text{area}} + (I_z)_{\text{area}} = \frac{8a^4}{3} + \frac{2a^4}{3} = \frac{10a^4}{3}$$

$$(I_{xy})_{\text{area}} = (I_{zx})_{\text{area}} = 0 \qquad \text{(the area lies in the } yz\text{-plane)}$$

$$I_{yz} = \frac{a^2(2a)^2}{4} = a^4$$

Substituting these results and $M = \rho At = \rho 2a^2 t$ into Eqs. (F.2) and (F.7), the mass properties of Plate 2 become

$$I_x = \rho t (J_O)_{\text{area}} = \rho t \frac{10a^4}{3} = \frac{5}{3} M a^2$$

$$I_y = \rho t (I_y)_{\text{area}} = \rho t \frac{8a^4}{3} = \frac{4}{3} M a^2$$

$$I_z = \rho t (I_z)_{\text{area}} = \rho t \frac{2a^4}{3} = \frac{1}{3} M a^2$$

$$I_{xy} = I_{zx} = 0$$

$$I_{yz} = \rho t (I_{yz})_{\text{area}} = \rho t a^4 = \frac{1}{2} M a^2$$

Assembly   The mass properties of the assembly are found by summing the properties of the two plates:

$$I_x = M a^2 \left(\frac{1}{3} + \frac{5}{3} \right) = 2 M a^2$$

$$I_y = M a^2 \left(\frac{1}{3} + \frac{4}{3} \right) = \frac{5}{3} M a^2$$

$$I_z = M a^2 \left(\frac{2}{3} + \frac{1}{3} \right) = M a^2$$

$$I_{xy} = 0 + 0 = 0$$

$$I_{yz} = 0 + \frac{1}{2} M a^2 = \frac{1}{2} M a^2$$

$$I_{zx} = 0 + 0 = 0$$

Substituting these values into Eq. (19.11), the inertia tensor at point O for the assembly becomes

$$\mathbf{I} = \begin{bmatrix} 2 & 0 & 0 \\ 0 & \dfrac{5}{3} & -\dfrac{1}{2} \\ 0 & -\dfrac{1}{2} & 1 \end{bmatrix} M a^2 \qquad\qquad \textit{Answer}$$

Part 2

Substituting the inertial properties of the assembly obtained in Part 1 into Eqs. (F.11) yields the following equations that must be solved for the principal moments of inertia and the direction cosines of the principal axes:

$$(2Ma^2 - I)\lambda_x \qquad\qquad + 0\lambda_y \qquad\qquad + 0\lambda_z = 0 \qquad\qquad \text{(a)}$$

$$0\lambda_x + \left(\frac{5}{3}Ma^2 - I\right)\lambda_y \qquad - \frac{Ma^2}{2}\lambda_z = 0 \qquad\qquad \text{(b)}$$

$$0\lambda_x \qquad - \frac{Ma^2}{2}\lambda_y + (Ma^2 - I)\lambda_z = 0 \qquad\qquad \text{(c)}$$

$$\lambda_x^2 + \lambda_y^2 + \lambda_z^2 = 1 \qquad\qquad \text{(d)}$$

The characteristic equation is obtained by setting the determinant of the coefficients in Eqs. (a) through (c) equal to zero:

$$\begin{vmatrix} 2Ma^2 - I & 0 & 0 \\ 0 & \frac{5}{3}Ma^2 - I & -\frac{Ma^2}{2} \\ 0 & -\frac{Ma^2}{2} & Ma^2 - I \end{vmatrix} = 0 \qquad\qquad \text{(e)}$$

Expanding the determinant and simplifying, we get

$$(2Ma^2 - I)\left(I^2 - \frac{8}{3}Ma^2 I + \frac{17}{12}M^2 a^4\right) = 0 \qquad\qquad \text{(f)}$$

Solving Eq. (f) and labeling the roots as I_1, I_2, and I_3 we obtain

$$I_1 = 0.7324 Ma^2 \qquad I_2 = 1.9343 Ma^2 \qquad I_3 = 2.0000 Ma^2 \qquad \textit{Answer}$$

as the principal moments of inertia (eigenvalues).

The direction cosines of a principal axis can be found by substituting the corresponding eigenvalue into Eqs. (a) through (d) and then solving these equations for the direction cosines λ_x, λ_y, and λ_z. Substituting $I = I_3 = 2.0Ma^2$, Eqs. (a) through (c) become

$$0 = 0$$

$$-\frac{1}{3}\lambda_y - \frac{1}{2}\lambda_z = 0$$

$$-\frac{1}{2}\lambda_y - \lambda_z = 0$$

Because the second and third equations are linearly independent, their solution is $\lambda_y = \lambda_z = 0$. Combining these results with Eq. (d), $\lambda_x^2 + \lambda_y^2 + \lambda_z^2 = 1$, we conclude that $\lambda_x = \pm 1$. Hence the unit vector in the direction of the axis associated with I_3—that is, the third eigenvector—is

$$\boldsymbol{\lambda}_3 = \pm \mathbf{i} \qquad\qquad \textit{Answer}$$

The plus-or-minus sign in this answer reflects the fact that an eigenvector defines only the direction of the principal axis; the sense of the unit vector along that axis is arbitrary. In other words, if $\boldsymbol{\lambda}$ is an eigenvector, then $-\boldsymbol{\lambda}$ is also an eigenvector.

With $I = I_2 = 1.9343Ma^2$, Eqs. (a) through (c) are

$$0.0657\lambda_x = 0$$

$$-0.2676\lambda_y - 0.5000\lambda_z = 0$$

$$-0.5000\lambda_y - 0.9343\lambda_z = 0$$

From the first equation we get $\lambda_x = 0$. The second and third equations are not independent, because both yield $\lambda_z = -0.5352\lambda_y$. Substituting these results into Eq. (d), we obtain

$$0^2 + \lambda_y^2 + (-0.5352\lambda_y)^2 = 1$$

which yields $\lambda_y = \pm0.8817$. Consequently, $\lambda_z = -0.5352(\pm0.8817) = \mp0.4719$, from which the second eigenvector is

$$\boldsymbol{\lambda}_2 = \pm0.882\mathbf{j} \mp 0.472\mathbf{k} \qquad\qquad \textit{Answer}$$

The direction of the first eigenvalue, corresponding to $I = I_1 = 0.7324\,Ma^2$, can be obtained in an analogous manner, with the result being

$$\boldsymbol{\lambda}_1 = \pm0.472\mathbf{j} \pm 0.882\mathbf{k} \qquad\qquad \textit{Answer}$$

The three principal axes are shown in Fig. (b). Note that these axes are mutually orthogonal, as they should be, because the principal moments of inertia are distinct. Furthermore, as shown in Fig. (b), the signs of the eigenvectors are usually chosen so that $\boldsymbol{\lambda}_1$, $\boldsymbol{\lambda}_2$, and $\boldsymbol{\lambda}_3$ form a right-handed triad.

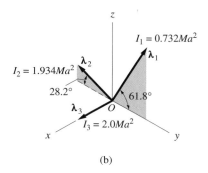

(b)

Problems

F.19   Use integration to determine the products of inertia of the homogeneous slender rod of mass m and length L about the axes shown.

F.20   Calculate I_{xy} for the slender curved rod in Prob. F.9 by integration.

F.21   The uniform slender wire of mass m is bent into the shape of a helix. Using integration, find the products of inertia about the axes shown. (Note that the equation of the helix is $z = h\theta/(2\pi)$, where θ is measured in radians.)

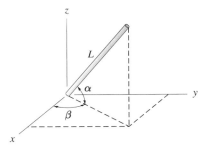

Fig. PF.19

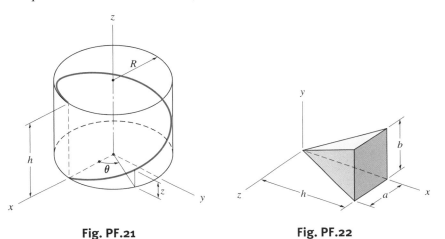

Fig. PF.21   **Fig. PF.22**

F.22   Use integration to determine the products of inertia of the homogeneous solid about the axes shown.

F.23   By integration, find the products of inertia of the homogeneous prism about the axes shown.

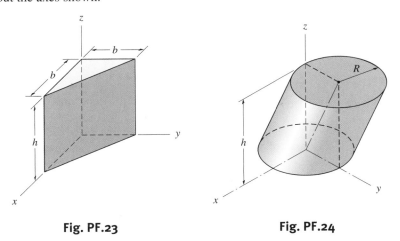

Fig. PF.23   **Fig. PF.24**

F.24   Determine the products of inertia of the homogeneous solid of mass m about the axes shown. Use integration.

F.25 The mass of the thin homogeneous plate is m. Determine its inertia tensor at point O with respect to the axes shown.

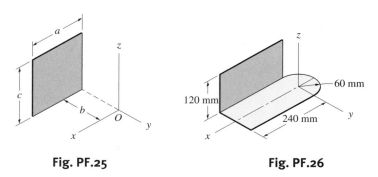

Fig. PF.25   **Fig. PF.26**

F.26 The steel bracket has a uniform thickness of 1.8 mm. Calculate its products of inertia about the axes shown. (For steel, $\gamma = 8000$ kg/m^3.)

F.27 Determine the moment of inertia of the 8-kg homogeneous disk about axis AB.

F.28 The part shown is made by slitting and bending a thin circular plate of mass m. Determine its products of inertia with respect to the axes shown.

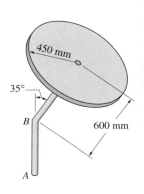

Fig. PF.27

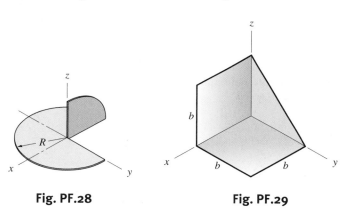

Fig. PF.28   **Fig. PF.29**

F.29 The body of mass m is fabricated from thin sheet metal of constant thickness. Calculate its products of inertia with respect to the axes shown.

F.30 The assembly is made by welding together three pieces of a uniform slender rod. If the mass of the assembly is m, determine its moment of inertia about the axis AB.

F.31 Calculate the moment of inertia of the homogeneous cone about the axis OA.

F.32 The mass of the uniform cube is m. (a) Determine the moment of inertia about the axis OA. (b) Show that OA is a principal axis of inertia at O. (Hint: Substitute your answer from part (a) into the characteristic equation for point O.)

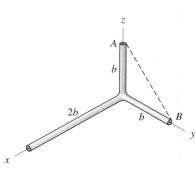

Fig. PF.30

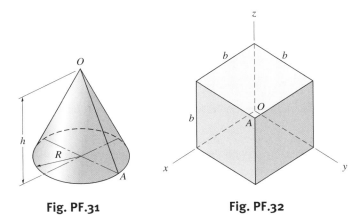

Fig. PF.31

Fig. PF.32

F.33   Determine the ratio h/R for the homogeneous cylinder so that the moments of inertia about all axes passing through its mass center are equal.

F.34   Find the moment of inertia for the homogeneous thin plate about the side AB from the properties of its area.

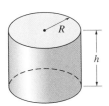

Fig. PF.33

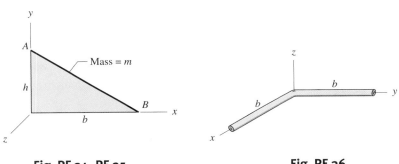

Fig. PF.34, PF.35

Fig. PF.36

F.35   The dimensions of the 0.36-kg thin uniform plate are $b = 240$ mm and $h = 150$ mm. For the mass center of the plate, (a) compute the principal moments of inertia; and (b) locate the principal axis corresponding to the smallest moment of inertia.

F.36   The mass of the slender bent bar is m. Determine the principal axes and the principal moments of inertia at the mass center of the bar. (Hint: The principal axes can be located by inspection from symmetry.)

F.37   The small masses are joined by a light rod. For point O, determine (a) the inertia tensor with respect to the axes shown; and (b) the principal moments of inertia and principal axes at O. Show the principal axes on a sketch of the body.

F.38   The two small masses are joined by a rigid rod of negligible mass. Determine the principal moments of inertia of the system at its mass center. (Hint: It is unnecessary to calculate the inertia tensor because the principal axes can be found by inspection.)

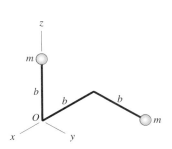

Fig. PF.37, PF.38

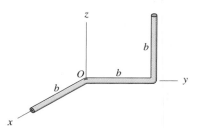

Fig. PF.39

F.39 The mass of the uniform bent rod is m. For point O, determine (a) the inertia tensor with respect to the axes shown; (b) the principal moments of inertia; and (c) the principal axis associated with the smallest moment of inertia.

F.40 The three small balls are joined by rods of negligible mass. For point O, determine (a) the inertia tensor with respect to the axes shown; and (b) the principal moments of inertia. Each rod is parallel to a coordinate axis.

F.41 The three small balls are joined by rods of negligible mass. Calculate the principal moments of inertia and the principal directions at the mass center of the system.

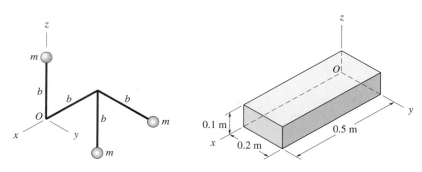

Fig. PF.40, PF.41   **Fig. PF.42**

F.42 The block is made of steel for which $\rho = 7850$ kg/m^3. For point O, find (a) the inertia tensor with respect to the axes shown; and (b) the principal moments of inertia and the principal axes.

Answers to Even-Numbered Problems

Chapter 11

11.2 104.51 N

11.6 (No answer)

11.8 (a) $[FL]$; (b) $[FT]$; (c) $[F]$

11.10 (a) $[FLT^2]$; (b) $[ML^2]$

11.12 $[c] = [FTL^{-1}]$, $[k] = [FL^{-1}]$, $[P_0] = [F]$, $[\omega] = [T^{-1}]$

11.14 (a) $[FL]$; (b) $[ML^2T^{-2}]$

11.16 6.00×10^{-10} N

11.18 (No answer)

11.20 5.93×10^{-3} N

Chapter 12

12.2 (a) $v = -gt + v_0$, $a = -g$; (b) $x_{max} = v_0^2/(2g)$, $t = 2v_0/g$; (c) $x_{max} = 31.89$ m, $t = 5.1$ s

12.4 $x = -40$ m, $v = 99$ m/s, $a = 42$ m/s^2, $s = 310$ m

12.6 (a) $v = v_0(1 - e^{-t/t_0})$; (b) $a = (v_0/t_0)e^{-t/t_0}$, $a = (v_0 - v)/t_0$

12.8 $v = 40.6$ mm/s, $a = 61.9$ mm/s^2

12.10 $v = v_0(v_0t - b)[(v_0t - b)^2 - b^2]^{-1/2}$, $a = -v_0^2 b^2 [(v_0t - b)^2 - b^2]^{-3/2}$

12.12 (a) $\mathbf{v} = -10\mathbf{j}$ m/s, $\mathbf{a} = -4\mathbf{i} + 2\mathbf{j}$ m/s^2; (b) $\mathbf{v} = -20\mathbf{i}$ m/s, $\mathbf{a} = -4\mathbf{i} + 2\mathbf{j}$ m/s^2

12.14 (a) $v_0\sqrt{1 + (2hx/b^2)^2}$; (b) $2hv_0^2/b^2 \downarrow$

12.16 $v = R\omega^2 t$

12.18 $v = 15.15$ m/s, $a = 189.5$ m/s^2

12.20 (a) $\mathbf{v} = (6t + 4)\mathbf{i} + (-8t + 3)\mathbf{j} - 6\mathbf{k}$ m/s, $\mathbf{a} = 6\mathbf{i} - 8\mathbf{j}$ m/s^2

12.22 (a) $\dot{x} = -2b\dot{\theta}\sin\theta$; (b) $\ddot{x} = -2b(\ddot{\theta}\sin\theta + \dot{\theta}^2\cos\theta)$

12.24 $\dot{x} = -R\omega\sin\theta(1 + \cos\theta/\sqrt{9 - \sin^2\theta})$

12.26 (a) 5770 m; (b) 263 m/s; (c) 18.86°

12.28 60.5 m

12.30 150.7 m

12.32 (a) $v_0 e^{-kx}$; (b) $v_0/(v_0 kt + 1)$

12.34 $\tan^{-1}(a/g)$

12.36 $0.894g$

12.38 4.75 m/s^2

12.40 $t = 0.685$ s, $v = 3$ m/s

12.42 (a) 4.88 m/s^2; (b) 3.64 m/s^2

12.44 7.17 m/s

12.46 1.417 m/s^2

12.48 1.373 m/s

12.50 6.17×10^5 N/m

12.52 $1.082\sqrt{Vq/m}$

12.54 $v_{max} = 1.75$ m/s at $x = 0.6434$ m

12.56 241 m

12.58 6.48 s

12.60 (a) $v_{max} = 19.5$ m/s at $x = 8882$ m

12.62 (a) $a = 6.25[(t + 1) + (t - 1)\,\text{sgn}\,(1 - t)] - 12.5x$ m/s^2; (b) $x_{max} = 1.554$ m, $v_{max} = 2.0$ m/s

12.64 $2.5\mathbf{i}$ N

12.66 (b) 7.5 N

12.68 0.524

12.70 9.02 m

12.72 99.6 m

12.74 (a) $y = -0.0436x^2 + 0.5311x$ m;
(b) Avoids net and ceiling, lands behind baseline

12.76 2.12 m

12.78 (a) $t = (m/c) \ln[1 + (cv_0/mg) \sin \alpha]$;
(b) $v = v_0 \cos \alpha/[1 + (cv_0/mg) \sin \alpha]$

12.80 (a) $a_x = 0.5x/(x^2 + y^2)^{3/2}$ m/s^2,
$a_y = 0.5y/(x^2 + y^2)^{3/2}$ m/s^2; Initial conditions:
$t = 0$, $x = 0.3$ m, $y = 0.4$ m, $v_x = 0$,
$v_y = -2$ m/s; (b) $x = 0.360$ m, $v = 1.795$ m/s

12.82 (a) $a_x = -0.13144v_x \left(v_x^2 + v_y^2\right)^{0.25}$ m/s^2,
$a_y = -0.13144v_y \left(v_x^2 + v_y^2\right)^{0.25} - 9.8$ m/s^2; Initial
conditions: $t = 0$, $x = 0$, $y = 1.8$ m, $v_x = 36$ m/s,
$v_y = 0$; (b) $R = 18.67$ m, $t = 0.652$ s

12.84 (a) $a_x = -40[1 - (0.5/R)]x$ m/s^2,
$a_y = -40[1 - (0.5/R)]y - 9.81$ m/s^2; Initial
conditions: $t = 0$, $x = 0.5$ m, $y = -0.5$ m,
$v_x = v_y = 0$

12.86 (a) $a_x = -0.05vv_x + 1.6v_y$ m/s^2,
$a_y = 0.05vv_y - 1.6v_x - 9.8$ m/s^2 where
$v = \sqrt{v_x^2 + v_y^2}$; (b) $t = 1.078$ s, $x = 10.59$ m

12.88 4038 m

12.90 16.4 m

12.92 $v_0 = 269.3$ m/s, $R = 2693$ m, $h = 490$ m

12.94 $t = 16.09$ s, $R = 3070$ m

12.96 84.2 s

12.98 203 m

12.100 (a) 0.4 N; (b) 0.294 m/s

12.102 $v_{max} = 97.6$ mm/s, $y_{max} = 63.5$ mm

12.104 32.0 mm/s^2

12.106 17.58 s

12.108 70 m

12.110 $-4 + (x/16)$ m/s^2

12.112 23.2 m/s

12.114 4230 m

12.116 (a) 1.635 N · s/m; (b) 25.0 s

12.118 765.7 m/s^2

12.120 0.48 m/s$^2 \rightarrow$

12.122 (a) $2mg$; (b) $g\sqrt{m/k}$

Chapter 13

13.2 57.3 km/h

13.4 112.7 km/h

13.6 9.25 m/s^2

13.8 1.4 m/s^2

13.10 (a) 0.64 m/s^2; (b) 4.0 m/s^2

13.12 (a) $\mathbf{v}_A = 4\mathbf{j}$ m/s, $\mathbf{v}_B = 8\mathbf{j}$ m/s;
(b) $\mathbf{a}_A = -32\mathbf{i} + 12\mathbf{j}$ m/s^2,
$\mathbf{a}_B = -64\mathbf{i} + 24\mathbf{j}$ m/s^2

13.14 (a) 0.2 m; (b) 22.4 m/s^2

13.16 72.5 m

13.18 5.18 m/s^2

13.20 76.4 s

13.22 1.19 m/s^2

13.24 $a_{max} = 2Av_0^2$, $x = -B/(2A)$

13.26 $v = 545$ m/s, $a = 101.4$ m/s^2

13.28 (a) $v_R = 3$ m/s, $v_\theta = 6$ m/s, $a_R = -9$ m/s^2,
$a_\theta = 9$ m/s^2; (b) $v_R = 0$, $v_\theta = 9$ m/s,
$a_R = -18$ m/s^2, $a_\theta = 0$

13.30 299.7 m/s

13.32 (a) $v_R = -0.255$ m/s, $v_\theta = 0.2$ m/s,
$a_R = -0.4$ m/s^2, $a_\theta = -1.018$ m/s^2;
(b) $v_R = -0.255$ m/s, $v_\theta = 0.333$ m/s,
$a_R = -0.667$ m/s^2, $a_\theta = -1.018$ m/s^2

13.34 $\mathbf{v} = 0.8\mathbf{e}_R + 0.96t^2\mathbf{e}_\theta$ m/s,
$\mathbf{a} = -1.152t^3\mathbf{e}_R + 2.88t\mathbf{e}_\theta$ m/s^2

13.36 $2.54b\omega^2$

13.38 (a) $2b\omega$; (b) $3b\omega^2$; (c) $b\omega$; (d) $b\omega^2$

13.40 $-3(v_0^2/b)\mathbf{e}_R$

13.42 $h = 1607$ m, $v = 115.6$ m/s

13.44 0.944 m/s

13.46 $v_{max} = 93.3$ mm/s at $\theta = n\pi/4$, $n = 0, 1, 2, \ldots$,
$a_{max} = 116$ mm/s^2 at
$\theta = (\pi/8) + (n\pi/4)$, $n = 0, 1, 2, \ldots$

13.48 $v = 3.66$ m/s, $a = 17.3$ m/s^2

13.50 (a) $h\omega\sqrt{1 + \theta^2 \sin^2 \beta}/(2\pi \cos \beta)$;
(b) $a_R = -(h\omega^2/2\pi)\theta \tan \beta$,
$a_\theta = (h\omega^2/\pi) \tan \beta$, $a_z = 0$

13.52 0.217

13.54 8.09 m/s

13.56 (a) 97.0 N; (b) -3.88 m/s^2

13.58 (a) $\ddot{\theta} = -4.905 \sin \theta$ rad/s^2;
(b) $6.26\sqrt{\cos \theta - 0.866}$ m/s

13.60 2.16 rad/s

13.62 0.75 N

13.64 (a) 3.46 rad/s; (b) 1.48 rad/s

13.66 (a) 1400 N; (b) 560 m/s^2

13.68 (a) $\ddot{R} = \omega^2 R$; (b) $4.12\omega R_0$

13.70 2.6 m/s

13.72 0.600

13.74 $F_R = -400$ N, $F_\theta = 480$ N

13.76 $F = 4.8 \tan \theta$ N

13.78 (a) $7.21\mathbf{e}_R + 8.0\mathbf{e}_\theta$ m/s; (b) 23.1 N

13.80 (a) 0.529 N; (b) 7.10 N

13.82 4.7 m/s

13.84 $F_R = -380.8$ N, $F_\theta = 230.9$ N,
$F_z = 5.4$ N

13.86 (b) 120.2°

13.88 (a) Initial conditions: $t = 0$, $z = 1.8$ m, $\dot{z} = 0$, $\dot{\theta} = 5/(3 \tan \beta)$ rad/s; (c) 0.64 m

13.90 (a) $\ddot{R} = R\dot{\theta}^2 - 40(R - 0.5) + 9.81 \sin \theta$ m/s^2, $\ddot{\theta} = (9.81 \cos \theta - 2\dot{R}\dot{\theta})/R$ rad/s^2; Initial conditions: $t = 0$, $R = 0.5$ m, $\dot{R} = \theta = \dot{\theta} = 0$; (c) $R_{max} = 1.282$ m, $\theta = 130.9°$

13.92 3330.8 N

13.94 39.2 m/s^2

13.96 0.621

13.98 42.7 N

13.100 29.9 m/s

13.102 $v = \sqrt{2gR(1 - \cos \theta)}$, $\theta = 48.2°$

13.104 (a) $\sqrt{2}b\omega$; (b) $\sqrt{5}b\omega^2$

13.106 $-160\mathbf{i} - 120\mathbf{j}$ mm/s

13.108 $\sqrt{Rg \tan \beta}$

Chapter 14

14.2 (a) $F_0 b/2$; (b) $F_0 b/2$

14.4 (a) $-0.414kR^2$; (b) WR

14.6 $-0.1186\mu kb^2$

14.8 5.7 m

14.10 19.5 km/h

14.12 1.860Ph

14.14 150.1 mm

14.16 98.1 N

14.18 29.6 N

14.20 1.181 m

14.22 3.6 m/s

14.24 59.7 m/s

14.26 $2W$

14.28 617 kN/m

14.30 168.7 N/m

14.32 2.36 m/s

14.34 4.90 m/s

14.36 8.41 N

14.38 25470.5 N

14.40 7.5 m/s

14.42 (a) $\sqrt{2gR(1 - \cos \theta)}$; (b) 48.2°

14.44 1.6 m/s

14.46 12.4 m/s

14.48 7.50×10^{10} J

14.50 4.57×10^6 m

14.52 109.5 m

14.54 46.9 hp

14.56 61.3%

14.58 $F^2 t/m$

14.60 (a) $47.7 \sin 2\pi t$ W; (b) $P_{max} = 47.7$ W, $t = 0.25$ s

14.62 0.815 m/s

14.64 (a) 4.98 m/s^2; (b) 2.43 m/s^2

14.66 (a) $36[t - (3/4)t^2 + (1/8)t^3]$ W; (b) $P_{max} = 13.86$ W, $t = 0.845$ s

14.68 3.84 N · s

14.70 $P = 4.9$ N at 30° ↖

14.72 3.92 s

14.74 147.7 N

14.76 $v_2 = 26.0$ m/s, $\theta_2 = 57.1°$

14.78 2271.5 N

14.80 $1.0\mathbf{i} + 0.463\mathbf{j}$ m/s

14.82 11.5 m/s

14.84 1.631 s

14.86 62.3 km/h

14.88 $P_0 L^2/(2v_0)$

14.90 $1.2\mathbf{i} - 0.4\mathbf{k}$ N · m · s

14.92 $2mvR \cos^2 \theta \mathbf{k}$

14.94 5.44 s

14.96 1.666 s

14.98 4340 m/s

14.100 21.2 N

14.102 (a) 2000 m/s; (b) 9950 km

14.104 (a) 32.0 rad/s; (b) 10.7 rad/s^2

14.106 (a) 4.336 m/s; (b) 22.7°

14.108 1682 days

14.110 $R_{max} = 4.04 \times 10^8$ m, $R_{min} = 3.62 \times 10^8$ m

14.112 1.736×10^{10} N · m

14.114 (a) hyperbolic; (b) 2.171×10^7 m (c) 5403 m/s

14.116 (b) 17.3°; (c) 152 km

14.118 At A: 976×10^3 N · s, at B: 562×10^3 N · s

14.120 82.7°, $v = 7.50 \times 10^3$ m/s

14.122 (a) $\ddot{R} = R\dot{\theta}^2 - (3.9866 \times 10^{14}/R^2)$, $\ddot{\theta} = -2\dot{R}\dot{\theta}/R$, Initial conditions: $t = 0$, $R_0 = 7.18 \times 10^6$ m, $\theta = 0$, $\dot{R} = 581.038$ m/s, $\dot{\theta} = 934.2 \times 10^{-6}$ rad/s; (c) 83.1°

14.124 8.26 m/s

14.126 92.6 N

14.128 (a) 1.612 m/s (b) 0.2149 N · m/s

14.130 8.1 m/s

14.132 6.02 s

14.134 $v_B = 9.8$ m/s, elongation rate $= 6.3$ m/s

14.136 4991 m/s

14.138 1000 N/m

14.140 6.4 m/s

Chapter 15

15.2 $a_{B/A} = -0.5$ m/s^2, $v_{B/A} = -0.5t + 5.0$ m/s, $x_{B/A} = -0.25t^2 + 5t + 4000$ m $x_{B/A}(t = 120$ s$) = 1000$ m

15.4 $v = 592$ km/h, $\theta = 5.03°$

15.6 40.9°

15.8 (a) $10.00\mathbf{i} + 51.96\mathbf{j}$ km/h;
(b) $10.00t\mathbf{i} + (51.96t - 3.46)\mathbf{j}$ km; (c) 655 m

15.10 (a) 100 m/s; (b) $160\mathbf{i} + 130.9\mathbf{j}$ m/s

15.12 1.92 m/s²

15.14 $-4.48\mathbf{i} + 5.47\mathbf{j}$ km/h

15.16 1.2 m/s $\longrightarrow$

15.18 2.0 m/s ↑

15.20 0.15 m/s ↑

15.22 2.81 m/s ↑

15.24 1.62 m/s ↑

15.26 $v_B = 0.1$ m/s ↓, $v_C = 0.5$ m/s ↑

15.28 (a) (4.33 m, 3.33 m); (b) $2.794\mathbf{i} + 0.560\mathbf{j}$ m/s²

15.30 12.24 m

15.32 $\mathbf{v}_A = 1.29\mathbf{i} + 3.44\mathbf{j}$ m/s, $\mathbf{v}_B = -3.61\mathbf{i}$ m/s

15.34 $x_B = 4390$ m, $y_B = 922$ m

15.36 $a = 3.49$ m/s², 2.83 N

15.38 18.24 N

15.40 $2g/3$

15.42 2.43 m

15.44 59.2 N

15.46 27.3 N

15.48 $v_A = 0.0960 \sin\theta$ m/s $\longrightarrow$,
$a_A = 0.230 \cos\theta$ m/s² $\longrightarrow$

15.50 2.55 m/s² $\longleftarrow$

15.52 (a) $a_1 = 2.2 - 140(x_1 - x_2)$ m/s²,
$a_2 = 140(x_1 - x_2)$ m/s², Initial conditions: $t = 0$,
$x_1 = x_2 = v_1 = v_2 = 0$; (b) $v_1 = 175.4$ mm/s,
$v_2 = 44.6$ mm/s, $P = 3.01$ N

15.54 (a) $a_A = (10/3)(v_B - v_A)$ m/s²,
$a_B = -2000(v_B - v_A)$ m/s², Initial conditions:
$t = 0$, $x_A = x_B = v_A = 0$, $v_B = 600$ m/s;
(b) $v_B = 81.8$ m/s, $x_A = 0.567$ mm

15.56 (a) $a_A = -2.943 \operatorname{sgn}(v_A - v_B) - 1000x_A$ m/s²,
$a_B = 1.4715 \operatorname{sgn}(v_A - v_B)$ m/s²

15.58 4.51 m/s

15.60 7.01 m/s

15.62 5.46 m/s $\longleftarrow$

15.64 $v_A = 0.7$ m/s $\longleftarrow$,
$v_B = 1.1$ m/s $\longrightarrow$

15.66 1.56 s

15.68 12.9 m/s ↑

15.70 1.05 m/s

15.72 2.57 m/s

15.74 24 000 N

15.76 $(8/3)m\omega L^2$ (CCW)

15.78 (a) 15.60 s; (b) 161.4 rev

15.80 Block A hits first with speed 1.1 m/s

15.82 1.198 m/s ↓

15.84 $\mathbf{v}_A = -(b\omega_0/4)(\mathbf{i} + 3\sqrt{3}\mathbf{j})$,
$\mathbf{v}_B = (b\omega_0/4)(-\mathbf{i} + \sqrt{3}\mathbf{j})$

15.86 (a) $\pm R_0\sqrt{(k/m) - 2\dot\theta_0^2}$; (b) $\dot\theta_0 > \sqrt{k/(2m)}$

15.88 (a) 1.5 m/s; (b) 1 m/s

15.90 $v_{\text{final}} = 1.440$ m/s $\longrightarrow$, energy lost $= 11.13\%$

15.92 1.67 m

15.94 (a) 0.65 m/s; (b) 81.22%

15.96 29.7°

15.98 $v_A = 21.8$ m/s, $v_B = 5.5$ m/s

15.100 588 N

15.102 0.933 m

15.104 (a) $v \cos\alpha$; (b) $0.5mv^2 \sin^2\alpha$

15.106 (a) $1.532\mathbf{i}$ m/s; (b) $1.532\mathbf{i} - 0.558\mathbf{j}$ m/s

15.108 (No answer)

15.110 $v_A = (13/64)\,v_0$, $v_B = (15/64)\,v_0$,
$v_C = (9/16)\,v_0$

15.112 v_0

15.114 0.268

15.116 $(\mathbf{v}_A)_2 = 2.93\mathbf{j}$ m/s, $(\mathbf{v}_B)_2 = 6\mathbf{i} + 4.73\mathbf{j}$ m/s

15.118 4.25 rad/s

15.120 6.54 m

15.122 $v_A = 7.29$ m/s, $v_B = 3.95$ m/s

15.124 (a) 5512.5 m/s; (b) 3934 m/s

15.126 $10713.7\mathbf{i} - 332.8\mathbf{j}$ m/s

15.128 $1.35\mathbf{i} - 5\mathbf{j}$ N

15.130 1.05×10^4 N $\longleftarrow$

15.132 $1447\mathbf{i} + 2584\mathbf{j}$ N

15.134 21.6 N

15.136 $\theta = 46.7°$, $a = 9.23$ m/s²

15.138 12.97 m/s² ↓

15.140 196.9 N/m²

15.142 (a) $0.8(6 - v)v$ N · m/s; (b) $v = 3$ m/s,
$P_{\max} = 72$ N · m/s

15.144 (a) $-4.713t\mathbf{i} + (8.604t - 120)\mathbf{j}$ m; (b) 57.8 m

15.146 4.0 km/h

15.148 197 m/s

15.150 5.72 rad/s

15.152 $v = 1981.5$ m/s, $\theta = 6°$

15.154 $v_0 m_h m_a (1 + e)/(m_h + m_a)$

15.156 $\mathbf{v}_A = 4.10\mathbf{i} - 1.88\mathbf{j}$ m/s, $\mathbf{v}_B = 6.58\mathbf{j}$ m/s

15.158 0.38 m

15.160 (a) 43.9 km/h; (b) 3.78%

Chapter 16

16.2 20 rad/s, 1.230 rev

16.4 (a) $\omega = -8$ rad/s, $\alpha = -8$ rad/s²; (b) 40 rad

16.6 (a) 9.0 s^{-2}; (b) 19 rad

16.8 48 rad

16.10 (a) $0.1667\omega^{1.5} + 8.0$ rad; (b) $4.0t^2$ rad/s;
       (c) $1.333t^3 + 8.0$ rad

16.12 80.1 m/s^2

16.14 (a) 5.0 rad/s^2; (b) 14.32 rev

16.16 $\mathbf{v}_B = -0.3021\mathbf{i} - 0.3625\mathbf{k}$ m/s,
       $\mathbf{a}_B = -1.605\mathbf{i} + 0.619\mathbf{j} - 0.812\mathbf{k}$ m/s^2

16.18 $\mathbf{v}_B = 1.9296\mathbf{i} + 1.5434\mathbf{j} + 2.573\mathbf{k}$ m/s,
       $\mathbf{a}_B = -75.02\mathbf{i} + 24.82\mathbf{j} + 41.37\mathbf{k}$ m/s^2

16.20 $\omega = 22.22$ rad/s (CW), $v_C = 9.33$ m/s $\longrightarrow$

16.22 $\omega = 5.833$ rad/s (CW), $v_C = 75.0$ mm/s $\longrightarrow$

16.24 $\omega_0(r_A + r_B)/r_B$

16.26 $\omega_B = 5$ rad/s (CCW), $\omega_{AB} = 2.22$ rad/s (CW)

16.28 $0.6\sin^2\theta$ rad/s

16.30 $\omega_{BD} = \omega_{DE} = 1.714$ rad/s (CCW)

16.32 4.54 m/s $\uparrow$

16.34 $\omega_{BD} = 20.6$ rad/s (CCW),
       $\omega_{DE} = 16.96$ rad/s (CCW)

16.36 4.57 rad/s (CCW)

16.38 (a) 5.31°; (b) 3.17 m/s

16.40 3.3 rad/s (CW)

16.42 721 mm/s at 33.7° $\nearrow$

16.44 (a) (0.12 m, .09 m); (b) (0, 1.08 m)

16.46 8 rad/s (CW)

16.48 7.80 rad/s (CCW)

16.50 $\omega = 1.0$ rad/s (CCW), $\mathbf{v}_G = -3.5\mathbf{j}$ m/s

16.52 $-1.420\mathbf{i}$ m/s

16.54 4.56 rad/s (CCW)

16.56 6.8 rad/s (CW)

16.58 $v_C = 13.11$ m/s $\longleftarrow$, $v_D = 13.11$ m/s $\uparrow$

16.60 0.447 m/s $\uparrow$

16.62 (a) $4.32\mathbf{i} + 6.48\mathbf{j}$ m/s^2; (b) 4.32 m/s^2

16.64 (a) $2\mathbf{i} + 2.16\mathbf{j}$ m/s^2; (b) 2.9i m/s^2;
       (c) $5.15\mathbf{i} - 5.4\mathbf{j}$ m/s^2

16.66 $a_B = 1.30$ m/s^2 $\swarrow$, $\alpha = 1.8$ rad/s^2 (CCW)

16.68 22.8 m/s^2 $\longrightarrow$

16.70 154.0 m/s^2 $\uparrow$

16.72 $\alpha_{AD} = 0.674$ rad/s^2 (CCW),
       $\alpha_{AE} = 0.818$ rad/s^2 (CW)

16.74 0.58 m/s^2 $\longrightarrow$

16.76 $\alpha_{AB} = 3.36$ rad/s^2 (CCW),
       $\alpha_{BC} = 20.4$ rad/s^2 (CCW)

16.78 $\alpha_{AB} = 3.0$ rad/s^2 (CCW), $a_B = 4.2$ m/s^2 $\uparrow$

16.80 5.10 m/s^2

16.82 (b) $-1.736\mathbf{i} - 0.768\mathbf{j}$ m/s^2

16.84 (a) 0.2 m/s^2 $\downarrow$; (b) 0.36 m/s^2 $\downarrow$

16.86 $\mathbf{v}_P = -0.519\mathbf{i} + 1.6\mathbf{j}$ m/s,
       $\mathbf{a}_P = -5.86\mathbf{i} - 3.36\mathbf{j}$ m/s^2

16.88 $v_P = -9.60$ m/s^2, $\mathbf{a}_P = -28.8\mathbf{i} - 76.8\mathbf{j}$ m/s^2

16.90 $-21.28\mathbf{i} - 16\mathbf{j}$ m/s^2

16.92 (a) 640 mm/s at 60° $\swarrow$; (b) 7.39 m/s^2 at 60° $\nearrow$

16.94 (a) $-54\mathbf{i} - 80\mathbf{j}$ m/s^2; (b) $48\mathbf{i} + 30\mathbf{j}$ m/s^2

16.96 (a) $-3.33\mathbf{i}$ m/s^2; (b) $-3.33\mathbf{i}$ m/s^2

16.98 2.88 rad/s (CW)

16.100 (a) 0.510 m/s $\downarrow$; (b) 1.312 m/s^2 $\downarrow$

16.102 1.045 rad/s (CCW)

16.104 (a) $(v_0/d)\sin^2\theta$; (b) $2(v_0/d)^2\sin^3\theta\cos\theta$

16.106 $2R\alpha_0(\sin\theta + 2\theta\cos\theta)$ $\downarrow$

16.108 0.64 rad/s (CCW)

16.110 4.66 rad/s (CCW), 2.62 rad/s (CW)

16.112 $11.54\omega^2$ (CW)

16.114 (a) $v_A\cos\theta/\cos(\alpha - \theta)$;
       (b) $-\left(v_A^2/L\right)\left[\sin^2\alpha/\cos^3(\alpha - \theta)\right]$

16.116 42.0 rad/s^2 (CCW)

16.118 $\mathbf{v}_C = -2.0\mathbf{i} + 3.464\mathbf{j}$ m/s,
       $\mathbf{a}_C = -5.43\mathbf{i} - 6.60\mathbf{j}$ m/s^2

16.120 $\mathbf{v}_C = -3.34\mathbf{i} - 5.01\mathbf{j}$ m/s,
       $\mathbf{a}_C = -8.4\mathbf{i} - 24.2\mathbf{j} + 35\mathbf{k}$ m/s^2

16.122 $\alpha_{AB} = 4.80$ rad/s^2 (CCW), $\alpha_{BC} = 0$

16.124 7.00 rad/s (CW)

16.126 $\omega_{BCD} = 74.3$ rad/s (CCW), $v_D = 54.1$ m/s

16.128 $\alpha_{AF} = 300$ rad/s^2 (CCW),
       $a_{F/BD} = 96$ m/s^2 $\longrightarrow$

16.130 (a) 5.21 m/s $\longrightarrow$; (b) 4.42 m/s^2 $\longleftarrow$

Chapter 17

17.2   2.5 kg · m^2

17.4 $\bar{z} = 50.33$ m, $\bar{I}_x = 1.37 \times 10^6$ kg · m^2

17.6 $mb^2/6$

17.8 $I_x = 13.98 \times 10^{-3}$ kg · m^2,
       $I_z = 70.9 \times 10^{-3}$ kg · m^2

17.10 $I_z = 0.1119$ kg m^2,
       $\bar{I}_z = 0.026$ kg · m^2

17.12 0.1015

17.14 $\bar{x} = 1.548$ m, $\bar{I} = 121.0$ kg · m^2

17.16 (No answer)

17.18 (a) 9.20 m/s^2; (b) 0.938

17.20 93.3 N

17.22 $a = 2$ m/s^2, $N_A = 77$ N

17.24 (b) 0.350

17.26 $1.5g$ $\downarrow$

17.28 $mg/7$

17.30 (No answer)

17.32 $\alpha = 10.15$ rad/s^2 (CCW), $T = 554$ N

17.34 27.7 N

17.36 $\alpha = 3.5$ rad/s^2, $T = 17.6$ N

17.38 $\alpha_A = 53.0$ rad/s^2, $\alpha_B = -13.24$ rad/s^2

17.40 $\alpha = 12.4$ rad/s^2 (CW),
$N = 364.2$ N $\uparrow$,
$F = 132.3$ N $\longrightarrow$

17.42 (a) 0.173 m; (b) 0.1134 m

17.44 204 N

17.46 1.01 N

17.48 (b) (i) to the right, (ii) to the left

17.50 1.826 m/s^2

17.52 (b) $\alpha = 5.45$ rad/s^2, $\bar{a} = 9.26$ m/s^2

17.54 (a) 0.283g/R; (b) 0.322

17.56 5.6 rad/s^2

17.58 $\alpha = 0.93$ rad/s^2, $T = 115.9$ N

17.60 (a) 8.0 m/s^2; (b) 0.8 m

17.62 $\alpha_C = 5.13$ rad/s^2 (CW);
$\alpha_{AB} = 17.58$ rad/s^2 (CCW)

17.64 (a) 9.05 rad/s^2; (b) 14.40 m/s^2 at 84.8° $\swarrow$

17.66 $(789 \sin\theta + 0.294)\cos\theta$

17.68 (b) $\sqrt{(3g/L)[2 - \cos\theta - 2\sqrt{2}\sin(\theta/2)]}$;
(c) $-\sqrt{3g/L}$

17.70 (a) 14.20 rad/s^2 (CCW); (b) 6.22 rev

17.72 (a) 0.5738 rad/s^2; (b) 73 s

17.74 (b) $\pm\sqrt{(3g/L)(\cos\theta - \cos\theta_0)}$;
(c) $(mg/2)(5 - 3\cos\theta_0)$ at $\theta = 0$

17.76 (a) $\alpha = 4.226 + 9.554\cos\theta$ rad/s^2; (b) $-707°$

17.78 (b) $\pm\sqrt{(3g/L)(\sin\theta + \cos\theta - 1)}$;
(c) $\theta_{max} = 90°$, $F = mg$

17.80 (b) $\pm\sqrt{20\theta - 19.6\sin\theta}$ rad/s

17.82 (a) $\bar{a} = 7.5 - 100x$ m/s^2,
$\bar{v} = \sqrt{15x - 100x^2}$ m/s;
(b) $\bar{v}_{max} = 0.750$ m/s at $x = 0.075$ m;
(c) 0.382

17.84 (a) $\omega = 3.13\sqrt{0.866 - \cos\theta}$ rad/s,
$\alpha = 4.905\sin\theta$ rad/s^2;
(b) $183.9(3\cos\theta - 1.732)\sin\theta$ N; (c) 54.7°

17.86 (b) 24.8°

17.88 (b) 2.3 rad/s

17.90 (b) rotation is counterclockwise

17.92 (a) 67.5 mm; (b) 111.6 mm; (c) 88.8 mm

17.94 $\bar{v} = 1.46$ m/s, $\bar{a} = 8.67$ m/s^2

17.96 0.283 mg

17.98 $\alpha = 0.6638$ rad/s^2 (CCW), $a_A = 3.58$ m/s^2 $\longleftarrow$

17.100 $\alpha = 3.02$ rad/s^2 (CW), $T = 140.8$ N

17.102 1.07 rad/s

17.104 $\bar{a} = 3.187$ m/s^2 $\longrightarrow$, $T = 610$ N

17.106 784 N

17.108 (a) $\alpha = [3\cos\theta/(2L)][(2P/m) - g]$,
$\omega = \sqrt{(3\sin\theta/L)[(2P/m) - g]}$;
(b) $(9\sin 2\theta/8)(2P - mg) \longrightarrow$

Chapter 18

18.2 -8.38 N $\cdot$ m

18.4 103.17 N $\cdot$ m

18.6 Case (a): 480 W, Case (b): 120 W

18.8 1.5 rad/s^2

18.10 (a) 63.41 N $\cdot$ m; (b) 38.05 N $\cdot$ m

18.12 (b) 22.5 hp at 200 rad/s

18.14 (a) $(17/48)mL^2\omega^2$; (b) $(25/96)mL^2\omega^2$

18.16 19 N $\cdot$ m

18.18 7.2 N $\cdot$ m

18.20 125 N $\cdot$ m

18.22 480 J

18.24 278.4 N $\cdot$ m

18.26 3.45 rad/s

18.28 5.99 N $\cdot$ m/rad

18.30 (a) $2\sqrt{ge/(R^2 + 2e^2)}$; (b) $R/\sqrt{2}$; (c) $1.189\sqrt{g/R}$

18.32 9.98 rad/s

18.34 0.966

18.36 (a) 3 m/s; (b) 6.26 m/s

18.38 (a) 2.97 m/s; (b) 98.9 mm

18.40 $\sqrt{3P/(mL)}$

18.42 $\omega = 2.86\sqrt{\cos\theta - 0.819}$ rad/s,
$\alpha = -4.1\sin\theta$ rad/s^2

18.44 (a) 5683 N $\cdot$ m; (b) 3.35 rad/s

18.46 2.36 m/s

18.48 $\sqrt{4m_2g/(3m_1R)}$

18.50 2.92 m/s

18.52 196.2 N

18.54 $P_0R/(r\sqrt{Ik})$

18.56 5.81 N $\cdot$ m

18.58 (a) $mb^2\omega/6$ (CW); (b) $2mb^2\omega/3$ (CW);
(c) $mb^2\omega/3$ (CCW)

18.60 $R/2$

18.62 1022 rad/s

18.64 24.3 rad/s

18.66 19.86 N

18.68 13.33 rad/s

18.70 (a) 9.13 rad/s (CW); (b) 9.6 rad/s (CCW)

18.72 (a) 3.50 rad/s; (b) 3.50 rad/s

18.74 $v_A = v_0/3 \uparrow$, $v_B = 2v_0/3 \uparrow$

18.76 14.02 rad/s

18.78 35.4 rad/s

18.80 166%

18.82 11.95 rad/s

18.84 (a) 35.25 N $\cdot$ s; (b) 24.38 N $\cdot$ s

18.86 (a) 2.56 rad/s; (b) 54.6°

18.88 (a) $2.40v_1$; (b) 1.430 m/s

18.90 3.84 rad/s (CW)

18.92 (a) v_1; (b) 0%

18.94   1.11 rad/s
18.96   310 m/s
18.98   (a) 0.4576 rad/s; (b) 3.95°
18.100 2.0ω
18.102 187.2 J
18.104 0
18.106 $t = 0.812$ s, $\omega = 30.0$ rad/s
18.108 3.29 m/s ↓
18.110 4.20 rad/s
18.112 3.82 rad/s
18.114 71.8°
18.116 44.5 rad/s (CW)
18.118 2 rad/s

Chapter 19

19.2 1.755 m/s^2
19.4   8.46 rad/s
19.6   (a) $2.22\mathbf{j}$ rad/s; (b) $2.04\mathbf{i}$ rad/s^2;
            (c) $\mathbf{v}_P = 0.205\mathbf{i}$ m/s,
            $\mathbf{a}_P = -0.188\mathbf{j} - 0.266\mathbf{k}$ m/s^2
19.8   (a) $1.80(\mathbf{j} + \mathbf{k})$ rad/s; (b) $2.37\mathbf{i}$ rad/s^2
19.10 (a) $36.7\mathbf{i} + 32.0\mathbf{k}$ rad/s; (b) $586\mathbf{j} + 400\mathbf{k}$ rad/s^2
19.12 (a) $(0.658\mathbf{j} - 0.940\mathbf{k})\omega_0$; (b) $4.17\omega_0\mathbf{j}$
19.14 (a) $6.0\mathbf{i} + 3.0\mathbf{j} - 4.0\mathbf{k}$ m/s;
            (b) $12.5\mathbf{i} - 16.0\mathbf{j} - 49.0\mathbf{k}$ m/s^2
19.16 (a) $\boldsymbol{\omega}_{CD} = 7.20\mathbf{j}$ rad/s, $\boldsymbol{\omega}_{BD} = \mathbf{0}$; (b) $34.6\mathbf{j}$ rad/s^2
19.18 $-0.866\mathbf{i} + 0.750\mathbf{j}$ rad/s^2
19.20 $\boldsymbol{\omega}_A = -2.40\mathbf{j}$ rad/s,
            $\boldsymbol{\omega}_C = 0.90\mathbf{i} - 2.40\mathbf{j} + 1.20\mathbf{k}$ rad/s
19.22 (a) $30\mathbf{j}$ rad/s^2; (b) $-18\,000\mathbf{i} - 8100\mathbf{j} + 30\,000\mathbf{k}$
            mm/s^2
19.24 (a) $108\mathbf{j}$ rad/s^2; (b) $159.6\mathbf{i} - 212.8\mathbf{j} - 59.8\mathbf{k}$ m/s^2
19.26 $\boldsymbol{\omega}_2 = 4.0\mathbf{i} - 3.14\mathbf{j} + 3.14\mathbf{k}$ rad/s, $T_2 = 1.997$ J
19.28 (a) $8.10\mathbf{j} + 1.35\mathbf{k}$ N · m · s; (b) 641 J
19.30 15.57 rad/s
19.32 0.410
19.34 $-12.69\mathbf{j} + 78.23\mathbf{k}$ N · m · s
19.36 $2.65\sqrt{g/L}$
19.38 15.63 rad/s
19.40 $(12/7)mb \int \hat{P}\, dt\, (\mathbf{i} - 5\mathbf{j})$
19.42 $1.02\mathbf{j} - 2.45\mathbf{k}$ rad/s
19.44 $0.776\mathbf{j} - 0.631\mathbf{k}$
19.46 $\cos^{-1}[3g/(2L\omega^2)]$
19.48 $\mathbf{C} = -(1/4)mR^2\omega_0^2 \sin\beta \cos\beta\mathbf{i}$, $\mathbf{R} = \mathbf{0}$
19.50 $0.5mRL\omega^2$
19.52 $\mathbf{R}_A = -648\mathbf{i}$ N, $\mathbf{R}_B = 648\mathbf{i}$ N
19.54 $\mathbf{R}_A = -1.5\mathbf{j}$ N, $\mathbf{R}_B = 1.5\mathbf{j}$ N
19.56 $\mathbf{C} = -6.91\mathbf{k}$ N · m

19.58 9.01 rad/s
19.60 360 N
19.62 $\mathbf{R}_A = -m\omega_0^2 b(\mathbf{j} + \mathbf{k})$, $\mathbf{C}_A = -(m\omega_0^2 b^2/8)\mathbf{j}$
19.64 123.6 N
19.66 26.5 N
19.68 $\dot{\boldsymbol{\psi}} = -4.98\mathbf{k}$ rad/s, $\boldsymbol{\omega} = 0.872\mathbf{j} + 4.980\mathbf{k}$ rad/s
19.70 2.13
19.72 11.38 N
19.74 (a) Initial conditions: $t = 0$, $\theta = 0.5236$ rad,
            $\dot{\theta} = 0$, $\phi = 0$, $\dot{\phi} = 200$ rad/s; (b) $\tau = 0.0628$ s,
            -86.6 rad/s $\le \dot{\theta} \le 86.6$ rad/s,
            50 rad/s $\le \dot{\phi} \le 200$ rad/s
19.76 (No answer)
19.78 (a) $\ddot{\theta} = (0.75\dot{\phi}^2 \cos\theta - 0.25\dot{\phi}\dot{\psi} + 117.70)\sin\theta$,
            $\ddot{\phi} = (-1.75\dot{\phi} \cos\theta + 0.25\dot{\psi})\dot{\theta}\csc\theta$,
            $\ddot{\psi} = (1.75\dot{\phi}\cos\theta - 0.25\dot{\psi})\dot{\theta}\cot\theta + \dot{\theta}\dot{\phi}\sin\theta - $
            $0.5(\dot{\phi}\cos\theta + \dot{\psi})$; (b) $\omega_z|_{t=0} = 123.8$ rad/s,
            $\omega_z|_{t=1.0\,s} = 75.1$ rad/s
19.80 $\dot{\phi} = 2Fb/(mR^2\dot{\psi})$
19.82 (a) $\theta = 20°$, $\dot{\psi} = 8.96$ rad/s, $\dot{\phi} = 3.18$ rad/s;
            (b) $15° \le \gamma \le 55°$
19.84 (No answer)
19.86 0.267 rad/s at 13.08° ↗

Chapter 20

20.2   (a) 24.5 Hz; (b) $x = 0.012 \sin(153.9t + 0.6228)$ m
20.4   1.858
20.6   0.747
20.8   (a) $\dot{\theta}_{max} = p\theta_0$, $\ddot{\theta}_{max} = p^2\theta_0$;
            (b) $\dot{\theta}_{max} = p\sqrt{2(1 - \cos\theta_0)}$, $\ddot{\theta}_{max} = p^2 \sin\theta_0$
20.10 $\ddot{x} + 2Tx/(mL) = 0$
20.12 1.256 s
20.14 (b) 0.2266 s
20.16 (No answer)
20.18 (b) 0.0659
20.20 31.0 N · s/m
20.22 -2.89 mm
20.24 0.435 s
20.26 5.2 m/s
20.28 1.419×10^{-4} N · s/m
20.30 (a) $\ddot{\theta} = -0.31\dot{\theta}\,|\dot{\theta}| - 19.6 \sin\theta$; (c) 30.5%
20.32 (a) $x = 0.02667(-0.5 \sin 50t + \sin 25t)$ m;
            (b) 1.333; (c) 34.6 mm
20.34 0.0489 m
20.36 (a) 2.5 rad/s; (b) 60 mm
20.38 $\omega < 41.16$ rad/s or $\omega > 52.57$ rad/s
20.40 (a) 400 rad/s; (b) 1.847 mm
20.42 (a) $\ddot{\theta} = -9.0 \sin\theta + 12.5 \sin 5t \cos\theta$

20.44 1568 rev/min.

20.46 12.01 N · s/m

20.48 (b) $x_{max} = 0.0393$ m, $x_{min} = 0.017285$ m

20.50 (a) $m\ddot{x} + c\dot{x} + kx = -c\omega Y \sin \omega t$;
(b) $|X| = 33.5$ mm, $\phi = 24.8°$

20.52 $x(t) = 19.95 \sin(600t + 0.499)$ mm,
$x(t)$ leads $y(t)$ by $28.6°$

20.54 $Z_{max} = 1.155Y$, $\omega/p = \sqrt{2}$

20.56 (a) 25.0 N · s/m; (b) $90°$

20.58 (a) $10\ddot{x} + 5\dot{x} + 150x = 0.5 \sin \omega t$;
(b) $x(t) = 25.82 \sin(3.873t + \pi/2)$ mm

20.60 (a) $\ddot{x} = 200 \sin 500t - 50\dot{x} |\dot{x}| - (250 \times 10^3)x$ m/s^2; (b) 4.4 mm

20.62 $0.276\sqrt{k/m}$ Hz

20.64 (b) $y = \bar{k}$, $p_{max} = \sqrt{g/(2\bar{k})}$

20.66 (a) $m\ddot{\theta}/3 + (c_1 + c_2/4)\dot{\theta} - k\theta = 0$, (b) 0.628

20.68 1.969 Hz

20.70 0.725 Hz

20.72 $9.51°$

20.74 3571.43 N · s/m

20.76 (c) $41°$

20.78 (a) 0.546; (b) 1.5663 N · m · s

20.80 (b) Reduction in amplitude in one cycle is $0.6q_0$, System stops at $(q = -0.1q_0, \dot{q}/p = 0)$

20.82 $5.13\sqrt{m/k}$

20.84 $0.936\sqrt{g/R}$

20.86 $2\pi(1 - R/L)\sqrt{L/g}$

20.88 0.0105 m, 0.4765 m

20.90 $2\pi\sqrt{3m/(5k)}$

20.92 0.60 s

20.94 (a) $\ddot{\theta} + 3(k/m - 2g/L)\theta = 0$; (b) 0.897 s;
(c) 30.0 kg

20.96 $0.1345\sqrt{g/(R - r)}$

20.98 (a) $\ddot{x} + 40x = 0$; (b) 1.007 Hz;
(c) $x(t) = 26.50 \sin(6.325t - 0.8550)$ mm

20.100 $0.0308 \sin(45.825t) - 0.0232 \sin(60t)$ m

20.102 (a) 32.7 rad/s; (b) 88.9 mm

20.104 $x(t) = 0.0577 \exp(-17.03t) \sin(24.497t + 1.0472)$ m

20.106 $2t \exp(-10t)$ m

20.108 (a) 1.23; (b) $(M.F.)_{max} = 1.23$,
$\omega = 12.93$ rad/s

20.110 5.58 kN · s/m

20.112 Underdamped

20.114 5.32 mm

Appendix F

F.2 $0.792mR^2$

F.4 $I_x = I_z = m(b^2 + 2a^2)/12$, $I_y = ma^2/3$

F.6 53.5 kg·m^2

F.8 (a) $(mL^2/12) \sin^2 \beta$; (b) $I_y = (mL^2/12) \cos^2 \beta$,
$I_z = mL^2/12$

F.10 (a) $(m/12)(L^2 + 3R^2)$; (b) $mR^2/2$

F.12 (a) $(m/5)(b^2/4 + 3h^2)$; (b) $mb^2/10$

F.14 0.304 kg · m^2

F.16 $0.0584mR^2$

F.18 6.22 kg · m^2

F.20 $3mR^2/(8\pi)$

F.22 $I_{xy} = 3mbh/10$, $I_{yz} = 3mab/20$, $I_{zx} = 3mah/10$

F.24 $I_{xy} = I_{zx} = 0$, $I_{yz} = mRh/3$

F.26 $I_{xy} = 2.986 \times 10^{-3}$ kg · m^2,
$I_{yz} = 1.493 \times 10^{-3}$ kg · m^2,
$I_{zx} = 2.986 \times 10^{-3}$ kg · m^2

F.28 $I_{xy} = mR^2/(8\pi)$, $I_{zx} = -mR^2/(8\pi)$, $I_{yz} = 0$

F.30 mb^2

F.32 (a) $mb^2/6$

F.34 $mb^2h^2(b^2 + h^2)/6$

F.36 $\lambda_1 = (\mathbf{i} + \mathbf{j})/\sqrt{2}$, $\lambda_2 = (-\mathbf{i} + \mathbf{j})/\sqrt{2}$, $\lambda_3 = \mathbf{k}$;
$\bar{I}_1 = mb^2/6$, $\bar{I}_2 = mb^2/24$, $\bar{I}_3 = 5mb^2/24$

F.38 $\bar{I}_1 = 0$, $\bar{I}_2 = \bar{I}_3 = 3mb^2/2$

F.40 (a) $mb^2 \begin{bmatrix} 3 & 1 & -1 \\ 1 & 4 & 0 \\ -1 & 0 & 3 \end{bmatrix}$; (b) $I_1 = 1.75mb^2$,
$I_2 = 4.80mb^2$, $I_3 = 3.45mb^2$

F.42 (a) $\begin{bmatrix} 1.308 & -1.963 & -0.981 \\ -1.963 & 6.803 & -0.393 \\ -0.981 & -0.393 & 7.588 \end{bmatrix}$ kg · m^2;
(b) $I_1 = 0.522$ kg · m^2, $I_2 = 7.416$ kg · m^2,
$I_3 = 7.762$ kg · m^2; $\lambda_1 = 0.941\mathbf{i} + 0.303\mathbf{j} + 0.148\mathbf{k}$, $\lambda_2 = -0.330\mathbf{i} + 0.919\mathbf{j} + 0.216\mathbf{k}$,
$\lambda_3 = -0.070\mathbf{i} - 0.252\mathbf{j} + 0.965\mathbf{k}$

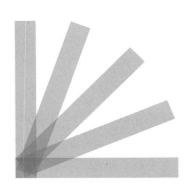

Index

Selected SI Units			Commonly used SI prefixes		
Quantity	Name	SI symbol	Factor	Prefix	SI symbol
Energy	joule	J ($1\ J = 1\ N \cdot m$)	10^9	giga	G
Force	newton	N ($1\ N = 1\ kg \cdot m/s^2$)	10^6	mega	M
Length	meter*	m	10^3	kilo	k
Mass	kilogram*	kg	10^{-3}	milli	m
Moment (torque)	newton meter	N $\cdot$ m	10^{-6}	micro	μ
Rotational frequency	revolution per second	r/s	10^{-9}	nano	n
	hertz	Hz ($1\ Hz = 1\ r/s$)			
Stress (pressure)	pascal	Pa ($1\ Pa = 1\ N/m^2$)			
Time	second*	s			
Power	watt	W ($1\ W = 1\ J/s$)			

* SI base unit

Selected Rules and Suggestions for SI Usage

1. Be careful in the use of capital and lowercase for symbols, units, and prefixes (e.g., m for meter or milli, M for mega).
2. For numbers having five or more digits, the digits should be placed in groups of three separated by a small space, counting both to the left and to the right of the decimal point (e.g., 61 354.982 03). The space is not required for four-digit numbers. Spaces are used instead of commas to avoid confusion—many countries use the comma as the decimal marker.
3. In compound units formed by multiplication, use the product dot (e.g., N $\cdot$ m).
4. Division may be indicated by a slash (m/s), or a negative exponent with a product dot (m $\cdot$ s^{-1}).
5. Avoid the use of prefixes in the denominator (e.g., km/s is preferred over m/ms). The exception to this rule is the prefix k in the base unit kg (kilogram).

Equivalence of U.S. Customary and SI Units (Asterisks indicate exact values; others are approximations.)

	U.S. customary to SI	SI to U.S. customary
1. Length	1 in. = 25.4* mm = 0.0254* m 1 ft = 304.8* mm = 0.3048* m	1 mm = 0.039 370 in. 1 m = 39.370 in. = 3.281 ft
2. Area	1 in.^{2} = 645.16* mm^2 1 ft^2 = 0.092 903 04* m^2	1 mm^2 = 0.001 550 in.^{2} 1 m^2 = 1550.0 in.^{2} = 10.764 ft^2
3. Volume	1 in.^{3} = 16 387.064* mm^3 1 ft^3 = 0.028 317 m^3	1 mm^3 = 0.000 061 024 in.^{3} 1 m^3 = 61 023.7 in.^{3} = 35.315 ft^3
4. Force	1 lb = 4.448 N 1 lb/ft = 14.594 N/m	1 N = 0.2248 lb 1 N/m = 0.068 522 lb/ft
5. Mass	1 lbm = 0.453 59 kg 1 slug = 14.593 kg	1 kg = 2.205 lbm 1 kg = 0.068 53 slugs
6. Moment of a force	1 lb $\cdot$ in. = 0.112 985 N $\cdot$ m 1 lb $\cdot$ ft = 1.355 82 N $\cdot$ m	1 N $\cdot$ m = 8.850 75 lb $\cdot$ in. 1 N $\cdot$ m = 0.737 56 lb $\cdot$ ft
7. Power	1 hp (550 lb $\cdot$ ft/s) = 0.7457 kW	1 kW = 1.3410 hp

Area Moments of Inertia

Rectangle	**Circle**	**Half parabolic complement**

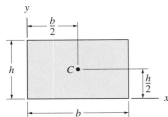

$$\bar{I}_x = \frac{bh^3}{12} \quad \bar{I}_y = \frac{b^3h}{12} \quad \bar{I}_{xy} = 0$$

$$I_x = \frac{bh^3}{3} \quad I_y = \frac{b^3h}{3} \quad I_{xy} = \frac{b^2h^2}{4}$$

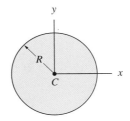

$$I_x = I_y = \frac{\pi R^4}{4} \quad I_{xy} = 0$$

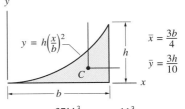

$$y = h\left(\frac{x}{b}\right)^2 \qquad \bar{x} = \frac{3b}{4}$$
$$\bar{y} = \frac{3h}{10}$$

$$\bar{I}_x = \frac{37bh^3}{2100} \quad I_x = \frac{bh^3}{21}$$
$$\bar{I}_y = \frac{b^3h}{80} \quad I_y = \frac{b^3h}{5}$$
$$\bar{I}_{xy} = \frac{b^2h^2}{120} \quad I_{xy} = \frac{b^2h^2}{12}$$

Right triangle	**Semicircle**	**Half parabola**

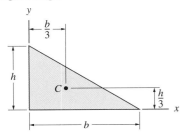

$$\bar{I}_x = \frac{bh^3}{36} \quad \bar{I}_y = \frac{b^3h}{36} \quad \bar{I}_{xy} = -\frac{b^2h^2}{72}$$

$$I_x = \frac{bh^3}{12} \quad I_y = \frac{b^3h}{12} \quad I_{xy} = \frac{b^2h^2}{24}$$

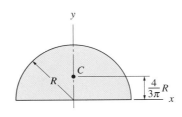

$$\bar{I}_x = 0.1098R^4 \quad \bar{I}_{xy} = 0$$

$$I_x = I_y = \frac{\pi R^4}{8} \quad I_{xy} = 0$$

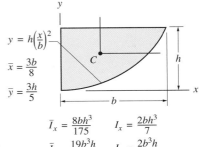

$$y = h\left(\frac{x}{b}\right)^2$$
$$\bar{x} = \frac{3b}{8}$$
$$\bar{y} = \frac{3h}{5}$$

$$\bar{I}_x = \frac{8bh^3}{175} \quad I_x = \frac{2bh^3}{7}$$
$$\bar{I}_y = \frac{19b^3h}{480} \quad I_y = \frac{2b^3h}{15}$$
$$\bar{I}_{xy} = \frac{b^2h^2}{60} \quad I_{xy} = \frac{b^2h^2}{6}$$

Isosceles triangle	**Quarter circle**	**Circular sector**

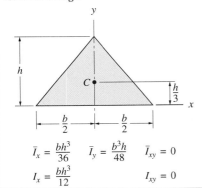

$$\bar{I}_x = \frac{bh^3}{36} \quad \bar{I}_y = \frac{b^3h}{48} \quad \bar{I}_{xy} = 0$$

$$I_x = \frac{bh^3}{12} \qquad\qquad I_{xy} = 0$$

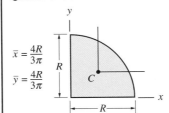

$$\bar{x} = \frac{4R}{3\pi}$$
$$\bar{y} = \frac{4R}{3\pi}$$

$$\bar{I}_x = \bar{I}_y = 0.05488R^4 \quad I_x = I_y = \frac{\pi R^4}{16}$$

$$\bar{I}_{xy} = -0.01647R^4 \quad I_{xy} = \frac{R^4}{8}$$

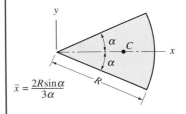

$$\bar{x} = \frac{2R\sin\alpha}{3\alpha}$$

$$I_x = \frac{R^4}{8}(2\alpha - \sin 2\alpha)$$
$$I_y = \frac{R^4}{8}(2\alpha + \sin 2\alpha)$$
$$I_{xy} = 0$$

Triangle	**Quarter ellipse**	

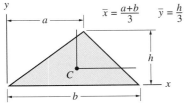

$$\bar{x} = \frac{a+b}{3} \qquad \bar{y} = \frac{h}{3}$$

$$\bar{I}_x = \frac{bh^3}{36} \qquad\qquad I_x = \frac{bh^3}{12}$$
$$\bar{I}_y = \frac{bh}{36}(a^2 - ab + b^2) \quad I_y = \frac{bh}{12}(a^2 + ab + b^2)$$
$$\bar{I}_{xy} = \frac{bh^2}{72}(2a - b) \quad I_{xy} = \frac{bh^2}{24}(2a + b)$$

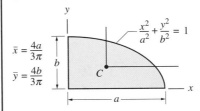

$$\bar{x} = \frac{4a}{3\pi}$$
$$\bar{y} = \frac{4b}{3\pi}$$

$$\bar{I}_x = 0.05488ab^3 \quad I_x = \frac{\pi ab^3}{16}$$
$$\bar{I}_y = 0.05488a^3b \quad I_y = \frac{\pi a^3b}{16}$$
$$\bar{I}_{xy} = -0.01647a^2b^2 \quad I_{xy} = \frac{a^2b^2}{8}$$

Mass Moments of Inertia

Slender rod

$m = AL\rho$

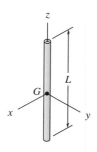

$$\bar{I}_x = \bar{I}_y = \frac{1}{12}mL^2 \qquad \bar{I}_z \approx 0$$

Cylinder

$m = \pi R^2 h\rho$

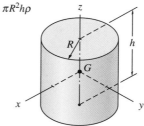

$$\bar{I}_x = \bar{I}_y = \frac{1}{12}m(3R^2 + h^2) \qquad \bar{I}_z = \frac{1}{2}mR^2$$

Hemisphere

$m = \frac{2\pi}{3}R^3\rho$

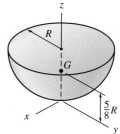

$$I_x = I_y = \frac{208}{320}mR^2 \qquad I_z = \bar{I}_z = \frac{2}{5}mR^2$$
$$\bar{I}_x = \bar{I}_y = \frac{83}{320}mR^2$$

Thin ring

$m = 2\pi AR\rho$

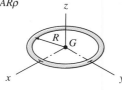

$$\bar{I}_x = \bar{I}_y = \frac{1}{2}mR^2 \qquad \bar{I}_z = mR^2$$

Sphere

$m = \frac{4\pi}{3}R^3\rho$

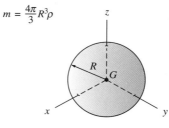

$$\bar{I}_x = \bar{I}_y = \bar{I}_z = \frac{2}{5}mR^2$$

Cylindrical shell

$m = 2\pi Rht\rho$

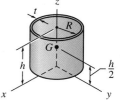

$$I_x = I_y = \frac{1}{6}m(3R^2 + 2h^2) \qquad I_z = \bar{I}_z = mR^2$$
$$\bar{I}_x = \bar{I}_y = \frac{1}{12}m(6R^2 + h^2)$$

Rectangular prism

$m = abc\rho$

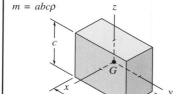

$$\bar{I}_x = \frac{1}{12}m(b^2 + c^2)$$
$$\bar{I}_y = \frac{1}{12}m(c^2 + a^2)$$
$$\bar{I}_z = \frac{1}{12}m(a^2 + b^2)$$

Cone

$m = \frac{\pi}{3}R^2 h\rho$

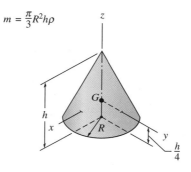

$$\bar{I}_x = \bar{I}_y = \frac{3}{80}m(4R^2 + h^2)$$
$$\bar{I}_z = \frac{3}{10}mR^2$$

Conical shell

$m = \pi Rt\rho\sqrt{R^2 + h^2}$

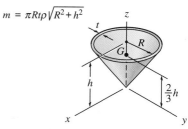

$$I_x = I_y = \frac{1}{2}m(R^2 + 2h^2) \qquad I_z = \bar{I}_z = \frac{1}{2}mR^2$$
$$\bar{I}_x = \bar{I}_y = \frac{1}{12}m(6R^2 + h^2)$$

Right tetrahedron

$m = \frac{1}{6}abc\rho$

$\bar{x} = \frac{1}{4}a$

$\bar{y} = \frac{1}{4}b$

$\bar{z} = \frac{1}{4}c$

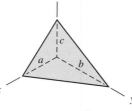

$$I_x = \frac{1}{10}m(b^2 + c^2) \qquad \bar{I}_x = \frac{3}{8}I_x$$
$$I_y = \frac{1}{10}m(c^2 + a^2) \qquad \bar{I}_y = \frac{3}{8}I_y$$
$$I_z = \frac{1}{10}m(a^2 + b^2) \qquad \bar{I}_z = \frac{3}{8}I_z$$

Semicircular rod

$m = \pi AR\rho$

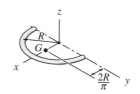

$$I_x = I_y = \frac{1}{2}mR^2 \qquad I_z = mR^2$$
$$\bar{I}_x = I_x \qquad \bar{I}_y = \left(\frac{1}{2} - \frac{4}{\pi^2}\right)mR^2$$
$$\bar{I}_z = \left(1 - \frac{4}{\pi^2}\right)mR^2$$

Hemispherical shell

$m = 2\pi R^2 t\rho$

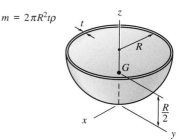

$$I_x = I_y = \frac{2}{3}mR^2 \qquad I_z = \bar{I}_z = \frac{5}{12}mR^2$$
$$\bar{I}_x = \bar{I}_y = \frac{5}{12}mR^2$$